Molecular Breeding and Genetics Research in Plants

Molecular Breeding and Genetics Research in Plants

Guest Editor

Shimeles Tilahun

MDPI

Basel • Beijing • Wuhan • Barcelona • Belgrade • Novi Sad • Cluj • Manchester

Guest Editor
Shimeles Tilahun
Agriculture and Life Sciences
Research Institute
Kangwon National
University
Chuncheon
Republic of Korea

Editorial Office
MDPI AG
Grosspeteranlage 5
4052 Basel, Switzerland

This is a reprint of the Special Issue, published open access by the journal *Current Issues in Molecular Biology* (ISSN 1467-3045), freely accessible at: https://www.mdpi.com/journal/cimb/special_issues/11VT7U3WC7.

For citation purposes, cite each article independently as indicated on the article page online and as indicated below:

Lastname, A.A.; Lastname, B.B. Article Title. *Journal Name* **Year**, *Volume Number*, Page Range.

ISBN 978-3-7258-4135-6 (Hbk)
ISBN 978-3-7258-4136-3 (PDF)
https://doi.org/10.3390/books978-3-7258-4136-3

Contents

About the Editor

Shimeles Tilahun

Shimeles Tilahun, Ph.D., is a senior researcher at the Agriculture and Life Science Research Institute of Kangwon National University. He has published more than 50 research articles in leading journals and presented his work at international conferences. His current research interest focuses on developing innovative solutions that combine postharvest physiology and technology with transcriptomics, metabolomics, and non-destructive quality prediction techniques to extend shelf life and maintain the quality of horticultural crops.

current issues in molecular biology

MDPI

Editorial

Editorial for the Special Issue 'Molecular Breeding and Genetics Research in Plants'

Shimeles Tilahun [1,2]

[1] Agriculture and Life Science Research Institute, Kangwon National University, Chuncheon 24341, Republic of Korea; shimeles@kangwon.ac.kr
[2] Department of Horticulture and Plant Sciences, Jimma University, Jimma 378, Ethiopia

Received: 27 December 2024
Accepted: 28 December 2024
Published: 29 December 2024

Citation: Tilahun, S. Editorial for the Special Issue 'Molecular Breeding and Genetics Research in Plants'. *Curr. Issues Mol. Biol.* **2025**, *47*, 11. https://doi.org/10.3390/cimb47010011

Despite significant advancements in plant breeding research, the challenges posed by a growing global population, the impact of abiotic and biotic stresses, and the uncertainties of climate change necessitate continued focus and innovation in plant breeding and genetic studies. In recent decades, advancements in molecular breeding and genetics research in plants have significantly accelerated the crop breeding process. For instance, recent studies have advanced rice yield and quality by regulating grain size [1], enhanced crop growth through improved nitrogen use efficiency [2], and developed strategies to make crops more tolerant to stresses like cold, salinity, and drought, ensuring consistent yields [3–5]. Among the traits influencing rice yield, studies suggest a strong connection between grain filling and starch metabolism [6,7]. However, achieving high yields alongside optimal grain filling ratios in cultivated rice varieties remains challenging and requires continued research leveraging modern molecular biology technologies to address this issue effectively [8,9]. Lee et al. [10] reported several quantitative trait loci (QTLs) related to rice grain filling. In addition to the yields, as living standards and consumer expectations continue to rise, enhancing the quality of rice has become increasingly important [11]. Gong et al. [12] reviewed the significant progress made in identifying genes and QTL linked to essential traits such as appearance, aroma, texture, and nutritional value in rice. These genetic discoveries have enabled the development of molecular markers, which serve as powerful tools for marker-assisted selection, for the improvement of rice quality.

Similarly, advances in molecular breeding and genetics have significantly transformed important horticultural crops such as tomatoes. In tomatoes, the development of molecular biological techniques, such as QTL mapping techniques and genome-wide association studies (GWASs), has facilitated the understanding of the genetic architecture of complex traits and germplasm management of both wild and cultivated tomatoes [13]. Breeding efforts have focused on addressing key challenges such as productivity, fruit sensorial and nutritional quality, resistance to biotic and abiotic stresses, and adaptation to new growth conditions [13,14]. These insights into molecular mechanisms driving better yield, product quality, nutrient efficiency, and stress adaptation in rice and tomato underscore the importance of further innovations in other crops as well.

This Special Issue, 'Molecular Breeding and Genetics Research in Plants', in *Current Issues in Molecular Biology* showcases advancements in molecular breeding, genomics, and gene characterization aimed at improving crop yield, stress tolerance, and disease resistance, providing valuable insights for sustainable agriculture. In this Special Issue, 19 manuscripts, comprising 16 research articles, 2 reviews, and 1 brief report are included after rigorous evaluation by the editors and expert reviewers. Five key topics are addressed: (i) advances in molecular breeding and genetics; (ii) stress tolerance and environmental adaptation;

(iii) plant health and resistance to weeds and diseases; (iv) anthocyanin biosynthesis and pigmentation; and (v) specialized functional studies.

Relating to the first topic (i), Tian et al. [15] developed a colorimetric loop-mediated isothermal amplification (LAMP) assay for rice to detect the dense and erect panicle 1 (dep1) allele, a gene critical for improving traits like yield, quality, and stress tolerance. Similarly, Tang et al. [16] identified and characterized 16 ZmFAR1 genes in maize (inbred line B73), highlighting their roles in light signaling and stress responses. Notably, ZmFAR1-14 and ZmFAR1-9 were linked to kernel row number and kernel weight, respectively, under shading conditions, emphasizing their potential for maize breeding. Furthering the genetic analysis of crops, Yan et al. [17] studied soluble solids content (SSC) in melon, identifying five candidate genes involved in sugar metabolism and accumulation, thereby offering valuable insights for breeding sweeter varieties. Meanwhile, Barbosa et al. [18] used whole-genome resequencing to uncover genetic markers for abaca varietal improvement, enhancing the identification and genotyping of this important fiber crop.

Regarding topic (ii), salinity, low-temperature tolerance, and environmental adaptability pose major challenges in agriculture. Li et al. [19] developed salt-tolerant rice varieties by incorporating Xian (*Indica*) alleles into Geng (*Japonica*) rice, leveraging the genetic diversity of Xian for high-yielding and salt-tolerant Geng varieties. In maize, Yu et al. [20] conducted a GWAS to identify SNPs associated with low-temperature tolerance, uncovering a key SNP on chromosome 5 that explained significant phenotypic variation. Addressing adaptation to low latitudes, Song et al. [21] used CRISPR-Cas9 to edit the EHD1 gene in *Japonica* rice, delaying flowering and improving adaptability while retaining superior grain quality and higher yields. Chen et al. [22] explored stress-specific miRNA expression in maize under salinity and alkalinity, revealing distinct molecular pathways linked to ion transport and metabolism. Further studies on stress resistance include Zhang et al. [23], who identified and characterized a novel gene, ClNUM1, and its splice variants in *Chrysanthemum lavandulifolium*, demonstrating varying levels of salt and cold tolerance. Guo et al. [24] identified the KvCHX gene in the halophyte *Kosteletzkya virginica*, which enhanced salinity tolerance and potassium homeostasis when overexpressed in Arabidopsis.

Concerning the third topic (iii), innovative strategies for combating biotic stressors are essential for sustainable agriculture. Gerakari et al. [25] studied broomrape resistance in tomatoes, identifying transcriptomic differences in introgression lines derived from *Solanum lycopersicum* and its wild relative *S. pennellii*, providing insights for molecular breeding and sustainable weed management in tomatoes and other crops. Hamidi et al. [26] focused on managing banded leaf and sheath blight in maize using salicylic acid (SA) and jasmonic acid (JA) as elicitors to boost resistance through induced host defense mechanisms. In viticulture, Kim et al. [27] developed an in vitro propagation method for 'Shine Muscat' grapevines, ensuring disease-free plants with confirmed genetic stability to support large-scale production.

Topic (iv) focuses on anthocyanin biosynthesis and pigmentation, as anthocyanins are fundamental to plant coloration. Fu et al. [28] identified IbERF1 as a key upstream regulator of the IbMYB1-4 promoter, which controls anthocyanin biosynthesis in purple-fleshed sweet potatoes. Wang et al. [29] reviewed black pigmentation in plants, emphasizing the significance of cyanidin in black pigments and the regulation of this coloration by R2R3-MYB transcription factors such as PeMYB7, PeMYB11, and CsMYB90, offering insights into plant color regulation and for the development of black-colored cultivars. In *Ficus virens*, Ma et al. [30] linked extended red-leaf periods to anthocyanin content and the MYB gene FvPAP1. Variations in FvPAP1 expression influence red leaf duration, providing molecular targets for breeding ornamental and woody plants with enhanced red-leaf traits.

Finally, regarding topic (v), focused investigations into plant-specific functions also contribute to agricultural advancements. Zhou et al. [31] reviewed the role of kinetochore proteins in plant meiosis, exploring their potential for genetic manipulation in plant breeding. Chen et al. [32] studied citric acid regulation in citrus, identifying citrate synthase (CS) and ATP citrate-pro-S-lyase (ACL) as key genes influencing acidity in low-acid and high-acid varieties. Sharma et al. [33] examined genetic and epigenetic stability in lingonberry during tissue culture and propagation, highlighting that while the plant retained genetic stability, it exhibited epigenetic variability under both in vitro and ex vitro conditions.

In conclusion, the papers in this Special Issue present innovative findings that advance our understanding of molecular breeding and genetics in plants. I extend my gratitude to all the authors who contributed their manuscripts, as well as to the *CIMB* editorial team, academic editors, and reviewers for their invaluable collaboration. I would also like to inform you that the 2nd edition of this Special Issue, 'Molecular Breeding and Genetics Research in Plants, 2nd Edition', has been launched (https://www.mdpi.com/journal/cimb/special_issues/6I87FSK9NV, accessed on 26 December 2024). We invite you to send your original articles and reviews to the new edition of this Special Issue of *CIMB*.

Acknowledgments: I thank all authors for submitting high-quality manuscripts and the reviewers for their careful evaluation with the common goal of improving the papers.

Conflicts of Interest: The author declares no conflicts of interest.

References

1. Ren, D.; Ding, C.; Qian, Q. Molecular Bases of Rice Grain Size and Quality for Optimized Productivity. *Sci. Bull.* **2023**, *68*, 314–350. [CrossRef] [PubMed]
2. Hou, M.; Yu, M.; Li, Z.; Ai, Z.; Chen, J. Molecular Regulatory Networks for Improving Nitrogen Use Efficiency in Rice. *Int. J. Mol. Sci.* **2021**, *22*, 9040. [CrossRef] [PubMed]
3. Li, J.; Zhang, Z.; Chong, K.; Xu, Y. Chilling Tolerance in Rice: Past and Present. *J. Plant Physiol.* **2022**, *268*, 153576. [CrossRef]
4. Qin, H.; Li, Y.; Huang, R. Advances and Challenges in the Breeding of Salt-Tolerant Rice. *Int. J. Mol. Sci.* **2020**, *21*, 8385. [CrossRef]
5. Oladosu, Y.; Rafii, M.Y.; Samuel, C.; Fatai, A.; Magaji, U.; Kareem, I.; Kamarudin, Z.S.; Muhammad, I.; Kolapo, K. Drought Resistance in Rice from Conventional to Molecular Breeding: A Review. *Int. J. Mol. Sci.* **2019**, *20*, 3519. [CrossRef]
6. Tang, T.; Xie, H.; Wang, Y.; Lü, B.; Liang, J. The Effect of Sucrose and Abscisic Acid Interaction on Sucrose Synthase and Its Relationship to Grain Filling of Rice (*Oryza sativa* L.). *J. Exp. Bot.* **2009**, *60*, 2641–2652. [CrossRef]
7. Jiang, Z.; Chen, Q.; Chen, L.; Yang, H.; Zhu, M.; Ding, Y.; Li, W.; Liu, Z.; Jiang, Y.; Li, G. Efficiency of Sucrose to Starch Metabolism Is Related to the Initiation of Inferior Grain Filling in Large Panicle Rice. *Front. Plant Sci.* **2021**, *12*, 732867. [CrossRef]
8. Zhang, W.; Cao, Z.; Zhou, Q.; Chen, J.; Xu, G.; Gu, J.; Liu, L.; Wang, Z.; Yang, J.; Zhang, H. Grain Filling Characteristics and Their Relations with Endogenous Hormones in Large- and Small-Grain Mutants of Rice. *PLoS ONE* **2016**, *11*, e0165321. [CrossRef]
9. Yang, J.; Zhang, J. Grain-Filling Problem in "super" Rice. *J. Exp. Bot.* **2010**, *61*, 1–5. [CrossRef]
10. Lee, S.M.; Kabange, N.R.; Kang, J.W.; Kwon, Y.; Cha, J.K.; Park, H.; Oh, K.W.; Seo, J.; Koh, H.J.; Lee, J.H. Identifying QTLs Related to Grain Filling Using a Doubled Haploid Rice (*Oryza sativa* L.) Population. *Agronomy* **2023**, *13*, 912. [CrossRef]
11. Rao, Y.; Li, Y.; Qian, Q. Recent Progress on Molecular Breeding of Rice in China. *Plant Cell Rep.* **2014**, *33*, 551–564. [CrossRef] [PubMed]
12. Gong, D.; Zhang, X.; He, F.; Chen, Y.; Li, R.; Yao, J.; Zhang, M.; Zheng, W.; Yu, G. Genetic Improvements in Rice Grain Quality: A Review of Elite Genes and Their Applications in Molecular Breeding. *Agronomy* **2023**, *13*, 1375. [CrossRef]
13. Wang, Y.; Sun, C.; Ye, Z.; Li, C.; Huang, S.; Lin, T. The Genomic Route to Tomato Breeding: Past, Present, and Future. *Plant Physiol.* **2024**, *195*, 2500–2514. [CrossRef] [PubMed]
14. Rothan, C.; Diouf, I.; Causse, M. Trait Discovery and Editing in Tomato. *Plant J.* **2019**, *97*, 73–90. [CrossRef] [PubMed]
15. Tian, Y.; Chen, X.; Xu, P.; Wang, Y.; Wu, X.; Wu, K.; Fu, X.; Chin, Y.; Liao, Y. Rapid Visual Detection of Elite Erect Panicle Dense and Erect Panicle 1 Allele for Marker-Assisted Improvement in Rice (*Oryza sativa* L.) Using the Loop-Mediated Isothermal Amplification Method. *Curr. Issues Mol. Biol.* **2024**, *46*, 498–512. [CrossRef]
16. Tang, H.; Jing, D.; Liu, C.; Xie, X.; Zhang, L.; Chen, X.; Li, C. Genome-Wide Identification and Expression Analyses of the FAR1/FHY3 Gene Family Provide Insight into Inflorescence Development in Maize. *Curr. Issues Mol. Biol.* **2024**, *46*, 430–449. [CrossRef]

17. Yan, H.; Wang, K.; Wang, M.; Feng, L.; Zhang, H.; Wei, X. QTL Mapping and Genome-Wide Association Study Reveal Genetic Loci and Candidate Genes Related to Soluble Solids Content in Melon. *Curr. Issues Mol. Biol.* **2023**, *45*, 7110–7129. [CrossRef]
18. Barbosa, C.F.C.; Asunto, J.C.; Koh, R.B.L.; Santos, D.M.C.; Zhang, D.; Cao, E.P.; Galvez, L.C. Genome-Wide SNP and Indel Discovery in Abaca (Musa Textilis Née) and among Other Musa Spp. for Abaca Genetic Resources Management. *Curr. Issues Mol. Biol.* **2023**, *45*, 5776–5797. [CrossRef]
19. Li, S.; Feng, T.; Zhang, C.; Zhang, F.; Li, H.; Chen, Y.; Liang, L.; Zhang, C.; Zeng, W.; Liu, E.; et al. Genetic Dissection of Salt Tolerance and Yield Traits of Geng (Japonica) Rice by Selective Subspecific Introgression. *Curr. Issues Mol. Biol.* **2023**, *45*, 4796–4813. [CrossRef]
20. Yu, T.; Zhang, J.; Cao, J.; Li, S.; Cai, Q.; Li, X.; Li, S.; Li, Y.; He, C.; Ma, X. Identification of Multiple Genetic Loci Related to Low-Temperature Tolerance during Germination in Maize (*Zea maize* L.) through a Genome-Wide Association Study. *Curr. Issues Mol. Biol.* **2023**, *45*, 9634–9655. [CrossRef]
21. Song, J.; Tang, L.; Fan, H.; Xu, X.; Peng, X.; Cui, Y.; Wang, J. Enhancing Yield and Improving Grain Quality in Japonica Rice: Targeted EHD1 Editing via CRISPR-Cas9 in Low-Latitude Adaptation. *Curr. Issues Mol. Biol.* **2024**, *46*, 3741–3751. [CrossRef] [PubMed]
22. Chen, Z.; Liu, Y.; Wang, Q.; Fei, J.; Liu, X.; Zhang, C.; Yin, Y. MiRNA Sequencing Analysis in Maize Roots Treated with Neutral and Alkaline Salts. *Curr. Issues Mol. Biol.* **2024**, *46*, 8874–8889. [CrossRef] [PubMed]
23. Zhang, W.; Wang, H.; Guo, Y.; Hao, X.; Li, Y.; He, W.; Zhao, X.; Cai, S.; Song, X. Functional Validation of Different Alternative Splicing Variants of the Chrysanthemum Lavandulifolium ClNUM1 Gene in Tobacco. *Curr. Issues Mol. Biol.* **2024**, *46*, 5242–5256. [CrossRef]
24. Guo, Y.; Zhu, C.; Tian, Z. Overexpression of KvCHX Enhances Salt Tolerance in Arabidopsis Thaliana Seedlings. *Curr. Issues Mol. Biol.* **2023**, *45*, 9692–9708. [CrossRef]
25. Gerakari, M.; Kotsira, V.; Kapazoglou, A.; Tastsoglou, S.; Katsileros, A.; Chachalis, D.; Hatzigeorgiou, A.G.; Tani, E. Transcriptomic Approach for Investigation of Solanum Spp. Resistance upon Early-Stage Broomrape Parasitism. *Curr. Issues Mol. Biol.* **2024**, *46*, 9047–9073. [CrossRef]
26. Hamidi, S.M.; Meshram, S.; Kumar, A.; Singh, A.; Yadav, R.; Gogoi, R. Biochemical and Molecular Basis of Chemically Induced Defense Activation in Maize against Banded Leaf and Sheath Blight Disease. *Curr. Issues Mol. Biol.* **2024**, *46*, 3063–3080. [CrossRef]
27. Kim, S.H.; Zebro, M.; Jang, D.C.; Sim, J.E.; Park, H.K.; Kim, K.Y.; Bae, H.M.; Tilahun, S.; Park, S.M. Optimization of Plant Growth Regulators for In Vitro Mass Propagation of a Disease-Free 'Shine Muscat' Grapevine Cultivar. *Curr. Issues Mol. Biol.* **2023**, *45*, 7721–7733. [CrossRef]
28. Fu, D.; Yang, S.; Liu, R.; Gao, F. Yeast One-Hybrid Screening to Identify Transcription Factors for IbMYB1-4 in the Purple-Fleshed Sweet Potato (*Ipomoea batatas* [L.] Lam.). *Curr. Issues Mol. Biol.* **2023**, *45*, 5765–5775. [CrossRef]
29. Wang, F.; Chen, J.; Tang, R.; Wang, R.; Ahmad, S.; Liu, Z.; Peng, D. Research Progress on Anthocyanin-Mediated Regulation of 'Black' Phenotypes of Plant Organs. *Curr. Issues Mol. Biol.* **2023**, *45*, 7242–7256. [CrossRef]
30. Ma, Q.; Zhong, S.; Ma, T.; Yue, Y.; Zou, S.; Sui, S.; Ai, L.; Guo, Y. Transcriptome Analysis Reveals That FvPAP1 Genes Are Related to the Prolongation of Red-Leaf Period in Ficus Virens. *Curr. Issues Mol. Biol.* **2024**, *46*, 5724–5743. [CrossRef]
31. Zhou, K.D.; Zhang, C.X.; Niu, F.R.; Bai, H.C.; Wu, D.D.; Deng, J.C.; Qian, H.Y.; Jiang, Y.L.; Ma, W. Exploring Plant Meiosis: Insights from the Kinetochore Perspective. *Curr. Issues Mol. Biol.* **2023**, *45*, 7974–7995. [CrossRef] [PubMed]
32. Chen, T.; Niu, J.; Sun, Z.; Chen, J.; Wang, Y.; Chen, J.; Luan, M. Transcriptome Analysis and VIGS Identification of Key Genes Regulating Citric Acid Metabolism in Citrus. *Curr. Issues Mol. Biol.* **2023**, *45*, 4647–4664. [CrossRef] [PubMed]
33. Sharma, U.; Sikdar, A.; Igamberdiev, A.U.; Debnath, S.C. Exploring Genetic and Epigenetic Changes in Lingonberry Using Molecular Markers: Implications for Clonal Propagation. *Curr. Issues Mol. Biol.* **2023**, *45*, 6296–6310. [CrossRef] [PubMed]

current issues in molecular biology

MDPI

Article

Rapid Visual Detection of Elite Erect Panicle *Dense and Erect Panicle 1* Allele for Marker-Assisted Improvement in Rice (*Oryza sativa* L.) Using the Loop-Mediated Isothermal Amplification Method

Yonghang Tian [1,2], Xiyi Chen [1,2], Peizhou Xu [3], Yuping Wang [3], Xianjun Wu [3], Kun Wu [4], Xiangdong Fu [4], Yaoxian Chin [1,2,*] and Yongxiang Liao [3,*]

1 College of Food Science and Engineering, Hainan Tropical Ocean University, No. 1 Yucai Road, Sanya 572022, China; areckting@hntou.edu.cn (Y.T.); mt48713651@163.com (X.C.)
2 Marine Food Engineering Technology Research Center of Hainan Province, No. 1 Yucai Road, Sanya 572022, China
3 State Key Laboratory of Crop Gene Exploration and Utilization in Southwest China, Rice Research Institute, Sichuan Agricultural University, No. 211 Huiming Road, Chengdu 611130, China; xpzhxj@163.com (P.X.); wyp-sc@163.net (Y.W.); wuxjsau@126.com (X.W.)
4 Institute of Genetics and Developmental Biology, Chinese Academy of Sciences, No. 1 West Beichen Road, Beijing 100101, China; kunwu@genetics.ac.cn (K.W.); xdfu@genetics.ac.cn (X.F.)
* Correspondence: chinyx1@hntou.edu.cn (Y.C.); 14387@sicau.edu.cn (Y.L.)

Abstract: Molecular-assisted breeding is an effective way to improve targeted agronomic traits. *dep1* (*dense and erect panicle 1*) is a pleiotropic gene that regulates yield, quality, disease resistance, and stress tolerance, traits that are of great value in rice (*Oryza sativa* L.) breeding. In this study, a colorimetric LAMP (loop-mediated isothermal amplification) assay was developed for the detection of the *dep1* allele and tested for the screening and selection of the heavy-panicle hybrid rice elite restorer line SHUHUI498, modified with the allele. InDel (Insertion and Deletion) primers (DEP1_F and DEP1_R) and LAMP primers (F3, B3, FIP, and BIP) for genotyping were designed using the Primer3 Plus (version 3.3.0) and PrimerExplore (version 5) software. Our results showed that both InDel and LAMP markers could be used for accurate genotyping. After incubation at a constant temperature of 65 °C for 60 min with hydroxynaphthol blue (HNB) as a color indicator, the color of the LAMP assay containing the *dep1* allele changed to sky blue. The SHUHUI498 rice line that was detected in our LAMP assay displayed phenotypes consistent with the *dep1* allele such as having a more compact plant architecture, straight stems and leaves, and a significant increase in the number of effective panicles and spikelets, demonstrating the effectiveness of our method in screening for the *dep1* allele in rice breeding.

Keywords: rice; *dep1*; InDel; LAMP; HNB; SHUHUI498

Citation: Tian, Y.; Chen, X.; Xu, P.; Wang, Y.; Wu, X.; Wu, K.; Fu, X.; Chin, Y.; Liao, Y. Rapid Visual Detection of Elite Erect Panicle *Dense and Erect Panicle 1* Allele for Marker-Assisted Improvement in Rice (*Oryza sativa* L.) Using the Loop-Mediated Isothermal Amplification Method. *Curr. Issues Mol. Biol.* **2024**, *46*, 498–512. https://doi.org/10.3390/cimb46010032

Academic Editor: Angel Llamas

Received: 30 November 2023
Revised: 28 December 2023
Accepted: 2 January 2024
Published: 4 January 2024

1. Introduction

The breeding of heavy-panicle hybrid rice to further increase rice yield has been gaining research attention [1]. Recently, researchers have incorporated molecular designs in breeding strategies for the development of improved heavy-panicle hybrid rice, leading to the discovery of various beneficial genes for the crop. One such discovery is the pleiotropic gene *DEP1*, which plays an important role in various physiological functions in rice plants, such as yield, nitrogen fertilizer utilization, grain quality, lodging resistance, disease resistance, cadmium tolerance, and drought tolerance [2–4]. The gene is located at chromosome 9 and encodes the G-protein gamma subunit in rice [5]. The mutant *DEP1* allele (*dep1*) is a gain-of-function variant with a 625 bp deletion and a 12 bp insertion in exon 5, which causes the formation of dense and erect panicles by allowing for an increased number of primary and secondary branches and, thus, grain number per panicle in rice. The allele is predominantly found in the *japonica* subspecies [5,6] and was introduced into the elite *indica* restorer line 9311, significantly increasing the yield and grain number per panicle [7]. The allele also enables denser planting and higher yields by improving canopy structure and increasing plant photosynthetic efficiency [8,9]. Research has shown that the *dep1* allele improves the harvest index and yield by improving the efficiency of nitrogen application [10], thereby helping to reduce the carbon footprint and promote a sustainable rice industry. In addition, dep1 interacts with LPA1 to activate the expression of *PIN1a*, which improves rice resistance to blights [4], minimizing the need for pesticides and, thus, ensuring stable yields. At the same time, the allele also improves plant tolerance to Cd (Cadmium) [11] and allows plants to survive harsh environments by improving their tolerance to drought, cold, and salinity [12,13]. Furthermore, *dep1* is able to modulate the balance of the carbon and nitrogen metabolism of rice, contributing to improved grain quality [3,14]. Therefore, the *dep1* allele has great potential in the development of green super rice [15,16] and warrants further research. As such, the development of convenient on-site detection for *dep1* would considerably enhance our ability to monitor the effects and success of *dep1* applications in rice breeding.

Several reports have been presented on the detection of the *dep1* allele using PCR (polymerase chain reaction) technology based on molecular markers such as InDel-E5 [17], *DEP1*S9 [18], *DEP1*E5ID [19], and H90 [20]. However, these methods require high temperatures and specialized devices for data visualization, rendering them impractical for on-site testing [21,22]. LAMP (loop-mediated isothermal amplification) is an isothermal nucleic acid detection technique [23,24] that is able to achieve an accuracy similar to PCR. Moreover, this technology does not require specialized or sophisticated instruments while also offering the rapid amplification of targeted DNA with high efficiency, specificity, and sensitivity. LAMP technology is also user-friendly, as it can be easily mastered and performed by laymen without the need for prior molecular experiences [25]. Recently, LAMP has been successfully used for the rapid detection of transgenes [26], blight-resistant genes [27], and authenticity [28] in rice plants.

In this study, we aim to establish a colorimetric LAMP assay for the visual detection of the *dep1* allele in the heavy-panicle elite hybrid rice restorer line SHUHUI498. The outcome of this study could provide a technical reference for the development and application of visual markers for other important functional genes. The LAMP molecular marker's design takes advantage of *dep1's* 625 bp deletion and a 12 bp insertion in exon 5, which is a conserved region of the *DEP1* gene and, therefore, enables specific differentiation between the two alleles.

2. Materials and Methods

2.1. Plant Material and Genomic DNA Isolation

The rice varieties used in the experiment consisted of 10 rice lines, namely, WYJ7, WYJ3, SN265, QCL2, N580, N130, Nip, SHUHUI498, ZH11, 9311, and the progenies of WYJ7 backcrossed with SHUHUI498, all provided by the Rice Research Institute of Sichuan Agricultural University.

Genomic DNA from rice leaves at the heading stage was extracted using the CTAB method [29]. The extracted DNA templates were stored at −20 °C for subsequent use.

2.2. The InDel and LAMP Marker Design

The DNA sequence for the rice *dep1* allele was downloaded from the RGAP (Rice Genome Annotation Project) [30].

After the alignment and analysis of the *dep1* allele sequence against the NCBI nucleotide database (https://blast.ncbi.nlm.nih.gov/Blast.cgi, accessed on 15 May 2019), the deletion and insertion variants [5] that include both upstream and downstream conserved sequences in exon 5 of *dep1* were taken as target sequences for the design of InDel primers using Primer3 [31]. The sequence was uploaded to PrimerExplorer V5 (http://primerexplorer.jp/lampv5e/index.html, accessed on 1 January 2024), and LAMP primers were designed with one of the primers spanning the 625 bp→12 bp mutated locus in exon 5 of the *dep1* variant to ensure specificity. After alignment was carried out in EnsemblPlants (https://plants.ensembl.org/Multi/Tools/Blast, accessed on 1 January 2024), generated primers with the highest specificity were selected. The primers were then synthesized by Sangon Biotech (Shanghai) Co., Ltd., Shanghai, China.

2.3. PCR Amplification and Detection

The PCR reaction was carried out in a reaction volume of 20 μL containing 2 μL of forward and reverse primer (10 μM), 4 μL of a 2 × Taq PCR mix, 10 μL of ddH_2O, and 2 μL of DNA templates. PCR amplification was performed on a Bio-Rad T100™ Thermal Cycler (Bio, Hercules, CA, USA) with the following conditions: initial denaturation at 94 °C (3 min); 30 cycles of denaturation (94 °C, 30 s), annealing (56–63 °C, 30 s), and extension (72 °C, 1 min); and final extension (72 °C, 5 min). The presence of the amplification products was verified via agarose gel electrophoresis (BIO-OI gel imaging system, Guangzhou Guangyi Biotechnology Co., Ltd., Guangzhou, China). DNA templates containing the *dep1* allele will generate smaller amplicons (409 bp), whereas templates containing the *DEP1* allele will produce larger amplicons (1034 bp).

2.4. LAMP Amplification and Colorimetric LAMP Assay

2.4.1. LAMP Amplification and Agarose Gel Electrophoresis Detection

The LAMP reaction system was prepared as described in the kit's manual (2× Lamp PCR Master Mix, B532455, Sangon Biotech (Shanghai) Co., Ltd., Shanghai, China) and adjusted for a 12.5 μL reaction volume: 6.25 μL of 2 × LAMP Mix Buffer, 10 μM F3/B3 primers (each 1 μL), 10 μM FIP/BIP primers (each 0.25 μL), 0.5 μL DNA template, 0.25 μL DNA polymerase (0.16 U/μL).

The LAMP reaction was carried out according to kit instructions with some modifications. Briefly, the target gene was denatured at 95 °C for 5 min, followed by incubation at a constant temperature (55–65 °C) for 1 h. The enzyme activity was then inactivated at 80 °C for 10 min, and the LAMP reaction was terminated by holding it at 12 °C for 5 min. The presence of LAMP products was verified with agarose gel electrophoresis, as described in Section 2.3.

2.4.2. Colorimetric LAMP Assay

The reaction system was prepared according to the manufacturer's instructions (LAMP HNB Amplification Kit, A3802, HaiGene Biotech Co., Ltd., Harbin, China), adjusted for a 12.5 μL reaction volume. The LAMP amplification protocol was performed as described in Section 2.4.1. After the completion of the amplification reaction, the color of the reaction solution was observed with the naked eye for changes [32]. The reaction solutions that contained the *dep1* allele changed to a sky-blue color, while the reaction solutions without it remained violet.

2.5. Application of LAMP Colorimetric Assay in the Improvement of the Heavy-Panicle Elite Hybrid Rice Restorer Line SHUHUI498

The potential application of our designed LAMP assay was tested via screening for heavy-panicle elite hybrid rice restorer line SHUHUI498 introduced with the *dep1* allele. Cross-breeding was first performed by using *dep1*-carrying WYJ7 [33] rice as the maternal parent and SHUHUI498 [34], a heavy-panicle elite hybrid rice restorer line, as the paternal parent. The resultant progenies of the cross were backcrossed twice (BC_1 and BC_2) with SHUHUI498 as the recurrent parent, followed by consecutive self-pollinations (F_n), during which elite progenies in each generation (n = 50) were screened and selected to establish a stable superior line. The screening and selection of superior individuals in generations of BC_1F_1 and BC_2F_1 were performed by observing phenotype characteristics supplemented by genotyping using the colorimetric LAMP assay described above. Individual plants carrying *dep1*, which is a dominant negative gene [5], are typically distinguished by plant architecture, panicle, leaf, and grain shape. As 87.5% of the WYJ7 genetic background would theoretically be eliminated in BC_2F_1 via self-pollination, an experienced breeder could achieve more than 95% clearance of the genetic background when carrying out phenotype selection in tandem with genotyping [35]. After progenies carrying the *dep1* allele in BC_2F_1 were verified with the LAMP assay, the successive selection of excellent individuals from each segregating generation was carried out until a stable line (tentatively named SHUHUI498^{dep1}) was established, and both its genotypes and phenotypes were analyzed.

Forceps emasculation (FE) was used to produce crosses and backcrosses [36]. Superior progenies in the backcross populations were screened and selected for hybridization using on-field performance, phenotype evaluation, and genotyping of *dep1* allele. Elite individuals that flowered during the heading stage were collected before 8 a.m. each day. Using flowering time as an indicator, both bloomed and immature florets were cut off and discarded, while the remaining florets were used for forceps emasculation on the next day [37]. After cutting off about 1/3 of the upper part of the spikelet diagonally with a pair of shears, the exposed stamens were wholly removed with tweezers before covering the panicles with hybridization bags. Panicles of SHUHUI498 that were emitting or on the verge of emitting pollen were collected around 10 a.m. to pollinate the emasculated spikelets, which were then bagged and recorded. Hybridized grains were collected at the end of 20 days with a grain count of approximately 100 grains, with sufficient grain counts obtained by re-hybridizing if necessary.

Twenty individuals of SHUHUI498 were grown per generation as controls, while 50 progenies were used for screening or backcrossing. Planting was carried out in ten plants per row at 30 cm × 15 cm spacing. WYJ7, SHUHUI498, and their progenies were planted in Lingshui District in Hainan Province during the winter and in Wenjiang District in Sichuan Province during the summer.

Agronomic traits such as PH (plant height), EPN (effective panicle number), PL (panicle length), GP (grains per panicle), SP (spikelets per panicle), and YP (yield per plant) were investigated in the stabilized lines according to the literature [5,10]. Genotyping was carried out according to the steps described in Section 2.4.

3. Results

3.1. InDel and LAMP Markers

The InDel primers (DEP1-F and DEP1-R) for genotyping the *dep1* allele were designed using the Primer3Plus (version 3.3.0, Figure 1, Table 1). The sizes of the PCR amplification products of the *dep1* and *DEP1* alleles using the InDel marker were 409 bp and 1034 bp, respectively. We also included an InDel marker that is reported elsewhere (Table 1) for result comparison and verification [17].

The LAMP primers F3, B3, FIP, and BIP were designed using PrimerExplorer V5 for genotyping the *dep1* allele (Table 1), specifically targeting six highly conserved regions of the *DEP1* gene (Figure 1). Except for differences in the 12 bp insertion and 625 bp deletion between the *DEP1* and *dep1* alleles at exon 5, the six gene segments were consistent across the approximately 600 published genomes (Supplementary Data S1.out and S2.out), revealing them as highly conserved regions. The distance between the 5′ end of B2 and the 5′ end of B1 (or the 5′ end of F2 and the 5′ end of F1) determined the size of the loop formed in the reaction, which, in turn, affected the LAMP amplification efficiency [38]. In this study, the most suitable distance was found to be 40~60 bp. Hence, the distance between B1 and B2 was set at 57 bp, which enabled the LAMP primers to accurately bind to the DNA template and amplify efficiently. The 625 bp→12 bp mutation between B2 and B1 in the *DEP1* allele severely impaired the loop formation in the reaction, thus preventing the LAMP reaction.

Table 1. Primers' sequences of InDel and LMP markers.

Primers	Sequences (5′-3′)	Usage	Product Size
DEP1-F	TAAGCCAAACTGCAGTGCG	Forward Primer for InDel Marker	*dep1* (409 bp), *DEP1* (1034 bp)
DEP1-R	GTTCAACCTCGTCTCATAGCT	Reverse Primer for InDel Marker	
F3	CATGCTGTAGTCCAGACTG	Forward Outer Primer for LAMP Marker	
B3	AAGCAACCACTGAGACAG	Backward Outer Primer for LAMP Marker	
FIP (F1C-F2)	TTCGGTTTGCAGCAAGAAGG-CTGCTCATGCTGTAAACCTA	Forward Inner Primer for LAMP Marker	
BIP (B1C-B2)	TGCGATACATCGTGCTGCAA-GGGCATCGACAACCCA	Backward Inner Primer for LAMP Marker	
InDel-E5-F [17]	TCCAGGGATGTAATCATCTTTGTT	Forward Primer for InDel Marker InDel-E5	*dep1* (733 bp), *DEP1* (1357 bp)
InDel-E5-R [17]	GGCTCCATATCTTCACGGTCTA	Forward Primer for InDel Marker InDel-E5	

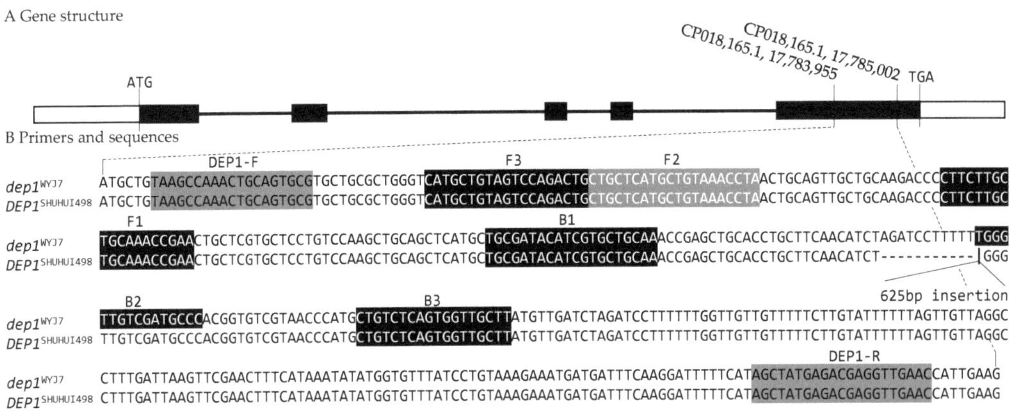

Figure 1. Schematic of the *DEP1* gene structure and primer locations. (**A**) Gene structure. ATG denotes the start codon, TGA denotes the stop codon position, the white box before ATG denotes the 5′UTR, the white box after TGA denotes the 3′UTR, and the black box in the middle denotes the exon. (**B**) Primers and sequences. *DEP1*^{WYJ7} indicates that the *dep1* gene sequence is from the rice variety WYJ7, while *DEP1*^{SHUHUI498} indicates a *DEP1* gene from the SHUHUI498 variety. "CP018,165.1, 17,783,955" is indicated as base "A" at positions 17, 783, and 955 bp of GenBank accession number CP018,165.1, "CP018,165.1, 17,785,002" is the same. The sequences shaded in dark gray under the InDel marker primers DEP1-F and DEP1-R denote the primer sequences and their positions on the genes. The shaded characters under F3, F2, F1, B1, B2, and B3 denote the six regions of the gene targeted by the LAMP primers.

3.2. Optimizing PCR and Detecting InDel Markers in Different Rice Varieties

The PCR amplification annealing temperatures of primers DEP1-F and DEP1-R were first optimized to ensure accuracy (Figure 2). The results showed that unique bands of approximately 1000 bp in size could be amplified via PCR using SHUHUI498's DNA as a template at different annealing temperatures ranging from 56.0 °C to 63.0 °C. The targeted bands produced in each lane were clear, indicating excellent primer specificity. An annealing temperature of 63.0 °C, which provided better amplification, was finally chosen as the optimized annealing temperature for PCR amplification.

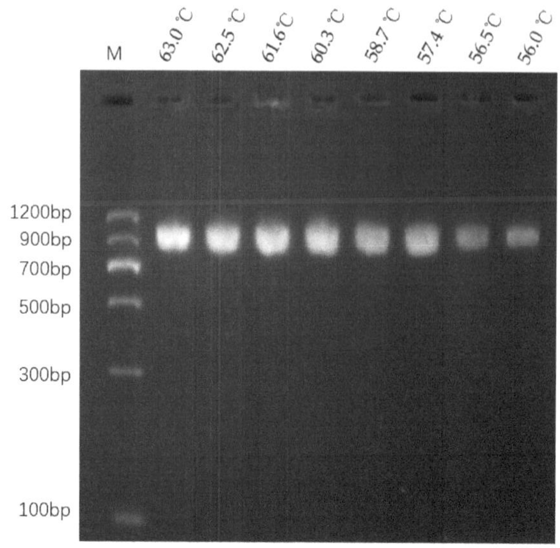

Figure 2. PCR amplification at different annealing temperatures. Lane M is a DNA ladder (MD102, Tiangen Biotech (Beijing) Co., Ltd., Beijing, China). Lanes 1 to 8 correspond, respectively, to annealing temperatures of 63.0 °C, 62.5 °C, 61.6 °C, 60.3 °C, 58.7 °C, 57.4 °C, 56.5 °C, and 56.0 °C, in that order.

Following the optimized PCR conditions, DNA materials from rice lines WYJ7, WYJ3, SN265, QCL2, N580, N130, Nip, SHUHUI498, ZH11, and 9311 were used as templates for PCR amplification, and the resultant amplicons were verified via electrophoresis on a 2% agarose gel (Figure 3). The results showed that the WYJ7, WYJ3, SN265, QCL2, N580, and N130 rice lines, all of which carry the *dep1* allele, produced smaller amplicons, while the *DEP1*-carrying rice lines Nip, SHUHUI498, ZH11, and 9311 produced larger amplicons. Therefore, the sizes of the PCR amplicons are in accordance with the genotypes.

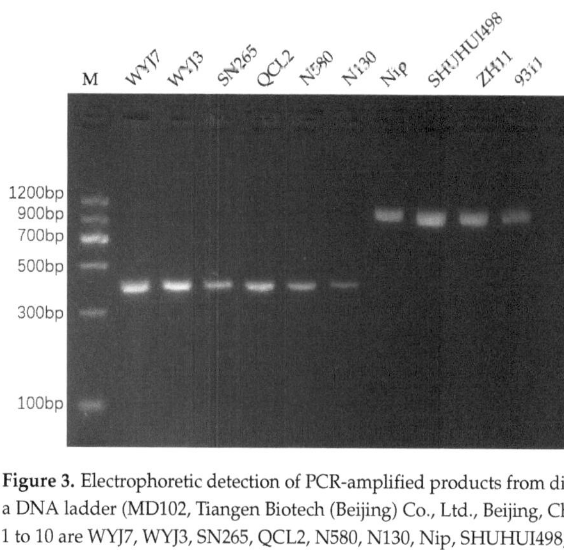

Figure 3. Electrophoretic detection of PCR-amplified products from different rice varieties. Lane M is a DNA ladder (MD102, Tiangen Biotech (Beijing) Co., Ltd., Beijing, China), and the varieties in lanes 1 to 10 are WYJ7, WYJ3, SN265, QCL2, N580, N130, Nip, SHUHUI498, ZH11, and 9311, in that order.

3.3. Optimization of LAMP Reaction Temperature

After the LAMP reaction was carried out at different temperatures ranging from 55 °C to 65 °C using the DNA template from WYJ7, the resultant products were verified using 2% agarose gel electrophoresis (Figure 4). The results showed typical ladder-like bands exhibited by LAMP amplification products. When the temperature reached 56.9 °C, the bands were blurred and dull, indicating few LAMP products. At 65.0 °C, 64.3 °C, 63.0 °C, 61.1 °C, and 58.8 °C, the bands were clearer and brighter, indicating more amplification products. Hence, 65.0 °C was chosen as the optimized LAMP reaction temperature for the following experiments.

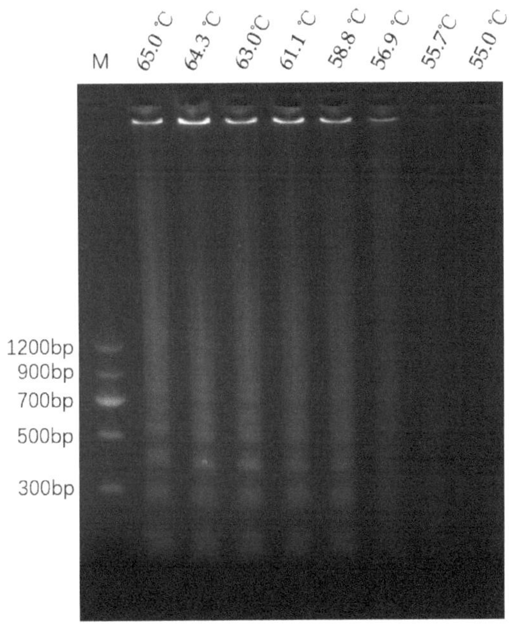

Figure 4. Electrophoresis of LAMP products at different temperatures. Lane M is a DNA ladder (MD104, Tiangen Biotech (Beijing) Co., Ltd., Beijing, China). Lanes 1 to 8 correspond to the respective LAMP reaction temperatures of 65.0 °C, 64.3 °C, 63.0 °C, 61.1 °C, 58.8 °C, 56.9 °C, 55.7 °C, and 55.0 °C, in that order.

3.4. LAMP Reaction and Colorimetric LAMP Assay

LAMP amplification and electrophoresis were performed under LAMP-optimized conditions using DNA from the aforementioned 10 different rice lines as templates (Figure 5A). The gel electrophoresis results showed characteristic ladder bands in the lanes of WYJ7, WYJ3, SN265, QCL2, N580, and N130, which are rice lines that carry the *dep1* allele. In contrast, no LAMP amplification products were found in the lanes representing the *DEP1*-containing varieties (Nip, SHUHUI498, ZH11, and 9311). These results were expected, as the LAMP markers were designed using the DNA sequence of the *dep1* allele, meaning only the *dep1*-containing DNA template would be amplified. Nip, SHUHUI498, ZH11, and 9311 are of the *DEP1* genotype, which has a large insertion on the DNA template corresponding to the primer's BIP-matching region, which impairs the formation of the loop in the LAMP reaction, thus resulting in no amplicons being produced.

A colorimetric LAMP assay was also performed with the aforementioned 10 rice varieties (Figure 5B). The results showed that the tubes containing DNA templates of WYJ7, WYJ3, SN265, QCL2, N580, and N130 successfully completed the LAMP reaction, turning the solution sky blue. In contrast, the tubes containing the DNA templates of

Nip, SHUHUI498, ZH11, and 9311 did not undergo a LAMP reaction, and the solution remained violet.

At present, the genomic information of the eight rice varieties used in this study, except for N580 and N130, are published in open databases. The genotypes of WYJ7, WYJ3, SN265, QCL2, N580, and N130 were all confirmed to be of *dep1*, while the genotypes of Nip, SHUHUI498, ZH11, and 9311 are *DEP1*. Hence, the results of the InDel markers, LAMP amplification, and LAMP colorimetric assay in this study were all consistent with the published genotypic data.

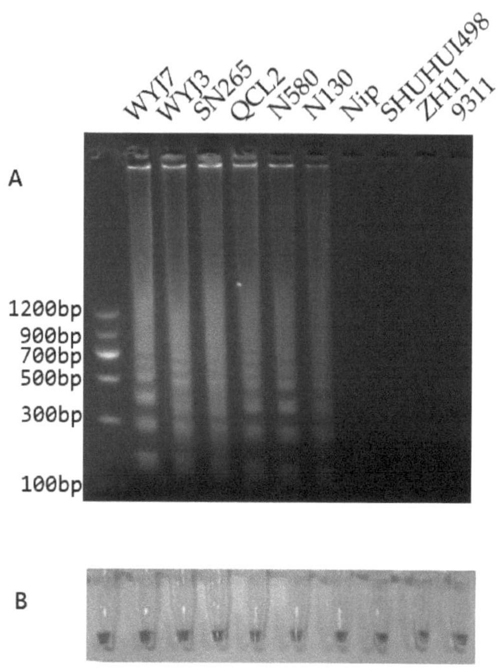

Figure 5. LAMP amplification and visual colorimetric assay for the detection of different rice varieties. (**A**) Electrophoretic detection of LAMP amplification products. (**B**) Visual colorimetric LAMP assay. Lane M is a DNA ladder (MD114, Tiangen Biotech (Beijing) Co., Ltd., Beijing, China). The DNA templates used for each of the lanes are the same as those in Figure 3.

3.5. Application of the Colorimetric LAMP Assay in Breeding an Improved Heavy-Panicle Elite Hybrid Rice Restorer Line

The SHUHUI498^{dep1} rice line was successfully bred by introducing the *dep1* allele into the heavy-panicle elite hybrid rice restorer line SHUHUI498 (Figures 6 and 7). SHUHUI498^{dep1} was compact, with upright leaves and stems, and had a reduced plant height of about 14 cm (Figures 6B and 7A). The plants grew uniformly in the field (Figure 6D) with shorter panicles and increased grain density (Figures 6B,C and 7C). Effective panicles per plant (Figures 6B and 7B) and spikelet numbers per panicle were significantly increased in SHUHUI498^{dep1} (Figure 7E), although the grain number per panicle and the yield per plant did not (Figure 7D,E). This is probably due to the low light and low day/night temperature difference in southwest China, which is unfavorable for organic matter accumulation, leading to insufficient photosynthetic products being produced to sustain the increase in the actual grain numbers per panicle and yield, despite the increase in effective panicles per plant and spikelet numbers per panicle in SHUHUI498^{dep1}. This may also be the reason why *dep1* is mainly found in *japonica* rice [5]. On the other hand, SHUHUI498^{dep1} has the potential to achieve high yields in high-light environments, which can boost its photosynthetic efficiency and allow for the increased planting density necessary to raise the yield per unit area [8,9].

Figure 6. Breeding plan and phenotypes of SHUHUI498^{dep1}. (**A**) Breeding plan of SHUHUI498^{dep1}. (**B**) Morphology comparison between SHUHUI498^{dep1} and SHUHUI498. (**C**) Comparison of panicle traits between SHUHUI498^{dep1} and SHUHUI498. (**D**) Field performance of SHUHUI498^{dep1} and SHUHUI498. Symbols: ×, cross; ⊗, self-cross.

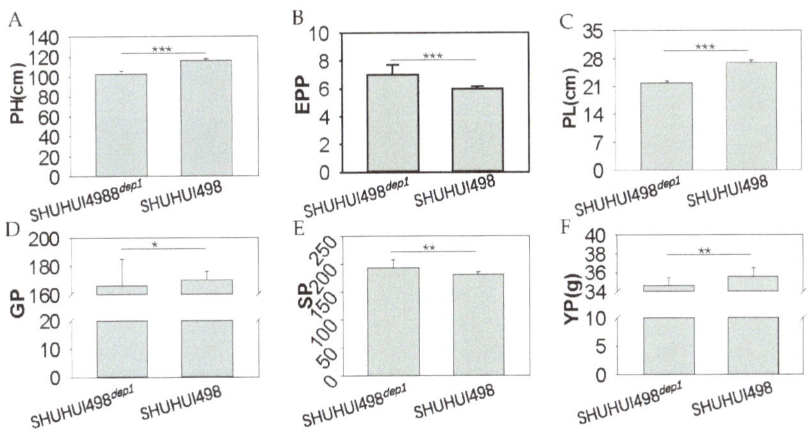

Figure 7. Analysis of key agronomic traits between SHUHUI498^{dep1} and SHUHUI498. (**A**) PH, plant height. (**B**) EPP, effective panicles per plant. (**C**) PL, panicle length. (**D**) GP, grain number per panicle. (**E**) SP, spikelet number per panicle. (**F**) YP, yield per plant. Analysis of the data was performed using a two-tailed Student's t-test. Significances are denoted as *** ($p < 0.001$), ** ($p < 0.01$), and * ($p < 0.05$).

SHUHUI498^{dep1}, SHUHUI498, and the progenies of SHUHUI498 were genotyped with the InDel markers (InDel-E5) reported elsewhere [17] (Figure 8A). The results show that the genotype of SHUHUI498 is *DEP1*, while SHUHUI498^{dep1}; individuals 3, 5, and 6 in BC_1F_1; and individuals 8 and 9 in BC_2F_1 carry the *dep1* allele. This is consistent with the results in both the gel electrophoresis and colorimetric assay of LAMP amplification (Figure 8A). Therefore, the colorimetric LAMP assay that we developed is able to accurately detect the targeted genotype in SHUHUI498^{dep1}, SHUHUI498, and their progenies (Figure 8B,C), which, when combined with the phenotypic characteristics of the *dep1* gene, can be used for the rapid on-site screening of target individuals in rice breeding programs (Figure 7).

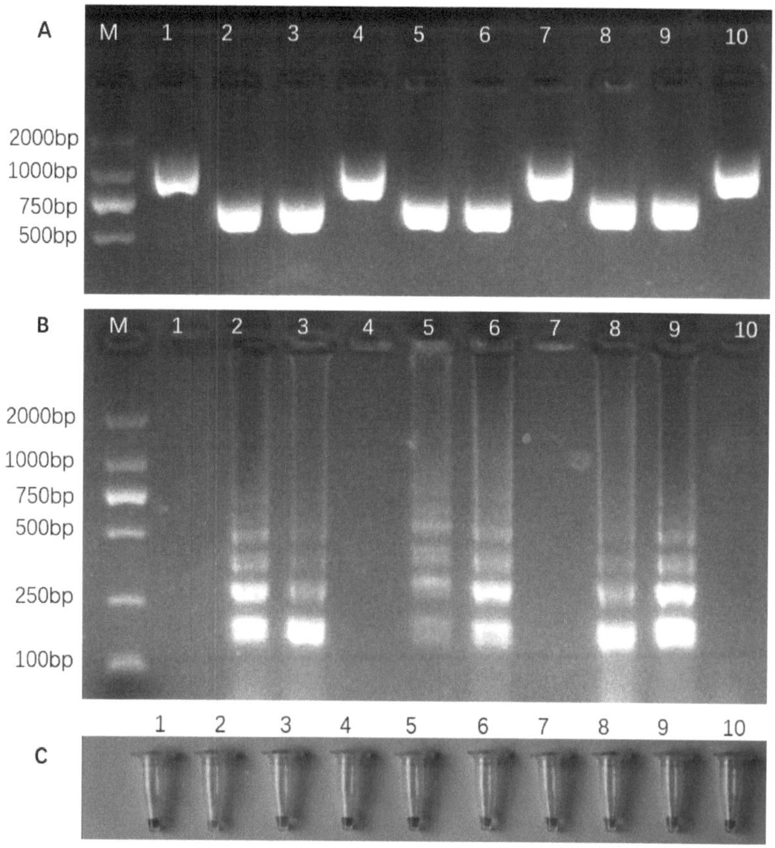

Figure 8. Genotyping of SHUHUI498^{dep1}, SHUHUI498, and their offspring. (**A**) Electrophoresis of the PCR amplification products with the InDel marker InDel-E5 [17]. (**B**) Electrophoresis of LAMP amplification products. (**C**) Visual colorimetric LAMP amplification. Lane M is a DNA ladder (MD114, Tiangen Biotech (Beijing) Co., Ltd., Beijing, China). The DNA templates in lanes 1 to 10 and tubes 1 to 10 are as follows: 1 to 2 are SHUHUI498 and SHUHUI498^{dep1}, 3 to 6 are individuals in BC_1F_1, and 7 to 10 are individuals in BC_2F_1.

4. Discussion

Molecular marker-assisted breeding (MAB) is a valuable technique for producing rice varieties with improved yield, quality, and disease resistance [39,40]. The technique has the advantages of being convenient, highly efficient, and unaffected by environmental conditions, making it possible to rapidly improve desired traits and develop superior varieties. The application of MAB has overcome many of the limitations faced in traditional breeding, enabling the development of high-quality improved lines in a relatively short period. Screening and selection in conventional breeding programs are usually achieved empirically by observing phenotypes, from the screened parental plants to the hybridized progenies, and success is more often than not due to the experience of the breeders. Hence, traditional breeding programs tend to require a long breeding cycle and, subsequently, have low efficiency. Following the booming development in functional genomics, many functional genes have been discovered [41]. The incorporation of these functional genes into commercial breeding is a crucial step in crop research [42]. With the aid of molecular markers, the screening and selection of desired traits in a breeding program can be greatly enhanced through the definite confirmation of genotypes even before phenotype observation is possible. In addition, MAB enables the simultaneous selection of targeted genes and their neighboring genes (both upstream and downstream), which improves the breeding accuracy, speeds up the removal of genetic background, and reduces linkage tedium, leading to a shortening of the breeding cycle. Thus, MAB is currently the most popular tool used by most breeders. Although molecular markers such as InDel, CAPs (Cleaved Amplified Polymorphic Sequences), and dCAPs (Derived Cleaved Amplified Polymorphic Sequences) are commonly used in both academic research and commercial breeding given their technical stability and precision, these methods are, nevertheless, complex and laborious, including multiple steps such as DNA extraction, PCR amplification, restriction enzyme digestion, gel electrophoresis, and imaging electrophoretic results. Furthermore, conventional PCR methods—such as RFLP (Restriction Fragment Length Polymorphism), AFLP (Amplified Fragment Length Polymorphism), SCAR (Sequence-Characterized Amplified Regions), SSR (Simple Sequence Repeats), RAPD (Random Amplified Polymorphic DNA), ISSR (Inter Simple Sequence Repeats), TRAP (Telomeric Repeat Amplification Protocol), ARMS (Amplification Refractory Mutation System PCR), KASP (Kompetitive Allele Specific PCR), and HRM (High-Resolution Melting curve)—require specialized equipment for thermal cycling and gel electrophoresis [43–45], which incurs extra cost. As such, isothermal methods such as LAMP represent attractive options for practical applications given their ease of use; sensitivity; reasonable accuracy; convenience; and, once developed, low cost [46]. Even though LAMP protocols are prone to false positives because of their high sensitivity, an established LAMP protocol, such as the one shown in this study, when combined with the rapid extraction of DNA templates [26,47] with commercially available kits, can enable a breeder to determine the targeted genotypes on-site with a simple and straightforward protocol at a reasonable cost. This is particularly appealing to rice breeders in developing countries, many of whom are farmers who are poorly educated with little to no knowledge of molecular biology and who are more concerned about cost than accuracy. This demand can be reflected by the fact that more than 60% of the molecular diagnostics kits available on the market are LAMP-based and are widely used in the identification of viruses, bacteria, and fungi, as well as medical diagnosis and food safety. The results of our study indicate that our developed LAMP assay has comparable precision and sensitivity to standard PCR using InDel markers (Figures 3 and 8A), further lending credence to the potential of colorimetric LAMP assays.

The need for global food security and environmental concerns have driven the research community to find the right balance between high yield and the use of nitrogen fertilizers [48]. In order to reduce the conflict between food resources and the environment, scientists have proposed the concept of green super rice, which focuses on the discovery and application of nitrogen-efficient genes [49]. Research has found that the *dep1* allele can increase the harvest index and yield by improving nitrogen use efficiency [10] and regulating rice quality, disease resistance, stress resistance, and so on. Hence, *dep1* is considered a key allele for breeding green super rice varieties [7,50]. The colorimetric LAMP assay for *dep1* allele designed in this study, combined with rapid DNA extraction methods [26,47], presents a convenient method for the on-site genotyping of *dep1*, contributing greatly to the breeding and development of green, high-yielding rice varieties.

Plant architecture is one of the key factors influencing yield and has received much attention in rice breeding. Numerous alleles, such as *dep1*, *sd1*, *IPA1*, *d14*, and *d61*, have been found to regulate plant architecture and, thus, have been widely cloned and used in rice breeding [51]. The *dep1* allele is widespread in *japonica* rice varieties, with more than half of the cultivated rice varieties in northeast China carrying this allele [6]. Rice lines WYJ7, WYG3, and LG31, all of which carry the *dep1i* allele, are among the most widely cultivated rice varieties. In addition, although *indica* rice varieties carry the *DEP1* allele natively, it has been reported that introducing the *dep1* allele into *indica* rice can significantly raise its yield [7]. The data from this study also showed that both the effective panicles per plant and the spikelet numbers per panicle were significantly increased after the successful introduction of the *dep1* allele into the heavy-panicle elite hybrid rice restorer line SHUHUI498 (Figures 6 and 7). Therefore, the colorimetric LAMP assay for the *dep1* allele developed in this study is a valuable screening tool for targeted gene improvement in MAB for rice.

Since global warming increases the threat of pests and diseases, breeding new varieties with improved stress and disease resistance is important to securing stable rice yields as global temperature rises [52]. The incorporation of newly discovered disease and stress resistance genes into rice production has shown satisfactory results and represents a viable strategy [53]. The colorimetric LAMP assay for *dep1* detection developed in this study can also be adapted and modified as a convenient way to screen and breed new disease- and stress-resistant rice varieties to ensure stable yields and food security.

It is important to note that, while the results of our study are in agreement with each other and are fully aligned with our expected outcome, the experiments were nevertheless conducted in a controlled environment with known genotypes. For the assay to be applicable for field use, which involves many unknown variables, internal and standard controls have to be developed to ensure accuracy, including the prevention of false-negative results.

5. Conclusions

In this study, we established a convenient colorimetric LAMP assay for genotyping the *dep1* allele in rice based on *dep1*-specific sequences available in the published literature and databases. The developed method proved to be accurate and specific, suitable for the on-site screening of the *dep1* allele. We further confirmed the feasibility of introducing the *dep1* allele into the heavy-panicle elite hybrid rice restorer line SHUHUI498 and showed its potential to increase yield. This study not only provides technical support for genotyping *dep1*, an allele for dense erect panicles and high yields in rice but also serves as a reference for the future development and application of LAMP-functional markers for other important genes. For further study, we will investigate the incorporation of additional loop primers into our LAMP assay in order to boost its efficiency.

Supplementary Materials: The following supporting information can be downloaded at: https://www.mdpi.com/article/10.3390/cimb46010032/s1, Data S1: Schematic comparison of the *DEP1* from Nipponbare_IRGSP_4.0_AP008215.2 with the *dep1* from 13 other varieties; Data S2: *DEP1* comparisons across nearly 600 different genomes.

Author Contributions: Y.T. and Y.L. conducted the bioinformatic analysis and primer design. Y.W., P.X. and X.W. confirmed the primers with experiments and breeding. X.C. conceptualized and wrote the first draft. K.W., Y.C. and X.F. designed the experiments and revised the manuscript. All authors discussed the results and contributed to the manuscript. All authors have read and agreed to the published version of the manuscript.

Funding: This study was supported by the National Natural Science Foundation of China, Grant No. 31960399; the Natural Science Foundation of Hainan Province, Grant No. 321MS050; and funding from the Chengdu Science and Technology Bureau, Grant No. 2022-YF09-00036-SN.

Institutional Review Board Statement: Not applicable.

Informed Consent Statement: Not applicable.

Data Availability Statement: Additional data are available from the corresponding authors upon request.

Conflicts of Interest: The authors declare no conflicts of interest.

References

1. Zhou, K.; Ma, Y.; Liu, T.; Shen, M. The breeding of subspecific heavy ear hybrid rice-exploration about super-high yield breeding of hybrid rice. *J. Sichuan Agric. Univ.* **1995**, *13*, 403–407. [CrossRef]
2. Zhang, A.; Wang, F.; Kong, D.; Hou, D.; Huang, L.; Bi, J.; Zhang, F.; Luo, X.; Wang, J.; Liu, G.; et al. Mutation of *DEP1* facilitates the improvement of plant architecture in *Xian* rice (*Oryza sativa*). *Plant Breed.* **2023**, *142*, 338–344. [CrossRef]
3. Huang, H.; Ye, Y.; Song, W.; Li, Q.; Han, R.; Wu, C.; Wang, S.; Yu, J.; Liu, X.; Fu, X.; et al. Modulating the C-terminus of DEP1 synergistically enhances grain quality and yield in rice. *J. Genet. Genom.* **2022**, *49*, 506–509. [CrossRef] [PubMed]
4. Liu, J.; Mei, Q.; Xue, C.; Wang, Z.; Li, D.; Zhang, Y.; Xuan, Y. Mutation of G-protein γ subunit *DEP1* increases planting density and resistance to sheath blight disease in rice. *Plant Biotechnol. J.* **2020**, *19*, 418–420. [CrossRef]
5. Huang, X.; Qian, Q.; Liu, Z.; Sun, H.; He, S.; Luo, D.; Xia, G.; Chu, C.; Li, J.; Fu, X. Natural variation at the *DEP1* locus enhances grain yield in rice. *Nat. Genet.* **2009**, *41*, 494–497. [CrossRef] [PubMed]
6. Xu, H.; Zhao, M.; Zhang, Q.; Xu, Z.; Xu, Q. The *DENSE AND ERECT PANICLE 1* (*DEP1*) gene offering the potential in the breeding of high-yielding rice. *Breed. Sci.* **2016**, *66*, 659–667. [CrossRef]
7. Qian, Q.; Guo, L.; Smith, S.M.; Li, J. Breeding high-yield superior-quality hybrid super-rice by rational design. *Natl. Sci. Rev.* **2016**, *3*, 283–294. [CrossRef]
8. Cheng, F.; Yu, J.; Xu, Z.; Xu, Q. Erect panicle architecture contributes to increased rice production through the improvement of canopy structure. *Mol. Breed.* **2019**, *39*, 128. [CrossRef]
9. Li, Y.; Lv, X.; Rui, M.; Hu, J.; Volkov, V.S.; Zeng, D.; Wang, Y. Rice *dep1* variety maintains larger stomatal conductance to enhance photosynthesis under low nitrogen conditions. *Crop Des.* **2023**, *2*, 100025. [CrossRef]
10. Sun, H.; Qian, Q.; Wu, K.; Luo, J.; Wang, S.; Zhang, C.; Ma, Y.; Liu, Q.; Huang, X.; Yuan, Q.; et al. Heterotrimeric G proteins regulate nitrogen-use efficiency in rice. *Nat. Genet.* **2014**, *46*, 652–656. [CrossRef]
11. Kunihiro, S.; Saito, T.; Matsuda, T.; Inoue, M.; Kuramata, M.; Taguchi-Shiobara, F.; Youssefian, S.; Berberich, T.; Kusano, T. Rice *DEP1*, encoding a highly cysteine-rich G protein γ subunit, confers cadmium tolerance on yeast cells and plants. *J. Exp. Bot.* **2013**, *64*, 4517–4527. [CrossRef] [PubMed]
12. Cui, Y.; Jiang, N.; Xu, Z.; Xu, Q. Heterotrimeric G protein are involved in the regulation of multiple agronomic traits and stress tolerance in rice. *BMC Plant Biol.* **2020**, *20*, 90. [CrossRef] [PubMed]
13. Zhang, D.; Zhou, Y.; Yin, J.; Yan, X.; Lin, S.; Xu, W.; Baluška, F.; Wang, Y.; Xia, Y.; Liang, G.; et al. Rice G-protein subunits *qPE9-1* and *RGB1* play distinct roles in abscisic acid responses and drought adaptation. *J. Exp. Bot.* **2015**, *66*, 6371–6384. [CrossRef] [PubMed]
14. Wu, Y.; Zhao, Y.; Yu, J.; Wu, C.; Wang, Q.; Liu, X.; Gao, X.; Wu, K.; Fu, X.; Liu, Q. Heterotrimeric G protein γ subunit DEP1 synergistically regulates grain quality and yield by modulating the TTP (TON1-TRM-PP2A) complex in rice. *J. Genet. Genom.* **2023**, *50*, 528–531. [CrossRef] [PubMed]
15. Zhao, M.; Wang, Q.; Sun, J.; Xu, Z.; Chen, W. Research progess of rice erect panicle gene *DEP1*. *Sci. Sin. Vitae* **2017**, *47*, 1036–1042. [CrossRef]
16. Xu, H.; Sun, J.; Xu, Q.; Pan, G.; Zhou, G.; Zhang, Z.; Sun, Y.; Xu, Z.; Chen, W. Research progress on optimizing panicle characters and quality improvement of *indica* pedigree in northern *japonica* rice. *Chin. Sci. Bull. Kexue Tongbao* **2022**, *67*, 135–142. [CrossRef]
17. Zhu, J.; Wang, J.; Yang, J.; Fan, F.; Yang, J.; Zhong, W. Development and application for a functional marker for erect panicle gene *qPE9-1* of rice. *Mol. Plant Breed.* **2012**, *10*, 583–588.
18. Pulindala, S.; Bommisetty, R.; Withanawasam, D.M.; Somagutta, S.; Kommana, M.; Kalluru, S.; Keerthi, I.; Chintala, S.; Vemireddy, L.R. Enhancement of yield through marker-assisted pyramiding of yield contributing genes in rice (*Oryza sativa* L.) var. BPT5204. *Euphytica* **2022**, *218*, 148. [CrossRef]
19. Li, P.; Li, Z.; Liu, X.; Zhang, H.; Wang, Q.; Li, N.; Ding, H.; Yao, F. Development and application of intragenic markers for 14 nitrogen-use efficiency genes in rice (*Oryza sativa* L.). *Front. Plant Sci.* **2022**, *13*, 891860. [CrossRef]
20. Wang, Y.; Zhang, N.; Chen, H.; Wang, F.; Huang, Y.; Jia, B.; Wang, S.; Wang, Y.; Xu, Z. Effects of *DEP1* on grain yield and grain quality in the background of two *japonica* rice (*Oryza sativa*) cultivars. *Plant Breed.* **2020**, *139*, 608–617. [CrossRef]
21. Soroka, M.; Wasowicz, B.; Rymaszewska, A. Loop-mediated isothermal amplification (LAMP): The better sibling of PCR? *Cells* **2021**, *10*, 1931. [CrossRef] [PubMed]
22. Becherer, L.; Borst, N.; Bakheit, M.; Frischmann, S.; Zengerle, R.; von Stetten, F. Loop-mediated isothermal amplification (LAMP)–review and classification of methods for sequence-specific detection. *Anal. Methods* **2020**, *12*, 717–746. [CrossRef]
23. Tomita, N.; Mori, Y.; Kanda, H.; Notomi, T. Loop-mediated isothermal amplification (LAMP) of gene sequences and simple visual detection of products. *Nat. Protoc.* **2008**, *3*, 877–882. [CrossRef] [PubMed]
24. Notomi, T.; Okayama, H.; Masubuchi, H.; Yonekawa, T.; Watanabe, K.; Amino, N.; Hase, T. Loop-mediated isothermal amplification of DNA. *Nucleic Acids Res.* **2000**, *28*, e63. [CrossRef]

25. Cuadros, J.; Pérez-Tanoira, R.; Prieto-Pérez, L.; Martin-Martin, I.; Berzosa, P.; González, V.; Tisiano, G.; Balcha, S.; Ramos, J.M.; Górgolas, M. Field evaluation of malaria microscopy, rapid malaria tests and loop-mediated isothermal amplification in a rural hospital in south western Ethiopia. *PLoS ONE* **2015**, *10*, e0142842. [CrossRef] [PubMed]
26. Narushima, J.; Kimata, S.; Soga, K.; Sugano, Y.; Kishine, M.; Takabatake, R.; Mano, J.; Kitta, K.; Kanamaru, S.; Shirakawa, N.; et al. Rapid DNA template preparation directly from a rice sample without purification for loop-mediated isothermal amplification (LAMP) of rice genes. *Biosci. Biotechnol. Biochem.* **2019**, *84*, 670–677. [CrossRef] [PubMed]
27. Prasannakumar, M.K.; Parivallal, B.P.; Manjunatha, C.; Pramesh, D.; Narayan, K.S.; Venkatesh, G.; Banakar, S.N.; Mahesh, H.B.; Vemanna, R.S.; Rangaswamy, K.T. Rapid genotyping of bacterial leaf blight resistant genes of rice using loop-mediated isothermal amplification assay. *Mol. Biol. Rep.* **2021**, *48*, 467–474. [CrossRef]
28. Kishine, M.; Okunishi, T. Rapid identification of rice cultivar "Koshihikari" using loop-mediated isothermal amplification (LAMP). *J. Jpn. Soc. Food Sci. Technol.* **2011**, *58*, 591–596. [CrossRef]
29. Porebski, S.; Bailey, L.G.; Baum, B.R. Modification of a CTAB DNA extraction protocol for plants containing high polysaccharide and polyphenol components. *Plant Mol. Biol. Report.* **1997**, *15*, 8–15. [CrossRef]
30. Kawahara, Y.; Bastide, M.d.l.; Hamilton, J.P.; Kanamori, H.; McCombie, W.R.; Ouyang, S.; Schwartz, D.C.; Tanaka, T.; Wu, J.; Zhou, S.; et al. Improvement of the *Oryza sativa* Nipponbare reference genome using next generation sequence and optical map data. *Rice* **2013**, *6*, 4. [CrossRef]
31. Untergasser, A.; Cutcutache, I.; Koressaar, T.; Ye, J.; Faircloth, B.C.; Remm, M.; Rozen, S.G. Primer3—new capabilities and interfaces. *Nucleic Acids Res.* **2012**, *40*, e115. [CrossRef]
32. Goto, M.; Honda, E.; Ogura, A.; Nomoto, A.; Hanaki, K.-I. Colorimetric detection of loop-mediated isothermal amplification reaction by using hydroxy naphthol blue. *Biotechniques* **2009**, *46*, 167–172. [CrossRef] [PubMed]
33. Liu, Q.; Han, R.; Wu, K.; Zhang, J.; Ye, Y.; Wang, S.; Chen, J.; Pan, Y.; Li, Q.; Xu, X.; et al. G-protein $\beta\gamma$ subunits determine grain size through interaction with MADS-domain transcription factors in rice. *Nat. Commun.* **2018**, *9*, 852. [CrossRef] [PubMed]
34. Du, H.; Yu, Y.; Ma, Y.; Gao, Q.; Cao, Y.; Chen, Z.; Ma, B.; Qi, M.; Li, Y.; Zhao, X.; et al. Sequencing and de novo assembly of a near complete *indica* rice genome. *Nat. Commun.* **2017**, *8*, 15324. [CrossRef] [PubMed]
35. Babu, N.N.; Krishnan, S.G.; Vinod, K.K.; Krishnamurthy, S.L.; Singh, V.K.; Singh, M.P.; Singh, R.; Ellur, R.K.; Rai, V.; Bollinedi, H.; et al. Marker aided incorporation of *Saltol*, a major QTL associated with seedling stage salt tolerance, into *Oryza sativa* 'Pusa Basmati 1121'. *Front. Plant Sci.* **2017**, *8*, 41. [CrossRef] [PubMed]
36. Berdrejo, V.J.; Magulama, E.E. Efficiency of five emasculation techniques in rice. *USM CA Res. J.* **1994**, *5*, 9–12.
37. Plant Genetics and Breeding Program, Biology Department, Wuhan University. Development and Flowering of Rice Panicles. In *Panicle Development, Flowering and Fruiting in Rice*; Hubei People's Publishing House: Wuhan, China, 1975; pp. 42–45.
38. Srividya, A.; Maiti, B.; Chakraborty, A.; Chakraborty, G. Loop mediated isothermal amplification: A promising tool for screening genetic mutations. *Mol. Diagn. Ther.* **2019**, *23*, 723–733. [CrossRef]
39. Gouda, G.; Gupta, M.K.; Donde, R.; Mohapatra, T.; Vadde, R.; Behera, L. Marker-assisted selection for grain number and yield-related traits of rice (*Oryza sativa* L.). *Physiol. Mol. Biol. Plants Int. J. Funct. Plant Biol.* **2020**, *26*, 885–898. [CrossRef]
40. Muhammad Gul, A.; Hameed, G. Application techniques of molecular marker and achievement of marker assisted selection (MAS) in three major crops rice, wheat and maize. *Int. J. Res. Appl. Sci. Biotechnol.* **2021**, *8*, 82–93. [CrossRef]
41. Gu, Z.; Gong, J.; Zhu, Z.; Li, Z.; Feng, Q.; Wang, C.; Zhao, Y.; Zhan, Q.; Zhou, C.; Wang, A.; et al. Structure and function of rice hybrid genomes reveal genetic basis and optimal performance of heterosis. *Nat. Genet.* **2023**, *55*, 1745–1756. [CrossRef]
42. Wei, S.; Jiang, J.; Liu, W.; Li, X.; Luo, J. Decade of crop science:Funding, outcomes and future prospects of national natural science foundation of China. *Bull. Natl. Nat. Sci. Found. China* **2022**, *36*, 972–981. [CrossRef]
43. Amiteye, S. Basic concepts and methodologies of DNA marker systems in plant molecular breeding. *Heliyon* **2021**, *7*, e08093. [CrossRef] [PubMed]
44. Yang, G.; Chen, S.; Chen, L.; Sun, K.; Huang, C.; Zhou, D.; Huang, Y.; Wang, J.; Liu, Y.; Wang, H.; et al. Development of a core SNP arrays based on the KASP method for molecular breeding of rice. *Rice* **2019**, *12*, 21. [CrossRef] [PubMed]
45. Li, R.; Ba, Y.; Song, Y.; Cui, J.; Zhang, X.; Zhang, D.; Yuan, Z.; Yang, L. Rapid and sensitive screening and identification of CRISPR/Cas9 edited rice plants using quantitative real-time PCR coupled with high resolution melting analysis. *Food Control* **2020**, *112*, 107088. [CrossRef]
46. Glökler, J.; Lim, T.S.; Ida, J.; Frohme, M. Isothermal amplifications—A comprehensive review on current methods. *Crit. Rev. Biochem. Mol.* **2021**, *56*, 543–586. [CrossRef] [PubMed]
47. Edwards, K.; Johnstone, C.; Thompson, C. A simple and rapid method for the preparation of plant genomic DNA for PCR analysis. *Nucleic Acids Res.* **1991**, *19*, 1349. [CrossRef] [PubMed]
48. Cai, S.; Zhao, X.; Pittelkow, C.M.; Fan, M.; Zhang, X.; Yan, X. Optimal nitrogen rate strategy for sustainable rice production in China. *Nature* **2023**, *615*, 73–79. [CrossRef]
49. Zhang, Q. Strategies for developing Green Super Rice. *Proc. Natl. Acad. Sci. USA* **2007**, *104*, 16402–16409. [CrossRef]
50. Yu, S.; Ali, J.; Zhou, S.; Ren, G.; Xie, H.; Xu, J.; Yu, X.; Zhou, F.; Peng, S.; Ma, L.; et al. From Green Super Rice to green agriculture: Reaping the promise of functional genomics research. *Mol. Plant* **2021**, *15*, 9–26. [CrossRef]
51. Li, P.; Chang, T.; Chang, S.; Ouyang, X.; Qu, M.; Song, Q.; Xiao, L.; Xia, S.; Deng, Q.; Zhu, X.-G. Systems model-guided rice yield improvements based on genes controlling source, sink, and flow. *J. Integr. Plant Biol.* **2018**, *60*, 1154–1180. [CrossRef]

52. Singh, B.K.; Delgado-Baquerizo, M.; Egidi, E.; Guirado, E.; Leach, J.E.; Liu, H.; Trivedi, P. Climate change impacts on plant pathogens, food security and paths forward. *Nat. Rev. Microbiol.* **2023**, *21*, 640–656. [CrossRef] [PubMed]
53. Li, S.; Feng, T.; Zhang, C.; Zhang, F.; Li, H.; Chen, Y.; Liang, L.; Zhang, C.; Zeng, W.; Liu, E.; et al. Genetic dissection of salt tolerance and yield traits of geng (*japonica*) rice by selective subspecific introgression. *Curr. Issues Mol. Biol.* **2023**, *45*, 4796–4813. [CrossRef] [PubMed]

current issues in molecular biology

MDPI

Article

Genome-Wide Identification and Expression Analyses of the *FAR1/FHY3* Gene Family Provide Insight into Inflorescence Development in Maize

Huaijun Tang [1,†], De Jing [2,†], Cheng Liu [1], Xiaoqing Xie [1], Lei Zhang [1], Xunji Chen [3,*] and Changyu Li [2,*]

1 Institute of Grain Crops, Xinjiang Academy of Agricultural Sciences, Urumqi 830091, China; tanghuaijun83@sina.com (H.T.); liuchengxj@126.com (C.L.); 13565990757@163.com (X.X.); 18999224030@163.com (L.Z.)
2 Key Laboratory of Genome Research and Genetic Improvement of Xinjiang Characteristic Fruits and Vegetables, Institute of Horticultural Crops, Xinjiang Academy of Agricultural Sciences, Urumqi 830091, China; 17608476967@163.com
3 Institute of Nuclear and Biotechnology, Xinjiang Academy of Agricultural Sciences, Urumqi 830091, China
* Correspondence: chenxunji@xaas.ac.cn (X.C.); licy@xaas.ac.cn (C.L.)
† These authors contributed equally to this work.

Abstract: As transcription factors derived from transposase, *FAR-RED IMPAIRED RESPONSE1* (*FAR1*) and its homolog *FHY3* play crucial roles in the regulation of light signaling and various stress responses by coordinating the expression of downstream target genes. Despite the extensive investigation of the FAR1/FHY3 family in *Arabidopsis thaliana* and other species, a comprehensive examination of these genes in maize has not been conducted thus far. In this study, we employed a genomic mining approach to identify 16 *ZmFAR1* genes in the maize inbred line B73, which were further classified into five subgroups based on their phylogenetic relationships. The present study characterized the predicted polypeptide sequences, molecular weights, isoelectric points, chromosomal distribution, gene structure, conserved motifs, subcellular localizations, phylogenetic relationships, and cis-regulatory elements of all members belonging to the ZmFAR1 family. Furthermore, the tissue-specific expression of the 16 *ZmFAR1* genes was analyzed using RNA-seq, and their expression patterns under far-red light conditions were validated in the ear and tassel through qRT-qPCR. The observed highly temporal and spatial expression patterns of these *ZmFAR1* genes were likely associated with their specific functional capabilities under different light conditions. Further analysis revealed that six *ZmFAR1* genes (*ZmFAR1-1*, *ZmFAR1-10*, *ZmFAR1-11*, *ZmFAR1-12*, *ZmFAR1-14*, and *ZmFAR1-15*) exhibited a response to simulated shading treatment and actively contributed to the development of maize ears. Through the integration of expression quantitative trait loci (eQTL) analyses and population genetics, we identified the presence of potential causal variations in *ZmFAR1-14* and *ZmFAR1-9*, which play a crucial role in regulating the kernel row number and kernel volume weight, respectively. In summary, this study represents the initial identification and characterization of ZmFAR1 family members in maize, uncovering the functional variation in candidate regulatory genes associated with the improvement of significant agronomic traits during modern maize breeding.

Keywords: *ZmFAR1*; transcription factors; far-red light; causal variation; maize

Citation: Tang, H.; Jing, D.; Liu, C.; Xie, X.; Zhang, L.; Chen, X.; Li, C. Genome-Wide Identification and Expression Analyses of the *FAR1/FHY3* Gene Family Provide Insight into Inflorescence Development in Maize. *Curr. Issues Mol. Biol.* **2024**, *46*, 430–449. https://doi.org/10.3390/cimb46010027

Academic Editor: Vijai Bhadauria

Received: 4 December 2023
Revised: 28 December 2023
Accepted: 31 December 2023
Published: 2 January 2024

1. Introduction

Light is crucial for plant growth [1,2], significantly affecting development under canopy shade. Plants use sophisticated photoreceptors to detect light changes [3] and adapt their growth strategies. The FAR1/FHY3 gene family plays a key role in plant light signaling and shade adaptation [4], with far-red light activating the *FAR1/FHY3* genes to enhance the phyA-mediated high light responses [5,6], aiding in light detection and growth. The FAR1/FHY3 proteins contain three domains: the C2H2 zinc finger, core transposase, and SWIM zinc finger domains [7,8]. Arabidopsis has 14 identified *FAR1/FHY3* genes, phylogenetically classified into five groups, encoding sequences of 531 to 851 amino acids, with specific functions across tissues and developmental stages [5,7].

FAR1 and *FHY3* are multifunctional genes studied in plants, known to upregulate ELF4 and influence flowering by connecting light signals to the miR156-SPL aging pathway, as well as to regulate starch synthesis by activating ISA2 [9–12]. In *far1* and *fhy3* mutants, the lack of *CCA1* expression suggests that these genes also affect light-induced processes [10]. Additionally, *FAR1/FHY3* interact with jasmonic acid pathways to balance growth and defense [12,13] and affect branching through *SPL9/SPL15* interactions and strigolactone signaling [14]. These genes also play roles in chlorophyll production, chloroplast division, ABA signaling, and delaying leaf senescence by integrating light and ethylene signals [15,16]. FHY3's extensive targets imply its broad impact on plant development [6].

In addition, the upregulation of *THERMO-SENSITIVE BARREN PANICLE* (*TAP*), a member of the FAR1 family in rice, might be induced by elevated temperatures. This up-regulation is believed to modulate the proper growth and development of rice inflorescences and spikelets through the formation of protein complexes with *OsYABBY4/5* [17]. Additionally, several *AhFAR1* genes have been detected in peanuts, suggesting their potential involvement in regulating pod development [18]. Moreover, the dispersed expansion of FAR1/FRS-like genes has been observed in *Rosa wichuraiana* 'Basye's Thornless', and their diverse expression patterns have been linked to the control of shoot growth and the flowering time [19]. Despite the numerous attributions assigned to these genes, there remains a dearth of systematic inquiries into this gene family in maize. Consequently, it is imperative to ascertain the constituents of the ZmFAR1 family in maize and comprehend their responses to shading (light conditions enriched in far-red light), along with their prospective functions in governing plant structural development.

Maize is a major global crop facing challenges such as reduced productivity due to shade avoidance syndrome (SAS) under high-density planting, which is further influenced by decreased red/far-red light [20,21]. In this study, a comprehensive analysis was conducted to identify ZmFAR1 family members in the maize B73 genome. The investigation included an examination of the physicochemical properties, sequence characteristics, conserved domains, and phylogenetic relationships of 16 *ZmFAR1* genes. Furthermore, the subcellular localization of maize genes was predicted, and analyses of the cis-regulatory elements and miRNA–mRNA regulatory networks were performed to gain insights into potential gene functions. The expression patterns of the *ZmFAR1* genes were analyzed using RNA-seq data and the expression levels were further validated through qRT-PCR experiments in the ear and tassel at various developmental stages under simulated shading conditions. In addition, we endeavored to ascertain the potential causal variations and favorable haplotypes for ZmFAR1 family members through the integration of eQTL analyses and population genetics. Our discoveries might establish a theoretical framework for future investigations into the biological functions of ZmFAR1 transcription factor family members and provide valuable perspectives on the genetic enhancement of maize inbred lines during modern maize breeding.

2. Materials and Methods

2.1. Identification of ZmFAR1 Genes in Maize B73

The FAR1 gene family's Hidden Markov Models (HMMs) PF03101, PF10551, and PF04434 were obtained from the Pfam protein family database (https://pfam.xfam.org/) (accessed on 2 April 2023). The maize B73 genome sequence, gff3 annotation, and protein sequence files were obtained from the Maize Genome Database (https://www.maizegdb.org/) (accessed on 2 April 2023). The protein sequences were scanned using hmmsearch in the HMMER (v3.3.2) software, and those with an E-value < 1×10^{-5} were treated as criteria for the screening of high-quality FAR1 proteins [22]. The FAR1 protein sequences of *Arabidopsis thaliana* were also retrieved from the TAIR database (https://www.arabidopsis.org/) (accessed on 3 April 2023). The BLASTP (v2.9.0) comparison tool was utilized for the identification of FAR1 homologous genes in maize [23]. The BLASTP results were integrated with the hmmsearch results, and the intersections were taken as candidate ZmFAR1 family members. Finally, protein domain prediction was performed via the Interpro (https://www.ebi.ac.uk/interpro, accessed on 7 February 2023) and SMART online tools (http://smart.embl-heidelberg.de/) to perform further verification (accessed on 5 April 2023). The physicochemical properties and subcellular localizations of *ZmFAR1* genes were analyzed through ExPASy (http://www.expasy.org) and WoLFPSORT (https://wolfpsort.hgc.jp/), respectively (accessed on 7 April 2023). The secondary structure was also predicted through SOPMA (https://npsa-prabi.ibcp.fr/cgi bin/npsa_automat.plpage=npsa_sopma.html) (accessed on 8 April 2023).

2.2. Phylogenetic Analyses of ZmFAR1 Genes

Multiple sequence alignment of the FAR1 protein sequences of maize B73, *Sorghum bicolor*, and *Arabidopsis thaliana* was performed using the muscle (v3.8.31) program with default parameters [24]. Based on the alignment results, the optimal model was validated, and the maximum-likelihood approach was used to create the phylogenetic tree using iqtree (v1.6.12); the bootstrap value was set to 1000 [25]. Using the web application iTOL, the phylogenetic tree was visualized and rendered (https://itol.embl.de/) (accessed on 10 April 2023).

2.3. Gene Structure and Conserved Motif Analysis of ZmFAR1 Genes

The *ZmFAR1* gene structures were investigated through GSDS (http://gsds.cbi.pku.edu.cn/) (accessed on 11 April 2023) [26]. Using web-based MEME, the conserved motifs in ZmFAR1 family proteins were predicted. The number of motifs was set to 15 (https://meme-suite.org/meme) (accessed on 12 April 2023) [27]. The evolutionary tree, exon–intron structures, and conserved motifs were all visualized by the TBtools program [28].

2.4. Chromosomal Location and Collinearity Analysis of ZmFAR1 Genes

Based on the genetic information and length information of maize chromosomes, all distributions of *ZmFAR1* genes were mapped using MapChart (v2.3.2). The collinearity of FAR1 family members among *Sorghum bicolor*, maize B73, and *Oryza sativa* was analyzed using MCScanX with default parameters and visualized by TBtools [28,29].

2.5. Cis-Acting Elements and Interaction Network Analysis of ZmFAR1 Genes

The upstream sequences (2000 bp) of all *ZmFAR1* genes were extracted to perform cis-acting element analysis by uploading to the PlantCARE database (http://bioinformatics.psb.ugent.be/webtools/plantcare/html/) (accessed on 13 April 2023). Using the R program ggplot2, the distributions and functions of all anticipated cis-regulatory elements in the promoters of all genes were studied and visualized [30]. The psRNAtarget online platform was used to predict the microRNAs (miRNAs) that target *ZmFAR1*. The expectation was set to 5 and no gaps were permitted (https://www.zhaolab.org/psRNATarget/) (accessed on 15 April 2023). Cytoscape was used to provide a visual representation of the interaction network.

2.6. Tissue-Specific Expression Analysis of ZmFAR1 Genes

Using the NCBI SRA database, we obtained transcriptome data for the ear, leaf tip, tassel, leaf base, leaf center, embryo, root, shoot, anther, and endosperm of maize B73 (http://www.ncbi.nlm.nih.gov/sra/) with the accession number ERR3773807-ERR3773827 (accessed on 20 April 2023). All raw sequencing data were first preprocessed using fastp (v0.20.1) [31], followed by mapping to the reference genome using HISAT2 [32]. Cufflinks and Cuffcompare were utilized to assemble and annotate the predicted transcripts, respectively [33]. After normalizing the gene counts, the expression levels of all expressed genes were obtained, from which the expression levels of *ZmFAR1* genes were extracted and visualized using the R package ggplot2 [30]. The transcriptome data for the far-red and shade treatment of maize B73 were obtained from the Maize RNA-seq Database (http://ipf.sustech.edu.cn/pub/zmrna/download/, accessed on 20 January 2023).

2.7. Plant Materials and Stress Treatments

The seedlings of the maize inbred line B73 were grown in an incubator at an ambient temperature of 27 °C. The intensity of white light was set to 330 μmol m^{-2} s^{-1} and the R:FR ratio was 14.3. The photoperiod was adjusted to 12 h light/12 h dark. The control group was defined as follows: the seedlings of maize B73 were grown under white light until the sampling period. The following criteria were used to define the treatment groups: the seedlings of maize B73 were grown in a simulated shade treatment after the seedlings were grown until stage V2 and then treated with 21 μmol m^{-2} s^{-1} far-red light for 30 min at the end of each day (EOD-FR treatment). The ear and tassel were collected at stages V8 to V11 and V8 to V12, respectively. After the samples were collected, they were promptly frozen in liquid nitrogen and stored in a −80 °C freezer until they were utilized for RNA extraction.

2.8. RNA Isolation and Quantitative qRT-PCR Analysis

Total RNA was extracted from samples using an RNAsimple Total RNA Kit (Tiangen, Beijing, China). First, cDNAs were reverse-transcribed from the total mRNAs using the Hifair® III 1st Strand cDNA Synthesis SuperMix (Yeasen Biotechnology, Shanghai, China). Then, quantitative real-time PCR (qRT-PCR) was performed using Hieff UNICON® qPCR SYBR Green Master Mix (Yeasen Biotechnology, Shanghai, China) on an Applied Biosystems 7500 real-time PCR system. The expression levels of *ZmFAR1* genes were calculated by the $2^{-\Delta CT}$ method and normalized against the reference gene ACTIN [34]. Values were calculated as the means ± standard deviations (SD) of three biological replicates. All primer sequences are listed in Supplementary Table S4.

2.9. Expression Quantitative Trait Loci (eQTL) Analysis

We performed eQTL analysis based on high-quality SNPs and gene expression levels in the ear in a previously reported study and using the LMM model in the EMMAX software emmax-intel64-20120205 [35,36]. The method of cis-eQTL identification was similar to a previous study, with minor modifications [35].

2.10. Statistical Analysis

Data processing with simple calculations (mean and standard deviation) and Microsoft Excel (version 16.01) were used to perform Student's *t*-tests. The p-value < 0.05 was considered statistically significant between the control and treatment (* p-value < 0.05, ** p-value < 0.01).

3. Results

3.1. Identification and Chromosomal Location of ZmFAR1 Genes in Maize

In the maize inbred line B73 genome, 16 putative *ZmFAR1* genes were found and named *ZmFAR1-1* to *ZmFAR1-16* based on their chromosomal locations. Based on the genomic annotation information of maize inbred line B73, all 16 *ZmFAR1* genes were unevenly distributed throughout the eight chromosomes (Figure 1). Specifically, *ZmFAR1-1*, *ZmFAR1-2*, and *ZmFAR1-3* were located on chromosome 1 (Chr1); *ZmFAR1-8*, *ZmFAR1-9*, and *ZmFAR1-10* were located on Chr5; and *ZmFAR1-11*, *ZmFAR1-12*, and *ZmFAR1-13* were located on Chr7. Chr3 and Chr10 each contained two gene members (*ZmFAR1-5* and *ZmFAR1-6*; *ZmFAR1-15* and *ZmFAR1-16*), and Chr2, Chr4, and Chr9 contained *ZmFAR1-4*, *ZmFAR1-7*, and *ZmFAR1-14*, respectively.

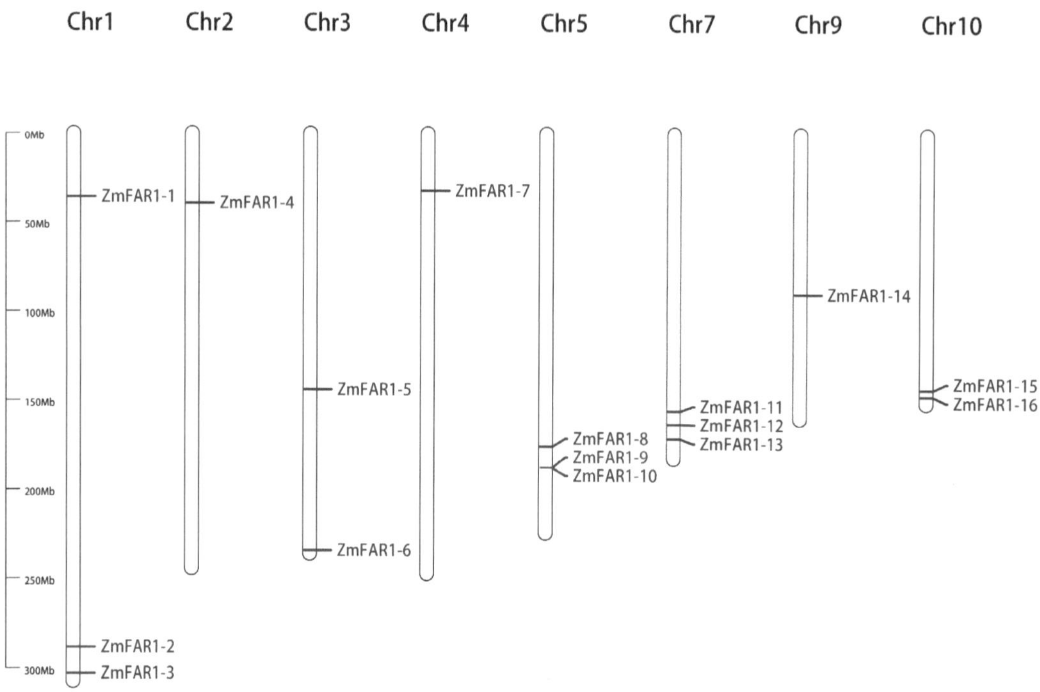

Figure 1. Distribution of ZmFAR1 family members in maize B73.

3.2. Phylogenetic Analysis of ZmFAR1 Genes

To evaluate the evolutionary relationships among the FAR1 proteins, an unrooted phylogenetic tree of 79 FAR1 protein sequences was constructed using the ML method (Supplementary Table S1). These 79 FAR1 proteins were derived from three species, 14 from *Arabidopsis thaliana* (*A. thaliana*), 49 from *Sorghum bicolor* (*S. bicolor*), and 16 from maize B73. All 79 *FAR1* genes were divided into five subgroups according to the classification of *FAR1* genes in *Arabidopsis*, designated I to V (Figure 2). There were 5 *AtFAR1* members in the I subgroup; 5 *ZmFAR1* members, 2 *AtFAR1* members, and 10 *SbFAR1* members in the II subgroup; 3 *ZmFAR1* members, 5 *AtFAR1* members, and 4 *SbFAR1* members in the III subgroup; 1 *ZmFAR1* member, 2 *AtFAR1* members, and 24 *SbFAR1* members in the IV subgroup; and 7 *ZmFAR1* members and 11 *SbFAR1* members in the V subgroup. However, subgroup I had no *ZmFAR1* members, and subgroup V contained the largest number of ZmFAR1 proteins.

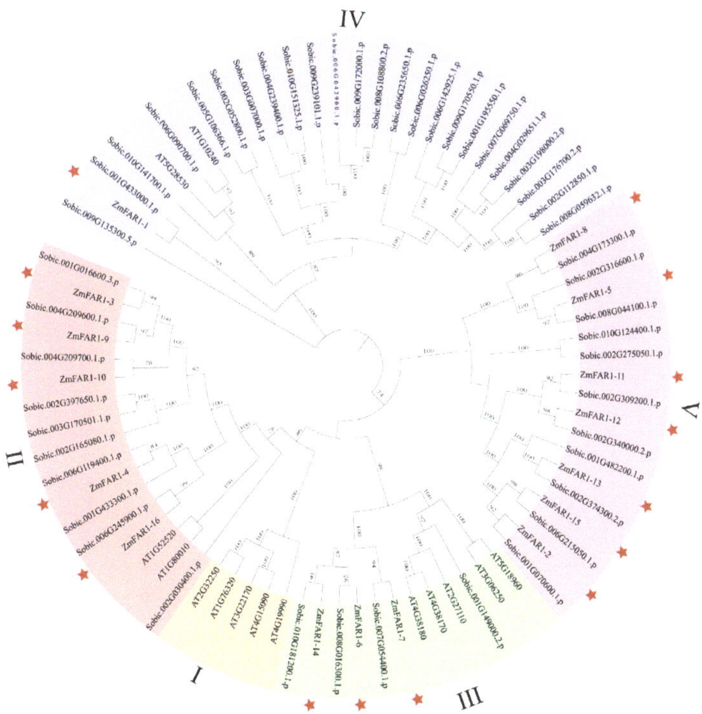

Figure 2. Phylogenetic analysis of maize B73, *Sorghum bicolor*, and *A. thaliana*. The genes labeled with "AT" represent the genes of *Arabidopsis*. The genes labeled with "Zm" represent the genes of maize inbred line B73. The genes labeled with "Sobic" represent the genes of *Sorghum bicolor*. Each position of the *ZmFAR1* gene is noted with an asterisk. All *FAR1* genes were subdivided into groups I through V.

3.3. Conserved Structure and Motif Analyses of ZmFAR1 Genes

To explore the distribution and structural diversification of the conserved motifs of the ZmFAR1 proteins, we analyzed their conserved motifs and intron–exon structure. The results revealed that all ZmFAR1 proteins in maize B73 featured an N-terminal C2H2 zinc finger domain at the N-terminus with the capability of binding DNA (Supplementary Figure S1). As shown in Figure 3A, the *ZmFAR1* genes with identical conserved motifs were clustered along the same evolutionary branch, suggesting that they might have functional similarities. Then, we further analyzed the conserved DNA motifs, and 15 conserved motifs were identified in the 16 ZmFAR1 proteins (Figures 3B and S2). This result illustrated that the highly conserved *ZmFAR1* genes might share the same structural characteristics. The structure motif3-motif9-motif8-motif6 (FAR1 DNA-binding domain) was found to be present in all family members. Interestingly, *ZmFAR1-1*, *ZmFAR1-11*, and *ZmFAR1-12* contained two C2H2 zinc finger domains, suggesting that they might have unique functions compared to other *ZmFAR1* genes. *ZmFAR1-2* and *ZmFAR1-15* contained all motifs, while *ZmFAR1-3* only contained four motifs. All subgroups except *ZmFAR1-1* and *ZmFAR1-3* contained motif 2, motif 4, and motif 7, while motif 10, motif 12, and motif 13 were only present in subgroups II and IV. In addition, motif 15 appeared only in the IV subgroup. Thus, *ZmFAR1* genes with comparable conserved motifs were clustered in the same evolutionary branch, and this finding was compatible with the phylogenetic relationships (Figure 3A,B).

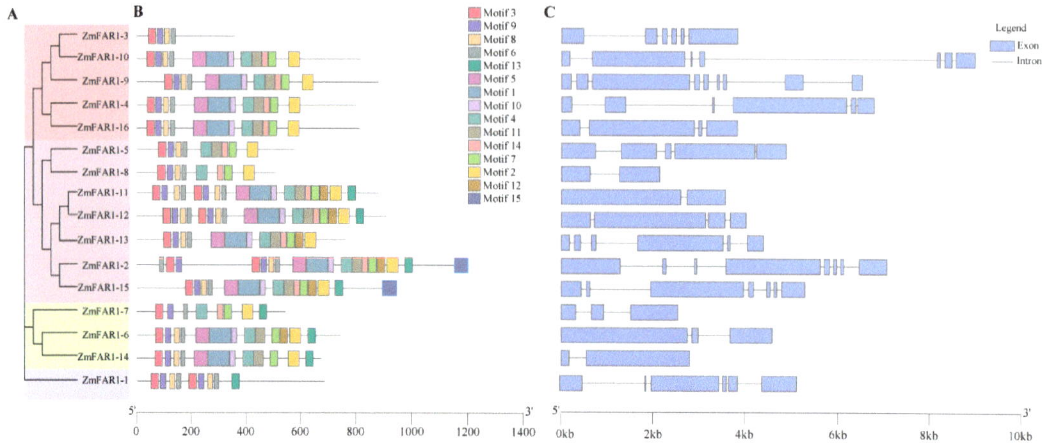

Figure 3. The gene structure and motifs of ZmFAR1 family members in maize B73. (**A**) Phylogenetic tree of *ZmFAR1* genes. (**B**) Conserved motifs of *ZmFAR1* proteins. The order of the motifs corresponds to their position within individual protein sequences. (**C**) The exon–intron organization of *ZmFAR1* genes. Blue boxes represent exons; black lines represent introns. The lengths of exons and introns for each *ZmFAR1* gene are proportionally displayed.

The exon–intron structures could reveal the lineage-specific features of eukaryotic genes, as they differ significantly among eukaryotic genomes [37]. Therefore, exon–intron organization analysis of the *ZmFAR1* genes was performed to understand the structural diversity and characteristics (Figure 3C). According to the data, *ZmFAR1-9* included the highest number of exons (9), followed by *ZmFAR1-2* (8), *ZmFAR1-10* (7), and *ZmFAR1-15* (7). In contrast, *ZmFAR1-8*, *ZmFAR1-11*, and *ZmFAR1-14* had only two exons and one intron, whereas *ZmFAR1-5* had five exons and four introns. *ZmFAR1-12* and *ZmFAR1-16* each had four exons and three introns. *ZmFAR1-6* and *ZmFAR1-7* both had three exons and two introns. The remaining genes all contained six exons and five introns.

3.4. Collinearity Analysis of ZmFAR1 Genes

Collinearity analysis could provide insights into the divergence and evolutionary history of the genome [38]. To further investigate the covariance of the *FAR1* gene family, we identified collinear relationships among *FAR1* in *S. bicolor*, maize B73, and *A. thaliana* (Figure 4). A total of 15 homologous gene pairs between maize B73 and *S. bicolor* were found, with *ZmFAR1-13* being homologous to two *S. bicolor* genes (*Sobic.0G482200.1* and *Sobic.0G374300.2*). The homologous genes of *ZmFAR1-4*, *ZmFAR1-8*, and *ZmFAR1-9* (*Sobic.0G119400.1*, *Sobic.0G173300.1*, and *Sobic.0G209600.1*) could only be found in *S. bicolor*. There was a total of 14 homologous gene pairs found between maize B73 and *O. sativa*. *ZmFAR1-6* and *ZmFAR1-16* were found to be homologous to *Sobic.0G016300.1* and *Sobic.006G245900.1*, respectively, but no homologous genes were found in *O. sativa*. Only *ZmFAR1-10* had no homologous genes in *S. bicolor* and *O. sativa*. Therefore, studying the evolutionary relationships of the FAR1 family members will aid in the study of *ZmFAR1* gene function.

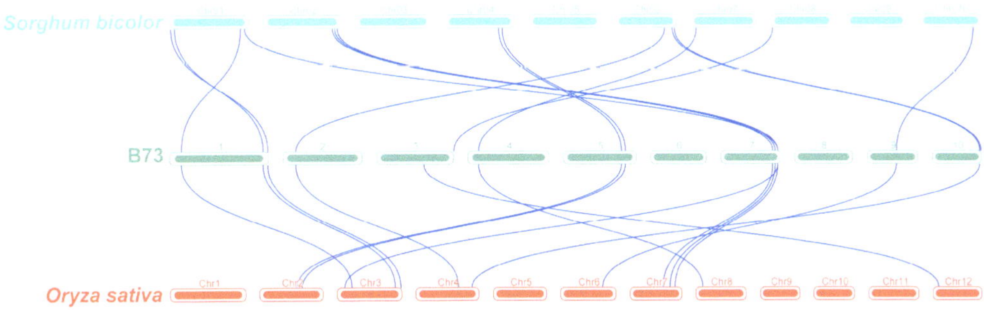

Figure 4. Synteny analysis of FAR1 genes between *Sorghum bicolor*, maize inbred line B73, and Oryza sativa. Gray lines represent all collinearity blocks between genomes, and the colinear *FAR1* gene pairs are highlighted by blue lines. The numbers in the figure represent chromosomes.

3.5. Cis-Acting Element Analysis of ZmFAR1 Genes

Cis-regulatory elements are specific DNA sequences that are located upstream of gene coding sequences, which are specific binding sites for the proteins involved in the initiation and regulation of transcription [39]. To investigate the probable roles and regulation of ZmFAR1 family members, we performed cis-acting element analysis based on the putative promoter sequences in the upstream 2000 bp region of each gene. In total, 374 cis-elements were identified across all *ZmFAR1* genes, which could be categorized into four primary groups, including hormone and stress responses, light responses, and plant growth and development (Figure 5). The majority of the 16 *ZmFAR1* genes included light-responsive cis-elements (40.11%, 150/374), followed by hormone-responsive elements (32.89%, 123/374), stress response elements (20.05%, 75/374), and the lowest proportion of cis-elements (6.95%, 26/374) associated with plant growth and development (Supplementary Table S2). The Gap-box, GA-motif, TCT-motif, GATA-motif, Sp1, TCCC-motif, GTGGC-motif, GT1-motif, chs-CMA1a, CAG-motif, I-box, MRE, ATCT-motif, ACE, Box 4, G-box, and AE-box were all identified as light-responsive cis-elements. Of these, the G-box was the most abundant (24%), followed by the GT1-motif (12.67%) and Sp1 (12%). Hormone-responsive cis-elements included growth hormone-responsive elements (AuxRR-core and TGA-element), gibberellin-responsive elements (GARE motif, TATC-box, and P-box), MeJA-responsive elements (TGACG-motif and CGTCA-motif), and salicylic acid responsiveness (TCA-element and ABRE). Among them, ABRE was the most abundant (25.2%, 31/123), and AuxRR-core and P-box were the least abundant (2.44%, 3/123) among the hormone-responsive cis-elements. The stress-responsive cis-elements comprised 46.67% ARE (anaerobic induction), 22.6% MBS (drought induction), 14.6% GC-motif (hypoxia-specific induction), 10.67% LTR (low-temperature response), and 5.3% TC-rich repeat sequences (defense and stress responsiveness). The plant growth and development cis-elements comprised 11.54% GCN4_motif (cis-regulation of endosperm expression), 53.85% CAT-box (meristem expression), 7.69% RY-elements (seed-specific regulation), and 26.92% O2-site maize alcoholic proteins (metabolic regulation). The investigated cis-elements had the potential to play vital roles in the process of regulating gene expression in response to a variety of stimuli, as well as in the process of plant growth and development. We also found differences in the number of cis-acting elements of each ZmFAR1 family member. *ZmFAR1-12* had the highest number of cis-elements (31 members), followed by *ZmFAR1-3* and *ZmFAR1-9* (28 members), while *ZmFAR1-7* had the lowest number of cis-elements (14 members). These results indicate that the functional differences between different subfamily members may be due to the presence of different numbers of cis-regulatory elements.

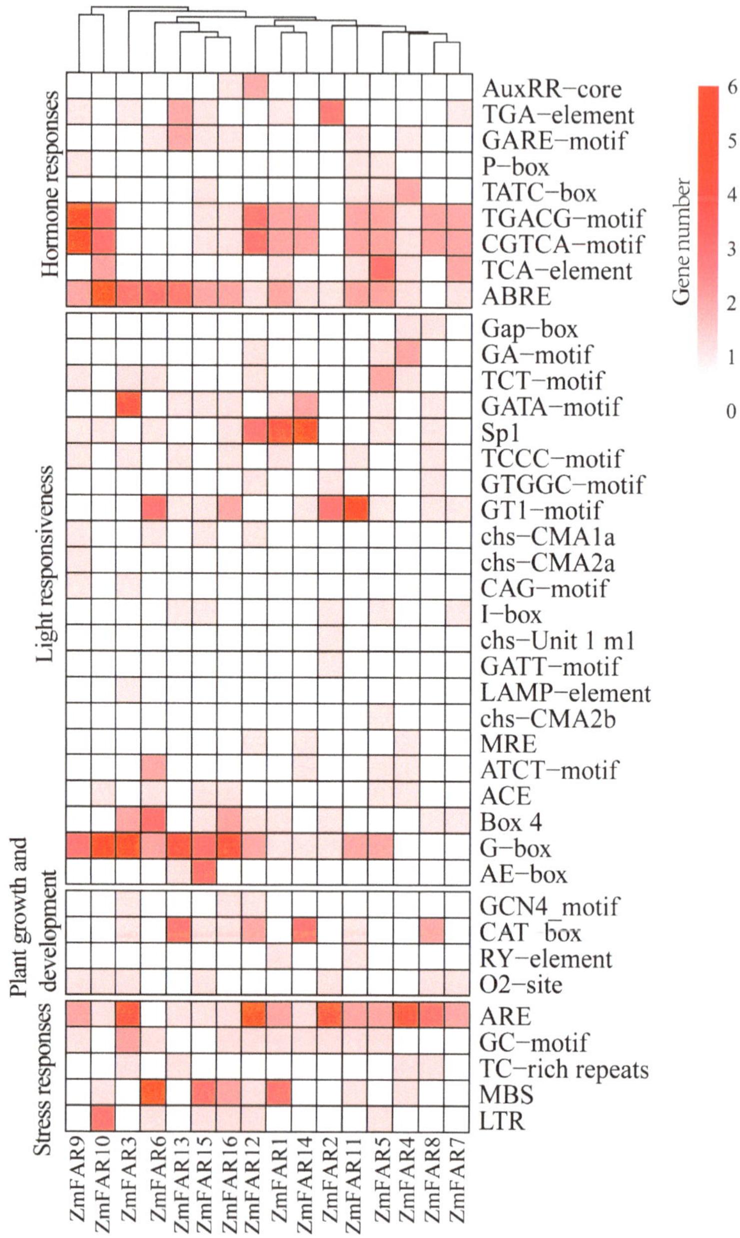

Figure 5. Cis-acting elements in the promoters of ZmFAR1 family members.

3.6. Target miRNA Prediction and Interactive Network Analysis of ZmFAR1 Genes

miRNAs control multiple cellular and biological processes by regulating gene expression. To predict the downstream targets of the *ZmFAR1* genes, we constructed a miRNA–mRNA coexpression network. The results showed that there was a potential targeting relationship between 179 miRNAs and 16 *ZmFAR1* genes (Supplementary Table S3). *ZmFAR1-4*, *ZmFAR1-5*, and *ZmFAR1-10* were the primary targets of 84 miRNAs from

20 miRNA families. In particular, *ZmFAR1-4* was targeted by the highest number of miRNAs (45 members) from 12 miRNA families (Figure 6). *ZmFAR1-10*, *ZmFAR1-5*, *ZmFAR1-2*, *ZmFAR1-15*, and *ZmFAR1-12* were targeted by 31, 26, 20, 20, and 19 miRNAs, respectively. From the miRNA viewpoint, the miR2275 family could target 10 *ZmFAR1* genes, followed by miR159 (eight target *ZmFAR1* genes) and miR395 (seven target *ZmFAR1* genes). Previous studies showed that miR156 could regulate flowering, and we found that the miR156 family targets six *ZmFAR1* genes, which may also regulate plant flowering. Furthermore, six miRNAs (miR396a-3p, miR396b-3p, miR396g-3p, miR169r-3p, miR171g-5p, miR397a-5p, and miR397b-5p) could only target *ZmFAR1-11*, suggesting that this gene might play a different role compared to the other *ZmFAR1* family members.

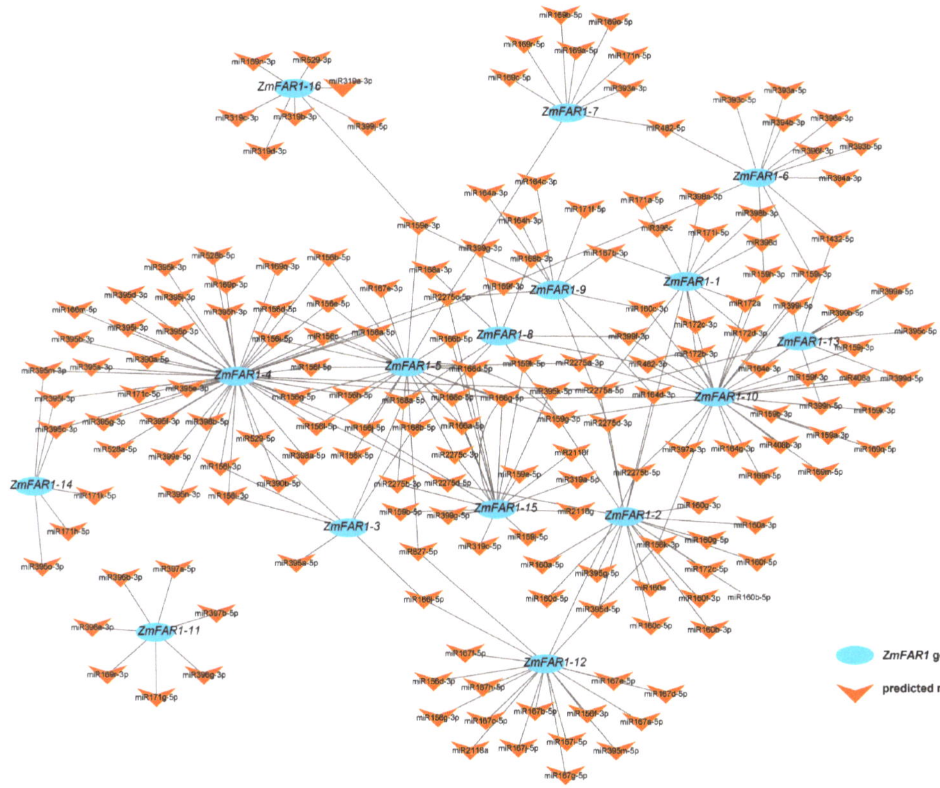

Figure 6. Predicted miRNA–mRNA regulatory network. Gray lines represent the interactions between miRNAs and *ZmFAR1* genes. The blue circles represent *ZmFAR1* genes, and the red arrowheads represent miRNAs.

3.7. Expression Analysis of ZmFAR1 Genes

Gene expression patterns provide important clues in understanding potential gene functions. To evaluate the transcriptome properties and biological activities of all *ZmFAR1* genes, we analyzed the expression patterns of these genes using public transcriptome data in different B73 tissues (ear, leaf tip, tassel, leaf base, leaf middle, embryo, root, shoot, anther, and endosperm), under simulated shade (low red/far-red ratio) and far-red (FR) light treatment (Figure 7). The results showed that the majority of the *ZmFAR1* family members were expressed in all tissues; however, their expression levels varied greatly among different tissues. Specifically, certain family members exhibited preferential expression in specific tissues. For instance, the highest expression levels of the *ZmFAR1-*

1 gene were found in the leaf tip, while the highest expression level of *ZmFAR1-6* was detected in the embryo. *ZmFAR1-15* showed considerably higher expression levels in the leaf tip and leaf middle than in other tissues. Intriguingly, the expression levels of all other *ZmFAR1* genes were highest in the ear and lower in both the anther and endosperm. In addition, all *ZmFAR1s* have different degrees of sensitivity to FR. *ZmFAR1-14* was induced under 1 h FR light treatment, and *ZmFAR1-7*, *ZmFAR1-9*, and *ZmFAR1-15* were induced under 6 h FR light treatment. *ZmFAR1-2* was reduced under 1 h and *ZmFAR1-12* was reduced under 6 h of shade treatment. These findings suggest that various members of the ZmFAR1 family might play diverse roles in separate tissues and suggest a potential role for certain members in ear growth and development and the far-red light response.

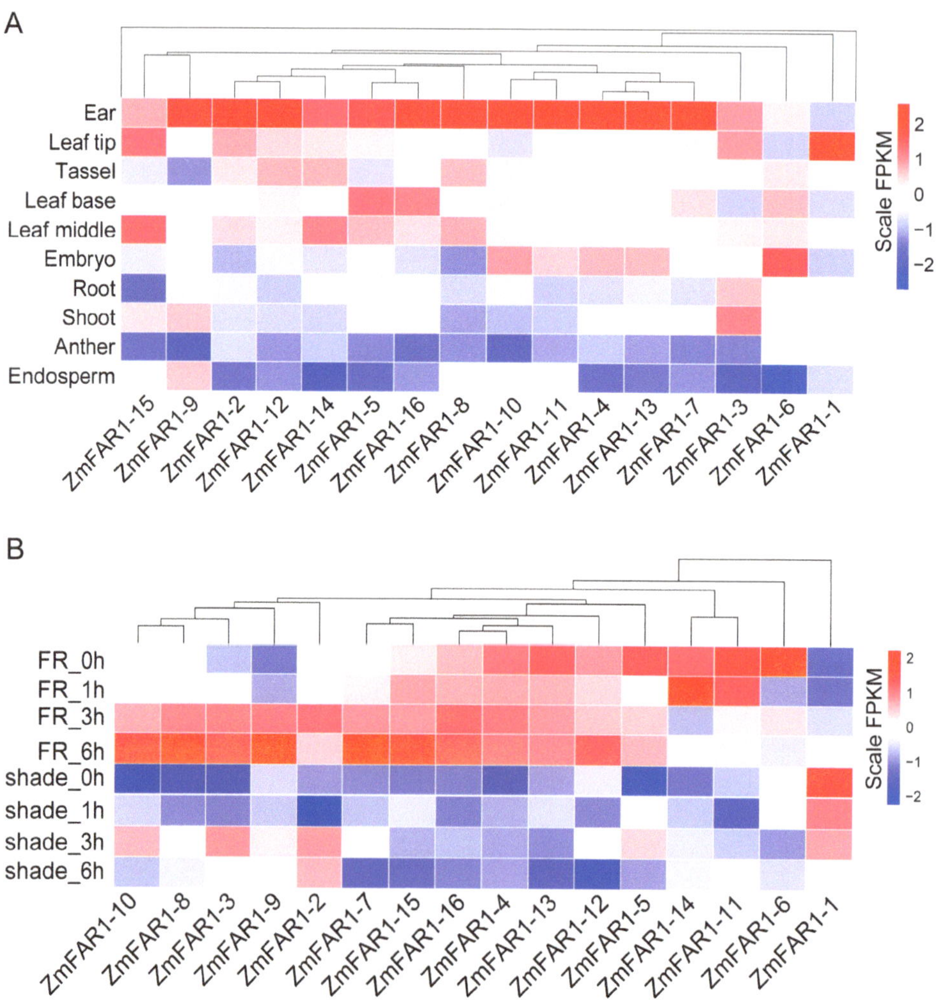

Figure 7. The expression profiles of ZmFAR1 family members in different tissues (**A**) and under different light treatment conditions (**B**). FR refers to far-red light treatment under 6 μmol/m^{2}s far-red light. Shade refers to the simulated shade treatment under blue, 15 μmol/m^{2}s; red, 12 μmol/m^{2}s; far-red, 105 μmol/m^{2}s.

3.8. Expression Analysis of ZmFAR1 Genes under Far-Red Light

Through the transcriptome expression analysis, we found that the *ZmFAR1* genes were highly expressed in maize inflorescence tissues and had a significant response to far-red light treatment. Consequently, we performed a quantitative reverse transcriptase polymerase chain reaction (RT-qPCR) to further verify the 16 *ZmFAR1* genes under far-red light, to determine their expression patterns in the ear and tassel at different developmental stages (V8–V12) (Figures 8 and 9). The results suggest that the peak expression of some *ZmFAR1* genes was significantly influenced by far-red light at various developmental stages and that the expression patterns of certain *ZmFAR1* genes might be limited to a specific period. In the ear, the peak expression of *ZmFAR1-10*, *ZmFAR1-11*, and *ZmFAR1-15* was advanced from V11 to V10, the peak expression of *ZmFAR1-1* was advanced from V10 to V9, and the peak expression of *ZmFAR1-12* was advanced from V9 to V8 (Figure 8).

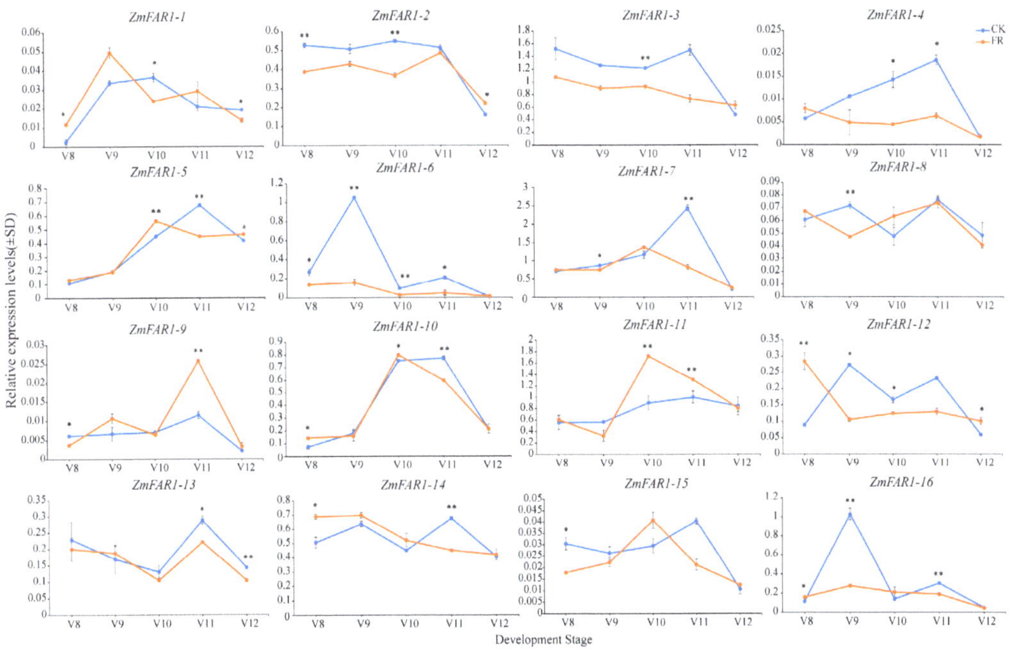

Figure 8. Expression patterns of 16 ZmFAR1 family members in the ear in response to far-red light. Error bars represent the means ± SD from three biological replicates. Statistically significant differences were indicated as follows: ** (p-value < 0.01) and * (p-value < 0.05), and the indicated p-value was determined by Student's t-test. CK (control light condition) and FR (EOD-FR treatment) are represented by blue lines and orange lines, respectively. The abscissa represents the consecutive developmental period; the ordinate represents the qRT-PCR expression levels.

During the V8–V12 stage in the tassel, the ZmFAR1 family members also showed varied expression patterns, with some being significantly up- or downregulated. We found that *ZmFAR1-1* and *ZmFAR1-12* were significantly upregulated under far-red light treatment during the V8 period, while *ZmFAR1-4*, *ZmFAR1-6*, and *ZmFAR1-7* were significantly downregulated under far-red light during the V11 period. No genes were differentially expressed during the V12 period. Compared to the ear, the expression patterns of the *ZmFAR1* genes were significantly different in the tassel (Figure 9). *ZmFAR1-2*, *ZmFAR1-3*, *ZmFAR1-5*, and *ZmFAR1-11* were significantly upregulated during the V8 period, and *ZmFAR1-2*, *ZmFAR1-9*, and *ZmFAR1-15* were significantly upregulated during the V9 period.

Far-red light treatment upregulated *ZmFAR1-6*, *ZmFAR1-9*, *ZmFAR1-12*, and *ZmFAR1-15* during the V11 period, while completely suppressing the expression of *ZmFAR1-4*.

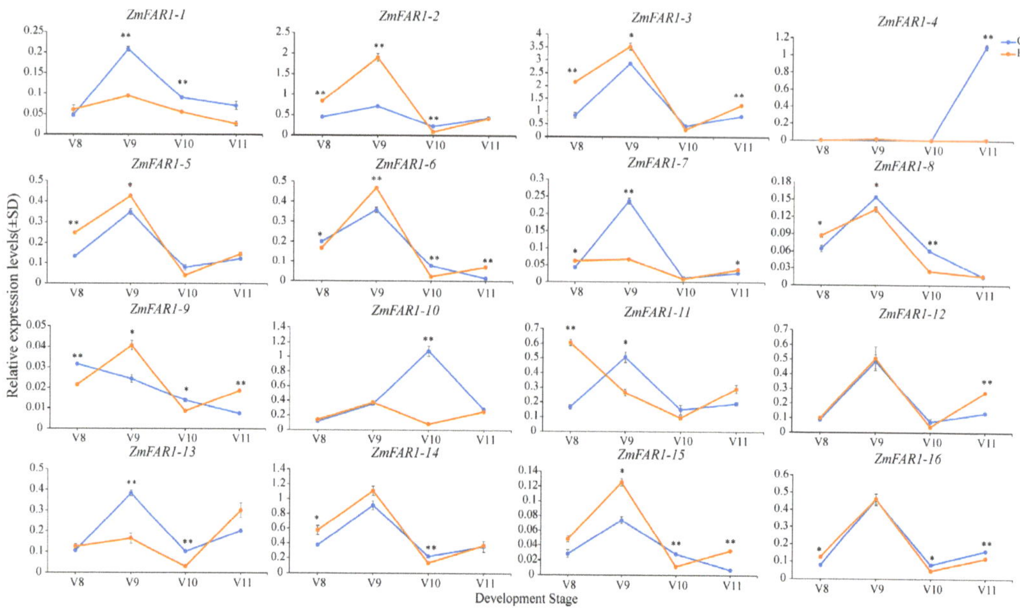

Figure 9. Expression patterns of 16 ZmFAR1 family members in the tassel in response to far-red light. Error bars represent the means ± SDs from three biological replicates. Statistically significant differences were indicated as follows: ** (*p*-value < 0.01) and * (*p*-value < 0.05), and the indicated *p*-value was determined by Student's *t*-test. CK (control light condition) and FR (EOD-FR treatment) are represented by blue lines and orange lines, respectively. The abscissa represents the consecutive developmental period; the ordinate represents the qRT-PCR expression levels.

Our findings demonstrated that the expression levels of *ZmFAR1* genes varied significantly between the ear and tassel during the same developmental stage, as well as between developmental periods within the same tissue. In our prior studies, one period in advance of ear development in the treatment group was observed under far-red light [40]. Combined with the analysis of the PCR results, we suggest that six genes with earlier peak expression in the ear (*ZmFAR1-1*, *ZmFAR1-10*, *ZmFAR1-11*, *ZmFAR1-12*, *ZmFAR1-14*, and *ZmFAR1-15*) might have roles in the growth and development process of the ear under far-red light conditions. However, the precise functions of these genes need to be explored further.

3.9. *Cis-eQTL Analyses Identify the Causal Variations Associated with Gene Expression Variation and Agronomic Traits*

To identify the genetic basis of *ZmFAR1* gene expression variation, we performed eQTL analysis based on published RNA-seq data for immature ears and the high-quality single-nucleotide polymorphisms (SNPs) of 137 ILs, using the linear mixed model in the EM-MAX software. We detected a cis-eQTL (B73_V3, chr9_87667000, *p*-value = 2.92×10^{-11}) associated with the expression level of *ZmFAR1-14* and a cis-eQTL (B73_V3, chr5_182810053, *p*-value = 6.42×10^{-9}) associated with the expression level of *ZmFAR1-9* (Figure 10). We next attempted to identify the causal variations of several key genes regulating important agronomic traits, taking advantage of the recently released de novo assembled genomes of 14 founder inbred lines (FILs) of maize [41–43]. We compared the genic region of *ZmFAR1-14* and 10 kb upstream and downstream sequences in the 14 FILs and found that variations in two indels (indel1(aaaaaga), indel2(ca)) located in the promoter were in high linkage disequilibrium (LD, r^2 = 0.58, r^2 = 0.43) with the cis-eQTL (Figure 10). Genotyping analysis of the 137 inbred lines identified two haplotypes (Hap1 and Hap2); the expression level of Hap1 was significantly higher than that in the ILs of Hap1, and the frequency of Hap1 continuously increased during modern maize breeding. We also found that a higher frequency of Hap1 was associated with a higher kernel row number. These results suggest that *ZmFAR1-14* likely plays a role in regulating the kernel row number in maize.

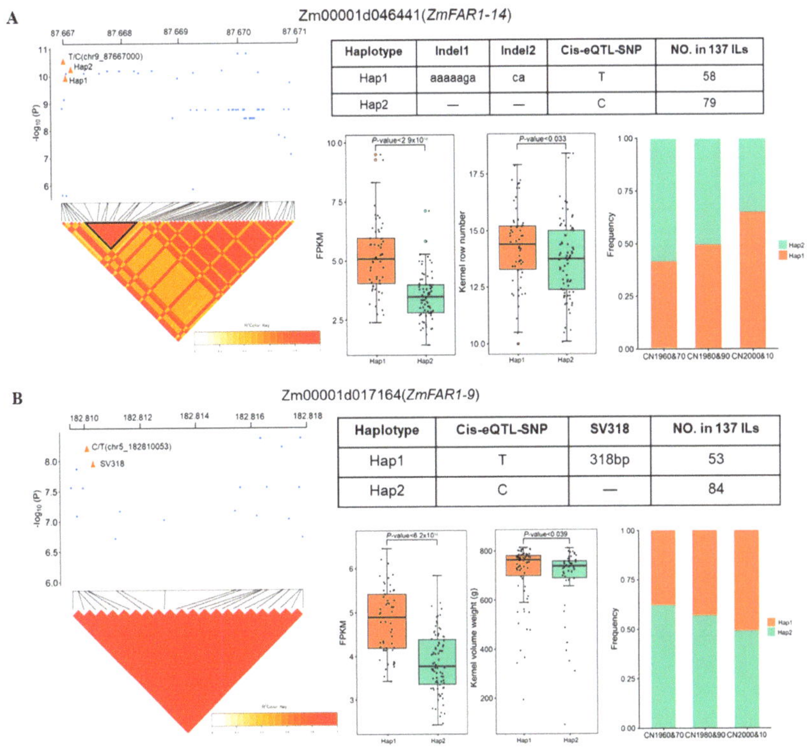

Figure 10. eQTL analyses identify the potential functional variations for expression variation of *ZmFAR1-14* and *ZmFAR1-9*. (**A**) Manhattan plot (upper) and LD heatmap (lower) for eQTLs and candidate association analysis of *ZmFAR1-14*. Two haplotypes of *ZmFAR1-14* based on the two SVs in the 137 ILs (right top). (d) Box plots for *ZmFAR1-14* expression levels and kernel row number for the two haplotypes in the 137 ILs. Bar chart for haplotype frequency changes during modern maize breeding. (**B**) Manhattan plot (upper) and LD heatmap (lower) for eQTLs and candidate association analysis of *ZmFAR1-9*. Two haplotypes of *ZmFAR1-9* based on the two SVs in the 137 ILs (right top). Box plots for *ZmFAR1-9* expression levels and kernel row number for the two haplotypes in the 137 ILs. Bar chart for haplotype frequency changes during modern maize breeding.

Similarly, by comparing the genic region of *ZmFAR1-9* and 10 kb upstream and downstream sequences in the 14 FILs, we found that variations in a 318 bp SV (SV318, located in the first intron) showed high LD with the lead SNP (r^2 = 0.61). The SV and cis-eQTL form two haplotypes in the 137 ILs. The expression level and kernel volume weight of Hap1 were significantly higher than those in the ILs of Hap2, and the frequency of Hap1 continuously increased during modern maize breeding. Together, these results suggest that Hap1 of *ZmFAR1-9* likely represents a favorable haplotype under selection for kernel weight during modern maize breeding.

4. Discussion

4.1. Characterization of the ZmFAR1 Family Members in Maize

FAR1/FHY3 is a novel transcription factor derived from transposase mutations and has evolved through long-term domestication and adaptation to form the *FAR1/FHY3* transcription factor family [7,44]. Recent studies revealed that *FAR1/FHY3* were crucial positive regulatory components in the *phyA* signaling pathway, primarily activating the transcription of light-induced target genes [7]. In *Arabidopsis*, more than 1000 possible *FHY3* binding genes have been identified, including *ERF4*, *ARC2*, *HEMB1*, and *COP1*, indicating that *FAR1/FHY3* might be involved in the control of several plant growth and development features [6]. However, there have been no reports of a thorough investigation of the *ZmFAR1* transcription factor family in maize. In this study, we identified a total of 16 *ZmFAR1* genes from the maize inbred line B73. The number of *ZmFAR1* genes was greater than that in *Arabidopsis* (14 members) but less than that in *Aarchis hypogaea* (246 members), *Camellia sinensis* (25 members), *Populus trichocarpa* (47 members), *Rosa wichuraiana 'Basye's Thornless'* (91 members), and *Rosa chinensis 'Old Blush'* (50 members) [8,19,44,45]. These findings indicate that the *FAR1/FHY3* gene family expanded throughout the evolution of these species. Despite a recent whole-genome duplication (WGD) event that occurred in maize after species formation, the number of genes belonging to the FAR1/FHY3 family did not vary dramatically. This suggests that the FAR1/FHY3 family is relatively conserved in maize.

4.2. Studies on the Structure, Evolutionary Characteristics, and Functions of Maize ZmFAR1 Genes

There were a total of 16 *ZmFAR1* genes that were unevenly distributed across eight chromosomes. The analysis of the physicochemical properties revealed that the majority of the *ZmFAR1* proteins in maize were unstable (with the exception of *ZmFAR1-3* and *ZmFAR1-4*), which was in line with the findings in *Arabidopsis thaliana* [44]. Based on the phylogenetic tree analysis results, all the *ZmFAR1* genes were divided into five subgroups according to the classification of the *FAR1* genes in *Arabidopsis*, with group I including no *ZmFAR1* genes and group V containing the highest number of *ZmFAR1* genes (Figure 1). Based on these findings, it seemed that the *ZmFAR1* genes in group I had been lost during evolution during maize breeding, while the *ZmFAR1* genes in group V might have evolved independently with unique functions. Gene structure analysis revealed that all *ZmFAR1* genes contained the FAR1 domain (Supplementary Figure S1), indicating that the C2H2 zinc finger domain was conserved in maize. Furthermore, the majority of the *ZmFAR1* genes in the same group had similar motifs (Figure 2B), and two FAR1 domains were found in *ZmFAR1-1*, *ZmFAR1-11*, and *ZmFAR1-12*, which was also similar to the results of *Arabidopsis* and poplar [5,8]. Based on the phylogenetic analysis and conserved domain analysis, the *ZmFAR1* genes were shown to be highly conserved at the protein level, and genes within the same subclade might have comparable roles. The significant changes in the number of exons and introns between the *ZmFAR1* genes might be associated with functional differences. Through binding to transcription factors, cis-acting regulatory elements can precisely activate gene transcription and influence transcriptional efficiency, and they have been implicated in several growth and development-related events in plants. Several cis-regulatory elements were found in the promoter regions of the *ZmFAR1* genes, such as the GT-1-motif, GATA-motif, Box-4, and G-box, suggesting that the *ZmFAR1* genes in maize might be involved in light-sensitive responses (Figure 5) [46–48]. There are regulatory relationships between the light response and various hormones [14,49]. For example, the *ZmFAR1-9* promoter contained six TGACG motifs and CGTCA motifs, which suggested a possible association with the MeJA signaling pathway [50,51]. By binding directly to the promoter of *ABA-insensitive 5* (*ABI5*) in *Arabidopsis*, FAR1 and FHY3 function as positive regulators of ABA signaling. The *ZmFAR1-10* promoter contained five ABRE motifs and might be related to ABA synthesis and signaling [15,52]. Many types of plant growth and development-related cis-regulatory elements, including the CAT-box, GCN4-motif, and O2-site (seed-specific regulation), were discovered in the promoter regions of several *ZmFAR1* genes, and these cis-elements had a significant role in regulating the expression of meristematic tissue genes. The RY promoter element was involved in controlling gene expression during late embryogenesis and seed development. In the light-induced starch of *Arabidopsis*, *FAR1* and *FHY3* might directly target the starch-debranching enzyme *isoamylase2* (*ISA2*) [11]. In addition, the *FAR1* genes in Arabidopsis could also regulate seed germination [15]. In the present study, RY-element motifs were identified in the promoters of *ZmFAR1-1* and *ZmFAR1-11*, suggesting possible similar functions. Ten and eight *ZmFAR1* genes were predicted to be targeted by miR2275 and miR159, respectively. Previous studies have demonstrated that the production of 24-nt phasiRNAs is mediated by miR2275, which was assoWciated with maize pollen fertility at various temperatures [53,54]. miR159 is very conserved in plants and several light, and stress response cis-elements, such as G-boxes, AREs, and CAAT-boxes, have been found in their promoters and are also involved in male reproductive development, seed development, flowering time, and growth [55–58]. These results show that miR2275 and miR159 might participate in maize growth and regulation by interacting with the *FAR1* genes, but further results have not been reported.

4.3. Functional Variations and Expression Patterns of ZmFAR1 Genes in Maize Ear Development and Phenotypic Variation

The expression patterns of the *ZmFAR1* genes in several organs of the maize inbred line B73 exhibited considerable variances, and the majority of the *ZmFAR1* genes were strongly expressed in the ear, indicating that these genes likely play a significant role in ear growth and development (Figure 7). Previous research had shown that the FAR1 family members *FRS7-FRS12* in *Arabidopsis* could regulate the flowering time and plant growth [37]; among them, *FRS9* was also a specific negative regulator of *PhyB* signaling that could mediate seedling de-etiolation [5]. Because of this, it was hypothesized that members of the same group, such as *ZmFAR1-1*, *ZmFAR1-10*, and *ZmFAR1-14*, could have functional similarities. Importantly, the qRT-PCR results demonstrated that the peak expression of three genes, *ZmFAR1-1*, *ZmFAR1-10*, and *ZmFAR1-14*, was advanced by one stage, which was consistent with the phenotype of ear development in our previous study (Figure 8) [40]. In addition, three expression patterns of the 16 *ZmFAR1* genes were found, with some genes exhibiting phase-advanced, upregulated, or downregulated peak expression under far-red light. Our analysis suggested that certain genes with earlier peak expression in the ear might play roles in growth and development under far-red light conditions, while nondifferentially expressed genes might be nonfunctional. These findings suggest a potential functional redundancy among ZmFAR1 family members, while further research is required to identify the particular activities of these genes. With the recent accumulation of high-density gene data and related high-throughput transcriptome data in maize, eQTL has been used to determine the mutual regulation of genes in the genome [59–61]. In this study, our analyses provided new insights into the expression regulation mechanism and phenotypic variation of *ZmFAR1* genes. In combination with eQTL mapping and association studies, we identified several SVs in the promoter of *ZmFAR1-14* as potential functional variations underlying its differential expression and likely regulating the kernel row number. We also found SVs in *ZmFAR1-9* in regulating gene expression and the kernel volume weight (Figure 10). These findings provide novel and important insights into the natural variations in the *ZmFAR1* genes at the transcriptome level and the influence of gene expression on phenotypic variations.

4.4. ZmFAR1 Genes' Role in Maize Yield and Trait Optimization

By uncovering the functional roles of the ZmFAR1 genes and their correlation with maize phenotypic traits such as ear development and the kernel row number, our research offers strategic insights for breeders aiming to enhance yields and environmental resilience. Utilizing genetic variants like SNVs in regulatory regions sheds light on the impact of ZmFAR1 expression on maize diversity, presenting opportunities in marker-assisted selection to foresee and promote desirable traits. This pivotal study not only sets the stage for the advanced functional investigation of the ZmFAR1 gene family but also charts a course for their deliberate application in optimizing agricultural outputs and conserving genetic variation.

5. Conclusions

In this study, 16 *ZmFAR1* genes were identified from the maize B73 genome and unevenly distributed over eight chromosomes. The cis-acting regulatory elements of 16 *ZmFAR1* genes were mainly related to the light response, hormone response, stress response, and plant growth and development. RNA-seq analysis showed that most *ZmFAR1* genes showed relatively higher expression levels in the ear and may play a pivotal role in the far-red (FR) light treatment of maize. The qRT-PCR study revealed the spatial and temporal specificity of *ZmFAR1* gene expression under far-red light conditions. Combining our previous phenotypic data, *ZmFAR1-1*, *ZmFAR1-10*, *ZmFAR1-11*, *ZmFAR1-12*, *ZmFAR1-14*, and *ZmFAR1-15* might have roles in ear development under far-red light treatment. Finally, we identified several SVs in *ZmFAR1-14* and *ZmFAR1-9* that contributed to the altered expression patterns and phenotypic differentiation. Overall, our study provided insights into understanding the functions of *ZmFAR1* genes during the light signaling pathway and identified the potential functional variations that may lead to increased yield gains during modern maize breeding.

Supplementary Materials: The following supporting information can be downloaded at: https://www.mdpi.com/article/10.3390/cimb46010027/s1, Figure S1: Multiple sequence alignment results for the ZmFAR1 family members. The FAR1 domains (motif3-motif9-motif8-motif6) are highlighted in red boxes; Figure S2: The sequence logos of all 15 motifs, where the base size at each point shows the likelihood that the base occurs there; Table S1: The protein sequences of 16 *ZmFAR1* genes in maize B73; Table S2: The cis-acting elements in the promoter regions of ZmFAR1 family members; Table S3: The interactions between miRNAs and putative target *ZmFAR1* genes; Table S4: The primer sequences used in qRT-PCR.

Author Contributions: C.L. (Changyu Li) and X.C. designed the project and wrote the manuscript; H.T. and D.J. conducted the experiments and analyzed the results; C.L. (Cheng Liu), X.X. and L.Z. helped with some molecular experiments. All authors have read and agreed to the published version of the manuscript.

Funding: This research was funded by a special project for basic scientific activities of non-profit institutes supported the government of Xin Jiang Uyghur Autonomous Region, funding number ky2021126.

Institutional Review Board Statement: Not applicable.

Informed Consent Statement: Not applicable.

Data Availability Statement: Existing datasets are available in a publicly accessible repository. Publicly available datasets were analyzed in this study. These data can be found at http://www.ncbi.nlm.nih.gov/sra/ with the accession numbers of ERR3773807 to ERR3773827.

Acknowledgments: The authors would like to express their appreciation to all members of the lab for their constant technical guidance and insightful discussions. The authors are appreciative of the feedback that they received from the manuscript's anonymous reviewers.

Conflicts of Interest: The authors declare that the research was conducted in the absence of any commercial or financial relationships that could be construed as a potential conflict of interest.

References

1. Casal, J.J. Photoreceptor Signaling Networks in Plant Responses to Shade. *Annu. Rev. Plant Biol.* **2013**, *64*, 403. [CrossRef] [PubMed]
2. Nilsson, H.; Rappaport, F.; Boussac, A.; Messinger, J. Substrate–water exchange in photosystem II is arrested before dioxygen formation. *Nat. Commun.* **2014**, *5*, 4305. [CrossRef] [PubMed]

3. Li, G.; Siddiqui, H.; Teng, Y.; Lin, R.; Wan, X.-Y.; Li, J.; Lau, O.-S.; Ouyang, X.; Dai, M.; Wan, J. Coordinated transcriptional regulation underlying the circadian clock in Arabidopsis. *Nat. Cell Biol.* **2011**, *13*, 616–622. [CrossRef] [PubMed]
4. Wang, H.; Wang, H. Multifaceted roles of FHY3 and FAR1 in light signaling and beyond. *Trends Plant Sci.* **2015**, *20*, 453–461. [CrossRef] [PubMed]
5. Lin, R.; Wang, H. Arabidopsis FHY3/FAR1 gene family and distinct roles of its members in light control of Arabidopsis development. *Plant Physiol.* **2004**, *136*, 4010–4022. [CrossRef]
6. Ouyang, X.; Li, J.; Li, G.; Li, B.; Chen, B.; Shen, H.; Huang, X.; Mo, X.; Wan, X.; Lin, R. Genome-wide binding site analysis of FAR-RED ELONGATED HYPOCOTYL3 reveals its novel function in Arabidopsis development. *Plant Cell* **2011**, *23*, 2514–2535. [CrossRef]
7. Ma, L.; Li, G. FAR1-related sequence (FRS) and FRS-related factor (FRF) family proteins in Arabidopsis growth and development. *Front. Plant. Sci.* **2018**, *9*, 692. [CrossRef]
8. Du, J.; Zhang, L.; Ge, X.; Xiang, X.; Cao, D.; Yang, H.; Hu, J. Genome-wide identification and characterization of the FAR1/FHY3 family in *Populus trichocarpa* Torr. & Gray and expression analysis in light response. *Forests* **2021**, *12*, 1385.
9. Liu, L.; Li, B.; Liu, X. FAR-RED ELONGATED HYPOCOTYL3 promotes floral meristem determinacy in Arabidopsis. *Plant. Signal. Behav.* **2016**, *11*, e1238545. [CrossRef]
10. Liu, Y.; Ma, M.; Li, G.; Yuan, L.; Xie, Y.; Wei, H.; Ma, X.; Li, Q.; Devlin, P.F.; Xu, X. Transcription factors FHY3 and FAR1 regulate light-induced CIRCADIAN CLOCK ASSOCIATED1 gene expression in Arabidopsis. *Plant Cell* **2020**, *32*, 1464–1478. [CrossRef]
11. Ma, L.; Xue, N.; Fu, X.; Zhang, H.; Li, G. *Arabidopsis thaliana* FAR-RED ELONGATED HYPOCOTYLS3 (FHY3) and FAR-RED-IMPAIRED RESPONSE1 (FAR1) modulate starch synthesis in response to light and sugar. *New Phytol.* **2017**, *213*, 1682–1696. [CrossRef] [PubMed]
12. Rea, A.C. Avoiding shade to grow taller but not always stronger: Phytochrome–jasmonic acid interplay. *Am. Soc. Plant Biol.* **2019**, *31*, 1941–1942. [CrossRef] [PubMed]
13. Liu, Y.; Wei, H.; Ma, M.; Li, Q.; Kong, D.; Sun, J.; Ma, X.; Wang, B.; Chen, C.; Xie, Y. Arabidopsis FHY3 and FAR1 regulate the balance between growth and defense responses under shade conditions. *Plant Cell* **2019**, *31*, 2089–2106. [CrossRef] [PubMed]
14. Xie, Y.; Liu, Y.; Ma, M.; Zhou, Q.; Zhao, Y.; Zhao, B.; Wang, B.; Wei, H.; Wang, H. Arabidopsis FHY3 and FAR1 integrate light and strigolactone signaling to regulate branching. *Nat. Commun.* **2020**, *11*, 1955. [CrossRef]
15. Tang, W.; Ji, Q.; Huang, Y.; Jiang, Z.; Bao, M.; Lin, W.R. FAR-RED ELONGATED HYPOCOTYL3 and FAR-RED IMPAIRED RESPONSE1 transcription factors integrate light and abscisic acid signaling in Arabidopsis. *Plant Physiol.* **2013**, *163*, 857–866. [CrossRef] [PubMed]
16. Xie, Y.; Ma, M.; Liu, Y.; Wang, B.; Wei, H.; Kong, D.; Wang, H. Arabidopsis FHY3 and FAR1 function in age gating of leaf senescence. *Front. Plant. Sci.* **2021**, *12*, 770060. [CrossRef] [PubMed]
17. Zhang, P.; Zhu, W.; He, Y.; Fan, J.; Shi, J.; Fu, R.; Hu, J.; Li, L.; Zhang, D.; Liang, W. THERMOSENSITIVE BARREN PANICLE (TAP) is required for rice panicle and spikelet development at high ambient temperature. *New Phytol.* **2023**, *237*, 855–869. [CrossRef]
18. Lu, Q.; Liu, H.; Hong, Y.; Liang, X.; Li, S.; Liu, H.; Li, H.; Wang, R.; Deng, Q.; Jiang, H. Genome-Wide Identification and Expression of FAR1 Gene Family Provide Insight Into Pod Development in Peanut (*Arachis hypogaea*). *Front. Plant. Sci.* **2022**, *13*, 893278. [CrossRef]
19. Zhong, M.-C.; Jiang, X.-D.; Cui, W.-H.; Hu, J.-Y. Expansion and expression diversity of FAR1/FRS-like genes provides insights into flowering time regulation in roses. *Plant Divers* **2021**, *43*, 173–179. [CrossRef]
20. Ruberti, I.; Sessa, G.; Ciolfi, A.; Possenti, M.; Carabelli, M.; Morelli, G. Plant adaptation to dynamically changing environment: The shade avoidance response. *Biotechnol. Adv.* **2012**, *30*, 1047–1058. [CrossRef]
21. Roig-Villanova, I.; Martínez-García, J.F. Plant responses to vegetation proximity: A whole life avoiding shade. *Front. Plant. Sci.* **2016**, *7*, 236. [CrossRef] [PubMed]
22. Mistry, J.; Finn, R.D.; Eddy, S.R.; Bateman, A.; Punta, M. Challenges in homology search: HMMER3 and convergent evolution of coiled-coil regions. *Nucleic Acids Res.* **2013**, *41*, e121. [CrossRef] [PubMed]
23. Jacob, A.; Lancaster, J.; Buhler, J.; Harris, B.; Chamberlain, R.D. Mercury BLASTP: Accelerating protein sequence alignment. *ACM Trans. Reconfigurable Technol. Syst.* **2008**, *1*, 1–44. [CrossRef] [PubMed]
24. Edgar, R.C. MUSCLE: Multiple sequence alignment with high accuracy and high throughput. *Nucleic Acids Res.* **2004**, *32*, 1792–1797. [CrossRef] [PubMed]
25. Nguyen, L.-T.; Schmidt, H.A.; Von Haeseler, A.; Minh, B.Q. IQ-TREE: A fast and effective stochastic algorithm for estimating maximum-likelihood phylogenies. *Mol. Biol. Evol.* **2015**, *32*, 268–274. [CrossRef] [PubMed]
26. Hu, B.; Jin, J.; Guo, A.-Y.; Zhang, H.; Luo, J.; Gao, G. GSDS 2.0: An upgraded gene feature visualization server. *Bioinformatics* **2015**, *31*, 1296–1297.
27. Bailey, T.L.; Johnson, J.; Grant, C.E.; Noble, W.S. The MEME suite. *Nucleic Acids Res.* **2015**, *43*, W39–W49. [CrossRef]
28. Chen, C.; Chen, H.; Zhang, Y.; Thomas, H.R.; Xia, R. TBtools: An Integrative Toolkit Developed for Interactive Analyses of Big Biological Data. *Mol. Plant* **2020**, *13*, 1194–1202. [CrossRef]
29. Yupeng, W.; Haibao, T.; Debarry, J.D.; Xu, T.; Jingping, L.; Xiyin, W.; Tae-Ho, L.; Huizhe, J.; Barry, M.; Hui, G. MCScanX: A toolkit for detection and evolutionary analysis of gene synteny and collinearity. *Nucleic Acids Res.* **2012**, *40*, e49.
30. Villanueva, R.A.M.; Chen, Z.J. *ggplot2: Elegant Graphics for Data Analysis*; Taylor & Francis: Oxford, UK, 2019.
31. Chen, S.; Zhou, Y.; Chen, Y.; Gu, J. fastp: An ultra fast all-in-one FASTQ preprocessor. *Bioinformatics* **2018**, *34*, i884–i890. [CrossRef]

32. Kim, D.; Langmead, B.; Salzberg, S.L. HISAT: A fast spliced aligner with low memory requirements. *Nat. Methods* **2015**, *12*, 357–360. [CrossRef] [PubMed]
33. Trapnell, C.; Roberts, A.; Goff, L.; Pertea, G.; Kim, D.; Kelley, D.R. Differential gene and transcript expression analysis of RNA-seq experiments with TopHat and Cufflinks. *Nat. Protoc.* **2012**, *7*, 562–578. [CrossRef] [PubMed]
34. Livak, K.J.; Schmittgen, T.D. Analysis of relative gene expression data using real-time quantitative PCR and the 2− ΔΔCT method. *Methods* **2001**, *25*, 402–408. [CrossRef] [PubMed]
35. Li, C.; Li, Y.; Song, G.; Yang, D.; Xia, Z.; Sun, C.; Zhao, Y.; Hou, M.; Zhang, M.; Qi, Z. Gene expression and expression quantitative trait loci analyses uncover natural variations underlying the improvement of important agronomic traits during modern maize breeding. *Plant J. Cell Mol. Biol.* **2023**, *115*, 772–787. [CrossRef] [PubMed]
36. Kang, H.M.; Sul, J.H.; Service, S.K.; Zaitlen, N.A.; Kong, S.Y.; Freimer, N.B.; Sabatti, C.; Eskin, E. Variance component model to account for sample structure in genome-wide association studies. *Nat. Genet.* **2010**, *42*, 348–354. [CrossRef] [PubMed]
37. Ritter, A.; Iñigo, S.; Fernández-Calvo, P.; Heyndrickx, K.S.; Dhondt, S.; Shi, H.; De Milde, L.; Vanden Bossche, R.; De Clercq, R.; Eeckhout, D. The transcriptional repressor complex FRS7-FRS12 regulates flowering time and growth in Arabidopsis. *Nat. Commun.* **2017**, *8*, 15235. [CrossRef] [PubMed]
38. Wang, J.; Sun, P.; Li, Y.; Liu, Y.; Yang, N.; Yu, J.; Ma, X.; Sun, S.; Xia, R.; Liu, X. An overlooked paleotetraploidization in Cucurbitaceae. *Mol. Biol. Evol.* **2018**, *35*, 16–26. [CrossRef]
39. Hernandez-Garcia, C.M.; Finer, J.J. Identification and validation of promoters and cis-acting regulatory elements. *Plant. Sci.* **2014**, *217*, 109–119. [CrossRef]
40. Kong, D.; Li, C.; Xue, W.; Wei, H.; Ding, H.; Hu, G.; Zhang, X.; Zhang, G.; Zou, T.; Xian, Y. UB2/UB3/TSH4-anchored transcriptional networks regulate early maize inflorescence development in response to simulated shade. *Plant Cell* **2023**, *35*, 717–737. [CrossRef]
41. Jiao, Y.; Peluso, P.; Shi, J.; Liang, T.; Stitzer, M.C.; Wang, B.; Campbell, M.S.; Stein, J.C.; Wei, X.; Chin, C.-S. Improved maize reference genome with single-molecule technologies. *Nature* **2017**, *546*, 524–527. [CrossRef]
42. Sun, S.; Zhou, Y.; Chen, J.; Shi, J.; Zhao, H.; Zhao, H.; Song, W.; Zhang, M.; Cui, Y.; Dong, X. Extensive intraspecific gene order and gene structural variations between Mo17 and other maize genomes. *Nat. Genet.* **2018**, *50*, 1289–1295. [CrossRef] [PubMed]
43. Wang, B.; Hou, M.; Shi, J.; Ku, L.; Song, W.; Li, C.; Ning, Q.; Li, X.; Li, C.; Zhao, B. De novo genome assembly and analyses of 12 founder inbred lines provide insights into maize heterosis. *Nat. Genet.* **2023**, *55*, 312–323. [CrossRef] [PubMed]
44. Wang, H.; Deng, X.W. Arabidopsis FHY3 defines a key phytochrome A signaling component directly interacting with its homologous partner FAR1. *EMBO J.* **2002**, *21*, 1339–1349. [CrossRef] [PubMed]
45. Liu, Z.; An, C.; Zhao, Y.; Xiao, Y.; Bao, L.; Gong, C.; Gao, Y. Genome-wide identification and characterization of the CsFHY3/FAR1 gene family and expression analysis under biotic and abiotic stresses in tea plants (*Camellia sinensis*). *Plants* **2021**, *10*, 570. [CrossRef] [PubMed]
46. Kumar, G.M.; Mamidala, P.; Podile, A.R. Regulation of Polygalacturonase-inhibitory proteins in plants is highly dependent on stress and light responsive elements. *Plant Omics* **2009**, *2*, 238–249.
47. Baek, K.; Lee, Y.; Nam, O.; Park, S.; Sim, S.J.; Jin, E. Introducing Dunaliella LIP promoter containing light-inducible motifs improves transgenic expression in Chlamydomonas reinhardtii. *Biotechnol. J.* **2016**, *11*, 384–392. [CrossRef] [PubMed]
48. Zhang, Z.; Ren, C.; Zou, L.; Wang, Y.; Li, S.; Liang, Z. Characterization of the GATA gene family in Vitis vinifera: Genome-wide analysis, expression profiles, and involvement in light and phytohormone response. *Genome* **2018**, *61*, 713–723. [CrossRef] [PubMed]
49. Liu, Y.; Jafari, F.; Wang, H. Integration of light and hormone signaling pathways in the regulation of plant shade avoidance syndrome. *Abiotech* **2021**, *2*, 131–145. [CrossRef]
50. Rouster, J.; Leah, R.; Mundy, J.; Cameron-Mills, V. Identification of a methyl jasmonate-responsive region in the promoter of a lipoxygenase 1 gene expressed in barley grain. *Plant J.* **1997**, *11*, 513–523. [CrossRef]
51. Chu, Z.; Wang, X.; Li, Y.; Yu, H.; Li, J.; Lu, Y.; Li, H.; Ouyang, B. Genomic organization, phylogenetic and expression analysis of the B-BOX gene family in tomato. *Front. Plant. Sci.* **2016**, *7*, 1552. [CrossRef]
52. Wang, X.; Guo, C.; Peng, J.; Li, C.; Wan, F.; Zhang, S.; Zhou, Y.; Yan, Y.; Qi, L.; Sun, K. ABRE-BINDING FACTORS play a role in the feedback regulation of ABA signaling by mediating rapid ABA induction of ABA co-receptor genes. *New Phytol.* **2019**, *221*, 341–355. [CrossRef] [PubMed]
53. Xia, R.; Chen, C.; Pokhrel, S.; Ma, W.; Huang, K.; Patel, P.; Wang, F.; Xu, J.; Liu, Z.; Li, J. 24-nt reproductive phasiRNAs are broadly present in angiosperms. *Nat. Commun.* **2019**, *10*, 627. [CrossRef] [PubMed]
54. Zhai, J.; Zhang, H.; Arikit, S.; Huang, K.; Nan, G.-L.; Walbot, V.; Meyers, B.C. Spatiotemporally dynamic, cell-type–dependent premeiotic and meiotic phasiRNAs in maize anthers. *Proc. Natl. Acad. Sci. USA* **2015**, *112*, 3146–3151. [CrossRef] [PubMed]
55. Luo, Y.; Guo, Z.; Li, L. Evolutionary conservation of microRNA regulatory programs in plant flower development. *Dev. Biol.* **2013**, *380*, 133–144. [CrossRef] [PubMed]
56. Li, B.; Tao, P.; Xu, F.; He, P.; Wang, J. Function of Soybean miR159 Family Members in Plant Responses to Low Phosphorus, High Salinity, and Abscisic Acid Treatment. *Agronomy* **2023**, *13*, 1798. [CrossRef]
57. Chen, J.-F.; Zhao, Z.-X.; Li, Y.; Li, T.-T.; Zhu, Y.; Yang, X.-M.; Zhou, S.-X.; Wang, H.; Zhao, J.-Q.; Pu, M. Fine-tuning roles of Osa-miR159a in rice immunity against Magnaporthe oryzae and development. *Rice* **2021**, *14*, 26. [CrossRef] [PubMed]

58. Zhao, Y.; Wang, S.; Wu, W.; Li, L.; Jiang, T.; Zheng, B. Clearance of maternal barriers by paternal miR159 to initiate endosperm nuclear division in Arabidopsis. *Nat. Commun.* **2018**, *9*, 5011. [CrossRef]
59. Wang, X.; Chen, Q.; Wu, Y.; Lemmon, Z.H.; Xu, G.; Huang, C.; Liang, Y.; Xu, D.; Li, D.; Doebley, J.F. Genome-wide analysis of transcriptional variability in a large maize-teosinte population. *Mol. Plant* **2018**, *11*, 443–459. [CrossRef]
60. Liu, S.; Li, C.; Wang, H.; Wang, S.; Yang, S.; Liu, X.; Yan, J.; Li, B.; Beatty, M.; Zastrow-Hayes, G. Mapping regulatory variants controlling gene expression in drought response and tolerance in maize. *Genome Biol.* **2020**, *21*, 163. [CrossRef]
61. Wu, X.; Feng, H.; Wu, D.; Yan, S.; Dai, M. Using high-throughput multiple optical phenotyping to decipher the genetic architecture of maize drought tolerance. *Genome Biol.* **2021**, *22*, 185. [CrossRef]

current issues in molecular biology

Article

QTL Mapping and Genome-Wide Association Study Reveal Genetic Loci and Candidate Genes Related to Soluble Solids Content in Melon

Honglang Yan *,†, Kang Wang †, Manman Wang, Lulu Feng, Huimin Zhang and Xiaoyun Wei

Jiangsu Yanjiang Institute of Agricultural Sciences, Jiangsu Academy of Agricultural Sciences, Nantong 226012, China; wangkang@jaas.ac.cn (K.W.); manmanwang@jaas.ac.cn (M.W.); fenglulu@jaas.ac.cn (L.F.); zhanghuimin1@jaas.ac.cn (H.Z.); 20042200@jaas.ac.cn (X.W.)

* Correspondence: yanhonglang@jaas.ac.cn

† These authors contributed equally to this work.

Abstract: Melon (*Cucumis melo* L.) is an economically important Cucurbitaceae crop grown around the globe. The sweetness of melon is a significant factor in fruit quality and consumer appeal, and the soluble solids content (SSC) is a key index of melon sweetness. In this study, 146 recombinant inbred lines (RILs) derived from two oriental melon materials with different levels of sweetness containing 1427 bin markers, and 213 melon accessions containing 1,681,775 single nucleotide polymorphism (SNP) markers were used to identify genomic regions influencing SSC. Linkage mapping detected 10 quantitative trait loci (QTLs) distributed on six chromosomes, seven of which were overlapped with the reported QTLs. A total of 211 significant SNPs were identified by genome-wide association study (GWAS), 138 of which overlapped with the reported QTLs. Two new stable, co-localized regions on chromosome 3 were identified by QTL mapping and GWAS across multiple environments, which explained large phenotypic variance. Five candidate genes related to SSC were identified by QTL mapping, GWAS, and qRT-PCR, two of which were involved in hydrolysis of raffinose and sucrose located in the new stable loci. The other three candidate genes were involved in raffinose synthesis, sugar transport, and production of substrate for sugar synthesis. The genomic regions and candidate genes will be helpful for molecular breeding programs and elucidating the mechanisms of sugar accumulation.

Keywords: melon; soluble solids content; linkage analysis; GWAS; candidate genes

Citation: Yan, H.; Wang, K.; Wang, M.; Feng, L.; Zhang, H.; Wei, X. QTL Mapping and Genome-Wide Association Study Reveal Genetic Loci and Candidate Genes Related to Soluble Solids Content in Melon. *Curr. Issues Mol. Biol.* **2023**, *45*, 7110–7129. https://doi.org/10.3390/cimb45090450

Academic Editor: Angel Llamas

Received: 1 August 2023
Revised: 21 August 2023
Accepted: 25 August 2023
Published: 26 August 2023

1. Introduction

Melon (*Cucumis melo* L.; $2n = 2x = 24$) is an economically important Cucurbitaceae crop grown around the globe. Based on differences of ovary pubescence, melon cultivars are divided into two subspecies: *C. melo* ssp. *agrestis* and *C. melo* ssp. *melo* [1,2]. The sweetness is an important quality trait of melons that affects consumer preferences, and the soluble solids content (SSC) is a key index of melon sweetness. Identifying genomic regions and genes controlling SSC will be helpful for molecular breeding.

Linkage analysis is an effective approach for detecting quantitative trait loci (QTLs) related to complex agronomic traits using biparental segregating populations, and several QTLs were found to regulate the accumulation of soluble solids content in melon [3–14]. However, a considerable number of QTLs for soluble solids content detected through diverse backgrounds have wide intervals, and these QTLs are difficult to apply to marker-assisted selection (MAS). Due to the rapid growth of sequencing technology, GWAS has proven a powerful approach in detecting genetic variations of important agronomic traits in melon [15–19]. Previously, some key loci related to soluble solids content were detected via GWAS in tomato [20,21], apricot [22], peach [23], and watermelon [24]. Leida et al. [25] identified a stable region on chromosome 3 associated with sugar content via association analysis using 251 single nucleotide polymorphism (SNP) markers. Wang et al. [26] detected 11 SNPs associated with soluble solids content in melon via GWAS using genotyping-by-sequencing (GBS).

Association mapping does not require a biparental segregating population and is useful for analyzing multiple alleles [27], but the population substructure and allele frequency could bring about false-positive results [28]. Combining association analysis and linkage analysis has been successfully used for dissecting complex quantitative traits in rice [29], maize [30], and soybean [31]. Wei et al. [29] identified eight candidate genes for rice seedlings in response to high temperature stress using linkage analysis and GWAS. Xu et al. [30] dissected the genetic basis of zinc deficiency tolerance in maize at the seedling stage via linkage mapping and GWAS, and three candidate genes were identified as being responsible for Zn transport. Hu et al. [31] identified a gibberellin 3b-hydroxylase gene related to soybean 100-seed weight through joint linkage mapping and GWAS which enhanced photosynthesis and increased seed yield in soybean. At present, association analysis and linkage analysis have been used to dissect the complex agronomic traits of melon. Nimmakayala et al. [32] detected several QTLs for fruit firmness through GWAS and QTL mapping. Perpiñá et al. [33] detected several stable QTLs related to SSC using linkage analysis with introgression lines and association analysis in backcross populations. Oren et al. [34] detected a major netting QTL on melon chromosome 2 by combining linkage analysis and association analysis. Du et al. [35] identified a candidate genomic region associate with the fruit surface groove of melon via linkage analysis and association analysis. These genetic loci identified by GWAS and linkage analysis will make MAS breeding more accurate and effective. Based on the rapid growth of sequencing technology, joint GWAS and linkage analysis will play an increasingly important role in gene mining for important agronomic traits in melon. Raffinose and stachyose are the core sugars of cucurbit crops transported in the phloem [36]. Based on the rapid advances of molecular biology and genomics, many genes related to sugar metabolism and accumulation during fruit growth have been reported such as alpha-galactosidase (AGA) [37,38], sugars will eventually be exported transporter (SWEET) [37–39], vacuolar sugar transporter (VST) [40], sucrose transporter (SUT) [39,41], tonoplast sugar transporter (TST) [42–44], hexose transporter [45], tonoplast H+/sugar antiporters [46], trehalose-6-phosphate (T6P) [47,48], sucrose synthase (SUS), sucrose phosphate synthase (SPS), and invertase [49].

Previous studies have shown that the integration of linkage analysis and GWAS provided a higher power and resolution to investigate complex quantitative traits. In the present study, we constructed a recombinant inbred line (RIL) population containing 146 lines derived from two oriental melon materials with different levels of sweetness. Then, a high-resolution genetic map of the RIL population was constructed to detect QTLs related to SSC in melon. In addition, a total of 213 melon accessions were utilized to detect the significance of SNPs associated with SSC via GWAS. Five candidate genes were identified by QTL mapping, GWAS, and qRT-PCR, which played important roles in synthesis and hydrolysis of raffinose, sucrose hydrolysis, and sugar transport. The genomic regions and candidate genes will be helpful for molecular breeding programs and elucidating the mechanisms of sugar accumulation.

2. Materials and Methods

2.1. Plant Materials and Phenotypic Data Analysis

The plant material contained an RIL population and a GWAS panel. The RIL population, consisting of 146 RILs, was derived from two oriental melon cultivars (1214 with low sweetness and 1228 with high sweetness). The RIL population was grown in a greenhouse at Xueyao Experiment Station (32°14′ N, 120°64′ E) of Jiangsu Yanjiang Institute of Agricultural Sciences (Nantong, China) in the spring of 2020, 2021, and 2022. The GWAS panel, consisting of 213 melon accessions that included 145 accessions of ssp. *agrestis*, 67 accessions of ssp. *melo*, and 1 Mapao melon accession, was planted in a greenhouse at Xueyao Experiment Station in the spring of 2020 and 2021 and the spring/autumn of 2021.

A randomized block design with three replicates was utilized to evaluate the plant materials. Five plants were grown for each melon germplasm. One fruit per plant was harvested at 30–35 days after pollination, and the SSC of fruit flesh was measured with a hand-held digital PAL-1 refractometer (Atago). ANOVA for SSC was calculated in R using the aov function. Based on the phenotypic data of two populations for multiple environments, the best linear unbiased predictions (BLUP) were calculated by the R package lme4.

2.2. DNA Extraction and Whole-Genome Resequencing

The young leaves of RIL population and GWAS panel were utilized to extract DNA using the CTAB method. The genomic DNA samples were fragmented to a size of 350 bp for sequencing on the Illumina HiSeq PE150 platform of Novogene, China. The melon reference genome HS_ZJU [50] and its annotation were downloaded online (http://www.ncbi.nlm.nih.gov, accessed on 19 July 2021). BWA [51] was utilized to map the clean reads to the melon reference genome using the command 'mem -t 4 -k 32 -M'. SNP calling was performed using SAMtools [52] with a Bayesian approach. The high-quality SNPs were performed using Genome Analysis Toolkit (GATK) software 4.0.4.0 [53]. ANNOVAR software (Version 20130520) [54] was utilized to annotate the SNPs identified in this study.

2.3. Linkage Analysis

The sequencing depths of parents (1214 and 1228) and RILs were 20-fold and 5-fold, respectively. The RILs were genotyped by polymorphic SNPs with an aa $\times$ bb segregation pattern between the parental materials. The recombinant breakpoints were evaluated by the sliding window approach [55]. The genotype was performed with a sliding window of 15 SNPs. The high-resolution genetic map of the RIL population was constructed with bin markers using JoinMap 4.1 (https://www.kyazma.nl/index.php/JoinMap/, accessed on 13 May 2022). The ALLMAPS program [56] was utilized to visualize the relationships between the physical and genetic locations. The composite interval mapping (CIM) method of Windows QTL Cartographer 2.5 software (https://brcwebportal.cos.ncsu.edu/qtlcart/WQTLCart.htm, accessed on 7 June 2022) was utilized to detect QTLs. The permutation test of MapQTL6.0 (https://www.kyazma.nl/index.php/MapQTL/, accessed on 7 June 2022) was utilized to calculate the LOD threshold values for the SSC.

2.4. Association Analysis

The sequencing depths of 211 melon accessions were 10-fold, and two accessions (1214 and 1228) were sequenced with a depth of 20-fold. The sequencing depth ≥ 6, missing rate ≤ 0.5, and minor allele frequency ≥ 0.05 were considered standard in identifying SNPs. Finally, a total of 1,681,775 SNPs were used for association analysis. The software TreeBeST 1.9.2 (https://treesoft.sourceforge.net/treebest.shtml, accessed on 2 April 2022) was utilized to create an individual-based neighbor-joining (NJ) tree using p-distance with 1000 bootstrap replications. The phylogenetic tree was visualized using MEGA6.0 (http://www.megasoftware.net/, accessed on 2 April 2022). The program ADMIXTURE (version 1.23) [57] was utilized to calculate the population genetics structure using an expectation maximization algorithm. To evaluate the LD decay in the association population, the software PopLDdecay 3.40 [58] was utilized to calculate the degree of linkage disequilibrium coefficient (r^2) between pairwise SNPs with the command '-n -dprime -minMAF 0.05'. The GEMMA 0.98.1 software package [59] was utilized to perform association analysis, and the significant P-value threshold was approximately 1×10^{-6}.

2.5. Candidate Genes Identification and qRT-PCR Analysis

The co-detected regions by linkage analysis and GWAS, and the $\pm$100 kb regions around the significant SNPs located within the reported QTL regions were considered significant regions to search for candidate genes. Annotations of these genes were analyzed following the melon reference genome 'HS_ZJU', and genes related to sugar metabolism and accumulation were considered candidate genes. The fruits of the two parents (1228 and 1214) harvested at 20DAP, 25DAP, and 30DAP were used to extract RNA for qRT-PCR analysis. The cDNA was synthesized using All-in-One First-Strand cDNA Synthesis SuperMix for qPCR (TransGen, Beijing, China). Gene expression was performed on the ABI7500 Real-Time PCR System (Applied Biosystems, Singapore) using Green qPCR SuperMix (TransGen, Beijing, China). The data of relative gene expression was analyzed according to the $2^{-\Delta\Delta Ct}$ method with the 18s rRNA used as the internal control. The primers used in this study are shown in Table S8.

3. Results

3.1. Phenotypic Variation of Soluble Solids Content in Two Populations

In this study, 146 RILs and 213 melon accessions were used. The SSC of RILs was evaluated in the Nantong greenhouse in the spring of 2020, 2021, and 2022 (Table S9). The SSC of the 213 melon accessions was evaluated in the spring of 2020 and 2022 and the spring/autumn of 2021 (Table S10). The descriptive statistics including the mean, range, standard deviation, kurtosis, skewness, and coefficient of variation (CV) for SSC of two panels are presented in Table 1.

The SSC of RILs ranged from 5.67% to 14.50% (with a mean of 10.43%) in spring 2020, from 5.83% to 14.07% (with a mean of 11.01%) in spring 2021, and from 7.15% to 14.80% (with a mean of 11.29%) in spring 2022. The SSC of 213 melon accessions ranged from 3.00% to 17.30% (with a mean of 10.01%) in spring 2020, from 3.00% to 17.43% (with a mean of 9.86%) in spring 2021, from 3.55% to 15.50% (with a mean of 8.97%) in autumn 2021, and from 3.80% to 18.98% (with a mean of 10.68%) in spring 2022. The CV of the RILs ranged from 12.93% to 14.96%, while the CV of the accessions ranged from 26.17% to 30.21%. The skewness and kurtosis in Table 1 and the histograms in Figure 1a,b show that the data of SSC have a normal distribution. The melon accessions (G), environment (E), and the genotype × environment interaction (G × E) had a significant effect ($p < 0.01$) on the SSC (Table 1). The broad-sense heritability (H^2) for SSC were 0.65 and 0.72 in two populations (Table 1).

The fruits of the two parents (1228 and 1214), were harvested at five growth stages to investigate the changes in the SSC (Figure 1c). With fruit development, the SSC of 1228 and 1214 accumulated continuously and peaked at 30 days after pollination (DAP), but the SSC of 1214 was significantly lower than that of 1228 at each stage (Figure 1c). The different levels of SSC led to the differences in sweetness between these two cultivars.

3.2. Population Sequencing and Linkage Map Construction

In order to construct a high-resolution genetic map, 146 RILs together with two parents were re-sequenced on the Illumina HiSeq PE150 platform. In total, 9.05 Gb (19.96-fold genome coverage) and 10.34 Gb (23.04-fold genome coverage) clean bases were obtained for two parents (1214 and 1228), respectively (Table S1). A total of 364 Gb clean bases were obtained for the 146 RILs with high quality (Q20 $\geq$ 95.42%, Q30 $\geq$ 88.72%), and the average sequencing depth for each individual was 6.7-fold (Table S1). A total of 380,864 SNPs were detected between the parents, and 111,500 SNPs were identified with the aa×bb segregation pattern. Then, a total of 1427 bin markers were obtained to construct a high-resolution genetic map (Figure 2a). The total length of the bin map was 1254.34 cM, and the average interval between the adjacent markers was 0.88 cM (Table 2). The genetic and physical positions of the bin markers on the 12 chromosomes were compared with each other, and a high level of collinearity was detected between the linkage map and the reference genome of melon cultivar 'HS_ZJU' (Figure 2b). However, there were several chromosomal locations of the bins that displayed inconsistencies with the genetic map.

Table 1. Descriptive statistics and variance analysis for SSC of the two panels in multiple environments.

Population	Environment [a]	Parents		Mean ± SE (%)	SD [b]	Range (%)	CV [c] (%)	Skewness	Kurtosis	G [d]	E [d]	G × E [d]	H^{2} [e]
		1214	1228										
RILs	2020 SPR	7.43	12.40	10.43 ± 0.13	1.56	5.67–14.50	14.96	−0.15	0.13	**	**	**	0.65
	2021 SPR	9.26	13.48	11.01 ± 0.13	1.55	5.83–14.07	14.07	−0.35	0.46				
	2022 SPR	9.62	14.35	11.29 ± 0.13	1.46	7.15–14.80	12.93	−0.34	−0.01				
Accessions	2020 SPR	-	-	10.01 ± 0.20	2.62	3.00–17.30	26.17	−0.39	−0.17	**	**	**	0.72
	2021 SPR	-	-	9.86 ± 0.20	2.87	3.00–17.43	29.01	−0.17	−0.57				
	2021 AUT	-	-	8.97 ± 0.21	2.71	3.55–15.50	30.21	0.16	−0.57				
	2022 SPR	-	-	10.58 ± 0.23	3.12	3.80–18.98	29.21	−0.03	−0.75				

[a] SPR represents spring and AUT represents autumn; [b] SD represents the standard deviation; [c] CV represents the coefficient of variation; [d] G, E, and G × E represent the effect for the genotype, environment, and genotype × environment interaction, respectively; [e] broad-sense heritability. ** Significant at $p \leq 0.01$.

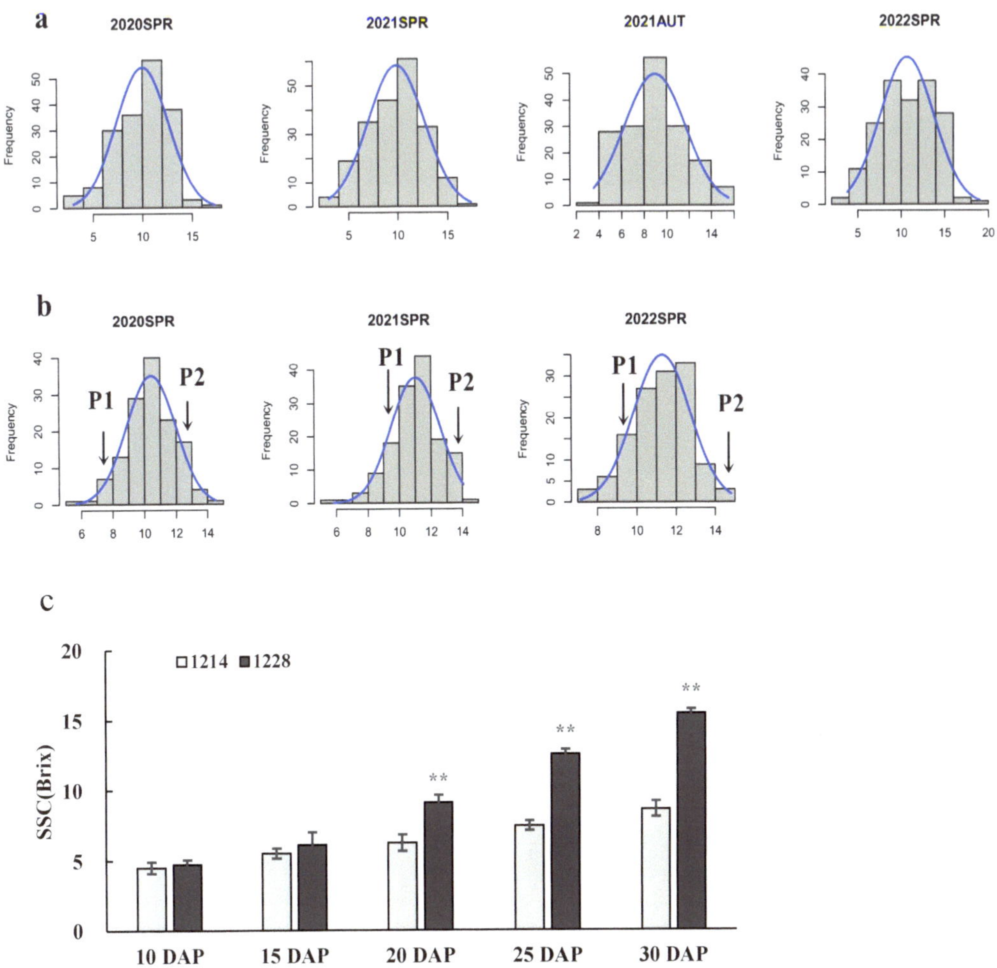

Figure 1. Frequency distribution and changes in soluble solids content. (**a**) Frequency distribution of SSC for 213 melon accessions in four environments. (**b**) Frequency distribution of SSC for RILs in three environments. SPR represents spring and AUT represents autumn. Arrows indicate the soluble solids content for the parents, 1214(P1) and 1228(P2). (**c**) SSC in the fruit during the five ripening stages. ** Significant at $p \leq 0.01$.

3.3. Identification of QTLs for Soluble Solids Content

Under three growth environments and BLUP, 14 QTLs related to SSC were identified using Windows QTL Cartographer 2.5 software with the bin map. These 14 QTLs were scattered on chromosomes 1, 2, 3, 4, 6 and 8, as follows: 1 QTL on chromosome 1, 4 QTLs on chromosome 2, 4 QTLs on chromosome 3, 2 QTLs on chromosome 4, 2 QTLs on chromosome 6, and 1 QTL on chromosome 8 (Figure 3, Table S3). The physical distances of these QTLs ranged from 0.39 to 3.15 Mb, and the average distances was 0.88 Mb. The LOD value of these QTLs ranged from 3.2 to 5.7, and the phenotypic variation explained (PVE) values ranged from 6.7% to 13.8%. All the QTL showed negative additive effects, indicating that alleles from the high-SSC parent 1228 contributed to the greater SSC. Based on the physical position of flanking markers, the 14 QTLs were clustered into 10 common QTLs. Of the 10 common QTLs, one QTL was repeatedly detected across three environments, and two QTLs were detected across two environments. In addition, seven common QTLs were mapped to the same loci of previously reported QTLs (Table S3) and confirmed the reliability of the genetic map.

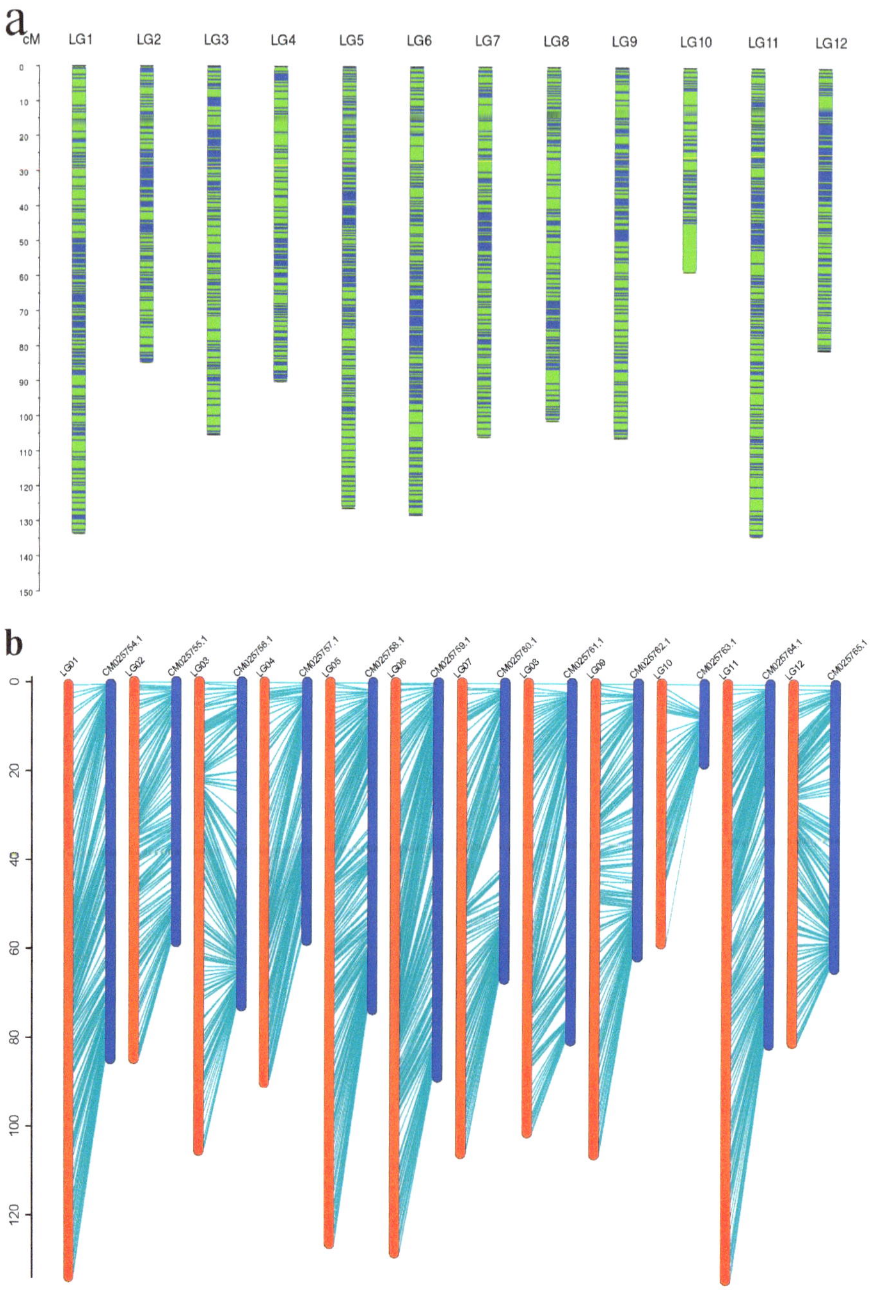

Figure 2. Construction of the linkage map (**a**) and collinearity analysis between the linkage map and melon genome 'HS_ZJU' (**b**).

Table 2. Information of the novel constructed genetic linkage map for melon.

LGs	Number of Bins	Map Length (cM)	Average Distance (cM)	Max Gap (cM)	<5 cM	>5 cM
LG01	148	133.75	0.90	4.27	148	0
LG02	109	84.73	0.78	3.00	109	0
LG03	119	105.39	0.89	3.84	119	0
LG04	96	90.16	0.94	3.84	96	0
LG05	152	126.17	0.83	3.42	152	0
LG06	158	128.21	0.81	4.27	158	0
LG07	114	105.84	0.93	3.84	114	0
LG08	109	101.16	0.93	4.27	109	0
LG09	118	105.97	0.90	3.00	118	0
LG10	43	58.46	1.36	14.26	42	1
LG11	149	133.92	0.90	3.84	149	0
LG12	112	80.59	0.72	3.84	112	0
total	1427	1254.34	0.88	14.26	1426	1

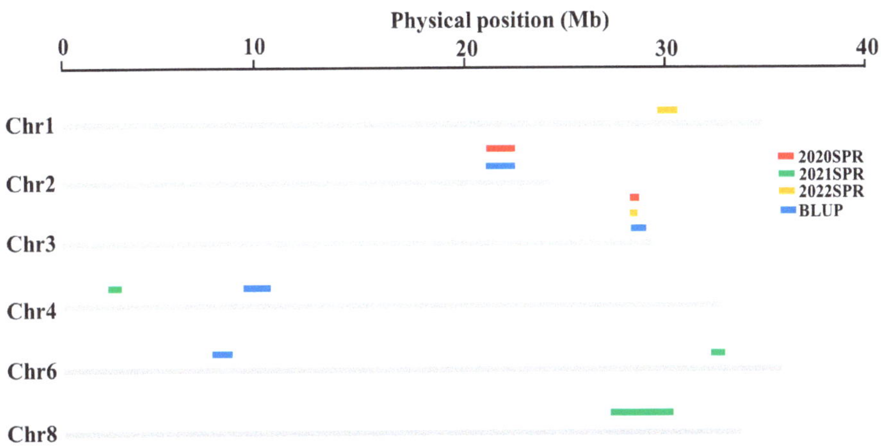

Figure 3. Distribution of QTLs related to SSC on the melon chromosomes. Gray bars represent the chromosomes and colored bars represent the QTLs for SSC. Bar length indicates the physical distance of QTLs.

3.4. GWAS for Soluble Solids Content

We re-sequenced 213 melon accessions at an average sequencing depth of 9.7× on the Illumina HiSeq PE150 platform, generating 941.6 Gb resequencing data of high quality (Q20 ≥ 94.43%, Q30 ≥ 86.92%, Table S2). The raw sequence data reported in this paper have been deposited in the Genome Sequence Archive at the National Genomics Data Center, China National Center for Bioinformation/Beijing Institute of Genomics, Chinese Academy of Sciences, under accession number CRA011939 and are publicly accessible at https://ngdc.cncb.ac.cn/gsa (released on 24 July 2023). The resequencing data were mapped to the melon genome 'HS_ZJU', and 1,681,775 high-confidence SNPs (missing rate ≤ 0.5; minor allele frequency ≥ 0.05) were detected. Of these, 99,547, 51,623, 304,931, 1,029,614, and 96,114 SNPs were located in upstream regions, exons, introns, intergenic regions, and downstream regions, respectively (Table S4). The markers were scattered on the melon genome, with the lowest (107,238) and highest (174,458) number of markers present on chromosome 9 and chromosome 1, respectively (Figure 4a). Additionally, the highest SNP frequency was 5.04 SNPs/kb on chromosome 7. The linkage disequilibrium (LD) level was calculated as the physical distance at which r^2 decreased to half of the maximum value, and LD decay was estimated at 103 kb (Figure 4b). From the phylogenetic tree constructed with the high-confidence SNPs (Figure 4c), the 213 accessions were divided into two subpopulations: subpopulation I primarily consisted of ssp. *agrestis* and subpopulation II consisted of ssp. *melo*. The result was further supported by a population structure plot (Figure 4d). In addition, the 'Mapao' melon accession was located within ssp. *agrestis*, and this result was consistent with a previous study [17].

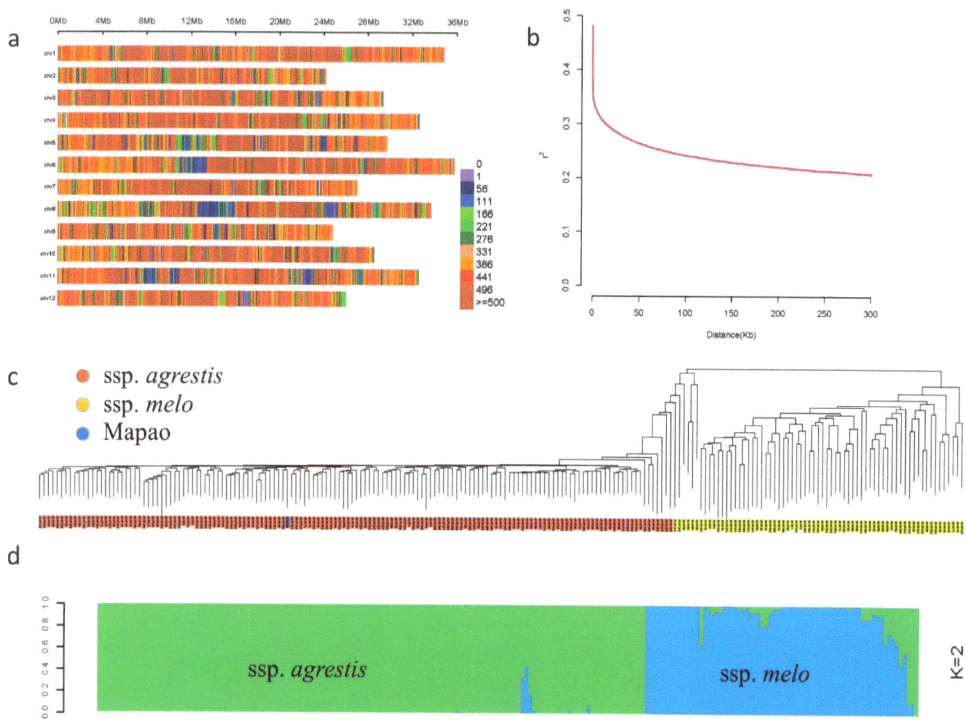

Figure 4. Distribution of the SNPs and the population structure of 213 melon accessions. (**a**) Distribution of SNPs and nucleotide diversity across the melon genome 'HS_ZJU'. Different colors indicate the number of SNPs within the 0.1 Mb window size. (**b**) Genome-wide average LD decay for 213 melon accessions. (**c**) Phylogenetic tree of 213 melon accessions. (**d**) Population structure of melon accessions estimated by ADMIXTURE with k = 2.

Based on the data of four growth environments and BLUP, GWAS of 213 melon accessions was performed with 1,681,775 SNPs using GEMMA software. At a significance level of $p < 10^{-6}$, a total of 211 SNPs significant associated with SSC was found (Figure 5; Table S5). Among these SNPs, s6-26722606 had the lowest p-value ($-\log10\ p = 10.44$), which was detected in 2022SPR. The two chromosomes with the highest number of markers were chromosome 9 with 61 SNPs and chromosome 12 with 34 SNPs. The intersections based on the SNPs detected across four environments and BLUP are presented in Figure 6 as the UpSet plot [60]. The UpSet plot showed the number of significant markers detected in each environment and BLUP, and also showed the number of the same markers detected across different environments (Figure 6). In Figure 6, 29 significant SNPs were detected coherently across three environments, 10 significant SNPs were detected coherently across four environments, and 1 significant SNP was detected coherently across all environments. Compared to the SNP loci detected by GWAS and the QTLs detected in previous studies, 138 SNPs overlapped with the reported QTLs (Table S6).

3.5. Co-Detected Regions by QTL Mapping and GWAS

Linkage analysis and association analysis could be complementary methods for QTLs detection. The results of linkage analysis and GWAS were analyzed together to identify stable regions significantly associated with SSC. Five QTLs identified by linkage analysis and 13 SNPs identified by GWAS were co-localized on chromosome 3, 4, and 6 (Table 3). QTL *qSSC3-1* and five SNPs, and *qSSC3-2* and four SNPs were co-localized on chromosome 3. QTL *qSSC4-1* and one SNPs, and *qSSC4-2* and two SNPs were co-localized on chromosome 4. QTL *qSSC6-2* and one SNP were co-localized on chromosome 6. The two co-localized regions on chromosome 3 were identified across multiple environments with a large PVE (>10%) and likely included genes affecting the accumulation of SSC.

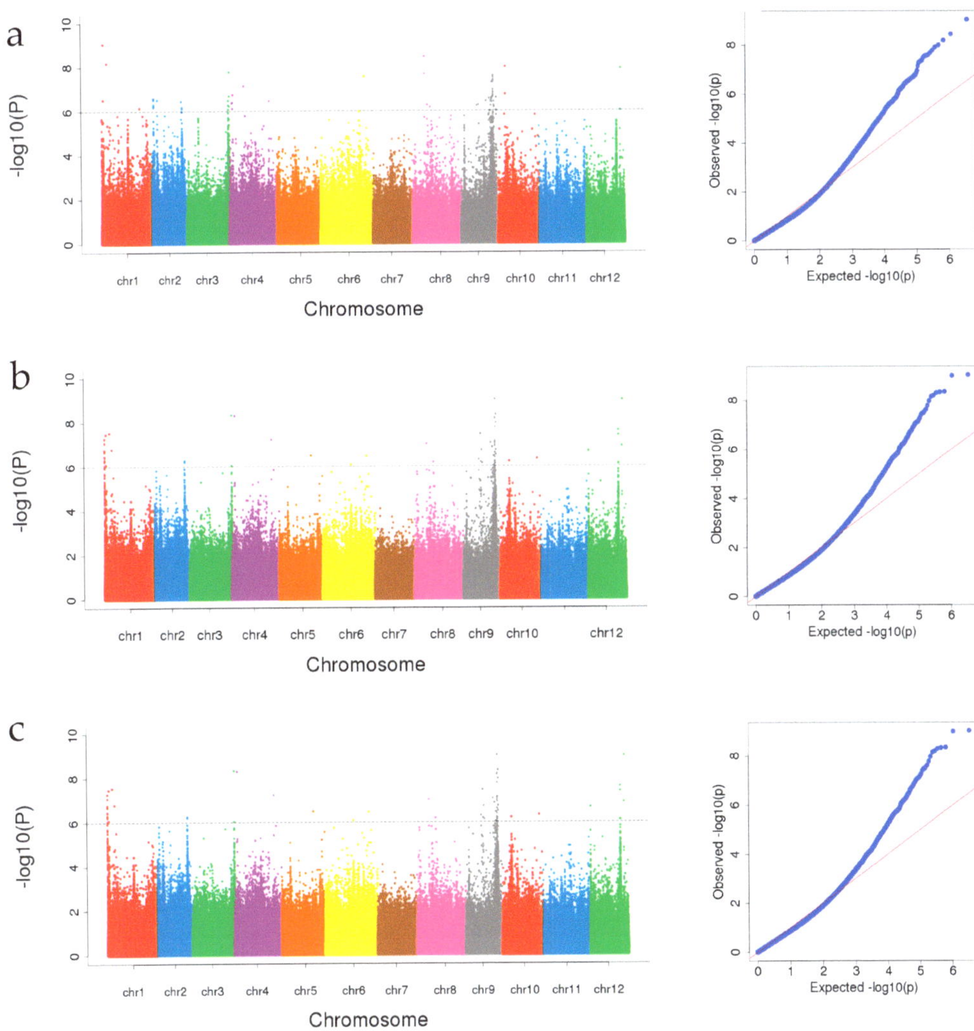

Figure 5. *Cont.*

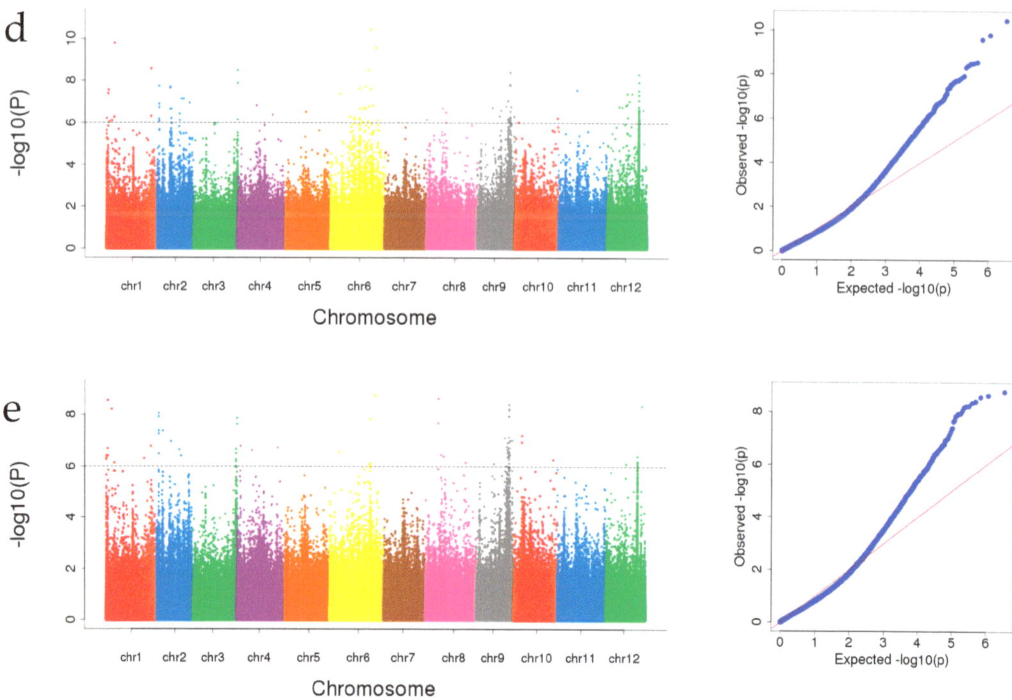

Figure 5. GWAS signals for SSC estimated across four environments and BLUP. Manhattan plot and quantile–quantile (Q–Q) plot estimated in (**a**) spring 2020 (E1), (**b**) spring 2021 (E2), (**c**) autumn 2021 (E3), (**d**) spring 2022 (E4), and (**e**) BLUP. The gray lines on the *Y*-axis denote the significance threshold ($-\log 10\, p > 6$).

3.6. Identification of Candidate Genes Related to SSC

The genes located within QTL intervals and the LD regions of significant SNPs, which were related to sugar metabolism and accumulation, were considered as candidate genes. As a result of *qSSC3-1* and *qSSC3-2* being co-detected by linkage analysis and GWAS across multiple environments, candidate genes related to SSC were searched for in these two stable regions on chromosome 3. A total of 113 genes were detected within the two co-localized intervals (28.5–28.8 Mb and 28.9–29.3 Mb) on chromosome 3, which consisted of 46 and 67 genes, respectively (Table S7). Of the 46 genes in *qSSC3-1*, the invertase gene *CWINV3* (*MELO 06759*) may be a candidate gene for SSC as a key enzyme for converting sucrose into glucose and fructose. The expression of *CWINV3* was significantly different between the high sweetness and low sweetness oriental melon accessions at different fruit growth stages [61]. Meanwhile, we found an alpha-galactosidase (*MELO 06810*) in *qSSC3-2*, which was related to the production of galactose and sucrose during phloem unloading [62]. Yang et al. [50] reported that *MELO 06810* had a parallel shift of amino acid in Cucurbitaceae via positive selection gene analysis, which may play an important role in phloem transport.

Based on the LD decay of 103 kb (Figure 4b), the ±100 kb regions around 138 significant SNPs that overlapped with the reported QTLs (Table S6) were considered significant regions to search for candidate genes. In these regions, nine candidate genes, which included sucrose phosphate synthase (SPS), trehalose 6-phosphate phosphatase (TPP), raffinose synthase (RS), sucrose synthase (SS), UDP-glucose 4-epimerase (UGE), sugar transport, and sugars will eventually be exported transporter (SWEET), were identified as being related to sugar metabolism and accumulation (Table 4). Quantitative real-time PCR was used to estimate the relative expression of 11 candidate genes at the stage of

SSC rapid accumulation (Figure 1c). The results showed that five candidate genes had different expression levels between the two parents at the stage of 20DAP, 25DAP, and 30DAP (Figure 7). The expression levels of *MELO21359*, *MELO06759*, *MELO22262*, and *MELO06810* were significantly higher in 1228 than 1214 at three stages. *MELO18502* had higher expression levels in 1228 than 1214 at 20DAP and 25DAP, while no significant difference was observed at 30DAP. *MELO22262* and *MELO06810* shared a pattern of higher expression level at 20DAP, and decreased expression at 25DAP and 30DAP. *MELO21359* had the higher expression level at 30DAP, while *MELO06759* had the higher expression levels at 25DAP and 30DAP. These five genes with different expression levels between two parents are the presumed candidates controlling SSC in melon.

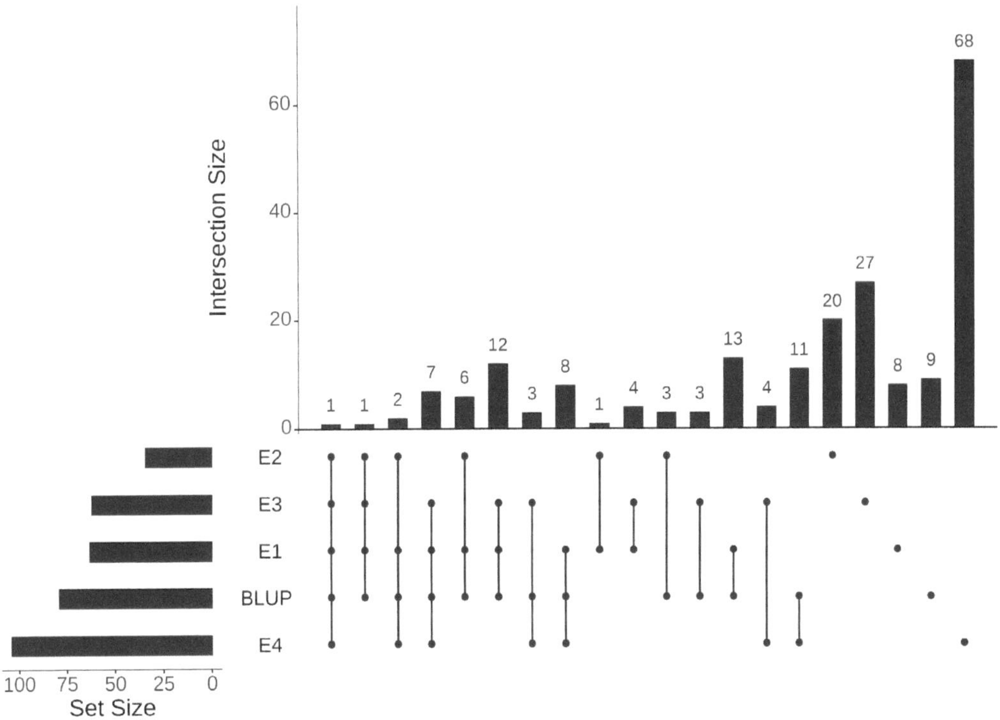

Figure 6. Intersection of significant SNPs across four different environments (E1, E2, E3, and E4) and BLUP.

Table 3. Co-detected loci by QTL mapping and GWAS.

Chr.	Environment	LOD	Position Interval (cM)	Physical Position (bp)	SNP Loci	−log10(p)				
						2020SPR	2021SPR	2021AUT	2022SPR	BLUP
3	2020SPR	4.5	95.2–104.3	28,509,689–28,965,705	28,672,844					6.01
3	2022SPR	5.5	93.7–99.5	28,446,470–28,845,998	28,694,392	6.27	6.05			6.27
3	BLUP	5.7	95.6–102.9	28,509,689–28,947,998	28,711,496	6.47				6.51
					28,714,701					6.19
					28,724,776	6.08				6.67
3	BLUP	5.2	102.9–105.1	28,947,998–29,341,849	29,267,782	6.68		6.04	8.49	7.65
					29,283,426				6.14	
					29,289,157	6.22				
					29,330,788	7.77		8.33	7.88	7.88
4	2021SPR	3.5	20.9–28.6	2,334,788–3,020,226	2,491,434	6.74				6.80
4	BLUP	3.4	51.7–54.2	8,961,531–10,360,279	10,141,036	7.13				6.63
					10,275,303		7.06			
6	2021SPR	3.5	95.5–101	32,217,543–32,935,488	32,656,543		6.56			

Table 4. Candidate genes related to SSC within the QTL/SNP regions.

Gene	Chromosome	Start	End	Functional Annotation
MELO03698	chr2	14,397,615	14,406,002	Sucrose-phosphate synthase
MELO06759	chr3	28,826,329	28,829,358	Beta-fructofuranosidase, insoluble isoenzyme CWINV3
MELO06810	chr3	29,119,436	29,126,262	Alpha-galactosidase
MELO12562	chr6	6,109,512	6,114,363	Trehalose 6-phosphate phosphatase
MELO18502	chr8	16,852,497	16,859,087	Raffinose synthase
MELO20547	chr9	14,794,455	14,800,212	Sucrose synthase
MELO21359	chr9	23,772,893	23,777,340	UDP-glucose 4′-epimerase
MELO22262	chr10	5,145,518	5,147,793	Sugar transport protein 10-like
MELO27354	chr12	20,598,706	20,606,584	Trehalose-6-phosphate synthase
MELO27634	chr12	23,004,024	23,006,015	Sugar transporter SWEET12-like
MELO27635	chr12	23,013,800	23,015,623	Sugar transporter SWEET12-like

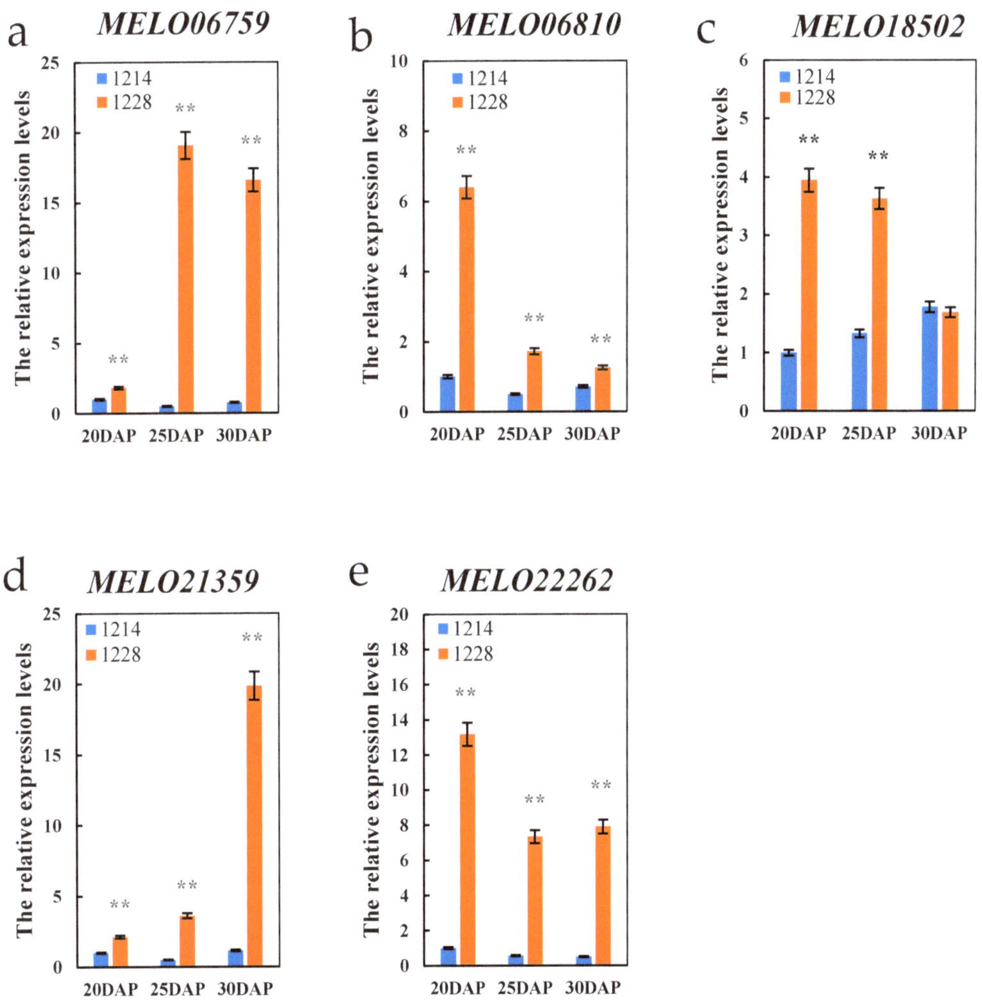

Figure 7. Relative expression levels of five candidate genes. (**a**) *MELO06759*; (**b**) *MELO06810*; (**c**) *MELO18502*; (**d**) *MELO21359*; and (**e**) *MELO22262*; ** Significant at $p \leq 0.01$.

4. Discussion

4.1. Phenotypic Data over Multiple Years and High-Density Markers Improved the Accuracy of QTL Mapping and Gwas

Linkage mapping based on biparental segregating populations and GWAS based on a natural population are complementary approaches to detect genetic variations for important agronomic traits. Linkage mapping can only evaluate limited variations between two parents in a single population. However, Association analysis can investigate all the allelic variations in natural populations. For QTL mapping and GWAS, phenotypic data over multiple years could reduce environmental factors to identify genetic loci that consistently affect the trait. The sweetness of melon has a significant effect on its quality and consumer appeal, and SSC is an important index of melon sweetness. In this study, phenotypic data of the two melon populations were investigated for three years. In addition, two cultivars with different SSC levels were utilized to understand the accumulation process. We found that SSC accumulated quickly from 20 DAP to 30 DAP, and a similar accumulation pattern for sugar has been observed in previous studies [50,61]. Higher marker density could also enhance the accuracy of linkage analysis and GWAS. In this study, 1427 bin markers generated via genome resequencing were used to construct a high-resolution genetic linkage map with an average interval between the adjacent markers of 0.88 cM. For GWAS, 1.68 million high-confidence SNPs were used in this study, leading to a high precision of GWAS to investigate the complex traits.

4.2. QTL Intervals and SNP Loci for SSC Identified via Linkage Mapping and GWAS

In this study, 10 QTLs (with an average distance of about 0.88 Mb) related to SSC were identified by linkage mapping using the high-resolution genetic linkage map. Seven QTLs were mapped to the same loci of previously reported QTLs (Table S3) and confirmed the reliability of our study. Based on genome resequencing, 1.68 million marks were utilized to detect the loci related to SSC via GWAS. A total of 211 SNPs significantly associated with SSC were found by GWAS with a significance level of $p < 10^{-6}$. Compared to the SNP loci detected by GWAS and the QTLs detected in previous studies, 138 SNPs overlapped with the reported QTLs (Table S6). These results indicate that the loci identified in the present study were reliable for candidate genes mining.

In this study, we found that 5 QTLs detected by linkage analysis were co-localized with 13 significant SNPs detected by GWAS. The genomic region overlapped by *qSSC4-2*, S4_10141036, and S4_10275303 on chromosome 4, and *qSSC6-2* and S6_32656543 on chromosome 6 (Table 3, Table S3 and Table S6) were also identified within the QTLs identified in previous studies [4,10]. Due to limited parental materials used in QTL mapping, not all of the genetic variations in melon have been identified, and populations derived from different materials, as in previous studies, could determine new QTLs or confirm previous QTLs [8]. In previous studies, the high sweetness parental materials utilized to construct segregated populations were almost always muskmelon. Saladie et al. [63] showed that the patterns of sugar accumulation were different between the oriental melon and muskmelon during fruit development. In this study, two oriental melon inbreeds with different soluble solids content were utilized to derive a segregated population for QTL mapping, and two stable QTLs, *qSSC3-1* and *qSSC3-2*, identified across multiple environments might be new QTLs associated with SSC.

4.3. Further Analysis of Candidate Genes Differentially Expressed between Parents at Different Fruit Growth Stages

Comparing the QTL intervals and SNP loci identified by linkage analysis and GWAS, the co-detected loci and ±100 kb regions around the significant SNPs within the reported QTLs were considered reliable regions to search for candidate genes. According to the annotation information, the genes related to sugar metabolism and accumulation were considered candidate gene controlling soluble solids content. The relative expression levels of eleven genes predicted to participate in the accumulation of SSC were investigated by qRT-PCR at different fruit stages. Finally, five genes with different expression levels between two parents during different stages were considered putative candidates controlling SSC.

In sweet melon, sucrose, fructose, and glucose are the major soluble sugars, and sucrose is the main factor affecting total sugar content [19,49,64]. The cell wall invertases (CWINV) hydrolyzing sucrose into fructose and glucose play an important role in sugar accumulation [65]. Zanor et al. [66] reported that silencing the cell wall invertase LIN5 could lead to a linear decrease in the total soluble solids content in tomato. Wang et al. [45] showed that a reduction in *LIN5* activity caused a decrease of fructose, glucose, and SSC in tomato. A previous study revealed that the expression of *CWINV3* was significantly higher in the high sweetness rather than low sweetness oriental melon accessions at different fruit growth stages [61]. In the present study, the same expression pattern of *CWINV3* was observed (Figure 7a) and the result indicated that *CWINV3* might be involved in SSC accumulation.

Alpha-galactosidases (AGA) that hydrolyze stachyose and raffinose play an important role in mediating source–sink communication during fruit development [62,67,68]. A previous study showed that *ClAGA2* controls fruit raffinose hydrolysis in watermelon, and that the sucrose content of fruits increased significantly in the overexpressed plants [67]. Based on the positive selection analysis in Cucurbitaceae genomes including *C. lanatus*, *C. maxima*, *C. pepo*, *C. sativus*, *L. siceraria*, and *C. melo* (HS, DHL92, Payzawat) with *A thaliana*, *S. lycopersicum*, *M. domestica*, and *O. sativa*, a parallel shift of an amino acid in alpha-galactosidase (*MELO 06810*) was identified in Cucurbitaceae, which may play an important role in raffinose hydrolysis [50]. In this study, the expression of *MELO 06810* was significantly higher in 1228 than 1214 at different fruit growth stages (Figure 7b). Therefore, the alpha-galactosidase (*MELO 06810*) located in *qSSC3-2* is considered a candidate gene related to SSC accumulation.

Raffinose could be hydrolyzed into sucrose and then transported and stored in tissues or cells. Raffinose synthase is a key enzyme for raffinose production, and overexpression of raffinose synthase significantly increased raffinose content [69]. A previous study revealed that the expression of raffinose synthase (*MELO18502*) was significantly higher in the high sweetness rather than low sweetness oriental melon accessions at different fruit growth stages [61]. In this study, the expression of *MELO18502* was significantly higher in 1228 than 1214 at 20DAP and 25DAP(Figure 7c).

Additionally, UDP-glucose 4′-epimerase (UGE) could catalyze the interconversion of UDP-glucose and UDP-galactose, and UDP-glucose may further serve as a substrate for the synthesis of sucrose. A previous study reported that overexpression of UGE could improve the accumulation soluble sugars [70]. In this study, the expression of *MELO21359* was significantly higher in 1228 than 1214 at three stages (Figure 7d).

Sugar transport proteins (STP) played an important role on sugar accumulation via the transport of glucose from the apoplast into plant cells [39,71]. A previous study showed that the soluble solids content increased in the overexpressed plants [45]. In this study, the expression of *MELO22262* was significantly higher in 1228 than 1214 at three stages (Figure 7e).

Sugar metabolism and accumulation are complex biological processes that require the synergistic action of key genes. In this study, five candidate genes related to SSC were identified by QTL mapping, GWAS, and qRT-PCR. These five genes play important roles in the synthesis and hydrolysis of raffinose, sucrose hydrolysis, sugar transport, and production of substrate for sugar synthesis. High gene expression levels of these five genes may collectively promote sugar accumulation.

5. Conclusions

The sweetness of melon is a significant factor in fruit quality and consumer appeal, and SSC is a key index of melon sweetness. In this study, a recombinant inbred population including 146 lines derived from two oriental melon materials with different sweetness levels and an association population consisting of 213 melon accessions were utilized to detect genomic regions influencing SSC. By comparing the results of linkage analysis and GWAS, we identified seven QTLs and 138 significant SNPs overlapped with the reported QTLs. In addition, we identified two new stable loci on chromosome 3 via QTL mapping and GWAS across multiple environments. Five candidate genes related to SSC were identified by QTL mapping, GWAS, and qRT-PCR, two of which were involved in raffinose and sucrose hydrolysis located in the new stable loci. The other three candidate genes were involved in raffinose synthesis sugar transport, and production of substrate for sugar synthesis. Synergistic action of these five candidate genes may cause the higher SSC. The identification of genomic regions and candidate genes that influence SSC will provide useful information and insights for molecular breeding and offer an opportunity for sugar accumulation research.

Supplementary Materials: The following supporting information can be downloaded at: https://www.mdpi.com/article/10.3390/cimb45090450/s1, Table S1: The sequence data summary for RILs; Table S2: Sequence information on the genomes of 213 melon accessions; Table S3: QTL analyses of soluble solids content in melon; Table S4: Results of filtered SNP annotation; Table S5: The SNPs with significant association signals ($-\log10(p) \geq 6$) for SSC; Table S6: GWAS signals overlapped with the reported QTLs; Table S7: Annotations of the genes underlying *qSSC3-1*, and *qSSC3-2* in melon; Table S8: Primer sequences for qRT-PCR analysis; Table S9: Phenotypic data of SSC for RILs; Table S10: Phenotypic data of SSC for the accessions.

Author Contributions: Conceptualization, H.Y. and K.W.; Formal analysis, H.Y. and M.W.; Validation, H.Y., M.W. and H.Z.; Investigation, H.Y., L.F. and X.W.; writing—original draft preparation, H.Y.; Writing—Review and editing, H.Y., M.W. and K.W.; visualization, H.Y. and L.F.; Supervision, K.W.; Project administration, M.W. and K.W. All authors have read and agreed to the published version of the manuscript.

Funding: This research was funded by the Natural Science Foundation of Jiangsu Province, China (BK20201205), the National Natural Science Foundation of China (31902037), Jiangsu Province Seed Industry Revitalization Project (JBGS [2021]075), and the Earmarked Fund for Jiangsu Agricultural Industry Technology System (JATS [2022]147).

Institutional Review Board Statement: Not applicable.

Informed Consent Statement: Not applicable.

Data Availability Statement: The datasets analyzed in this study are available in the manuscript text and Supplementary Materials. The raw sequence data reported in this paper have been deposited in the Genome Sequence Archive at the National Genomics Data Center, China National Center for Bioinformation/Beijing Institute of Genomics, Chinese Academy of Sciences, under accession number CRA011939 and are publicly accessible at https://ngdc.cncb.ac.cn/gsa (released on 24 July 2023).

Conflicts of Interest: The authors declare no conflict of interest.

References

1. Jeffrey, C. A review of the Cucurbitaceae. *Bot. J. Linn. Soc.* **1980**, *81*, 233–247. [CrossRef]
2. Pitrat, M.; Hanelt, P.; Hammer, K. Some Comments on Infraspecific Classification of Cultivars of Melon. *Acta Hortic.* **2000**, *510*, 29–36. [CrossRef]
3. Monforte, A.J.; Oliver, M.; Gonzalo, M.J.; Alvarez, J.M.; Dolcet-Sanjuan, R.; Arús, P. Identification of quantitative trait loci involved in fruit quality traits in melon (*Cucumis melo* L.). *Theor. Appl. Genet.* **2003**, *108*, 750–758. [CrossRef] [PubMed]
4. Eduardo, I.; Arús, P.; Monforte, A.J.; Obando, J.; Fernández-Trujillo, J.P.; Martínez, J.A.; Alarcón, A.L.; Álvarez, J.M.; van der Knaap, E. Estimating the Genetic Architecture of Fruit Quality Traits in Melon Using a Genomic Library of Near Isogenic Lines. *J. Am. Soc. Hortic. Sci.* **2007**, *132*, 80–89. [CrossRef]
5. Paris, M.K.; Zalapa, J.E.; McCreight, J.D.; Staub, J.E. Genetic dissection of fruit quality components in melon (*Cucumis melo* L.) using a RIL population derived from exotic × elite US Western Shipping germplasm. *Mol. Breed.* **2008**, *22*, 405–419. [CrossRef]
6. Harel-Beja, R.; Tzuri, G.; Portnoy, V.; Lotan-Pompan, M.; Lev, S.; Cohen, S.; Dai, N.; Yeselson, L.; Meir, A.; Libhaber, S.E.; et al. A genetic map of melon highly enriched with fruit quality QTLs and EST markers, including sugar and carotenoid metabolism genes. *Theor. Appl. Genet.* **2010**, *121*, 511–533. [CrossRef]

7. Diaz, A.; Fergany, M.; Formisano, G.; Ziarsolo, P.; Blanca, J.; Fei, Z.; Staub, J.E.; Zalapa, J.E.; Cuevas, H.E.; Dace, G.; et al. A consensus linkage map for molecular markers and Quantitative Trait Loci associated with economically important traits in melon (*Cucumis melo* L.). *BMC Plant Biol.* **2011**, *11*, 111. [CrossRef]
8. Ramamurthy, R.K.; Waters, B.M. Identification of fruit quality and morphology QTLs in melon (*Cucumis melo*) using a population derived from flexuosus and cantalupensis botanical groups. *Euphytica* **2015**, *204*, 163–177. [CrossRef]
9. Baloch, A.; Baloch, A.W.; Liu, S.; Gao, P.; Baloch, M.J.; Wang, X.; Davis, A.; Ali, M.; Luan, F. Linkage map construction and qtl analysis of fruit traits in melon (*Cucumis melo* L.) based on caps markers. *Pak. J. Bot.* **2016**, *48*, 1579–1584.
10. Argyris, J.M.; Díaz, A.; Ruggieri, V.; Fernández, M.; Jahrmann, T.; Gibon, Y.; Picó, B.; Martín-Hernández, A.M.; Monforte, A.J.; Garcia-Mas, J. QTL Analyses in Multiple Populations Employed for the Fine Mapping and Identification of Candidate Genes at a Locus Affecting Sugar Accumulation in Melon (*Cucumis melo* L.). *Front. Plant Sci.* **2017**, *8*, 1679. [CrossRef]
11. Amanullah, S.; Liu, S.; Gao, P.; Zhu, Z.; Zhu, Q.; Fan, C.; Luan, F. QTL mapping for melon (*Cucumis melo* L.) fruit traits by assembling and utilization of novel SNPs based CAPS markers. *Sci. Hortic.* **2018**, *236*, 18–29. [CrossRef]
12. Galpaz, N.; Gonda, I.; Shem-Tov, D.; Barad, O.; Tzuri, G.; Lev, S.; Fei, Z.; Xu, Y.; Mao, L.; Jiao, C.; et al. Deciphering genetic factors that determine melon fruit-quality traits using RNA -Seq-based high-resolution QTL and eQTL mapping. *Plant J.* **2018**, *94*, 169–191. [CrossRef] [PubMed]
13. Pereira, L.; Ruggieri, V.; Pérez, S.; Alexiou, K.G.; Fernández, M.; Jahrmann, T.; Pujol, M.; Garcia-Mas, J. QTL mapping of melon fruit quality traits using a high-density GBS-based genetic map. *BMC Plant Biol.* **2018**, *18*, 324. [CrossRef] [PubMed]
14. Pereira, L.; Domingo, M.S.; Argyris, J.; Mayobre, C.; Valverde, L.; Martín-Hernández, A.M.; Pujol, M.; Garcia-Mas, J. A novel introgression line collection to unravel the genetics of climacteric ripening and fruit quality in melon. *Sci. Rep.* **2021**, *11*, 11364. [CrossRef] [PubMed]
15. Gur, A.; Tzuri, G.; Meir, A.; Sa'ar, U.; Portnoy, V.; Katzir, N.; Schaffer, A.A.; Li, L.; Burger, J.; Tadmor, Y. Genome-Wide Linkage-Disequilibrium Mapping to the Candidate Gene Level in Melon (*Cucumis melo*). *Sci. Rep.* **2017**, *7*, 9770. [CrossRef] [PubMed]
16. Zhao, G.; Lian, Q.; Zhang, Z.; Fu, Q.; He, Y.; Ma, S.; Ruggieri, V.; Monforte, A.J.; Wang, P.; Julca, I.; et al. A comprehensive genome variation map of melon identifies multiple domestication events and loci influencing agronomic traits. *Nat. Genet.* **2019**, *51*, 1607–1615. [CrossRef]
17. Liu, S.; Gao, P.; Zhu, Q.; Zhu, Z.; Liu, H.; Wang, X.; Weng, Y.; Gao, M.; Luan, F. Resequencing of 297 melon accessions reveals the genomic history of improvement and loci related to fruit traits in melon. *Plant Biotechnol. J.* **2020**, *18*, 2545–2558. [CrossRef]
18. Pavan, S.; Marcotrigiano, A.; Ciani, E.; Mazzeo, R.; Zonno, V.; Ruggieri, V.; Lotti, C.; Ricciardi, L. Genotyping-by-sequencing of a melon (*Cucumis melo* L.) germplasm collection from a secondary center of diversity highlights patterns of genetic variation and genomic features of different gene pools. *BMC Genom.* **2017**, *18*, 59. [CrossRef]
19. Oren, E.; Dafna, A.; Tzuri, G.; Halperin, I.; Isaacson, T.; Elkabetz, M.; Meir, A.; Saar, U.; Ohali, S.; La, T.; et al. Pan-genome and multi-parental framework for high-resolution trait dissection in melon (*Cucumis melo*). *Plant J.* **2022**, *112*, 1525–1542. [CrossRef]
20. Tripodi, P.; Soler, S.; Campanelli, G.; Díez, M.J.; Esposito, S.; Sestili, S.; Figàs, M.R.; Leteo, F.; Casanova, C.; Platani, C.; et al. Genome wide association mapping for agronomic, fruit quality, and root architectural traits in tomato under organic farming conditions. *BMC Plant Biol.* **2021**, *21*, 481. [CrossRef]
21. Wang, Y.; Shi, C.; Ge, P.; Li, F.; Zhu, L.; Wang, Y.; Tao, J.; Zhang, X.; Dong, H.; Gai, W.; et al. A 21-bp InDel in the promoter of *STP1* selected during tomato improvement accounts for soluble solid content in fruits. *Hortic. Res.* **2023**, *10*, uhad009. [CrossRef] [PubMed]
22. Ferik, F.; Ates, D.; Ercisli, S.; Erdogan, A.; Orhan, E.; Tanyolac, M.B. Genome-wide association links candidate genes to fruit firmness, fruit flesh color, flowering time, and soluble solid content in apricot (*Prunus armeniaca* L.). *Mol. Biol. Rep.* **2021**, *49*, 5283–5291. [CrossRef] [PubMed]
23. Cao, K.; Zhou, Z.; Wang, Q.; Guo, J.; Zhao, P.; Zhu, G.; Fang, W.; Chen, C.; Wang, X.; Wang, X.; et al. Genome-wide association study of 12 agronomic traits in peach. *Nat. Commun.* **2016**, *7*, 13246. [CrossRef]
24. Guo, S.; Zhao, S.; Sun, H.; Wang, X.; Wu, S.; Lin, T.; Ren, Y.; Gao, L.; Deng, Y.; Zhang, J.; et al. Resequencing of 414 cultivated and wild watermelon accessions identifies selection for fruit quality traits. *Nat. Genet.* **2019**, *51*, 1616–1623. [CrossRef]
25. Leida, C.; Moser, C.; Esteras, C.; Sulpice, R.; Lunn, J.E.; De Langen, F.; Monforte, A.J.; Picó, B. Variability of candidate genes, genetic structure and association with sugar accumulation and climacteric behavior in a broad germplasm collection of melon (*Cucumis melo* L.). *BMC Genet.* **2015**, *16*, 28. [CrossRef] [PubMed]
26. Wang, X.; Ando, K.; Wu, S.; Reddy, U.K.; Tamang, P.; Bao, K.; Hammar, S.A.; Grumet, R.; McCreight, J.D.; Fei, Z. Genetic characterization of melon accessions in the U.S. National Plant Germplasm System and construction of a melon core collection. *Mol. Hortic.* **2021**, *1*, 11. [CrossRef]
27. Yang, X.; Yan, J.; Zheng, Y.; Yu, J.; Li, J. Reviews of association analysis for quantitative traits in plants. *Acta Agron. Sin.* **2007**, *33*, 523.
28. Yu, J.; Buckler, E.S. Genetic association mapping and genome organization of maize. *Curr. Opin. Biotechnol.* **2006**, *17*, 155–160. [CrossRef]
29. Wei, Z.; Yuan, Q.; Lin, H.; Li, X.; Zhang, C.; Gao, H.; Zhang, B.; He, H.; Liu, T.; Jie, Z.; et al. Linkage analysis, GWAS, transcriptome analysis to identify candidate genes for rice seedlings in response to high temperature stress. *BMC Plant Biol.* **2021**, *21*, 85. [CrossRef]

30. Xu, J.; Ni, Z.; Chen, F.; Fu, X.; Yu, F. Integrated linkage mapping and genome-wide association study to dissect the genetic basis of zinc deficiency tolerance in maize at seedling stage. *Crop. J.* **2022**, *10*, 1807–1818. [CrossRef]
31. Hu, D.; Li, X.; Yang, Z.; Liu, S.; Hao, D.; Chao, M.; Zhang, J.; Yang, H.; Su, X.; Jiang, M.; et al. Downregulation of a gibberellin 3β-hydroxylase enhances photosynthesis and increases seed yield in soybean. *New Phytol.* **2022**, *235*, 502–517. [CrossRef] [PubMed]
32. Nimmakayala, P.; Tomason, Y.R.; Abburi, V.L.; Alvarado, A.; Saminathan, T.; Vajja, V.G.; Salazar, G.; Panicker, G.K.; Levi, A.; Wechter, W.P.; et al. Genome-Wide Differentiation of Various Melon Horticultural Groups for Use in GWAS for Fruit Firmness and Construction of a High Resolution Genetic Map. *Front. Plant Sci.* **2016**, *7*, 1437. [CrossRef] [PubMed]
33. Perpiñá, G.; Esteras, C.; Gibon, Y.; Monforte, A.J.; Picó, B. A new genomic library of melon introgression lines in a cantaloupe genetic background for dissecting desirable agronomical traits. *BMC Plant Biol.* **2016**, *16*, 154. [CrossRef]
34. Oren, E.; Tzuri, G.; Dafna, A.; Meir, A.; Kumar, R.; Katzir, N.; Elkind, Y.; Freilich, S.; Schaffer, A.A.; Tadmor, Y.; et al. High-density NGS-based map construction and genetic dissection of fruit shape and rind netting in *Cucumis melo*. *Theor. Appl. Genet.* **2020**, *133*, 1927–1945. [CrossRef] [PubMed]
35. Du, X.; Liu, H.; Zhu, Z.; Liu, S.; Song, Z.; Xia, L.; Zhao, J.; Luan, F.; Liu, S. Identification of Candidate Chromosome Region Related to Melon (*Cucumis melo* L.) Fruit Surface Groove Trait Through Biparental Genetic Mapping and Genome-Wide Association Study. *Front. Plant Sci.* **2022**, *13*, 828287. [CrossRef]
36. Zhang, C.; Yu, X.; Ayre, B.G.; Turgeon, R. The Origin and Composition of Cucurbit "Phloem" Exudate. *Plant Physiol.* **2012**, *158*, 1873–1882. [CrossRef] [PubMed]
37. Ko, H.-Y.; Ho, L.-H.; Neuhaus, H.E.; Guo, W.-J. Transporter SlSWEET15 unloads sucrose from phloem and seed coat for fruit and seed development in tomato. *Plant Physiol.* **2021**, *187*, 2230–2245. [CrossRef]
38. Li, Y.; Liu, H.; Yao, X.; Wang, J.; Feng, S.; Sun, L.; Ma, S.; Xu, K.; Chen, L.-Q.; Sui, X. Hexose transporter CsSWEET7a in cucumber mediates phloem unloading in companion cells for fruit development. *Plant Physiol.* **2021**, *186*, 640–654. [CrossRef]
39. Wen, S.; Neuhaus, H.E.; Cheng, J.; Bie, Z. Contributions of sugar transporters to crop yield and fruit quality. *J. Exp. Bot.* **2022**, *73*, 2275–2289. [CrossRef]
40. Ren, Y.; Sun, H.; Zong, M.; Guo, S.; Ren, Z.; Zhao, J.; Li, M.; Zhang, J.; Tian, S.; Wang, J.; et al. Localization shift of a sugar transporter contributes to phloem unloading in sweet watermelons. *New Phytol.* **2020**, *227*, 1858–1871. [CrossRef]
41. Peng, Q.; Cai, Y.; Lai, E.; Nakamura, M.; Liao, L.; Zheng, B.; Ogutu, C.; Cherono, S.; Han, Y. The sucrose transporter MdSUT4.1 participates in the regulation of fruit sugar accumulation in apple. *BMC Plant Biol.* **2020**, *20*, 191. [CrossRef] [PubMed]
42. Ren, Y.; Guo, S.; Zhang, J.; He, H.; Sun, H.; Tian, S.; Gong, G.; Zhang, H.; Levi, A.; Tadmor, Y.; et al. A Tonoplast Sugar Transporter Underlies a Sugar Accumulation QTL in Watermelon. *Plant Physiol.* **2017**, *176*, 836–850. [CrossRef] [PubMed]
43. Cheng, J.; Wen, S.; Xiao, S.; Lu, B.; Ma, M.; Bie, Z. Overexpression of the tonoplast sugar transporter CmTST2 in melon fruit increases sugar accumulation. *J. Exp. Bot.* **2017**, *69*, 511–523. [CrossRef] [PubMed]
44. Huang, W.; Hu, B.; Liu, J.; Zhou, Y.; Liu, S. Identification and Characterization of Tonoplast Sugar Transporter (TST) Gene Family in Cucumber. *Hortic. Plant J.* **2020**, *6*, 145–157. [CrossRef]
45. Wang, Z.; Wei, X.; Yang, J.; Li, H.; Ma, B.; Zhang, K.; Zhang, Y.; Cheng, L.; Ma, F.; Li, M. Heterologous expression of the apple hexose transporter Md HT 2.2 altered sugar concentration with increasing cell wall invertase activity in tomato fruit. *Plant Biotechnol. J.* **2019**, *18*, 540–552. [CrossRef] [PubMed]
46. Zhu, L.; Li, B.; Wu, L.; Li, H.; Wang, Z.; Wei, X.; Ma, B.; Zhang, Y.; Ma, F.; Ruan, Y.-L.; et al. MdERDL6-mediated glucose efflux to the cytosol promotes sugar accumulation in the vacuole through up-regulating TSTs in apple and tomato. *Proc. Natl. Acad. Sci. USA* **2020**, *118*, e2022788118. [CrossRef]
47. Lunn, J.E.; Delorge, I.; Figueroa, C.M.; Van Dijck, P.; Stitt, M. Trehalose metabolism in plants. *Plant J.* **2014**, *79*, 544–567. [CrossRef]
48. Zhang, Z.; Deng, Y.; Song, X.; Miao, M. Trehalose-6-phosphate and SNF1-related protein kinase 1 are involved in the first-fruit inhibition of cucumber. *J. Plant Physiol.* **2014**, *177*, 110–120. [CrossRef]
49. Dai, N.; Cohen, S.; Portnoy, V.; Tzuri, G.; Harel-Beja, R.; Pompan-Lotan, M.; Carmi, N.; Zhang, G.; Diber, A.; Pollock, S.; et al. Metabolism of soluble sugars in developing melon fruit: A global transcriptional view of the metabolic transition to sucrose accumulation. *Plant Mol. Biol.* **2011**, *76*, 1–18. [CrossRef]
50. Yang, J.; Deng, G.; Lian, J.; Garraway, J.; Niu, Y.; Hu, Z.; Yu, J.; Zhang, M. The Chromosome-Scale Genome of Melon Dissects Genetic Architecture of Important Agronomic Traits. *iScience* **2020**, *23*, 101422. [CrossRef]
51. Li, H.; Durbin, R. Fast and accurate short read alignment with Burrows—Wheeler transform. *Bioinformatics* **2009**, *25*, 1754–1760. [CrossRef] [PubMed]
52. Li, H.; Handsaker, B.; Wysoker, A.; Fennell, T.; Ruan, J.; Homer, N.; Marth, G.; Abecasis, G.; Durbin, R. Genome Project Data Processing S: The Sequence Alignment/Map format and SAMtools. *Bioinformatics* **2009**, *25*, 2078–2079. [CrossRef] [PubMed]
53. McKenna, A.; Hanna, M.; Banks, E.; Sivachenko, A.; Cibulskis, K.; Kernytsky, A.; Garimella, K.; Altshuler, D.; Gabriel, S.; Daly, M.; et al. The Genome Analysis Toolkit: A MapReduce framework for analyzing next-generation DNA sequencing data. *Genome Res.* **2010**, *20*, 1297–1303. [CrossRef]
54. Wang, K.; Li, M.; Hakonarson, H. ANNOVAR: Functional annotation of genetic variants from high-throughput sequencing data. *Nucleic Acids Res.* **2010**, *38*, e164. [CrossRef] [PubMed]
55. Huang, X.; Feng, Q.; Qian, Q.; Zhao, Q.; Wang, L.; Wang, A.; Guan, J.; Fan, D.; Weng, Q.; Huang, T.; et al. High-throughput genotyping by whole-genome resequencing. *Genome Res.* **2009**, *19*, 1068–1076. [CrossRef] [PubMed]

56. Tang, H.; Zhang, X.; Miao, C.; Zhang, J.; Ming, R.; Schnable, J.C.; Schnable, P.S.; Lyons, E.; Lu, J. ALLMAPS: Robust scaffold ordering based on multiple maps. *Genome Biol.* **2015**, *16*, 3. [CrossRef] [PubMed]
57. Alexander, D.H.; Novembre, J.; Lange, K. Fast model-based estimation of ancestry in unrelated individuals. *Genome Res.* **2009**, *19*, 1655–1664. [CrossRef]
58. Zhang, C.; Dong, S.-S.; Xu, J.-Y.; He, W.-M.; Yang, T.-L. PopLDdecay: A fast and effective tool for linkage disequilibrium decay analysis based on variant call format files. *Bioinformatics* **2019**, *35*, 1786–1788. [CrossRef]
59. Zhou, X.; Stephens, M. Efficient multivariate linear mixed model algorithms for genome-wide association studies. *Nat. Methods* **2014**, *11*, 407–409. [CrossRef]
60. Lex, A.; Gehlenborg, N.; Strobelt, H.; Vuillemot, R.; Pfister, H. UpSet: Visualization of Intersecting Sets. *IEEE Trans. Vis. Comput. Graph.* **2014**, *20*, 1983–1992. [CrossRef]
61. Cheng, H.; Kong, W.; Tang, T.; Ren, K.; Zhang, K.; Wei, H.; Lin, T. Identification of Key Gene Networks Controlling Soluble Sugar and Organic Acid Metabolism During Oriental Melon Fruit Development by Integrated Analysis of Metabolic and Transcriptomic Analyses. *Front. Plant Sci.* **2022**, *13*, 830517. [CrossRef]
62. Ohkawa, W.; Kanayama, Y.; Daibo, N.; Sato, T.; Nishiyama, M.; Kanahama, K. Metabolic process of the 14C-sugars on the translocation pathways of cucumber plants. *Sci. Hortic.* **2010**, *124*, 46–50. [CrossRef]
63. Saladié, M.; Cañizares, J.; Phillips, M.A.; Rodriguez-Concepcion, M.; Larrigaudière, C.; Gibon, Y.; Stitt, M.; Lunn, J.E.; Garcia-Mas, J. Comparative transcriptional profiling analysis of developing melon (*Cucumis melo* L.) fruit from climacteric and non-climacteric varieties. *BMC Genom.* **2015**, *16*, 440. [CrossRef]
64. Shin, A.-Y.; Kim, Y.-M.; Koo, N.; Lee, S.M.; Nahm, S.; Kwon, S.-Y. Transcriptome analysis of the oriental melon (*Cucumis melo* L. var. *makuwa*) during fruit development. *PeerJ* **2017**, *5*, e2834. [CrossRef] [PubMed]
65. Jin, Y.; Ni, D.-A.; Ruan, Y.-L. Posttranslational Elevation of Cell Wall Invertase Activity by Silencing Its Inhibitor in Tomato Delays Leaf Senescence and Increases Seed Weight and Fruit Hexose Level. *Plant Cell* **2009**, *21*, 2072–2089. [CrossRef]
66. Zanor, M.I.; Osorio, S.; Nunes-Nesi, A.; Carrari, F.; Lohse, M.; Usadel, B.; Kühn, C.; Bleiss, W.; Giavalisco, P.; Willmitzer, L.; et al. RNA Interference of LIN5 in Tomato Confirms Its Role in Controlling Brix Content, Uncovers the Influence of Sugars on the Levels of Fruit Hormones, and Demonstrates the Importance of Sucrose Cleavage for Normal Fruit Development and Fertility. *Plant Physiol.* **2009**, *150*, 1204–1218. [CrossRef] [PubMed]
67. Ren, Y.; Li, M.; Guo, S.; Sun, H.; Zhao, J.; Zhang, J.; Liu, G.; He, H.; Tian, S.; Yu, Y.; et al. Evolutionary gain of oligosaccharide hydrolysis and sugar transport enhanced carbohydrate partitioning in sweet watermelon fruits. *Plant Cell* **2021**, *33*, 1554–1573. [CrossRef]
68. Liu, H.; Liu, X.; Zhao, Y.; Nie, J.; Yao, X.; Lv, L.; Yang, J.; Ma, N.; Guo, Y.; Li, Y.; et al. *Alkaline α-galactosidase 2* (*CsAGA2*) plays a pivotal role in mediating source–sink communication in cucumber. *Plant Physiol.* **2022**, *189*, 1501–1518. [CrossRef] [PubMed]
69. Liu, Y.; Li, T.; Zhang, C.; Zhang, W.; Deng, N.; Dirk, L.M.A.; Downie, A.B.; Zhao, T. Raffinose positively regulates maize drought tolerance by reducing leaf transpiration. *Plant J.* **2023**, *114*, 55–67. [CrossRef]
70. Tang, Y.; Wang, M.; Cao, L.; Dang, Z.; Ruan, N.; Wang, Y.; Huang, Y.; Wu, J.; Zhang, M.; Xu, Z.; et al. *OsUGE3*-mediated cell wall polysaccharides accumulation improves biomass production, mechanical strength, and salt tolerance. *Plant Cell Environ.* **2022**, *45*, 2492–2507. [CrossRef]
71. Bavnhøj, L.; Paulsen, P.A.; Flores-Canales, J.C.; Schiøtt, B.; Pedersen, B.P. Molecular mechanism of sugar transport in plants unveiled by structures of glucose/H+ symporter STP10. *Nat. Plants* **2021**, *7*, 1409–1419. [CrossRef] [PubMed]

current issues in molecular biology

MDPI

Article

Genome-Wide SNP and Indel Discovery in Abaca (*Musa textilis* Née) and among Other *Musa* spp. for Abaca Genetic Resources Management

Cris Francis C. Barbosa [1,2], Jayson C. Asunto [1], Rhosener Bhea L. Koh [3], Daisy May C. Santos [2], Dapeng Zhang [4], Ernelea P. Cao [2] and Leny C. Galvez [1,*]

[1] Philippine Fiber Industry Development Authority (PhilFIDA), PCAF Building, Department of Agriculture (DA) Compound, Quezon City 1101, Philippines; jcasunto.philfida@gmail.com (J.C.A.)

[2] Institute of Biology, College of Science, University of the Philippines Diliman, Quezon City 1101, Philippines; daconstantino@up.edu.ph (D.M.C.S.); epcao@up.edu.ph (E.P.C.)

[3] National Institute of Molecular Biology and Biotechnology, University of the Philippines Diliman, Quezon City 1101, Philippines; k.rbhea@gmail.com

[4] Sustainable Perennial Crops Laboratory, United States Department of Agriculture—Agricultural Research Service, Beltsville, MD 20705, USA; dapeng.zhang@usda.gov

* Correspondence: lcgalvez@huskers.unl.edu

Abstract: Abaca (*Musa textilis* Née) is an economically important fiber crop in the Philippines. Its economic potential, however, is hampered by biotic and abiotic stresses, which are exacerbated by insufficient genomic resources for varietal identification vital for crop improvement. To address these gaps, this study aimed to discover genome-wide polymorphisms among abaca cultivars and other *Musa* species and analyze their potential as genetic marker resources. This was achieved through whole-genome Illumina resequencing of abaca cultivars and variant calling using BCFtools, followed by genetic diversity and phylogenetic analyses. A total of 20,590,381 high-quality single-nucleotide polymorphisms (SNP) and DNA insertions/deletions (InDels) were mined across 16 abaca cultivars. Filtering based on linkage disequilibrium (LD) yielded 130,768 SNPs and 13,620 InDels, accounting for 0.396 ± 0.106 and 0.431 ± 0.111 of gene diversity across these cultivars. LD-pruned polymorphisms across abaca, *M. troglodytarum*, *M. acuminata* and *M. balbisiana* enabled genetic differentiation within abaca and across the four *Musa* spp. Phylogenetic analysis revealed the registered varieties Abuab and Inosa to accumulate a significant number of mutations, eliciting further studies linking mutations to their advantageous phenotypes. Overall, this study pioneered in producing marker resources in abaca based on genome-wide polymorphisms vital for varietal authentication and comparative genotyping with the more studied *Musa* spp.

Keywords: *Musa textilis*; whole-genome resequencing; polymorphism discovery; single-nucleotide polymorphisms (SNPs); DNA insertions/deletions (InDels)

Citation: Barbosa, C.F.C.; Asunto, J.C.; Koh, R.B.L.; Santos, D.M.C.; Zhang, D.; Cao, E.P.; Galvez, L.C. Genome-Wide SNP and Indel Discovery in Abaca (*Musa textilis* Née) and among Other *Musa* spp. for Abaca Genetic Resources Management. *Curr. Issues Mol. Biol.* **2023**, *45*, 5776–5797. https://doi.org/10.3390/cimb45070365

Academic Editor: Quan Zou

Received: 23 June 2023
Revised: 5 July 2023
Accepted: 7 July 2023
Published: 12 July 2023

1. Introduction

Abaca (*Musa textilis* Née) is a fiber crop endemic to the Philippines [1]. Being the center of origin of abaca, the Philippines contributes about 86.1% of the global fiber and fiber-based product requirement, generating an average annual income of USD 111.9M from 2010 to 2019 [2]. In many countries, abaca planting materials have been sourced from the Philippines, such as those planted in Borneo and islands in the Indies [3]. This makes the abaca industry an extremely important livelihood for Filipinos, wherein a total of 126,436 are abaca farmers and 160,400 hectares of land are planted [2]. Abaca is mainly harvested for its fiber, the Manila hemp. It has higher tensile strength than synthetic fibers, such as nylon and rayon, and has equivalent flexural strength with fiberglass [3–5]. Abaca also possesses long fiber length and high resistance to saltwater damage; thus, it is widely used for making marine cordages, paper, money, furnishings, insulators and textiles and environmentally friendly materials for automobile manufacturing [3,5–8]. A total of 170 varieties and 773 accessions have been documented in the Philippines [9,10]. To preserve genetic resources for breeding programs, germplasms are being maintained across the country by the Philippine Fiber Industry Development Authority (PhilFIDA). Varieties possessing advantageous characteristics can also be found in the wild, such as Ihalas in the Visayas, which was observed to have good tolerance to abaca diseases. There are also varieties that have been identified to be recommended for planting in particular regions, such that the fiber quality and yield obtained are optimal. In the Bicol region, *Musa* tex 51, Tinawagan Puti and Abuab are recommended for cultivation; Linawaan, Laylay and Inosa in the Visayas; and Bongolanon, Maguindanao and Tangongon in Mindanao [11]. Among these, Abuab, Inosa and Tangongon are registered under the National Seed Industry Council (NSIC) [12].

Abaca fiber production, however, is hampered by viral diseases such as bunchy top disease, mosaic disease and bract mosaic disease [13,14]. Abiotic stresses such as typhoons, severe flooding and drought also serve as important constraints to optimal abaca fiber productivity. In 2018, the abaca production was documented to decline by 8.0% as a result of El Niño, which consequently led to losses of 11.5% in fiber exports and 14.0% in abaca fiber shipment [15]. The reduced abaca fiber production is further aggravated by the absence of identified resistant varieties to both biotic and abiotic stresses. Regarding management of abaca germplasms and plantations, variety misidentification due to mislabeling during early stages of domestication, misidentifying morphologies and phenotypically variable progenies produced from seed-derived abaca plants also contribute to the retardation of the progress of the Philippine abaca industry [9]. These errors in identification of varieties have long been problems that impede the economic potential of abaca due to fluctuating qualities of fibers released in the market. Moreover, the lack of a well-curated germplasm collection produces uncertainties in the identity of plant materials used for breeding programs, leading back to the problem of the lack of highly proven resistant varieties to date. Furthermore, currently there is no information on genetic polymorphisms among abaca and other *Musa* spp. instrumental in genetic studies for advantageous or deleterious polymorphisms. Breeding for biotic and abiotic stress resistance hence requires information on abaca polymorphisms as a resource for developing genetic markers for accurate varietal identification. At the same time, information on the genetic differences among abaca and other *Musa* spp. can serve as a resource for mining of genes and polymorphisms linked to biotic and abiotic stress resistance.

To maintain the integrity and stability of the Philippine abaca industry, there is a dire need for genomic resources for the establishment of methods of identification of varieties based on their genetic nature instead of their morphologies or phenotypes. This will not only solve varietal misidentifications and labeling but will also be useful in marker-assisted selection in breeding programs, tracking movement of planting materials and authenticating varietal source of fibers released in the market. Establishing varietal identity can be best achieved using genetic markers such as single-nucleotide polymorphisms (SNPs) and DNA insertions/deletions (InDels), which are the top two most abundant DNA polymorphisms/variants in the plant genome [16,17].

SNPs are the smallest unit of genetic differences and are the most abundant sequence variations in the genomes of most organisms [18]. They have a wide distribution across the genome, with loci located within genes such as SNPs among exons, introns or promoters, and between genes [19]. The high abundance of SNPs (indicating a higher chance of finding markers linked to traits of interest), their biallelic nature leading to lower error rate in allele calls and their higher amenability to high-throughput genotyping methods such as genotyping arrays [20] and genotyping-by-sequencing (GBS; [21]) have led to the preference of using SNP markers for genetics studies [22].

InDels, which are the second most abundant genetic variation next to SNPs, are another group of markers of choice due to their codominant inheritance, wide genome distribution and multiallelic nature [18]. Compared to SNPs, InDels have higher functional consequences in organisms because they often result in frameshift mutations, disrupting the reading frames and gene expression among coding genes [18].

As of this date, only two studies [23,24] have been able to publish genetic markers for abaca genetic differentiation, which were based on banana-based simple sequence repeats (SSRs) and SNPs, respectively. These SSR markers were able to distinguish among 150 abaca accessions and revealed that the diversity within *M. textilis* is very rich, as evidenced by the calculated Shannon's Diversity Index of 0.68 [23]. The aforementioned SNP markers were also able to distinguish among 62 abaca accessions [24]. Randomly Amplified Polymorphic DNA (RAPD) markers were also reported years earlier [25] during an international symposium. In this work, the RAPD markers were able to distinguish abaca from other *Musa* species, but the identification of varieties within *M. textilis* was not performed. Aside from varietal identification, banana-based SSR markers associated with abaca bunchy top virus (ABTV) resistance have also been used to screen potential resistance to said virus among 57 accessions [26]. Since the aforementioned markers described in these two studies were based in bananas, these findings still need further confirmation. Moreover, the genotyping accuracy of these SNPs can have limitations due to differences in the genome sequence of abaca and banana. The SNP markers were assayed using microarray technology, which is expensive and not sustainable for routine genotyping. One disadvantage of SNP arrays for variant discovery is the bias of SNP arrays for preselected SNPs that capture only a fraction of the total diversity, especially if cross-species adaptation of SNPs is attempted. The use of SNP arrays will produce genotypes only based on the preselected SNPs. In cases where a different set of variants exist in the same loci across the samples being genotyped, possible errors in allele calls could occur. This is an instance of an ascertainment bias of SNP panels [27].

To facilitate the uncovering of true genome-wide polymorphisms that will be extremely important resources for abaca varietal identification and efficient breeding programs, the declining cost and increasing accessibility of next-generation sequencing (NGS) can be taken advantage of. NGS allows the generation of huge datasets containing genome-wide variants across individuals, hence providing a highly abundant resource of genetic markers for varietal identification and comparative genotyping across individuals, populations or species [28,29]. A reference genome for a species can be first generated through high-coverage de novo whole-genome sequencing and assembly, followed by sequencing of individuals within that population. This approach is often referred to as resequencing [28].

Recently, the abaca genome has been assembled [30], thus providing a reference sequence for discovery of SNPs and InDels present in abaca itself through next-generation sequencing (NGS). With the availability of an abaca reference genome, this study aimed to mine abaca genome-wide SNPs and InDels among abaca cultivars and other *Musa* and analyze their potential as genetic marker resources through genetic diversity and phylogenetic analyses.

Specifically, this study aimed at (1) whole-genome resequencing of 11 abaca varieties and accessions, (2) mapping of sequence reads to the abaca reference genome, (3) mining and analysis of SNPs and InDels and (4) determining genetic variation and evolutionary relationships within *M. textilis* and among *Musa* spp.

2. Materials and Methods

2.1. Plant Materials

Nine varieties housed at the Mindoro Fiber Experimental Station and Seedbank (MFESS), a PhilFIDA germplasm collection located in Socorro, Oriental Mindoro, Philippines, were used for the determination of genome-wide variants in abaca in the form of SNPs and InDels. The varieties are Abuab, Inosa, Kutay-kutay, Laylay, Tangongon, Hagbayanon, Luno, Socorro and Tinawagang Puti (T-puti). The Abuab, Inosa and Tangongon varieties are among the only three abaca varieties that are currently registered under the National Seed Industry Council (NSIC) of the Philippines. NSIC registration indicates that these varieties qualified for the criteria set for commercially propagated varieties, such that they must: (1) yield at least 800 kg/hectare dried fiber, (2) have a minimum fiber recovery of 1.5%, (3) be currently planted commercially at the event of registration and (4) have been reported and described technically in reputable publications [31]. Moreover, the T-puti and Laylay are varieties recommended for planting in Bicol and Visayas areas in the Philippines [11]. Furthermore, the Kutay-kutay, Hagbayanon, Luno and Socorro varieties are varieties with good observed agronomic performance in Zamboanga. There are also other accessions present in MFESS and included in this study, such as Samoro and 'Luno Green'; however, they are only present as a single hill. One sample from each of the 11 varieties and accessions was randomly collected at the MFESS and was flash frozen in liquid nitrogen prior to storage at −80 °C. The descriptions of these samples are listed in Table 1.

Table 1. Abaca and *Musa* spp. accessions involved in the study.

Accession	Species	Description	Source of Sequence Reads	Accession Number
Abuabref	*M. textilis*	Reference genome	[30]	N/A
Abuab	*M. textilis*	Commercial abaca variety	This study	SRR22906090
Hagbayanon	*M. textilis*	Commercial abaca variety	This study	SRR22906084
Ihalas	*M. textilis*	Wild abaca accession	[32]	N/A
Inosa	*M. textilis*	Commercial abaca variety	This study	SRR22906089
Kutay	*M. textilis*	Commercial abaca variety	This study	SRR22906087
Laylay	*M. textilis*	Commercial abaca variety	This study	SRR22906086
Luno	*M. textilis*	Commercial abaca variety	This study	SRR22906083
LunoGreen	*M. textilis*	Luno with green inflorescence	This study	SRR22906080
Samoro	*M. textilis*	Present as a single hill in MFESS	This study	SRR22906088
Socorro	*M. textilis*	Commercial abaca variety	This study	SRR22906082
Tangongon	*M. textilis*	Commercial abaca variety	This study	SRR22906085
Tputi	*M. textilis*	Commercial abaca variety	This study	SRR22906081
Unknown cultivar	*M. textilis*	Abaca accession	[33]	SRR9696635
Unknown cultivar	*M. textilis*	Abaca accession	[33]	SRR8989639
Unknown cultivar	*M. textilis*	Abaca accession	[33]	SRR9850642
Banana	*M. acuminata*	Wild banana accession	[33]	SRR8989638
Banana	*M. acuminata*	Wild banana accession	[33]	SRR8989629
Banana	*M. acuminata*	Wild banana accession	[33]	SRR8989632
Banana	*M. balbisiana*	Wild banana accession	[33]	SRR8989633
Banana	*M. balbisiana*	Wild banana accession	[33]	SRR9734079
Banana	*M. balbisiana*	Wild banana accession	[33]	SRR6147592
Banana	*M. balbisiana*	Wild banana accession	[33]	SRR9850640
Banana	*M. troglodytarum*	Wild banana accession	[33]	SRR8989640
Banana	*M. troglodytarum*	Wild banana accession	[33]	SRR9734080
Banana	*M. troglodytarum*	Wild banana accession	[33]	SRR9850641

2.2. DNA Extraction and Sequencing

DNAs were extracted from all samples using the SEPa Plant DNA Isolation Reagent Kit (1st Base, Selangor, Malaysia). For next-generation sequencing, libraries were prepared using the Nextera DNA Library Prep Kit (Illumina, San Diego, CA, USA), and PE 150 sequencing was performed using the Illumina Hiseq 4000 platform (Illumina, San Diego, CA, USA).

2.3. Quality Control and Trimming of Sequence Reads

To ensure the quality of bases used for variant calling, the program Trimmomatic [34] RRID:SCR_011848 was run to remove low-quality bases and contaminating adapters. Trimming of low-quality bases was employed using the parameters LEADING:3 TRAILING:3 SLIDINGWINDOW:4:20 MINLEN:80 in Trimmomatic. To determine the quality of the sequence reads before and after quality trimming, the FastQC tool [35] RRID:SCR_014583 was utilized using the raw sequence reads and trimmed paired-end reads, respectively. MultiQC [36] RRID:SCR_014982 was then used to compile and further process the FastQC results in one report. Adapter removal was also conducted through Trimmomatic using the parameter ILLUMINACLIP:NexteraPE-PE.fa:2:30:10, which removed adapters from the Nextera DNA Library Prep Kit.

2.4. Mapping of Reads to the Reference Genome

The quality-filtered sequence reads were mapped to a polished version of the Abuab reference genome [30] using the BWA version 0.7.17-r1188 [37] RRID:SCR_010910. The reference genome was first indexed using the bwa index command, followed by alignment of the reads to the indexed reference genome using the bwa mem command. The product of this step is a Sequence Alignment Map (SAM) file, which is expected to contain the read alignments. Each of the SAM files was compressed into a binary format called Binary Alignment Map (BAM) using the SAMtools version 1.15.1 RRID:SCR_002105 view, collate, fixmate, sort, markdup and index commands to reduce the complexity and size of the alignment file for downstream procedures [38]. The SAMtools view command converts the complex SAM file into a simpler but unsorted BAM file. The SAMtools collate ensures that sequence reads with the same names are grouped together, while the SAMtools fixmate adds mate coordinates as well as fields for insert size [39,40]. The SAMtools sort then sorts the resulting coordinate sorted BAM file. To ensure that duplicate alignments resulting from the sequencing process are marked, the SAMtools markdup function was used. Finally, the SAMtools index command was used to enable indexing of the BAM files to make these files simpler and to allow its features to be searchable when they are used in downstream bioinformatics analysis [39,40].

The bamqc command by the Qualimap tool v.2.2.2-dev [41] RRID:SCR_001209 was then used to evaluate the generated alignments and produce statistics, such as the percentage of mapped reads, mean coverage, duplication rate, general error rate and mean mapping quality.

To serve as additional sources of polymorphisms, quality-filtered reads previously produced for the Ihalas accession [32] were included in read-mapping and variant calling steps. Sequence reads from three unidentified abaca accessions [33] deposited under the SRA accession numbers SRR8989639, SRR9696635 and SRR9850642 were included as well.

Sequence reads from *M. acuminata*, *M. balbisiana* and *M. troglodytarum* were also included in the mapping step to determine the polymorphisms present and evolutionary relationships between *M. textilis* and the other *Musa* spp. These will be important resources and information for identification of polymorphisms associated with disease resistance and advantageous agronomic traits. A summary of all sequence reads mapped to the reference genome is shown in Table 1.

2.5. Calling of Genome-Wide SNPs and InDels

Raw variants between the reference genome and each of the varieties were called through BCFtools version 1.15.1 RRID:SCR_005227 mpileup and call -vc commands using the sorted BAM files as input [40]. The mpileup command ensures that variants are called based on uniquely mapped reads, while the call command performs the actual calling of the SNPs/InDels from the compressed BCF file [40,42]. The -vc option was used to generate variants using the consensus caller in BCFtools [40]. The called variants were printed using the BCFtools view command, followed by initial filtering of the variants using the vcfutils.pl varFilter option. This final step produced VCF files containing the variants called between each variety and the reference genome. This variant calling pipeline was conducted for each of the abaca varieties and the additional accessions.

2.6. Quality Filtering of SNPs and InDels

The VCF files generated from the variant calling step were indexed and merged into one VCF file using the BCFtools commands index and merge, respectively. The SNPs and InDels were distributed into two separate VCF files using VCFtools version 0.1.16 [43] RRID:SCR_001235. Using the merged VCF as input for VCFtools, the --remove-indels option was used to generate a VCF containing SNPs only, while the --keep-only indels option was used to generate a VCF with indels only.

To filter high-quality variants, the VCF files were used as input for VCFtools. The parameters --min-alleles 2 and --max-alleles 2 were used to select only biallelic variants, while the --minQ 40 parameter was used to select only the variants with a mapping quality higher than 40.

After quality filtering in VCFtools, the variants were further filtered using the PLINK 2.0 software (http://pngu.mgh.harvard.edu/purcell/plink/, accessed on 11 November 2022) [44] RRID:SCR_001757 executed in the R program. Highly associated variants loci were pruned to ensure absence of linkage disequilibrium (LD) between the variant positions using the parameters --indep-pairwise 50 2 0.2. LD occurs when alleles are non-randomly associated at a genetic loci pair and is observed when the observed haplotype frequencies and expected haplotype frequencies deviate, meaning the assumption that alleles at a pair of loci should have independent association was not met [45]. As a consequence, when a pair of selected variants is in LD, the information provided by one variant is redundant with the other variant. As a result, the informativeness of a set of variants containing variants in LD is reduced. This step removed pairs of variant loci within a 50 bp window with the correlation $r^2 > 0.2$. The r^2 [46] refers to the squared correlation between the presence of an allele at a locus and another allele at another locus, and is represented by the function:

$$r^2 = \frac{\left(pA_iB_j - pA_ipB_j\right)^2}{pA_i(1 - pA_i)pB_j(1 - pB_j)}$$

where the ith allele frequency of the locus A and the jth allele frequency of the locus B are denoted by pA_i and pB_j, respectively, and the haplotype frequency of A_iB_j is represented by pA_iB_j [45,47].

Pruned variants were further filtered using the parameters --geno 0.10 and --maf 0.05. The former filters out all variants with missing call rates exceeding 10% and the latter filters out all variants with minor allele frequency (MAF) below 5%, respectively. This step yielded high-quality genome-wide SNPs and InDels.

The retained high-quality polymorphisms were analyzed for their informativeness using metrics such as heterozygosity (*He*); minor allele frequency (MAF) and polymorphism information content (PIC) were also calculated. *He* provides information on the likelihood of particular loci being heterozygous; hence, *He* is a fundamental metric for estimating the genetic diversity in a population in terms of gene diversity at a locus [48,49]. *He* is calculated using the following formula:

$$He = 1 - \sum_{i=1}^{I} Pi^2$$

wherein I represents the distinct number of alleles at a particular locus and Pi represents the frequency of allele I within the population [50]. MAF is a measurement of the frequency of the second most common allele in the population [51], and is calculated using the formula:

$$\text{MAF} = \frac{minor\ allele\ count\ in\ the\ population}{Total\ allele\ count\ in\ the\ population}$$

Usually, MAF < 0.05 are excluded. PIC, on the other hand, is useful for measuring the polymorphism detection capability of markers within a population, and is calculated as follows:

$$PIC = 1 - \sum_{i=1}^{n} Pi^2 - \left(\sum_{i=1}^{n} Pj^2\right)^2 - \sum_{i=1}^{n} Pi^4$$

where in Pi and Pj represent the frequencies of a selected marker's ith and jth alleles, respectively [49,52]. Polymorphisms with PIC higher than 0.50 are considered very informative, while polymorphisms with PIC between 0.25–0.50 are considered somewhat informative [49].

2.7. *Analysis of Genome-Wide Variation and Phylogenetic Relationships*

For downstream genetic variation and phylogenetic analyses, the VCF files produced after variant pruning were preprocessed using the vcfR package [53] RRID:SCR_023453 in R. Principal component analysis (PCA) was performed using the Poppr package [54] RRID:SCR_023452 and visualized using ggplot2 [55] RRID:SCR_014601 to project the variation among the samples involved in terms of these genome-wide polymorphisms. To quantitatively assess genetic diversity, the expected heterozygosity (*He*) of each locus was calculated. Phylogenetic analysis was conducted using the Poppr package through a neighbor-joining algorithm [56] with 1000 bootstrapping. Analysis using the Poppr package was executed following the publicly available workflows and code structures [57].

3. Results

3.1. *DNA Extracts and Quality of Generated Sequence Reads*

The extracted DNA from the nine abaca varieties and two additional accessions, namely 'Luno Green' and Samoro, were of sufficient purity, with A260/A280 ratios ranging from 2.161 to 3.056; A260/A230 ratios ranging from 0.373 to 0.738; and DNA concentration ranging from 266.50 to 641.35 ng/µL (Table S1). The DNA quality and quantity were sufficient to pass the quality control procedures for next-generation sequencing.

Whole-genome resequencing through the Illumina Hiseq platform generated a total of 646,842,305 sequences, ranging from 37,019,179 to 83,388,873 sequences across the varieties, and equivalent to a total of 1,053,306,114 sequence reads (Table S2). The reads were determined to have sufficient base quality scores (Figure S1) but were highly contaminated with Nextera sequencing adapters (Figure S2). After base quality and adapter trimming with Trimmomatic, the mean base quality scores (>35) improved (Figure S3) and all adapters were successfully removed (Figure S4). This step retained 64–90% of the bases, which were then used for mapping to the reference genome.

3.2. *Mapping Quality Statistics*

The quality-filtered sequence reads were used for mapping to the polished version (through the PILON v1.22 program) of the reference genome [30]. This step also included the mapping of *Musa* reads [33] for *M. textilis* (T genome, 2n = 2x = 20) as well as for other diploid relatives, namely, *M. acuminata* (A genome, 2n = 2x = 22), *M. balbisiana* (B genome, 2n = 2x = 22) and *M. troglodytarum* (T genome, 2n = 2x = 20) [58,59]. The BWA program was used for this step, resulting in mapping of the reads at 22–46× coverage and the calculated mean mapping qualities (MQ) having scores ranging from 35.8–38.4 across the resequenced samples (Figure 1).

Mapping of the *Musa* sequence reads [33] to the abaca reference genome revealed most accessions to map to the reference at >20× coverage except for SRR9734079 (*M. balbisiana*, 18.4×), SRR9850640 (*M. balbisiana*, 11.0×) and SRR9850641 (*M. troglodytarum*, 5.8×). Mean MQ values across these accessions also ranged between 17.9–37.1.

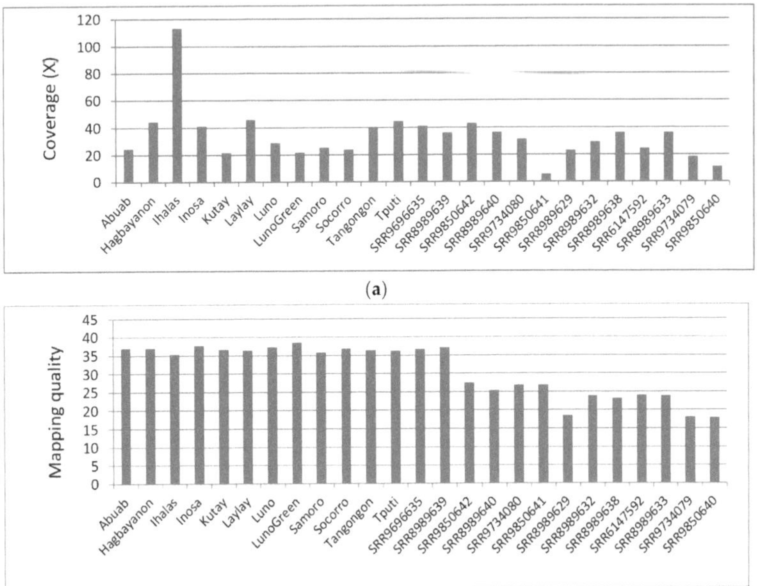

Figure 1. Sequencing coverage (**a**) and mapping quality evaluation (**b**) results generated through Qualimap across abaca cultivars and *Musa* accessions. The graph was generated using Microsoft Excel 2010s.

3.3. Variant Calling and Filtering Statistics

To mine and analyze the SNP and InDel polymorphisms within *M. textilis* and among *Musa* spp., variant calling was performed using the SAMtools and BCFtools programs. Variant calling produced a VCF file for polymorphisms within *M. textilis* containing 19,189,434 SNPs and 1,400,947 InDels (Table 2).

Table 2. Variant calling and filtering statistics.

	Within *M. textilis*		**Among *Musa* spp.**	
	SNPs	**InDels**	**SNPs**	**InDels**
Total	19,189,434	1,400,947	42,647,249	2,466,646
High quality (MQ > 40), biallelic	15,410,778	1,109,789	34,643,663	1,933,417
High quality (MQ > 40), multiallelic	388,338	135,230	1,814,554	2,31,670
Number of transitions (Ts)	108,067	N/A	18,106	N/A
Number of transversions (Tv)	22,701	N/A	13,138	N/A
Ts/Tv ratio	4.76	N/A	1.38	N/A
Retained after LD pruning	635,945	84,711	2,130,711	192,835
Number of loci with at most 10% missing genotypes	130,768	13,620	31,244	577

Among these, the number of high-quality (MQ > 40) and biallelic SNPs and InDels was 15,410,778 and 1,109,789, respectively. Another VCF containing polymorphisms between and among *Musa* spp. (*M. textilis*, *M. acuminata*, *M. balbisiana* and *M. troglodytarum*) was produced and contained 34,643,663 and 1,933,417 high-quality and biallelic SNPs and InDels, respectively.

Despite the huge number of polymorphisms, a total of 15,799,911 high-quality polymorphisms within *M. textilis* and 34,253,534 high-quality polymorphisms among *Musa* spp. were found to be correlated in terms of linkage ($r^2 > 0.2$). To avoid capturing loci with redundant genetic information, these polymorphisms were pruned, retaining 635,945 SNPs and 84,711 InDels within *M. textilis*, and 2,130,711 SNPs and 192,835 InDels among *Musa* spp. To ensure the accuracy of downstream multivariate and phylogenetic analysis, only loci with at most 10% missing genotypes were selected, retaining 130,768 SNPs and 13,620 InDels within *M. textilis* and 31,244 SNPs and 577 InDels among *Musa* spp.

A small portion of multiallelic variants within *M. textilis* was also discovered, consisting of 2.5% and 10.9% of the total SNPs and InDels, respectively (Figure 2). Among *Musa* spp., multiallelic variants consist of 5.0% and 10.7% of the total SNPs and InDels, respectively. While SNPs are known to be mostly biallelic in nature, up to four SNP alleles were detected within *M. textilis*, wherein 380,772 were triallelic and 7566 were tetra-allelic. Likewise, among *Musa* spp., up to four SNP alleles were found, comprising 1,748,160 triallelic SNPs and 142,145 tetra-allelic SNPs. For InDels, up to 18 alleles were detected, wherein 75,949 triallelic InDels (6.1% of total InDels) comprise most of the multiallelic InDels within *M. textilis*, and 142,145 triallelic InDels (6.6% of total InDels) comprising most of the multiallelic InDels among *Musa* spp. (Figure 2).

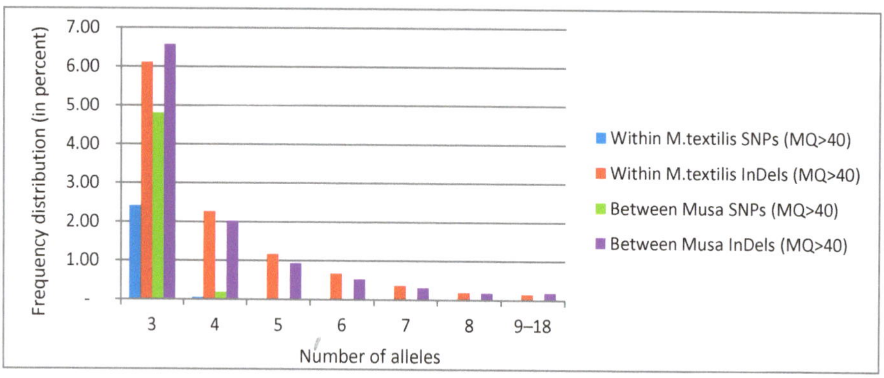

Figure 2. Frequency distribution of multiallelic SNPs and InDels within *M. textilis* and among *Musa* spp. The graph was generated using Microsoft Excel 2010.

Calculation of transitions (Ts) and transversions (Tv) among the mined high-quality and pruned SNPs within *M. textilis* and among *Musa* spp. revealed a higher frequency of occurrence of transitions over transversions (Figure 3).

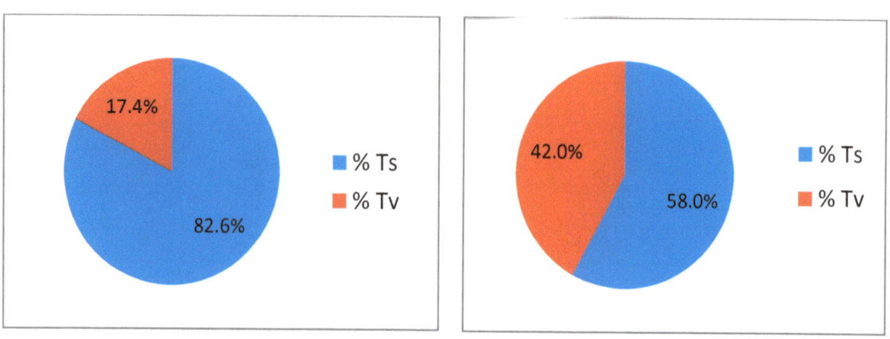

(**a**) (**b**)

Figure 3. Percentage distribution of transitions and transversions among the discovered SNPs within *M. textilis* (**a**) and among *Musa* spp. (**b**). The pie chart was generated using Microsoft Excel 2010.

Within the *M. textilis* species, the Ts/Tv ratio was calculated to be 4.76 (Table 2). Among *Musa* spp., however, the Ts/Tv ratio was calculated to be 1.38.

To compare this dataset with existing literature, SNPs were mined within *M. acuminata* only using the accession numbers SRR8989629, SRR8989632 and SRR8989638 [33]. The mined SNPs were found to have a Ts/Tv ratio of 1.77 (Table S3), which is not far from the published findings [60] for *M. acuminata* in East Africa (Ts/Tv = 1.37).

3.4. Genome-Wide Variation and Phylogenetic Relationships within Musa textilis

The retained 130,768 SNPs and 13,620 InDels within *M. textilis*, and 31,244 SNPs and 577 InDels among *Musa* spp. in Table 2, represent the variants that passed the quality control procedures, and represent the genome-wide SNPs and InDels considered to be at linkage equilibrium.

The genome-wide variation between abaca varieties and accessions was visualized through multivariate analysis by PCA (Figure 4).

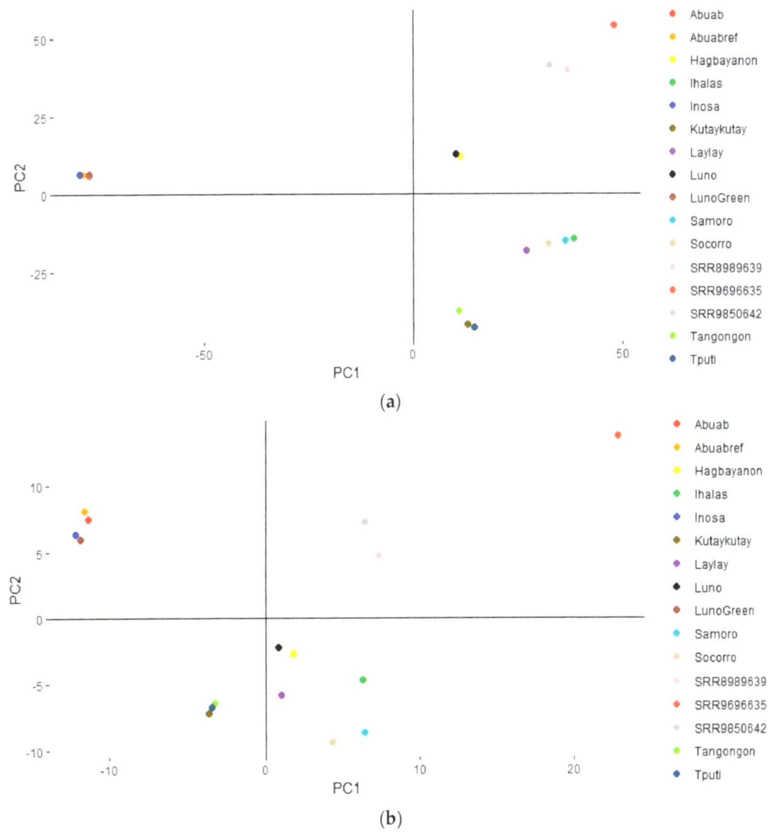

Figure 4. PCA plot displaying genetic variation among 16 abaca varieties and accessions in terms of their genome-wide SNPs (**a**) and InDels (**b**). These are represented by 130,768 and 13,620 LD-pruned loci, respectively, with at most 10% missing genotypes. The first and second components in the SNP PCA plot were able to explain 39.4% and 14.4% of the variance, respectively. The first and second components in the InDel PCA plot were able to explain 17.8% and 11.4% of the variance, respectively. Plots were generated using RStudio 2022.02.3.

The Abuab, Inosa and 'Luno Green' varieties were observed to have coinciding and indistinguishable points in PCA, while the Luno and Hagbayanon varieties were observed to have partially coinciding points. The rest of the samples have points that are clearly distinguishable from each other in PCA.

Calculation of PIC, *He* and MAF across the 130,768 SNPs and 13,620 InDels revealed quantitative estimates of the genetic diversity among the abaca varieties and accessions

(Table 3). In this study, the diversity among the abaca cultivars was found to be equivalent across SNPs and InDels. The mean PIC across SNPs and InDels was found to be moderate (0.312 and 0.332, respectively). Gene diversity, represented by *He*, was calculated to be higher in this study than the cross-species markers. MAF calculations were likewise found to be higher in this study than the cross-species SNPs. This indicates that the marker set in this study is more capable in detecting rare variants in populations.

Table 3. Mean polymorphism information content, expected heterozygosity, minor allele frequency and nucleotide diversity across SNPs and InDels scored on 16 *M. textilis* varieties and accessions.

Diversity Metrics	This Study		[24]
	SNPs	InDels	SNPs
PIC	0.312 ± 0.068	0.332 ± 0.076	-
He	0.396 ± 0.106	0.431 ± 0.111	0.281 ± 0.135
MAF	0.310 ± 0.126	0.362 ± 0.124	0.196 ± 0.132

A neighbor-joining (NJ) tree was also generated using Hamming distance (through the bitwise.dist function) for calculation of pairwise genetic distances, i.e., the proportion of dissimilar loci between varieties/accessions [61] (Figure 5).

The NJ tree generated using the genetic distances generated for the SNP genotypes grouped the varieties and accessions into three main monophyletic groups. The accessions [33], namely, SRR9696635, SRR8989639 and SRR9850642, as well as the abaca varieties in the current study, namely Laylay, Tangongon, Tputi, Kutaykutay, Hagbayanon, Luno, 'Luno Green', Abuab and Inosa, were grouped within the first main clade. The Samoro and Socorro varieties made up the second main clade, and the Ihalas accession is the lone constituent of the third clade (Figure 5a). Within the first clade, the Abuab, Inosa and 'Luno Green' varieties clustered together and were deeply separated from the other varieties and accessions. The branch length that immediately precedes their divergence was found to be significantly longer than the other varieties. On the other hand, the Socorro, Samoro and Ihalas varieties experienced the fewest mutations among the other varieties and accessions.

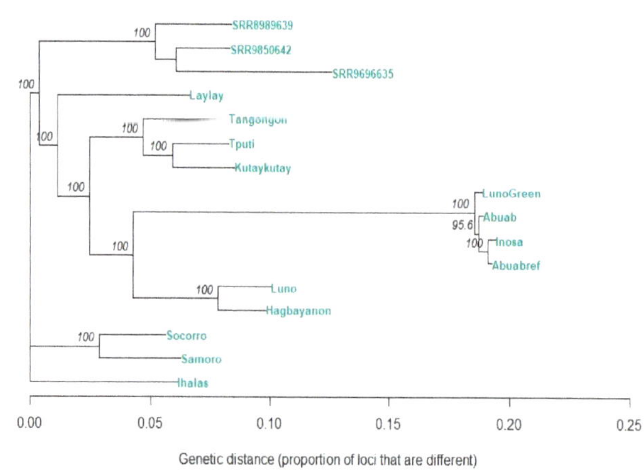

(a)

Figure 5. *Cont.*

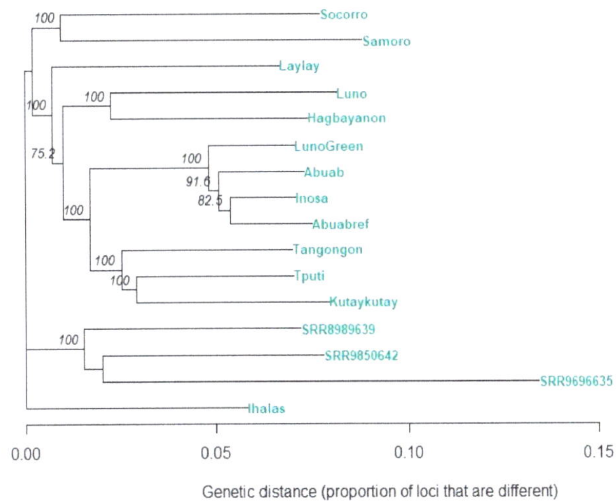

(**b**)

Figure 5. Unrooted neighbor-joining tree displaying phylogenetic relationships among 16 abaca varieties and accessions in terms of their genome-wide SNPs (**a**) and InDels (**b**). These are represented by 130,768 and 13,620 LD-pruned loci, respectively, with at most 10% missing genotypes. Values on nodes represent bootstrap support out of 1000 NJ bootstrap sampling using Hamming distance for genetic distance calculation. The accessions SRR9696635, SRR8989639 and SRR9850642 were not identified in terms of their varietal identity [33], and hence were labeled as accession numbers. Trees were generated using RStudio 2022.02.3.

3.5. Genome-Wide Variation and Phylogenetic Relationships among Musa Species

PCA of the SNP and InDel genotypes using the VCF file containing polymorphisms among *Musa* spp. distinguished *M. textilis* and *M. troglodytarum* species, and distinguished each of the *Musa* accessions, indicating the resolving power of InDels for genetic differentiation (Figure 6).

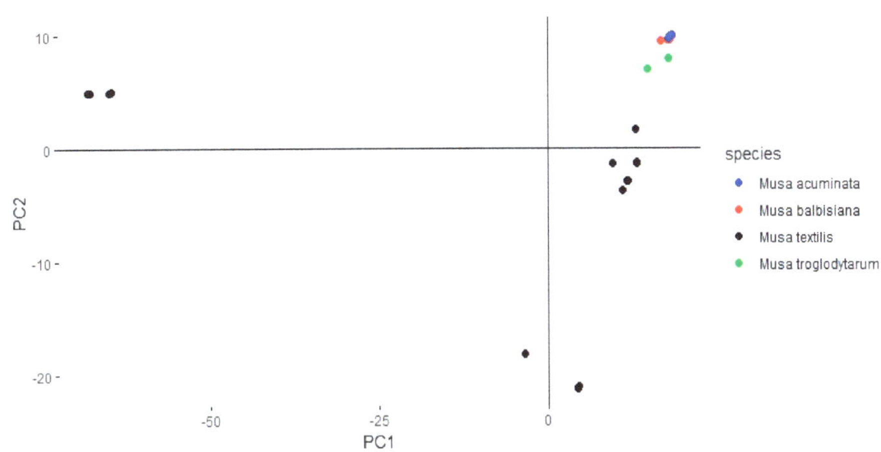

(**a**)

Figure 6. *Cont.*

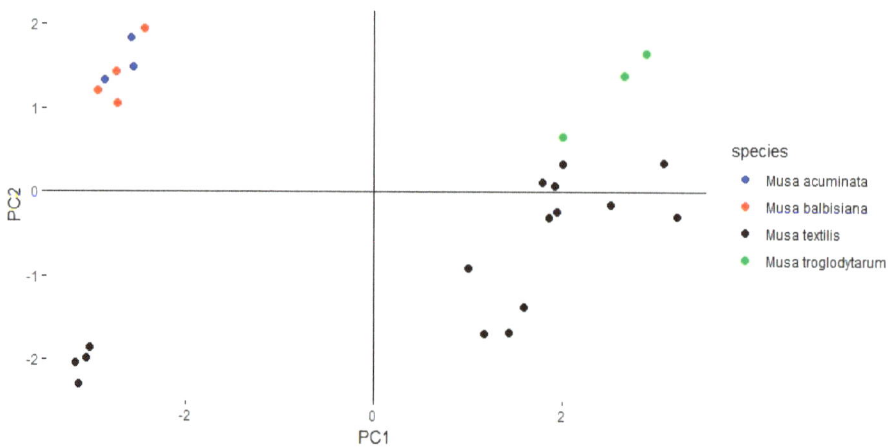

(**b**)

Figure 6. PCA plot displaying genetic variation among *Musa* species in terms of their genome-wide SNPs (**a**) and InDels (**b**). These are represented by 31,244 and 577 LD-pruned loci, respectively, with at most 10% missing genotypes. The first and second PCA components in the SNP PCA plot were able to explain 69.3% and 9.4% of the variance. The first and second PCA components in the InDel PCA plot were able to explain 27.0% and 7.9% of the variance. Species labels 'Musa balbisiana', 'Musa troglodytarum', 'Musa acuminata' and 'Musa *textilis*' represent the species *Musa balbisiana*, *M. troglodytarum*, *M. acuminata* and *M. textilis*, respectively. Plots were generated using RStudio 2022.02.3.

The *M. balbisiana* and *M. acuminata* accessions, however, did not form separate clusters. Nevertheless, the SNP and InDels were able to group the accessions into *M. textilis*, *M. troglodytarum* and *M. acuminata/balbisiana*. In addition, PCA analysis of the InDel genotypes also enabled separation among all samples involved (Figure S5).

Like the polymorphisms within *M. textilis*, genetic distances were likewise calculated among all *Musa* accessions involved in this study (Table S5). The rooted unweighted pair group method with arithmetic mean (UPGMA) analysis using the genetic distances generated for the SNP genotypes (Figure 7a) showed that all abaca varieties and accessions formed a monophyletic group and are closely related with *M. troglodytarum* among other *Musa* spp. UPGMA analyses of the InDel genotypes (Figure 7b), however, showed clustering of the *M. troglodytarum* accessions within the *M. textilis* clade, particularly the subclade containing SRR9850642, SRR8989639, Luno, Hagbayanon, Tputi, Tangongon, Kutaykutay, Laylay, Ihalas, Socorro, Samoro and SRR9696635. This subclade is distinct from the other subclade that contains the other varieties of *M. textilis*, which includes Inosa, Abuab and 'Luno Green'.

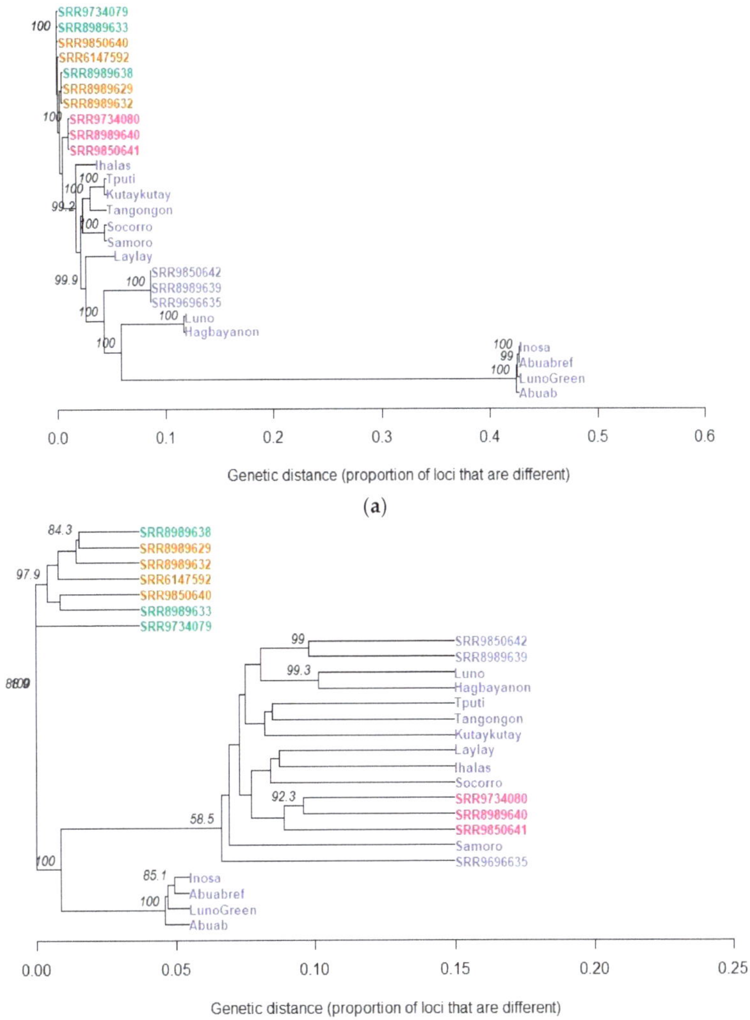

(b)

Figure 7. Rooted unweighted pair group method with arithmetic mean (UPGMA) tree displaying phylogenetic relationships among *Musa* accessions in terms of their genome-wide SNPs (**a**) and InDels (**b**). These are represented by 31,244 and 577 LD-pruned loci, respectively, with at most 10% missing genotypes. Values on nodes represent bootstrap support out of 1000 UPGMA bootstrap sampling using Hamming distance for genetic distance calculation. *M. textilis* accessions are labeled in purple, *M. troglodytarum* accessions are labeled in pink, *M. acuminata* accessions are labeled in green and *M. balbisiana* accessions are labeled in orange. For both trees, the SRR9734079 *M. acuminata* accession was used as outgroup from where the trees were rooted. Trees were generated using RStudio 2022.02.3.

3.6. Genetic Characterization of the Musa Accessions

Homozygosity statistics calculated through the VCFtools program across all accessions revealed Abuab, Inosa and 'Luno Green' to have a significantly low homozygosity, indicating a significantly high heterozygosity among all accessions (Figure 8).

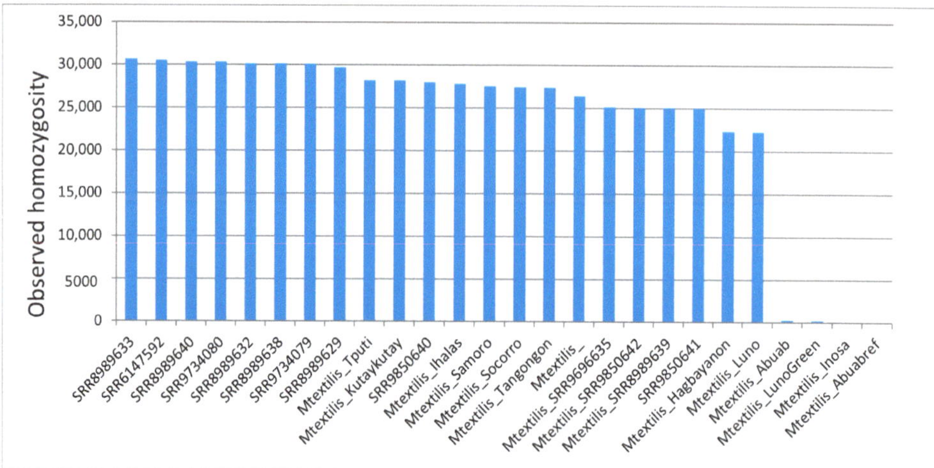

Figure 8. Graphical comparison of homozygosity statistics per individual (variety and accession) based on genome-wide SNPs. Labels starting with 'Mbalbisiana', 'Mtroglodytarum', 'Macuminata' and 'Mtextilis' represent varieties/accessions under the *Musa balbisiana*, *M. troglodytarum*, *M. acuminata* and *M. textilis*, respectively. The specific varieties/accessions are indicated following the aforementioned labels. The pie chart was generated using Microsoft Excel 2010.

The inbreeding coefficient calculations further support these findings, wherein Abuab, Inosa and 'Luno Green' were ranked last and highly negative in terms of inbreeding coefficient (F) (Table 4).

Table 4. Inbreeding coefficient (F) statistics across *Musa* accessions.

Accession/Variety	Species	N_SITES	F
SRR8989629	*M. acuminata*	30,123	0.93086
SRR6147592	*M. balbisiana*	30,963	0.928
SRR9850640	*M. balbisiana*	28,416	0.92696
SRR9734079	*M. balbisiana*	30,554	0.9252
SRR8989632	*M. acuminata*	30,643	0.91983
SRR8989633	*M. balbisiana*	31,192	0.91946
SRR8989638	*M. acuminata*	30,648	0.91891
SRR9850641	*M. troglodytarum*	25,747	0.86544
SRR8989640	*M. troglodytarum*	31,205	0.8618
SRR9734080	*M. troglodytarum*	31,204	0.85997
Ihalas	*M. textilis*	30,374	0.57407
Kutaykutay	*M. textilis*	31,079	0.53542
Tputi	*M. textilis*	31,165	0.52615
Samoro	*M. textilis*	30,835	0.46154
Socorro	*M. textilis*	30,819	0.45137
Tangongon	*M. textilis*	31,072	0.40478
Laylay	*M. textilis*	30,220	0.35956
SRR9696635	*M. textilis*	31,032	0.04446
SRR8989639	*M. textilis*	31,185	0.02363
SRR9850642	*M. textilis*	31,214	0.02259
Luno	*M. textilis*	30,726	−0.37979
Hagbayanon	*M. textilis*	30,791	−0.38391
Abuab	*M. textilis*	29,625	−3.893
LunoGreen	*M. textilis*	29,761	−3.90269
Inosa	*M. textilis*	31,039	−3.97664
Abuabref	*M. textilis*	31,180	−3.99309

4. Discussion

The persisting problems of destructive abaca diseases and abiotic stresses and the slow development of abaca breeding programs for disease resistance and climate resilience, aggravated by confusions in the true identity of abaca varieties and unavailability of abaca genome-based genetic marker resources, impede the Philippine abaca industry.

The whole-genome resequencing approach is a useful method for identification of genome-wide polymorphisms vital for plant genetic resources management [62]. While a high read depth would provide assurance in genotype calling accuracy, applying this for a large number of samples can be expensive. A cost-effective solution in mining true polymorphisms is the low-depth resequencing of several individuals, which is based on the information that true variants are expected to appear multiple times upon mapping of sequence reads to the reference [28]. In *Musa*, this resequencing approach has been previously employed [63] for detection of hidden diversity within the *Musa itinerans* species by mining genome-wide variation using sequencing data with 15.5× coverage. This approach was employed in another study (through resequencing of at least 20× coverage) for multiple *Musa* species to generate and analyze genomic resources for polymorphism detection usable for studying *Musa* evolution, diversity and breeding [33].

The current study utilized the resequencing approach in an attempt to capture genetic diversity and relationships among abaca cultivars and among *Musa* spp. That are important for addressing the scarcity of genetic information and resources vital for efficient abaca breeding programs. Specifically, these are the lack of abaca genome-based genetic markers usable for accurate identification of abaca varieties used for mass cultivation and as breeding materials, and the lack of genetic marker resources for identification of polymorphisms across *Musa* spp. that are associated with agronomically and economically important traits.

The first steps undertaken were the resequencing of 11 abaca varieties and accessions at a minimum of 20× coverage and mapping the reads to the abaca reference genome [30]. The calculated base quality scores (>35) indicate high confidence in the base calls from the sequencing (Figure S3). The calculated coverages (22–46×) matched the coverage used in a related study [33], hence indicating the sufficiency of the reads for variant calling using the abaca reference genome previously assembled at high (65×) coverage [30] (Figure 1). Using the MQ formula [64], the MQ scores among the resequenced abaca samples (35.8–38.4) indicate a very low 0.014–0.026% probability that the reads were incorrectly mapped to the reference genome. The reads, therefore, were uniquely mapped to the reference genome, indicating the high accuracy of the mapping step (Figure 1).

Mapping of the published *Musa* sequence reads [33] to the abaca reference genome [30] produced similar coverages (except for three accessions) and lower MQ values. Calculation of the mismapping probability reveals 0.02–1.62% probability that the reads were incorrectly mapped. Hence, even though the MQ values for these accessions were lower than the resequenced abaca samples and the coverage obtained for SRR9850640 and SRR9850641 are significantly lower than the other accessions, their mapping accuracy to the abaca reference genome is nevertheless satisfactory.

Mining of genome-wide SNPs and InDels was achieved using a bioinformatics pipeline featuring SAMtools and BCFtools. This step produced 20,590,381 variants within *M. textilis*, consisting of 93.2% SNPs and 6.8% InDels, and 47,580,541 variants among *M. textilis*, *M. acuminata*, *M. balbisiana* and *M. troglodytarum*, consisting of 94.8% SNPs and 5.2% InDels. This number of polymorphisms among *Musa* spp. is significantly higher compared to those within *M. textilis* polymorphisms owing to the higher expected interspecific (i.e., between-species) genetic diversity than infraspecific (i.e., within-species) genetic diversity. Among the mined polymorphisms, the high-quality (MQ > 40) SNPs and InDels were selected, and accounted for 80.3% and 79.2% of their respective total number within *M. textilis*, and 76.7% and 78.3% of their respective total number among the *Musa* spp. Even though mean MQ values around 17 were regarded as acceptable based on the equivalent mismapping probability, filtering of individual polymorphisms was based on more stringent MQ values to ensure optimal accuracy of these mined polymorphisms.

Since the majority of the polymorphisms have MQ above 40, this indicates that most of the genomic loci have accurately mapped reads. The accuracy of the identified polymorphisms is further assured by the fact that most of the polymorphisms within *M. textilis* have MQ values within the range of 57.5–60.0 (Figure S6). This MQ range indicates a 1/100,000 probability that the reads were incorrectly mapped, therefore giving information that the reads were indeed uniquely mapped to the reference genome [28,63].

While both SNPs and InDels are predominantly biallelic, multiallelic loci were also identified. For SNPs, the percentage of multiallelic loci is justifiably small (~2.3–4.7% of the total number of SNPs), owing to their supposedly biallelic nature [65]. The percentage of multiallelic SNPs is notably higher among the four *Musa* spp. than within *M. textilis*. SNP loci with greater than two alleles, such as those stated for triallelic loci [66], enable better genetic discrimination than biallelic loci. This highlights the possible role of multiallelic SNPs in genetic and phenotypic differentiation among the *Musa* spp. For InDels, which are naturally multiallelic, a total of 18 alleles were identified and have similar proportions within *M. textilis* and among *Musa* spp. Even though a high number of alleles were identified, setting the mapping quality threshold to MQ > 40 ensures accuracy of these mined InDels.

Further analysis of the SNPs revealed the transition/transversion ratio to be higher within *M. textilis* (4.76) than among *Musa* spp. (1.38). Transitions are single-base mutations resulting in the nucleotide with the same number of rings as the wild type (e.g., purine to purine mutation), while transversions occur as a result of mutation of the wildtype nucleotide base into a base with different number of rings (e.g., purine to pyrimidine mutation) [67]. Since transversions have a higher potential in altering the primary structure of encoded proteins and are more likely to alter the shape of the DNA backbone, transversions have higher impacts on gene regulatory elements and consequentially on gene expression [67]. Since SNPs have lower Ts/Tv ratios among *Musa* spp., there is a higher rate of transversion; this could be another source of phenotypic differences among the *Musa* spp.

The current study thus suggests that the higher percentage of multiallelic loci and higher transversion rate among the four *Musa* spp. than within the *M. textilis* species contributed to the genetic differentiation across the four *Musa* spp., and these findings can serve as a basis for further research on the phenotypic implications of these genetic variations in *Musa*.

Multivariate analysis (through PCA) of the high-quality polymorphisms that were pruned for linkage disequilibrium revealed genetic variation within *M. textilis* and among *Musa* spp. PCA of the SNP genotypes within *M. textilis* revealed a high similarity between the Abuab reference and the resequenced Abuab variety. It also enabled differentiation among abaca varieties based on their SNP profiles, except for two sets of varieties in which their SNP profiles were projected by coinciding points (Figure 4a). The overlapping points by Abuab, Inosa and 'Luno Green' varieties indicate their high genetic similarity in terms of SNPs. This provides evidence that the uncharacteristic morphology of the 'Luno Green' accession is a result of mislabeling in the field as a Luno variety. This is supported by initial morphological characterization studies by PhilFIDA revealing some Inosa collections possess the green male flower coloration observed for 'Luno Green'. Using the genome-wide InDel genotypes, PCA resulted in improved resolution among the abaca samples, as represented by the absence of coinciding points (Figure 4b). Calculation and comparison of mean PIC, however, show equivalent mean PIC between the InDel loci (0.33 ± 0.08) and SNP loci (0.31 ± 0.07) (Table 3), both of which display moderate polymorphism detection capability.

Calculation of *He* for each locus provided information on genetic diversity among the *M. textilis* varieties and accession. In comparison with the only published literature on abaca SNPs, i.e., the banana genome-based SNP markers cross-species adapted into abaca [18], this current study produced a significantly higher *He* than the cross-species markers. This therefore indicates the higher capability of the abaca genome-based SNPs mined in this study to capture genetic variation within the *M. textilis* species compared to the cross-species markers. The mean MAF calculations, which are also genetic diversity metrics since they measure the frequency of the second most common allele in the population [51], were also found to be higher across the SNP loci in the current study than those in the cross-species adapted SNPs [24]. These metrics therefore indicate sufficient capability of the polymorphism markers mined in this study to capture genetic diversity within the *M. textilis* species, and thus are usable for genetic differentiation applications such as varietal identification.

To elucidate phylogenetic relationships within *M. textilis*, a neighbor-joining tree generation approach was conducted. The calculated pairwise genetic distances among the abaca varieties and accessions are listed in Table S4. Since the branch length indicates the degree of evolutionary divergence such as the number of substitutions along a particular branch, NJ analysis using Hamming distances revealed extensive mutations experienced by Abuab, Inosa and 'Luno Green' compared to the rest of the varieties and accessions, as evidenced by their long branch length and deep separation from the rest of the varieties and accessions [68]. This is supported by the PCA plot in Figure 4a, wherein these varieties are closely clustered with each other and away from the other varieties and accessions. The Abuab reference genome was found to be $0.7 \pm 0.1\%$ dissimilar (on average) with Inosa and 'Luno Green' in terms of the genome-wide SNPs and $4.2 \pm 0.3\%$ dissimilar in terms of the genome-wide InDels. In contrast, the Abuab reference genome was calculated to be $18.1 \pm 6.3\%$ and $14.2 \pm 3.1\%$ dissimilar with the rest of the abaca varieties and accessions in terms of the genome-wide SNPs and InDels, respectively. The extensive mutations experienced by the Abuab and Inosa varieties are explained by their extensive domestication due to their popularity as commercial varieties. Since seeds have been rampantly used as planting materials during the 1900s and the abaca reproductive cycle consists of alternating emergence of male and female flowers (thereby preventing selfing events), genetic variability arising from continuous cross-pollination within the population most likely occurred [9]. In contrast, the Socorro, Samoro and Ihalas varieties were observed to form a separate cluster away from the rest of the abaca samples and shared the oldest common ancestor. The separate clades formed by these three abaca varieties are explained by genetic isolation by distance (IBD). IBD occurs due to the limit in gene flow as a result of geographical isolation [69]. The Socorro and Samoro varieties are popularly planted only in the Mindoro province, while the Ihalas variety is a wild variety known to exist only in the Leyte province.

Both the NJ tree for the SNP and InDel loci clustered together the pairs Samoro and Socorro, Luno and Hagbayanon and Tangongon and Tputi (Figure 5b). The branch lengths within these pairs, however, have higher length differences observed in the InDel NJ tree compared to those observed in the SNP tree, indicating a higher degree of separation within these pairs of varieties due to differences in the number of InDel mutations, which typically arise from polymerase slippage and unequal meiotic crossing-over [70]. To establish phylogenetic relationships among *Musa* spp., UPGMA analysis was performed. Within *M. textilis*, the Ihalas was observed to be the abaca accession closest to *M. troglodytarum*, and the clusters that were formed were in agreement with Figure 5a. Since Ihalas was reported by farmers in Leyte to display tolerance to abaca viral disease, its close relationship with a banana species could have contributed to its disease tolerance. Banana species, especially *M. balbisiana*, were widely documented to possess degrees of tolerance to viral diseases, e.g., [71]. These data, hence, support the use of Ihalas as a potential source of disease resistance genes and as a candidate for intraspecific hybridization.

Since SNPs are the most abundant genetic variation and are significantly more abundant than InDels (Table 2), the phylogenetic clustering based on SNPs observed in Figure 7a indicates the phylogenetic relationships across *Musa* spp., and then the insertion/deletion events in Figure 7b may likely contribute to enhancement of genetic diversity among *Musa* spp. and within *M. textilis*. These findings, together with the homozygosity data in Figure 8 and inbreeding coefficient calculations in Table 4, complement the NJ analysis results in Figure 5a, further supporting that the Abuab and Inosa varieties resulted from abaca being highly cross-pollinated due to alternating emergence of male and female flowers and extensive domestication exacerbated by the utilization of seeds as planting materials during the early times [9]. The results, therefore, indicate that the Abuab, Inosa and 'Luno Green' varieties have accumulated the most mutations among varieties and accessions in this study, and could be interesting sources of possibly advantageous mutations leading to disease resistance, abiotic stress resistance (including climate change adaptability) and good agronomic traits. Since Abuab and Inosa are NSIC-registered varieties, the advantageous characteristics of these varieties, hence, could have arisen from advantageous phenotypes caused by accumulation of advantageous mutations over time.

Overall, this study successfully mined genome-wide SNPs and InDels in abaca, which served as highly important resources for genetic marker discovery and downstream applications in genetic resources management such as varietal identification. Their potential as genetic marker resources is exemplified by their capability to detect variation within the *M. textilis* species and between *Musa textilis*, *M. acuminata*, *M. balbisiana* and *M. troglodytarum*. This is further supported by the elucidated genome-wide variations, genetic diversity and evolutionary relationships among abaca varieties and other *Musa* spp. While markers for abaca genetics studies were developed in the past few years, these markers were based on the banana genome. This study, hence, is the first report on genome-wide polymorphisms across abaca varieties based on the genome of abaca itself. Consequently, this is a pioneering work elucidating the genome-wide variation and evolutionary information across abaca varieties and among the four *Musa* species.

The genetic markers with sufficient diversity and polymorphism detection capability generated in this study will be impactful for the development of abaca genome-based assays for routine identification of abaca varieties and accessions. The mined polymorphisms among abaca and the more extensively studied banana will be important resources for identifying genes (through comparative genotyping) that may be important for abaca crop improvement and climate change adaptability. Altogether, these will lead to an efficient and effective abaca breeding program that will highly benefit the Philippine abaca industry and its global stakeholders.

For future studies, more phenotypically associated markers will need to be mined from the between-*Musa* polymorphisms identified in this study to enable the discovery of polymorphisms in the widely studied banana species that are linked to important traits. Since phylogenetic analysis using the genome-wide SNPs revealed the Abuab and Inosa varieties to have experienced a significant number of mutations, further characterization studies of these two varieties are recommended. This will identify agronomically important traits that resulted from accumulation of possibly advantageous characteristics.

Supplementary Materials: The following supporting information can be downloaded at: https://www.mdpi.com/article/10.3390/cimb45070365/s1.

Author Contributions: Conceptualization, L.C.G., E.P.C. and C.F.C.B.; methodology, C.F.C.B.; software, D.Z., C.F.C.B. and J.C.A.; validation, D.Z., E.P.C., L.C.G. and C.F.C.B.; formal analysis, C.F.C.B., L.C.G., D.Z., E.P.C., R.B.L.K. and J.C.A.; investigation, L.C.G., E.P.C. and C.F.C.B.; resources, L.C.G. and R.B.L.K.; data curation, C.F.C.B.; writing—original draft preparation, C.F.C.B.; writing—review and editing, L.C.G., E.P.C., D.M.C.S., R.B.L.K. and J.C.A.; visualization, C.F.C.B.; supervision, L.C.G., E.P.C. and D.M.C.S.; project administration, L.C.G.; funding acquisition, L.C.G. All authors have read and agreed to the published version of the manuscript.

Funding: This work was carried out as a part of the PhiFIDA project DABIOTECH R-1806 funded by the Department of Agriculture (DA) Bureau of Agricultural Research of the Philippines through the DA Biotechnology Program Office. This work was also supported through the Department of Science and Technology (DOST) Accelerated Science and Technology Human Resource Development Program scholarship.

Institutional Review Board Statement: Not applicable.

Informed Consent Statement: Not applicable.

Data Availability Statement: The sequence reads generated in the current study are available in the NCBI repository (BioProject ID PRJNA916350). The following accession numbers were generated for the paired-end sequence reads produced for each sample: SRR22906090 (Abuab), SRR22906084 (Hagbayanon), SRR22906089 (Inosa), SRR22906087 (Kutaykutay), SRR22906086 (Laylay), SRR22906083 (Luno), SRR22906080 ('Luno Green'), SRR22906088 (Samoro), SRR22906082 (Socorro), SRR22906085 (Tangongon) and SRR22906081 (T.puti). The generated VCF files are stored and are accessible through the Digital Object Identifier DOI: 10.17632/27kywgfynh.1 stored under Mendeley Data.

Acknowledgments: We are grateful to the DOST Advanced Science and Technology Institute for their high-performance computing facility administered through the Computing and Archiving Research Environment (DOST-ASTI-CoARE). We also thank Maria Magdalena Damo for her assistance in the collection of the abaca samples.

Conflicts of Interest: The authors have no conflict of interest and relevant financial or non-financial interest to disclose.

References

1. Lasalita-Zapico, F.; Aguilar, C.H. Elucidating plant genetic diversity and evolution through bioinformatics. *Int. Proc. Chem. Biol. Environ. Eng.* **2014**, *63*, 28–34.
2. *Philippine Fiber Industry Development Authority (PhilFIDA) Annual Report*; Philippine Fiber Industry Development Authority: Quezon City, Philippines, 2020.
3. Spencer, J.E. The abacá plant and its fiber, Manila Hemp. *Econ. Bot.* **1953**, *7*, 195–213. [CrossRef]
4. Bledzki, A.K.; Mamun, A.A.; Faruk, O. Abaca fibre reinforced PP composites and comparison with jute and flax fibre PP composites. *Express Polym. Lett.* **2007**, *1*, 755–762. [CrossRef]
5. Barba, B.J.; Madrid, J.; Penaloza, D.J.P. A review of abaca fiber-reinforced polymer composites: Different modes of preparation and their applications. *J. Chil. Chem. Soc.* **2020**, *65*, 4919–4924. [CrossRef]
6. Liu, K.; Takagi, H.; Yang, Z. Dependence of tensile properties of abaca fiber fragments and its unidirectional composites on the fragment height in the fiber stem. *Compos. Part A Appl. Sci.* **2013**, *45*, 14–22. [CrossRef]
7. *Philippine Fiber Industry Development Authority (PhilFIDA) Abaca Sustainability Manual*; PhilFIDA: Quezon City, Philippines; Deutsche Gesellschaft für Internationale Zusammenarbeit (GIZ) GmbH: Makati City, Philippines, 2016.
8. Karthik, T.; Prabha Karan, C.; Rathinamoorthy, R. *Nonwovens: Process, Structure, Properties and Applications*; Woodhead Publishing India PVT: Delhi, India, 2017.
9. Halos, S.C. *The Abaca*; Department of Agriculture-Biotechnology Program Office: Quezon City, Philippines, 2008.
10. Altoveros, N.C.; Borromeo, T.H. *The State of Plant Genetic Resources for Food and Agriculture in the Philippines, A Country Report*; Bureau of Plant Industry, Department of Agriculture: Malate, Philipines, 2007. Available online: https://www.fao.org/pgrfa-gpa-archive/phl/docs/countryreport2.pdf (accessed on 28 September 2022).
11. Philippine Fiber Industry Development Authority (PhilFIDA) (2021) Abaca Technoguide. 2021. Available online: https://philfida.da.gov.ph/images/Publications/Technoguides/abaca-technoguide.pdf (accessed on 29 December 2022).
12. Galvez, L.C.; Catalla, J.L.; Borromeo, T.H.; Altoveros, N.C. *Abaca Germplasm Collection*; Philippine Fiber Industry Development Authority: Quezon City, Philippines, 2018.
13. Bajet, N.B.; Magnaye, L. *Virus Diseases of Banana and Abaca in the Philippines*; PARRFI: Laguna, Philippines, 2000.
14. Galvez, L.C.; Barbosa, C.F.C.; Koh, R.B.L.; Aquino, V.M. Loop-mediated isothermal amplification (LAMP) assays for the detection of abaca bunchy top virus and banana bunchy top virus in abaca. *Crop Prot.* **2020**, *131*, 105101. [CrossRef]
15. Arcalas, J.Y. El Niño, Storms Slashed Abaca Output in 2019–PhilFida. El Niño Ready Nations. 2020. Available online: https://elninoreadynations.com/el-nino-storms-slashed-abaca-output-in-2019-philfida/ (accessed on 22 January 2022).
16. Vishwakarma, M.K.; Kale, S.M.; Sriswathi, M.; Naresh, T.; Shasidhar, Y.; Garg, V.; Pandey, M.K.; Varshney, R.K. Genome-wide discovery and deployment of insertions and deletions markers provided greater insights on species, genomes, and sections relationships in the genus Arachis. *Front. Plant Sci.* **2017**, *8*, 2064. [CrossRef]
17. Xia, W.; Luo, T.; Zhang, W.; Mason, A.S.; Huang, D.; Huang, X.; Tang, W.; Dou, Y.; Zhang, C.; Xiao, Y. Development of high-density SNP markers and their application in evaluating genetic diversity and population structure in *Elaeis guineensis*. *Front. Plant Sci.* **2019**, *10*, 130. [CrossRef]
18. Pan, Z.; Li, Z.; Zhang, J.; Bai, S.; Zhao, W.; Tong, C. Investigation of genome-wide InDel distribution and segregation in *Populus* with restriction-site associated DNA sequencing data. *Trop. Plant Biol.* **2022**, *15*, 171–180. [CrossRef]
19. Korir, N.K.; Han, J.; Shangguan, L.; Wang, C.; Kayesh, E.; Zhang, Y.; Fang, J. Plant variety and cultivar identification: Advances and prospects. *Crit. Rev. Biotechnol.* **2013**, *33*, 111–125. [CrossRef]
20. Garrido-Cardenas, J.A.; Mesa-Valle, C.; Manzano-Agugliaro, F. Trends in plant research using molecular markers. *Planta* **2018**, *247*, 543–557. [CrossRef] [PubMed]
21. Kim, M.; Jung, J.-K.; Shim, E.-J.; Chung, S.-M.; Park, Y.; Lee, G.P.; Sim, S.-C. Genome-wide SNP discovery and core marker sets for DNA barcoding and variety identification in commercial tomato cultivars. *Sci. Hortic.* **2021**, *276*, 109734. [CrossRef]

22. Fang, W.-P.; Meinhardt, L.W.; Tan, H.-W.; Zhou, L.; Mischke, S.; Zhang, D. Varietal identification of tea (*Camellia sinensis*) using nanofluidic array of single nucleotide polymorphism (SNP) markers. *Hortic. Res.* **2014**, *1*, 14035. [CrossRef] [PubMed]
23. Yllano, O.B.; Diaz, M.G.Q.; Lalusin, A.G.; Laurena, A.C.; Tecson, E.M. Genetic analyses of abaca (*Musa textilis* Née) germplasm from its primary center of origin, the Philippines, using Simple Sequence Repeat (SSR) markers. *Philipp. Agric. Sci.* **2020**, *103*, 12.
24. Galvez, L.C.; Meinhardt, L.W.; Goenaga, R.; Zhang, D. Accurate identification of abaca (*Musa textilis* Née) cultivars using single nucleotide polymorphisms (SNP) markers developed for banana (*Musa acuminata* Colla). *Int. J. Plant Biol. Res.* **2021**, *9*, 1125.
25. Ratnadewi, D. Identification of Musa textilis by using RAPD markers. In *Resource Management: Public-Private Partnership and Knowledge Sharing*; Adhi, A.K., Pambudy, R., Burhanuddin, Eds.; SEAG IPB: Bogor, Indonesia, 2001; pp. 75–80. Available online: https://repository.ipb.ac.id/jspui/bitstream/123456789/64194/1/karil1.pdf (accessed on 13 September 2022).
26. Boguero, A.P.B.; Parducho, M.A.L.; Mendoza, M.R.; Abustan, M.A.M.; Lalusin, A.G. Molecular screening of abaca (*Musa textilis* Née) accessions using microsatellite markers associated with resistance to bunchy top virus. *Philipp. J. Crop Sci.* **2016**, *41*, 13–19.
27. Kumar, S.; Banks, T.W.; Cloutier, S. SNP Discovery through Next-Generation Sequencing and Its Applications. *Int. J. Plant Genom.* **2012**, *2012*, e831460. [CrossRef]
28. Hahn, M.W. *Molecular Population Genetics*; Oxford University Press: Cary, NC, USA, 2018.
29. Li, J.; Jew, B.; Zhan, L.; Hwang, S.; Coppola, G.; Freimer, N.B.; Sul, J.H. ForestQC: Quality control on genetic variants from next-generation sequencing data using random forest. *PLoS Comput. Biol.* **2019**, *15*, e1007556. [CrossRef]
30. Galvez, L.C.; Koh, R.B.L.; Barbosa, C.F.C.; Asunto, J.C.; Catalla, J.L.; Atienza, R.G.; Costales, K.T.; Aquino, V.M.; Zhang, D. Sequencing and de novo assembly of abaca (*Musa textilis* Née) var. Abuab genome. *Genes* **2021**, *12*, 1202. [CrossRef]
31. Philippine Fiber Industry Development Authority (PhilFIDA). 3 Abaca Varieties Now Registered with the National Seed Industry Council. 2017. Available online: http://www.philfida.da.gov.ph/index.php/news-articles/116-3-abaca-varities-now-regsitered-with-nsic (accessed on 29 December 2022).
32. Koh, R.B.L.; Barbosa, C.F.C.; Aquino, V.M.; Galvez, L.C. Extraction of high molecular weight DNA suitable for next-generation sequencing from the fiber crop abaca. *Ind. Crops Prod.* **2021**, *161*, 113194. [CrossRef]
33. Sambles, C.; Venkatesan, L.; Shittu, O.M.; Harrison, J.; Moore, K.; Tripathi, L.; Grant, M.; Warmington, R.; Studholme, D.J. Genome sequencing data for wild and cultivated bananas, plantains and abacá. *Data Brief* **2020**, *33*, 106341. [CrossRef]
34. Bolger, A.M.; Lohse, M.; Usadel, B. Trimmomatic: A flexible trimmer for Illumina sequence data. *Bioinformatics* **2014**, *30*, 2114–2120. [CrossRef] [PubMed]
35. Andrews, S. FastQC: A Quality Control Tool for High Throughput Sequence Data. 2010. Available online: http://www.bioinformatics.babraham.ac.uk/projects/fastqc/ (accessed on 18 August 2022).
36. Ewels, P.; Magnusson, M.; Lundin, S.; Käller, M. MultiQC: Summarize analysis results for multiple tools and samples in a single report. *Bioinformatics* **2016**, *32*, 3047–3048. [CrossRef]
37. Li, H.; Durbin, R. Fast and accurate short read alignment with Burrows-Wheeler transform. *Bioinformatics* **2009**, *25*, 1754–1760. [CrossRef] [PubMed]
38. Li, H.; Handsaker, B.; Wysoker, A.; Fennell, T.; Ruan, J.; Homer, N.; Marth, G.; Abecasis, G.; Durbin, R.; 1000 Genome Project Data Processing Subgroup. The Sequence Alignment/Map format and SAMtools. *Bioinformatics* **2009**, *25*, 2078–2079. [CrossRef] [PubMed]
39. Bonfield, J.K.; Marshall, J.; Danecek, P.; Li, H.; Ohan, V.; Whitwham, A.; Keane, T.; Davies, R.M. HTSlib: C library for reading/writing high-throughput sequencing data. *GigaScience* **2021**, *10*, giab007.
40. Danecek, P.; Bonfield, J.K.; Liddle, J.; Marshall, J.; Ohan, V.; Pollard, M.O.; Whitwham, A.; Keane, T.; McCarthy, S.A.; Davies, R.M.; et al. Twelve years of SAMtools and BCFtools. *GigaScience* **2021**, *10*, giab008. [CrossRef] [PubMed]
41. García-Alcalde, F.; Okonechnikov, K.; Carbonell, J.; Cruz, L.M.; Götz, S.; Tarazona, S.; Dopazo, J.; Meyer, T.F.; Conesa, A. Qualimap: Evaluating next-generation sequencing alignment data. *Bioinformatics* **2012**, *28*, 2678–2679. [CrossRef]
42. Lefouili, M.; Nam, K. The evaluation of Bcftools mpileup and GATK HaplotypeCaller for variant calling in non-human species. *Sci. Rep.* **2022**, *12*, 1. [CrossRef]
43. Danecek, P.; Auton, A.; Abecasis, G.; Albers, C.A.; Banks, E.; DePristo, M.A.; Handsaker, R.E.; Lunter, G.; Marth, G.T.; Sherry, S.T.; et al. The variant call format and VCFtools. *Bioinformatics* **2011**, *27*, 2156–2158. [CrossRef]
44. Purcell, S.; Neale, B.; Todd-Brown, K.; Thomas, L.; Ferreira, M.A.R.; Bender, D.; Maller, J.; Sklar, P.; de Bakker, P.I.W.; Daly, M.J.; et al. PLINK: A tool set for whole-genome association and population-based linkage analyses. *Am. J. Hum. Genet.* **2007**, *81*, 559–575. [CrossRef] [PubMed]
45. Kang, J.T.L.; Rosenberg, N.A. Mathematical properties of linkage disequilibrium statistics defined by normalization of the coefficient D=pAB−pApB. *Hum. Hered.* **2019**, *84*, 127–143. [CrossRef] [PubMed]
46. Hill, W.G.; Robertson, A. Linkage disequilibrium in finite populations. *Theor. Appl. Genet.* **1968**, *38*, 226–231. [CrossRef] [PubMed]
47. Qu, J.; Kachman, S.D.; Garrick, D.; Fernando, R.L.; Cheng, H. Exact Distribution of Linkage Disequilibrium in the Presence of Mutation, Selection, or Minor Allele Frequency Filtering. *Front. Genet.* **2020**, *11*, 362. [CrossRef] [PubMed]
48. Nei, M. Estimation of average heterozygosity and genetic distance from a small number of individuals. *Genetics* **1978**, *89*, 583–590. [CrossRef]
49. Serrote, C.M.L.; Reiniger, L.R.S.; Silva, K.B.; Rabaiolli, S.M.D.S.; Stefanel, C.M. Determining the Polymorphism Information Content of a molecular marker. *Gene* **2020**, *726*, 144175. [CrossRef]

50. Harris, A.M.; DeGiorgio, M. An unbiased estimator of gene diversity with improved variance for samples containing related and inbred individuals of any ploidy. *Genes Genomes Genet.* **2017**, *7*, 671–691. [CrossRef]
51. Chanock, S.J.; Ostrander, E.A. 22—Discovery and Characterization of Cancer Genetic Susceptibility Alleles. In *Abeloff's Clinical Oncology*, 5th ed.; Niederhuber, J.E., Kastan, M.B., Tepper, J.E., Eds.; Elsevier: Amsterdam, The Netherlands, 2014; pp. 309–321.
52. Botstein, D.; White, R.L.; Skolnick, M.; Davis, R.W. Construction of a genetic linkage map in man using restriction fragment length polymorphisms. *Am. J. Hum. Genet.* **1980**, *32*, 314–331.
53. Knaus, B.J.; Grünwald, N.J. VCFR: A package to manipulate and visualize variant call format data in R. *Mol. Ecol. Resour.* **2017**, *17*, 44–53. [CrossRef]
54. Kamvar, Z.N.; Tabima, J.F.; Grünwald, N.J. Poppr: An R package for genetic analysis of populations with clonal, partially clonal, and/or sexual reproduction. *PeerJ* **2014**, *2*, e281. [CrossRef]
55. Wickham, H. ggplot2: Elegant Graphics for Data Analysis. 2016. Available online: https://ggplot2.tidyverse.org (accessed on 18 August 2022).
56. Saitou, N.; Nei, M. The neighbor-joining method: A new method for reconstructing phylogenetic trees. *Mol. Biol. Evol.* **1987**, *4*, 406–425. [CrossRef] [PubMed]
57. Tabima, J.F.; Knaus, B.J.; Grünwald, N.J. (n.d) GBS Analysis. Available online: https://grunwaldlab.github.io/Population_Genetics_in_R/gbs_analysis.html#:~:text=GBS%20is%20one%20of%20several,recommend%20reading%20Baird%20et%20al (accessed on 28 December 2022).
58. Simmonds, N.W. Notes on Banana Varieties in Hawaii. *Pac. Sci.* **1954**, *8*, 226–229.
59. Li, Z.; Wang, J.; Fu, Y.; Jing, Y.; Huang, B.; Chen, Y.; Wang, Q.; Wang, X.B.; Meng, C.; Yang, Q.; et al. The *Musa troglodytarum* L. genome provides insights into the mechanism of non-climacteric behaviour and enrichment of carotenoids. *BMC Biol.* **2022**, *20*, 186. [CrossRef] [PubMed]
60. Nyine, M.; Uwimana, B.; Akech, V.; Brown, A.; Ortiz, R.; Doležel, J.; Lorenzen, J.; Swennen, R. Association genetics of bunch weight and its component traits in East African highland banana (*Musa* spp. AAA group). *Theor. Appl. Genet.* **2019**, *132*, 3295–3308. [CrossRef] [PubMed]
61. Anselmo, C.A.F.; Pinheiro, A. Phylogenetic trees via Hamming distance decomposition tests. *J. Stat. Comput. Simul.* **2012**, *82*, 1287–1297. [CrossRef]
62. Volk, G.M.; Byrne, P.F.; Coyne, C.J.; Flint-Garcia, S.; Reeves, P.A.; Richards, C. Integrating Genomic and Phenomic Approaches to Support Plant Genetic Resources Conservation and Use. *Plants* **2021**, *10*, 2260. [CrossRef]
63. Wu, W.; Ng, W.-L.; Yang, J.-X.; Li, W.-M.; Ge, X.-J. High cryptic species diversity is revealed by genome-wide polymorphisms in a wild relative of banana, *Musa itinerans*, and implications for its conservation in subtropical China. *BMC Plant Biol.* **2018**, *18*, 194. [CrossRef]
64. Li, H.; Ruan, J.; Durbin, R. Mapping short DNA sequencing reads and calling variants using mapping quality scores. *Genome Res.* **2008**, *18*, 1851–1858. [CrossRef]
65. Foster, J.T.; Bull, R.L.; Keim, P. Ricin forensics: Comparisons to microbial forensics. In *Microbial Forensics*, 3rd ed.; Budowle, B., Schutzer, S., Morse, S., Eds.; Elsevier: Amsterdam, The Netherlands, 2020; pp. 241–250.
66. Phillips, C.; Amigo, J.; Tillmar, A.O.; Peck, M.A.; de la Puente, M.; Ruiz-Ramírez, J.; Bittner, F.; Idrizbegović, Š.; Wang, Y.; Parsons, T.J.; et al. A compilation of tri-allelic SNPs from 1000 Genomes and use of the most polymorphic loci for a large-scale human identification panel. *Forensic Sci. Int. Genet.* **2020**, *46*, 102232. [CrossRef]
67. Guo, C.; McDowell, I.C.; Nodzenski, M.; Scholtens, D.M.; Allen, A.S.; Lowe, W.L.; Reddy, T.E. Transversions have larger regulatory effects than transitions. *BMC Genom.* **2017**, *18*, 394. [CrossRef]
68. Choudhuri, S. Chapter 9—Phylogenetic Analysis. In *Bioinformatics for Beginners*; Choudhuri, S., Ed.; Academic Press: Cambridge, MA, USA, 2014; pp. 209–218. [CrossRef]
69. Jensen, J.L.; Bohonak, A.J.; Kelley, S.T. Isolation by distance, web service. *BMC Genet.* **2005**, *6*, 13. [CrossRef] [PubMed]
70. Jain, A.; Roorkiwal, M.; Kale, S.; Garg, V.; Yadala, R.; Varshney, R.K. InDel markers: An extended marker resource for molecular breeding in chickpea. *PLoS ONE* **2019**, *14*, e0213999. [CrossRef] [PubMed]
71. Dela Cueva, F.M.; Perez, N.A.M.; Benjamin, A.F.A.; Yanos, L.A.; Gueco, L.S.; Thomas, J.E. *Resistance of Musa balbisiana Accessions of the Philippines to Banana Bunchy Top Virus*; The American Phytopathological Society (APS): St. Paul, MN, USA, 2023. [CrossRef]

current issues in molecular biology

MDPI

Article

Genetic Dissection of Salt Tolerance and Yield Traits of *Geng* (*japonica*) Rice by Selective Subspecific Introgression

Simin Li [1,†], Ting Feng [1,†], Chenyang Zhang [1], Fanlin Zhang [1], Hua Li [1], Yanjun Chen [1], Lunping Liang [1], Chaopu Zhang [1], Wei Zeng [1], Erbao Liu [1], Yingyao Shi [1], Min Li [1,*] and Lijun Meng [2,*]

1 College of Agronomy, Anhui Agricultural University, Hefei 230036, China; lisimin@stu.ahau.edu.cn (S.L.); 20720156@stu.ahau.edu.cn (T.F.); zhangchenyang@stu.ahau.edu.cn (C.Z.); 21721689@stu.ahau.edu.cn (F.Z.); 2550361252@stu.ahau.edu.cn (H.L.); chen21720131@stu.ahau.edu.cn (Y.C.); liang2505568381@stu.ahau.edu.cn (L.L.); zchaopu@163.com (C.Z.); isaac777@ahau.edu.cn (W.Z.); liuerbao@ahau.edu.cn (E.L.); shiyingyao@ahau.edu.cn (Y.S.)

2 Kunpeng Institute of Modern Agriculture at Foshan, Foshan 528200, China

* Correspondence: limin@ahau.edu.cn (M.L.); menglijun@caas.cn (L.M.)

† These authors contributed equally to this work.

Abstract: Salinity is a major factor limiting rice productivity, and developing salt-tolerant (ST) varieties is the most efficient approach. Seventy-eight ST introgression lines (ILs), including nine promising lines with improved ST and yield potential (YP), were developed from four BC_2F_4 populations from inter-subspecific crosses between an elite *Geng* (*japonica*) recipient and four *Xian* (*indica*) donors at the Institute of Crop Sciences, Chinese Academy of Agricultural Sciences. Genome-wide characterization of donor introgression identified 35 ST QTLs, 25 of which harbor 38 cloned ST genes as the most likely QTL candidates. Thirty-four are *Xian-Geng* differentiated ones with the donor (*Xian*) alleles associated with ST, suggesting differentiated responses to salt stress were one of the major phenotypic differences between the two subspecies. At least eight ST QTLs and many others affecting yield traits were identified under salt/non-stress conditions. Our results indicated that the *Xian* gene pool contains rich 'hidden' genetic variation for developing superior *Geng* varieties with improved ST and YP, which could be efficiently exploited by selective introgression. The developed ST ILs and their genetic information on the donor alleles for ST and yield traits would provide a useful platform for developing superior ST and high-yield *Geng* varieties through breeding by design in the future.

Keywords: salt tolerance; selection; correlated phenotypic responses; QTL clusters; rice

Citation: Li, S.; Feng, T.; Zhang, C.; Zhang, F.; Li, H.; Chen, Y.; Liang, L.; Zhang, C.; Zeng, W.; Liu, E.; et al. Genetic Dissection of Salt Tolerance and Yield Traits of *Geng* (*japonica*) Rice by Selective Subspecific Introgression. *Curr. Issues Mol. Biol.* **2023**, *45*, 4796–4813. https://doi.org/10.3390/cimb45060305

Academic Editors: Shimeles Tilahun and Vijai Bhadauria

Received: 30 April 2023
Revised: 13 May 2023
Accepted: 22 May 2023
Published: 31 May 2023

1. Introduction

Rice is one of the most important staple crops feeding half the world's population. It is expected that the world population will continue to grow and exceed nine billion by 2050, which demands a nearly 70% increase in food production [1]. Primarily grown in Asia, rice crops suffer a wide range of abiotic stresses such as salt, submerge, and extreme temperatures [2]. Among these abiotic stresses, soil salinization is an alarming issue, as it affects more than 6% of the agricultural land worldwide, particularly along the coastal areas of rice-growing countries. Furthermore, the threat of salinity to rice production is expected to increase as the sea level rises from global warming [3,4]. The damaging effects of salt stress on rice plants usually include decreased photosynthesis, inhibited plant growth, biomass loss, partial sterility, and thus varied degrees of yield losses [5,6]. Most rice varieties are sensitive to salt, and their sensitivities change with age, most acute at the seedling (2–3 leaf) age, moderately tolerant at the tillering stage, and highly sensitive at the reproductive stage [7,8]. Different rice accessions are known to vary considerably in their salt tolerance (ST). The use of salinity-resistant rice varieties is an important topic, especially in organic farming, with regard to the reduction of synthetic products and chemical fertilizers and to increase product quality, such as the 'Green Super Rice (GSR)' which is based on the results of functional genomics research, improving the new varieties with less pesticide and fertilizer, water saving and drought/salt tolerance, high quality and yield were used to realize the fundamental transformation of crop production mode and promote the green development of agriculture [9]. Thus, developing ST rice varieties is the most effective approach to increasing rice productivity, particularly in the coastal areas of Asia.

In the past decades, most breeding efforts for developing ST rice varieties have used relatively few ST accessions as ST donors in their crossing schemes with limited success. This is largely because ST in rice is a quantitative trait controlled by multiple genes and involves complex biological and genetic mechanisms [10,11]. To facilitate the genetic improvement of rice ST, considerable signs of progress have been made in identifying genes and quantitative trait loci (QTLs) controlling ST and its components in rice using bi-parental segregating populations or random germplasm populations in the past two decades, resulting in the identification of hundreds of ST QTLs across the rice genome [11,12]. However, past research on the genetic and molecular dissection of rice ST has been primarily focused on ST at the seedling stage. This is largely because young rice seedlings tend to be more sensitive to salt [13] and partially because it is easier to evaluate rice ST at the seedling stage. Although rice suffers more yield reduction from salt stress at the reproductive stage, phenotyping rice ST at the reproductive stage in genetic and breeding research has been considered to be more time-consuming and laborious [14,15]. More importantly, rice ST at the seedling stage is poorly correlated with its ST at the tillering and reproductive stages, indicating different sets of genes involved in ST at different developmental stages [8,16]. Among the few reported studies on rice ST at the reproductive stage, Mohammadi et al. (2013) reported the identification of 35 QTLs for 11 traits associated with ST at the reproductive stage in an F_2 population of cross Sadri/FL478, most of which were found to be novel [17]. Mondal et al. (2022) also reported the identification of 40 ST QTLs in a large BC_1F_2 population [18]. In both cases, ST QTLs were identified in the F_2 or BC_1F_2 population, in which the phenotypic effects of the ST QTLs were determined based on their associations with yield-related traits under salt stress. Thus, it remains unknown if these QTLs affect the same traits under non-stress conditions, which is of great interest to breeders. *Saltol*, a major QTL on chromosome 1 for ST at the seedling stage [19], is known to be associated with an ST component, the Na^+/K^+ ratio, at the reproductive stage [20]. An additional ST QTL gene, SKC1, was reported to function by maintaining K^+ homeostasis in rice [16,21]. Researchers identified several QTLs on rice chromosomes 1, 7, 8 and 12 for ST at the reproductive stage using a recombinant inbred population. Furthermore, (Takehisa et al., 2004) reported 17 ST QTLs identified in BC_1F_9 to BC_1F_{12} populations of Nipponbare/Kasalath [22]. Two of these QTLs showed strong epistatic effects. Moreover, (Ammar et al., 2009) reported 25 QTLs for 17 traits, including seedling salt-injury score, Na^+, K^+, Cl^- concentrations and Na^+/K^+ ratio in the leaf and stem at the vegetative and reproductive stages using an $F_{2:3}$ population of CSR27/MI48. Meanwhile, considerable progress has been made in the functional genomic research of rice, which has resulted in the cloning and functional characterization of at least 50 genes related to rice ST [1,20]. In contrast, theoretical advances in genome mapping and functional genomic research have not yet been widely utilized in the genetic improvement of rice ST, except for the successful transfer of *Saltol* into some mega-rice varieties by marker-assisted introgression [11], in which FL478 was almost exclusively used as the ST donor. However, the introgression of single large-effect genes/QTLs from a few highly tolerant donors for improving rice ST is not the best choice to resolve the salinity problem, given the large numbers of ST loci involved and possible epistasis among the identified ST loci [23]. Lack of information on the genetic control of rice ST at the reproductive stage also limits our efforts in developing ST rice varieties. Fortunately, an efficient strategy, "selective introgression", that integrates introgression of useful genes/alleles from diverse germplasm accessions by BC breeding and strong phenotypic selection with marker-assisted tracking and characterization of donor alleles for target and non-target traits in the selected BC progeny has been well-demonstrated for simultaneous improvement and allelic mining of abiotic stresses in rice [24].

We report here the presence of a rich but largely 'hidden' genetic variation for rice ST at the vegetative and reproductive stages, yield potential in the rice subspecific gene pools, and demonstrate a powerful strategy to efficiently exploit and dissect this valuable source of genetic variation using the strategy of selective introgression, a complex process involving random introgression of diverse donor segments into an elite genetic background through backcrossing, strong phenotypic selection plus replicated progeny testing to identify promising BC progenies with desirable target traits and genetic tracking of donor segments and gene/QTL identification of selected BC progeny by DNA markers [24].

2. Results

2.1. Selection Efficiencies for Improving ST

The salt stress during the initial screening severely hindered the growth and development of the recipient, CY1, and caused a severe reduction in SF (17.1%) and GYP (2.4 g). However, we were able to select 126 BC_2F_4 plants which had significantly higher SF (>32.3%) and GYP (>4.7 g) than CY1, from the 2500 plants of the four BC_2F_4 populations, including 24 plants from population CY1/Bg90-2, 30 plants from population CY1/X21, 30 plants from population CY1/X22, 29 plants from CY1/Q5, and 13 plants from population CY1/SN265, respectively (Supplementary Table S1).

ANOVA of the data from the replicated progeny testing of the 126 selected BC_2F_5 ILs under salt stress and non-stress conditions indicated that the variance components between the salt treatments, among different populations, among ILs within each of the populations and ILs × stress treatment were all highly significant for all measured traits. In the progeny testing, CY1 (check) suffered severe reductions in all yield traits, including 19.1 g (84.5%) reduced GYP, 23.5% (38.3%) reduced SF, 65.0 (56.7%) reduced FGN, 39.7 (29.9%) reduced SNP, 1.6 g (16.1%) reduced TGW, 3.6 (50.0%) reduced PN, 38.8 cm (31.2%) reduced height, and 11.7 days (11.1%) delayed heading. Comparatively, the 126 selected ILs suffered much less with an average 18.5 g (79.0%) reduced GYP, 27.3% (32.5%) reduced SF, 65.0 (49.3%) reduced FGN, 39.7 reduced SNP (25.2%), 1.6 g (6.7%) reduced TGW, 2.8 reduced PN (41.0%), 36.9 cm (27.2%) reduced height, and only three-day (2.9%) delayed heading.

Of the 126 selected BC_2F_5 ILs, 80 lines showed significantly ($p < 0.005$) higher GYP than CY1 under salt stress in the replicated progeny testing (Table S2), including 17 ILs from population CY1/Bg90-2 with an average GYP of 10.3 g, 21 ILs from population CY1/X21 with an average GYP of 9.1 g, 18 ILs from population CY1/X22 with an average GYP of 9.1 g, 22 ILs from population CY1/Q5 with an average GYP of 9.4 g, and only two ILs from the intra-subspecific population of CY1/SN265, which out-yielded CY1 by 48.6–194.3% (Table S2). Notably, the selection efficiency for improving ST was 9.75 times higher in the four inter-subspecific populations with *Xian* donors (3.9%) than in the intra-subspecific population (0.4%), indicating that the *Xian* gene pool contained rich useful genetic diversity for improving ST of *Geng* cultivars.

2.2. Phenological Mechanisms Underlying Rice ST

When the results of the progeny testing were carefully examined (Table S2), three interesting observations were noted regarding the phenotypic responses in the yield traits of ST ILs to salt stress. First, when compared to CY1, the significantly improved ST (GYP under salt stress) of most ST ILs was associated with significantly higher SF/PH/TGW and less delayed heading under salt stress, though the higher SF/PH and heading delay varied considerably in magnitude among ST ILs from different populations. Secondly, more dramatically reduced SNP was observed in most ST ILs, except those from population CY1/Bg90-2, compared to CY1. Thirdly, ST ILs from population CY1/Q5 showed less reduced SF and PN but were accompanied by more dramatically reduced height under salt stress compared with CY1. These results suggested the presence of different physiological and genetic mechanisms for the improved ST of different ILs. Under salt stress, GYP was positively correlated with PH and its direct components, SNP, PN, and SF, but not with TGW and negatively correlated with HD (Table S3). A positive correlation also existed between PH and SNP and PH and HD, while HD negatively correlated with SF in the ST ILs. Under normal conditions, a positive correlation was observed between GYP and PN or SNP, between HD and PH or PN, while a negative correlation was observed between HD and SF.

2.3. Identification and Mapping of ST QTLs

Based on the genotypic data of 210 polymorphic markers, the overall introgression (heterozygosity) in the initially selected 126 BC_2F_4 plants was 0.108 (0.013), 0.137 (0.069), and 0.122 (0.027) in populations CY1/Bg90-2, CY1/X21, and CY1/X22, respectively, largely fitting the Mendelian expectations (Table S4), but was 0.216 (0.086) in population CY1/Q5, significantly higher than the Mendelian expectations. To understand the genetic basis of ST in the selected ILs, we adopted the strategy of selective introgression (Zhang et al., 2021 [22]) to detect the donor genomic segments that were strongly associated with the significantly improved ST in the 78 ST ILs from the four inter-subspecific populations. Because only two lines from cross CY1/SN265 were confirmed to have significantly improved ST, this population was not used for genetic mapping because of the low power in QTL detection.

Based on a threshold of $p < 0.001$ in the χ^2 tests against the average whole genome introgression of respective populations, we were able to detect 35 ST QTLs on 11 rice chromosomes, except for chromosome 12 (Figure 1 and Supplementary Table S4). These included 15 ST QTLs from population CY1/Bg90-2, which showed an average donor introgression of 0.483, 4.5 times as much as the whole genome introgression of 0.108; nine ST QTLs from population CY1/X21, which showed an average donor introgression of 0.465, 3.4 times as much as the whole genome introgression of 0.137; 17 ST QTLs from population CY1/X22 which showed an average introgression of 0.482 or four times as much as the whole genome introgression of 0.122; and 17 ST QTLs from population CY1/Q5 which showed an average introgression of 0.494 or 2.3 times as much as the whole genome introgression of 0.216. Important ST QTLs included *qST3.4* and *qST7.1*, which were detected in all four populations, plus *qST2.5* and *qST3.2*, which were detected in three populations. Thirteen additional ST QTLs were detected in two of the four populations, while the remaining 17 were detected in a single population.

We searched the genomic regions 5 cM flanking each (the peak marker) of the identified ST QTL regions and found that 25 of the 35 ST QTL regions harbor 38 cloned genes reported to be associated with ST and/or other abiotic stress tolerances (Tables S4 and S7). Clearly, these are potential candidate genes for the identified ST QTLs. To find additional evidence supporting these genes as candidates for the identified ST QTLs, we performed comprehensive gene CDS haplotype (gcHap) analyses for 31 ST genes in the 3000 diverse rice accessions (Wang et al. 2018 [25]) and the five parents (Figures 2 and S2). Table S7 shows the gcHap diversity (E_H) and the number of major gcHaps (gcHapN) in different rice populations and the parents, which varied considerably for different ST genes. Of these ST genes, only four genes (*DSM3, OsCIPK15, OsHDT1*, and *OsCIPK15*) are conserved ones with relatively low E_H and gcHaoN, while the remaining 34 ST genes are all of *Xian-Geng* differentiated type with each subspecies having one predominant (allelic frequency >70%) but different gcHap. For 31 of the ST loci with the parental gcHap information, the recipient (CY1) and donors have different gcHaps at 28 ST genes, except for *SNAC3, OsMYB48-1*, and *OsAKT2*. At all 28 ST gene loci, the donor (*Xian*) gcHaps were associated with ST, while the CY1 (*Geng*) alleles were associated with salt susceptibility. However, it was noted that in 17 of these cases, the same donor alleles (gcHaps) at the ST genes were not detectable in all populations, even though >2 of the donors had the same gcHaps. For example, all four donors have Hap3 at *OsHAK5*, but it was detected as an ST QTL only in CY1/X21. In nine cases (*OsMAPK33, OsAPX8, GnT1, OsPEX11-1, OsSRO1C, OsAPX1, OsHAK16, qSE3*, and *OsWRKY08*), more than one donor gcHap was associated with ST.

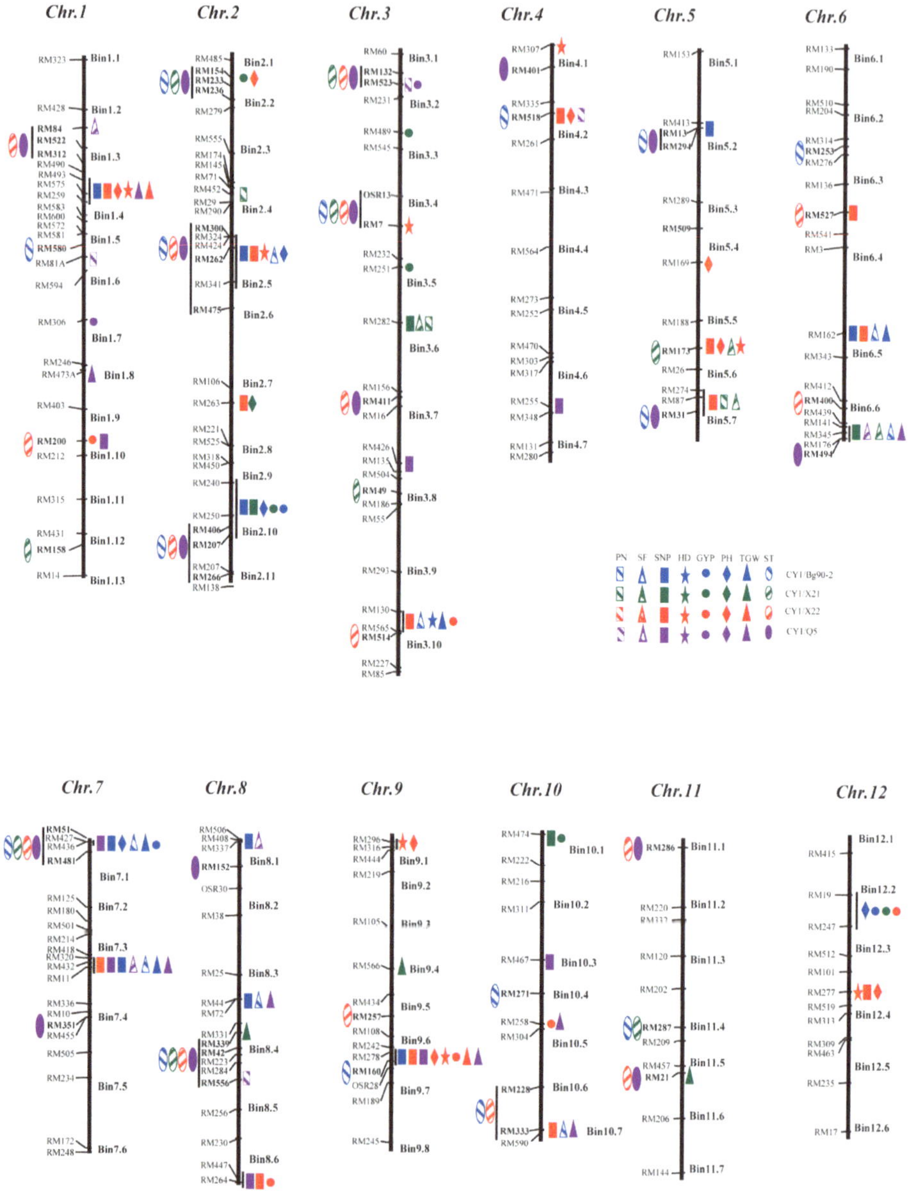

Figure 1. Genomic distribution of 35 loci (on the left side of rice chromosomes) for salt tolerance (ST) and 102 QTLs (on the right side of chromosomes) identified by positive selection in 78 ST introgression lines (ILs) and those ILs selected for high-trait values of seven yield traits from four BC_2F_4 populations derived from inter-subspecific crosses between a *Geng* (*japonica*) variety, Chaoyou1 (CY1), and four *Xian* (*indica*) donors. At all these QTLs, the donor alleles were associated with reduced trait values.

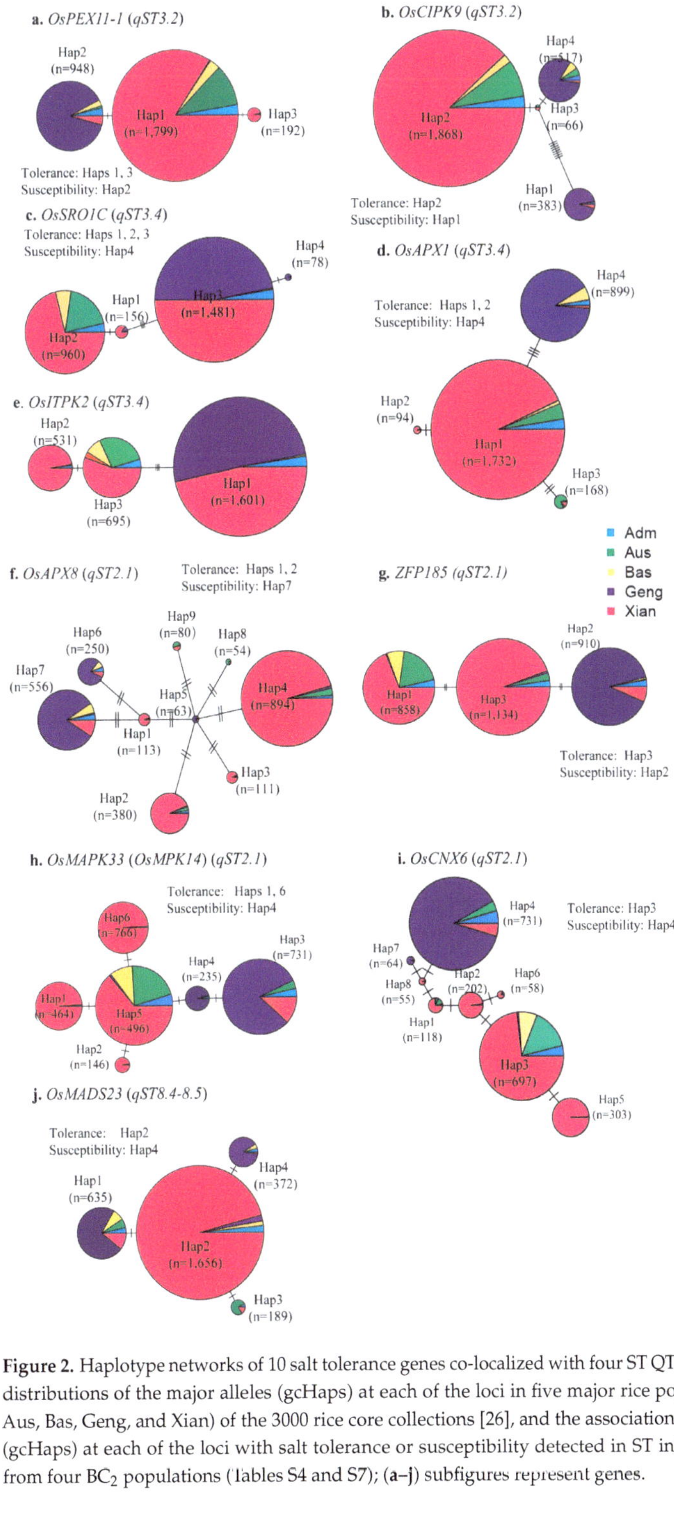

Figure 2. Haplotype networks of 10 salt tolerance genes co-localized with four ST QTLs, the frequency distributions of the major alleles (gcHaps) at each of the loci in five major rice populations (Adm, Aus, Bas, Geng, and Xian) of the 3000 rice core collections [26], and the associations of major alleles (gcHaps) at each of the loci with salt tolerance or susceptibility detected in ST introgression lines from four BC_2 populations (Tables S4 and S7); (**a**–**j**) subfigures represent genes.

Figure 2 shows the haplotype networks of the nine most likely candidate genes for four important ST QTLs identified in ≥3 of the BC populations. For *qST3.2*, we had two candidates, *OsPEX11-1* and *OsCIPK9*. *OsPEX11-1* encodes a downstream gene in the ABA-mediated pathway for enhanced ST in rice [27]. Our gcHap analyses identified three major gcHaps at *OsPEX11-1* in rice populations, with Hap2 predominant in subspecies *Geng* and CY1 as the susceptibility allele. At the same time, Hap1 and Hap3 were predominant in subspecies *Xian* as the candidates for *qST2.1* detected in all four *Xian* donors (Figure 2a, Supplementary Table S7). *OsCIPK9* encodes calcineurin B-like protein-interacting protein kinase-9, whose non-functional mutant has enhanced ST [28]. Four major gcHaps at *OsCIPK9* are present in rice populations. CY1 carries the susceptibility Hap4 predominant in subspecies *Geng*. Differing from Hap4 by four and five synonymous mutations in the *OsCIPK9* CDS region, Hap1 and Hap2 were predominant in subspecies *Xian* and detected as the candidates for *qST2.1* in the four *Xian* donors (Supplementary Table S7). A detailed comparison between Hap4 and Haps1, 2, and 3 indicated these nonsynonymous mutations occur at exon12 of the gene, causing aa substitutions from Tyrosine to Histidine in exon12 (Figure 1).

For *qST3.4*, we identified two candidate genes, *OsSRO1C* and *OsAPX1*. *OsSRO1C* encodes a rice SRO protein regulated by *SNAC1*, and the loss-of-function mutant of *OsSRO1C* showed increased stomatal aperture and sensitivity to drought [29]. In contrast, *OsAPX1* encodes a rice cytoplasm ascorbate peroxidase, which, when transformed into Tobacco plants, causes enhanced ST [30]. We identified four major gcHaps at *OsSRO1C*, with Hap2 present only in population *Xian* and Hap4 only in population *Geng*. Haps 1, 2 and 3 were detected as ST QTLs in the four donors, while Hap4 was associated with salt susceptibility in CY1 (Figure 2c). Similarly, four major gcHaps were identified at *OsAPX1*, with Hap1 and Hap2 predominant in subspecies *Xian*, while Hap3 is present only in populations *Aus* and *Xian* for ST. Carried by CY1 for salt susceptibility and predominant in subspecies *Geng*, Hap4 differs from Hap1, Hap2 and Hap3 by three, four and five synonymous mutations of exons 7 on the *OsAPX1* CDS region (Figure 2d), which cause an amino acid (aa) substitution from Arginine to Cysteine in exon7 (Figure 1).

We identified four ST genes, *OsAPX8*, *ZFP185*, *OsMAPK33* (*OsMPK14*), and *OsCNX6*, in the region of *qST2.1* (Figure 1, Supplementary Table S4). *OsAPX8* encodes a chloroplast ascorbate peroxidase, and reduced expression of *OsAPX8* leads to enhanced ST in rice [31]. Eight major gcHaps at *OsAPX8* are present in rice populations (Figure 2e). CY1 carries Hap7, which is present in most *Geng* accessions, while Haps 1 and 2 at *OsAPX8* were candidates for *qST2.1* detected in three donors (Q5, X21, and X22) (Supplementary Table S7). ST Haps 1 and 2 were present in only 493 *Xian* and *Aus* accessions and differed from the susceptibility Hap7 by two and six nonsynonymous mutations in the *OsAPX8* CDS region. *ZFP185* encodes a C2H2 zinc-finger protein which interacts with *OsMAPK3* to enhance salt tolerance in rice [27]. We found three major gcHaps at *ZFP185* in rice populations, with Hap2 predominant in subspecies *Geng* and CY1 for salt susceptibility. Differing from Hap2 by only two nonsynonymous mutations in the gene CDS region, Hap3 is predominant in subspecies *Xian* and one candidate of *qST2.1* detected in donors Q5, X21, and X22 (Figure 2f, Supplementary Table S7). *OsMAPK33* encodes *a* mitogen-activated protein kinase whose overexpression leads to enhanced rice sensitivity to salt [32]. Six major gcHaps are present at *OsMAPK33* in rice populations, with Hap3 and Hap4 predominant in subspecies *Geng*, while Hap1, Hap 5, and Hap 6 are predominant in subspecies *Xian* (Figure 2f, Supplementary Table S7). As one candidate for *qST2.1*, ST alleles Hap1 and Hap6 in donors Q5, X21, and X22 differ from the salt susceptibility CY1 Hap4 by only two nonsynonymous mutations in the *OsMAPK33* CDS region. Encoding homologs of *MoaE* and *MoeA* for ST in rice [33], *OsCNX6* has eight major gcHaps in rice populations, with Hap3 predominant in subspecies *Xian* and Hap4 predominant in subspecies *Geng*. As one candidate for *qST2.1*, ST Hap3 in donors Q5, X21, and X22 differs from the susceptibility CY1 Hap4 by four nonsynonymous mutations in the *OsCNX6* CDS region.

Only a single candidate ST gene, *OsMADS23*, was found in the *qST8.4-8.5* region. *OsMADS23* is a MADS-box transcription factor gene reportedly to confer rice osmotic stress tolerance by regulating ABA biosynthesis [34]. With four major gcHaps at *OsMADS23* in rice populations (Figure 2f), the tolerance, Hap2, was predominant in subspecies *Xian*. Differing from Hap2 by only a single nonsynonymous mutation, the susceptibility Hap4 was present in more than 350 *Geng* accessions. Interestingly, Hap3 was predominant only in the population *Aus*, but it remains unknown for its effect on ST. Taken together, these results strongly suggested that the donor gcHaps at many of these ST genes were most likely to be the candidate genes for the identified ST QTLs.

2.4. Correlated Phenotypic Responses from Selection for ST and Mapping of QTLs for 'Non-Target' Traits

Under non-stress conditions of the progeny testing, selection for ST (GYP under salt stress) resulted in correlated responses of significantly increased GYP in 40 (51.3%) of the 78 ST ILs from the four populations by an average of 8.5 g (37.6%) over CY1 under normal conditions. Most of the correlated responses in GYP could be attributed to significantly increased SNP in 57 (73.1%) of the ST ILs, with an average SNP of 44.9 (33.5%) over CY1. This was in contrast to only eight ST ILs that showed significantly reduced GYP by an average of 7.0 g (31.0%), six of which could be attributed to significantly reduced SF under normal conditions. Similarly, correlated responses for significantly increased PH by 19.6 cm occurred in 53 (74.6%) of the ST ILs from populations CY1/X21, CY1/X22, and CY1/Q56, in contrast to only 5 (7.0%) ST ILs showing significantly reduced average PH by 10.7 cm under normal conditions. In 35 of these cases, the correlated increases in PH were associated with delayed heading (Supplementary Table S2). In the same three populations, the correlated responses by an average 7.6-day delayed heading were observed in the 48 (67.6%) ST ILs, in contrast to only 3 (4.2%) ST ILs showing significantly early heading by 4.3 days under normal conditions. Taken together, we were able to identify nine promising ILs which produced significantly higher GYP under both salt stress and non-stress normal conditions with very small changes in PH and HD (Table 1). The progeny testing identified the two best lines, A-12 and C-30, from populations CY1/Bg90-2 and CY1/X22. With yield increases of 211.4% and 77.4% under salt stress and normal conditions, line A-12 had significant improvements in all yield traits except for slightly reduced TGW by 1.7 g, plus 6.6 cm reduced height and three days of delayed heading as compared to CY1 under normal conditions. With yield increases of 297.1% and 50.9% under salt stress and normal conditions, line C-30 had significant improvements in all yield traits except for slightly reduced SF of 4.4%, 18.6 cm shorter and two days of early heading compared to CY1 (Table 1).

Table 1. Performances of nine promising CY1 introgression lines for seven yield traits under salt stress and non-stress conditions in the replicated progeny testing.

Lines	Donor	GYP (g) [1]	SF(%)	FGN	SNP	TGW (g)	PN	PH (cm)	HD (Days)
A-11	Bg90-2	29.3	81.7	150.5	184.3	22.6	7.7	126.1	108.0
A-12	Bg90-2	40.1	92.9	163.5	176.1	22.5	9.7	117.7	108.0
A-21	Bg90-2	25.9	92.5	153.4	165.9	24.3	6.7	124.5	95.0
A-3	Bg90-2	27.3	95.3	153.6	161.2	26.2	7.0	127.3	112.0
B-5	X21	29.5	92.0	120.3	130.8	23.8	9.7	115.0	108.0
C-2	X22	28.0	87.0	132.5	152.3	26.7	8.3	127.1	104.0
C-30	X22	34.1	83.1	161.0	193.7	26.6	7.7	105.7	103.0
D-10	Q5	32.9	92.5	149.6	161.7	25.4	8.7	130.8	108.0
D-9	Q5	20.0	93.0	106.8	114.8	22.5	7.7	110.6	102.0
Mean		29.7	90.0	143.5	160.1	24.5	8.1	120.5	105.3
CY1		22.6	87.5	124.0	134.2	24.2	7.2	124.3	105.3

Table 1. *Cont.*

Lines	Donor	GYP (g) [1]	SF(%)	FGN	SNP	TGW (g)	PN	PH (cm)	HD (Days)
A-11	Bg90-2	12.8	63.0	473.0	187.8	23.5	4.0	96.0	107.0
A-12	Bg90-2	10.9	68.6	386.0	140.8	26.5	4.0	89.0	108.0
A-21	Bg90-2	14.6	73.5	476.0	162.0	27.5	4.0	93.0	107.0
A-3	Bg90-2	8.4	83.9	224.0	133.5	27.0	2.0	85.0	108.0
B-5	X21	9.4	66.0	285.0	72.0	23.3	6.0	76.0	118.0
C-2	X22	11.5	78.4	373.0	95.2	22.0	5.0	84.0	115.0
C-30	X22	13.9	73.2	583.0	132.7	23.3	6.0	118.0	106.0
D-10	Q5	11.1	81.1	344.0	84.8	23.3	5.0	98.0	110.0
D-9	Q5	11.7	63.1	352.0	93.0	26.3	6.0	91.0	112.0
Mean		11.6	72.3	388.4	122.4	24.7	4.7	92.2	110.1
CY1		3.5	54	48.8	99.4	20.3	3.6	85.5	117.0

[1] GYP = grain yield per plant, SF = spikelet fertility, SNP = spikelet number per panicle, PN = panicle number per plant, TGW = 1000 grain weight, PH = plant height and HD = heading date, respectively.

2.5. Identification and Mapping of QTLs Affecting the Yield Traits under Normal Conditions

The significant amounts of residual genetic variation for the measured yield traits in the 113 ILs initially selected from the four populations under non-stress normal conditions in the progeny testing allowed us to select ILs showing significantly ($p < 0.01$) lower (negative selection) or higher (positive selection) trait values than CY1 for each of the measured traits. These included 11 low GYP ILs and 17 high GYP ILs from all four populations, eight low SF ILs from populations CY1/X22 and CY1/Q5, 11 high SF ILs from populations CY1/Bg90-2, CY1/X21, and CY1/Q5, 32 high SNP ILs from all four populations, 13 low-PN ILs populations CY1/Bg90-2, CY1/X21, and CY1/Q5, five high PN ILs from populations CY1/X21 and CY1/Q5, 11 low TGW ILs from populations CY1/Bg90-2, CY1/X21, and CY1/X22, 15 high TGW ILs from all four populations, seven low PH ILs from populations CY1/Bg90-2 and CY1/X22, 67 high-PH ILs from all four populations, six low HD ILs and 11 high HD ILs from populations CY1/Bg90-2 and CY1/X22, respectively. Using these low- and high-trait value ILs and the approach of selective introgression (Supplementary Table S4), we were able to identify and map 160 QTLs responsible for the associated responses in the seven yield traits in the selected ILs. These QTLs were mapped to 49 genomic regions across all 12 rice chromosomes, including 102 QTLs in 43 genomic regions with the donor alleles associated with increased trait values detected by positive selection and 58 QTLs in 35 genomic regions with the donor alleles associated with reduced trait values detected by negative selection (Figure 1 and Supplementary Table S5). The number of identified QTLs was 50 in population CY1/Bg90-2, 74 in population CY1/X21, 69 in population CY1/X22, and 41 in population CY1/Q5 (Figure 1; Supplementary Table S5 and Supplementary Figure S1). Nineteen of these QTLs were identified in 2–3 of the populations.

QTLs affecting GYP: Twenty-three QTLs affecting GYP were identified and mapped to 10 rice chromosomes except for chromosomes 4 and 11 (Supplementary Table S5). These included eight QTLs detected in 11 low GYP ILs with the donor alleles causing low GYP (negative selection) and 15 QTLs in 17 high GYP ILs with the donor alleles causing high GYP (positive selection) from the four populations. Of these GYP QTLs, $qGY12.2^{P}$ was detected in three populations, $qGY1.4^{N}$ and $qGY7.1^{P}$ were detected in two populations, and the remaining 22 GYP QTLs were detected in one of the populations.

QTLs affecting SF: Seventeen QTLs affecting SF were identified and mapped to nine rice chromosomes except for chromosomes 9, 11 and 12 (Supplementary Table S5). These included four QTLs detected in eight low SF ILs from populations CY1/X22 and CY1/Q5, with the donor alleles causing low SF (negative selection) and 13 QTLs in 11 high SF ILs with the donor alleles causing high SF (positive selection) from three of the four populations except for CY1/X22. Of these QTLs, $qSFY6.6^{P}$ was detected in three populations, and qSF3.2^{N} and qGY7.3^{P} were detected in two populations. The remaining 14 QTLs were detected in one of the populations.

QTLs affecting SNP: Twenty-eight QTLs affecting SNP were identified and mapped to 11 of the 12 rice chromosomes except for chromosome 11, all of which were detected in 32 high-SNP ILs from the four populations with the donor alleles increasing SNP (positive selection) at all 29 loci (Supplementary Table S5). Of these QTLs, qSNP7.3^{P} and qSNP9.6^{P} were detected in three populations. Five additional QTLs, qSNP1.4^{P}, qSNP2.5^{P}, qSNP6.5^{P}, qSNP7.1^{P}, and qSNP8.6^{P}, were detected in two of the populations, while the remaining 21 SNP QTLs were detected in one of the populations.

QTLs affecting PN: Twenty-one QTLs affecting PN were identified and mapped to 11 rice chromosomes except for chromosome 11 (Supplementary Table S5). These included 14 QTLs detected in 13 low PN ILs from populations CY1/Bg90-2, CY1/X22, and CY1/Q5 with the donor alleles causing reduced PN (negative selection) and seven QTLs in five high PN ILs with the donor alleles causing increased PN in populations CY1/X21 and CY1/Q5. In addition, two QTLs, qPN5.7^{N} and qPN10.7^{N}, were detected in two populations, while the remaining 19 PN QTLs were identified in one.

QTLs affecting TGW: Thirty QTLs affecting TGW were identified and mapped to the 12 rice chromosomes, including 16 QTLs detected in 11 low TGW ILs from populations CY1/Bg90-2, CY1/X21, and CY1/X22 plus 14 QTLs detected in 15 high TGW ILs from the four populations with the donor alleles increasing TGW (Supplementary Table S5). In addition, four QTLs, qGW1.4^{P}, qGW7.3^{P}, and qGW9.6^{P}, were detected in two populations, while the remaining 27 TGW QTLs were detected in one.

QTLs affecting PH and HD: Twenty-four QTLs affecting PH were identified and mapped to 11 of the 12 rice chromosomes except for chromosome 11, including 10 QTLs detected in seven short ILs (negative selection) plus 15 QTLs detected in 14 tall (positive selection) ILs from populations CY1/Bg90-2 and CY1/X22 (Supplementary Table S5). In addition, 57 of the 59 ST ILs from populations CY1/X21 and CY1/Q5 showed significantly increased PH. Thus all 26 ST QTLs (nine in population CY1/X21 and 17 in population CY1/Q5) showed significant over-introgression in the 57 tall ILs. Of these ST/PH QTLs identified in populations CY1/X21 and CY1/Q5, five (qPH2.1^{P}, qPH2.5^{P}, qPH2.10^{P}, qPH5.6^{P}, and qPH7.1^{P}) were also detected in populations CY1/Bg90-2 and/or CY1/X22. Fifteen QTLs affecting HD were identified and mapped to nine of the 12 rice chromosomes except for chromosomes 6, 7 and 11, including six HD QTLs detected in six early heading ILs (negative selection) plus nine QTLs detected in 11 late heading (positive selection) ILs from populations CY1/Bg90-2 and CY1/X22 (Supplementary Table S5). All these HD QTLs were detected in either population CY1/Bg90-2 or population CY1/X22.

Taken together, several observations were noted regarding the genetic bases of ST and non-target yield traits identified under normal conditions, as summarized in (Supplementary Table S6). First, ILs with increased trait values were found in 158 cases (75.9% of which also showed confirmed ST), resulting in the identification of 136 QTLs at which the donor alleles were associated with increased values of the yield traits under normal conditions. In contrast, ILs with reduced trait values were found in 56 cases (53.6% of which showed confirmed ST), from which the donor alleles were associated with reduced values of the yield traits at 64 QTLs under normal conditions. Second, PN was the only exception in which most ST ILs showed unchanged or fewer PN, with few PN QTLs detected.

It is well-known that the responses of rice plants to salinity vary at different developmental stages, but salt stress causes more severe yield losses at the reproductive stage [35]. Most previous studies focused on rice ST at either the germination and/or seedling stages, primarily because of the difficulty in phenotyping ST at the vegetative and reproductive stages. Meanwhile, breeding progress for improved ST at the reproductive stage has been slow largely because of relatively few germplasm accessions showing high levels of ST. In this study, we tried to improve the ST of CY1, a high-yield commercial but salt-sensitive *Geng* variety, using BC breeding. We started the salt stress at the tillering stage and continued until maturity in both the initial screening and progeny testing because this type of salt stress represented the real cases of most coastal saline areas in Asia where transplanted rice is the predominant way for growing rice crops. Following the previous success in improving rice tolerances to abiotic stresses such as drought, submergence, anaerobic germination, salinity at the seedling stage, heat and cold [36–42], we tried to test the hypothesis for the existence of rich genetic variation for ST in different sub-specific gene pools of rice. As expected, the development of 78 ST ILs and nine promising lines with greatly improved ST and yield potential, plus the identification of more than 200 QTLs underlying ST and yield traits achieved in this study, again demonstrated the power and efficiency of selective introgression for simultaneous improvement and genetic dissection of multiple complex traits. Specifically, our results revealed several phenological and genetic mechanisms underlying rice ST at the vegetative and reproductive stages that merit further discussion.

3. Discussion

3.1. The Phenological Mechanisms of Rice ST at the Vegetative and Reproductive Stages

Our results suggest that ST at the vegetative and reproductive stages involved complex phenological and genetic mechanisms, evidenced by the following observations. First, the improved ST of most ILs were reflected by their significantly higher PH/PN and SF/TGW than CY1 under salt stress. The former two traits indicated that these ST ILs were able to maintain relatively normal growth/development at the tillering stage under salt stress. In comparison, the latter two traits indicated that most ST ILs had relatively normal fertilization and grain filling at the reproductive stage [43,44]. Second, most ST ILs showed accelerated heading under salt stress, as compared with CY1, suggesting the possible presence of salt escape [45]. This was because the level of salt stress in the concrete ponds increased with plant development. This was due to continuous irrigation with salty water and water evaporation during plant growth. Thus, early maturing lines were expected to suffer less from salt stress. Thirdly, most ST ILs showed dramatically increased height under both salt stress and non-stress conditions, which suggested the presence of two additional ST mechanisms. One was the possible salt dilution [46] because tall ILs from populations CY1/Bg90-2, CY1/X21, and CY1/X22 had larger plant sizes under both stress and non-stress conditions, even though they showed the same height reduction under salt stress as CY1 (Supplementary Table S2). The other was reflected by the more dramatically reduced height of most tall ILs from population CY1/Q5 under salt as compared with the non-stress conditions, suggesting these ILs might have benefited from a possible ST mechanism by salt compartment in which more salt was deposited in the vegetative parts (leaves and stems) than in the reproductive parts (spikelets and grains) in addition to salt dilution. Fourthly, we observed significantly increased SNP in 32 (41%) of the ST ILs from all four populations; most also showed significantly increased height and GYP under normal conditions. In other words, improved ST in these ILs would have been achieved through their significantly increased SNP and GYP under normal conditions. This would obviously be the preferred ST mechanism by plant breeders because most of these ILs also had significantly improved yield potential under non-stress conditions. Of course, one may perceive that many ST ILs have more than one of the above-mentioned ST mechanisms [47].

3.2. Genetic Basis of Rice ST and Its Associated Yield Traits

The most important discovery of this study was the identification of 35 ST QTLs in each of which the donor (*Xian*) allele was associated with significantly improved ST in the selected ILs. These ST QTLs could be categorized into two major groups. The first group involved 25 ST QTLs, each of *which* one or more candidate genes with confirmed functions on ST were found. Thus, many of the identified ST QTLs, particularly those important ones detected in multiple populations (Figure 2), may each have resulted from the accumulated effects of multiple ST genes. Notably, 34 (90%) of the candidate ST genes showed the maximum *Xian-Geng* subspecific differentiation with one allele (gcHap) predominant in one subspecies and a different one in the other. In fact, approximately one-fifth of all rice genes belong to this type of *Xian-Geng* subspecific differentiated gene [24]. More importantly, at virtually all candidate genes with confirmed functions on ST, it is the CY1 (*Geng*) alleles that were associated with salt susceptibility, while the donor *Xian* alleles were associated with ST, consistent with the QTLs results. Moreover, the differences between the tolerance *Xian* and susceptibility *Geng* alleles involved only one to a few aa substitutions. Thus, the current knowledge that the *Xian-Geng* subspecific differentiation of rice is primarily reflected by their respective adaptations to different temperature and water environments is incomplete. In other words, the gradual adaptation of wild and ancient *Geng*-type accessions to drier and colder environments was also accompanied by reduced ST. Then, an important question arises regarding how such small allelic variation involving few *aa* substitutions between the *Xian* and *Geng* alleles at the ST gene loci would cause significant phenotypic differences in ST, which should be addressed in the future.

The second group involved QTL clusters at bins 1.4, 2.1, 2.5, 2.10, 5.7, 6.6, 7.1, and 9.6, at each of which the donor alleles for improved ST were also associated with increased GYP, SNP, SF and PH, responsible for the correlated responses to grain yield traits observed in many of the selected ST ILs. As discussed above, the donor alleles at these QTLs should be more important because their contributions to ST were achieved by improved growth rate and yield under salt. Still, they also contributed to improved yield under non-stress conditions. Empirical evidence from past rice QTL cloning efforts suggests that these QTL clusters with large effects on multiple related traits tend to harbor important regulatory genes, which should be the targets in future rice functional genomics research.

Another important result of this study was the identification of an additional 160 QTLs for seven yield traits in the ST ILs under non-stress conditions, including 58 (in 65 cases) QTLs with the donor alleles for reduced trait values detected by negative selection and 102 QTLs (in 119 cases) with the donor alleles for increased trait values detected by positive selection (Figure 1 and Supplementary Table S5). In this study, at least three factors could have contributed to the high power and efficiency in identifying QTLs affecting complex traits. First, the 500 individuals in the original BC_2F_4 populations from subspecific introgression contained very large genetic variation for ST and the measured yield traits. Thus, the 113 ILs selected from the initial salt screening of 2000 BC_2F_4 plants were still segregating at large numbers of loci for the seven yield traits. Notably, most of the identified QTLs were clustered in 31 genomic regions (including 160 QTL clusters affecting ST and yield traits), each of which affected three or more of the yield traits, including several large QTL clusters in bins 1.4, 2.10–2.11, 3.10, 5.6–5.7, 6.6, 7.1, 7.3, 9.6, and 10.7, each of which was associated with >5 of the yield traits (Figure 1; Supplementary Table S5 and Supplementary Figure S1). Genetic pleiotropy was suggested to be responsible for QTL clusters at bins 1.4, 2.9, 2.10, 2.11, 3.2, 3.6, 3.10, 7.1, 7.3, 8.6, 9.6, and 12.2, each of which a GYP QTL was detected with QTLs for one or more of its components (PN, SNP, SF, and TGW) by the same marker in the same population, and in eight additional cases where a QTL for reduced height co-localized with a HD QTL for early heading, or vice versa in the same population. Possible physiological pleiotropy was suggested in seven cases of negative associations between QTLs for yield components (SNP, SF, and TGW) at bins 1.4, 2.7, 3.6, 3.8, 4.2, 5.6, and 12.4 identified in the same populations. The linkage could be responsible for the majority of the remaining detected QTL clusters. Among the QTL clusters identified, we noted six cases ($qPN2.4^P$ and $qPN2.4^N$, $qPH2.5^P$ and $qPH2.5^N$, $qGW3.10^P$ and $qGW3.10^N$, $qPN5.7^P$ and $qPN5.7^N$, $qSF6.6^P$ and $qSF6.6^N$, and $qGW3.10^P$ and $qGW3.10^N$) at each of which, QTLs affecting the same traits were detected by both positive and negative selection with the same or tightly linked markers but most in different populations (Supplementary Table S5 and Supplementary Figure S1). Genetically, this could be due to linkage drag or the possible presence of multiple alleles from different donors at the same QTLs. We believe linkage drag could be more likely responsible for most of these cases because the presence of linkage drag in the *Xian* genomes from the tight repulsive linkage of QTLs with opposite effects on rice yield traits could be relatively easily detected in the same population by positive and negative selection [2,48,49].

3.3. Implications for Improving Rice ST at The Vegetative and Reproductive Stages

Our results have important implications in breeding for improved ST at the vegetative and reproductive stages. First, all five donors in our BC breeding populations were sensitive to salt. Still, many more ST BC progeny were identified in the four populations with *Xian* donors than the one with *Geng* donors (SN265), indicating that the *Xian* gene pool contains rich, valuable genetic variation for improving ST, and much of this variation is 'hidden' because none of the donors had high levels of ST. Similar results were obtained in previous efforts to improve tolerances to cold, heat and drought [38–41]. All these results indicate that the subspecific gene pools, *Xian* or *Geng*, contain rich, valuable genetic variation not only for improving abiotic stress tolerances but also for improving yield potential for each other. Exploiting this rich 'hidden' diversity can be easily achieved by subspecific introgression through BC breeding. Notably, different donors did differ in their contributing genetic, physio- and morphological mechanisms underlying ST and yield traits, which could readily be discovered and exploited through selective introgression but not through phenotyping the parental lines. Secondly, strong phenotypic selection in the initial screening for ST using SF as the primary target trait followed by replicated progeny testing under both salt stress and non-stress conditions using yield as the target trait was powerful and essential for simultaneous improvement of ST and yield potential of rice. Moreover, large segregating breeding populations should be used in the first round of selection for ST to ensure a high selection efficiency for improving ST and yield potential, to break the possible genetic drag, to take advantage of correlated responses for increased yield traits and to overcome hybrid breakdown commonly observed in most segregating populations from inter-subspecific crosses of rice [48]. This was consistent with previously reported breeding efforts for improving tolerances to drought, cold and heat [35,36,38–41,47]. Finally, the development of nine promising lines with short stature, early heading, large SNP and GYP, in addition to excellent ST (Table 1), strongly suggested that the undesirable phenotypic association between ST and high PH and/or delayed heading observed in this study could be broken, even though it remains unknown whether tight linkage or genetic pleiotropy was the underlying genetic basis for the association. More importantly, the developed ST ILs plus their genetic/phenological constitutions in ST and yield traits provide useful information for breeders to develop better varieties in the future using novel breeding strategies such as breeding by design or designed QTL pyramiding [24,49–52].

4. Materials and Methods

4.1. Plant Materials

The materials used in this study included an elite *Geng* (*japonica*) variety, Chaoyou 1 (CY1) used as the recipient, four *Xian* (*indica*) donors from Vietnam (X21, X22, and Q5) and Sri Lanka (Bg90-2), and one *Geng* donor, Liao*Geng*265 from Northeast China. Crosses were made between CY1 and the donors to create F_1 plants, which were then backcrossed to CY1 to produce BC_1F_1 plants. Then, approximately 25 BC_1F_1 plants of each cross were backcrossed to CY1 to produce 25 BC_2F_1 lines. Selfed seeds from all ~25 BC_2F_1 lines were bulk harvested to produce a bulk BC_2F_2 population. This selfing process (without selection) continued for two more generations for all BC populations until large amounts of BC_2F_4 seeds were obtained for each BC population (Table 1).

4.2. Screening BC_2F_4 Progeny for Salt Tolerance at the Vegetative and Reproductive Stages

In the normal rice growing season (May to October), seeds of the BC_2F_4 segregating populations were sown in the seedling nursery, and 500 seedlings (30-day-old) of each BC_2F_4 population were transplanted into the concrete tanks (3 × 6 m) at a spacing of 15 × 20 cm with two rows of CY1 as the check in each tank in the screening house at the Rice Research Institute, Tianjin Academy of Agricultural Sciences (TAAS). The transplanted seedlings were allowed to grow under normal freshwater irrigation for 15 days. For the salt treatment, we mixed approximately equal amounts of salt water from two wells with a water salt (NaCl) content of 1.6% in one well and 0.3% in the other. The salt content of the mixed water was adjusted to ~0.8% (~140 mmol·L^{-1} NaCl). Then, the mixed water was used to irrigate all the concrete tanks every 3–5 days from 15 days after transplanting until maturity. Under this type of salt treatment, the soil salt level in the ponds was expected to increase slowly from evaporation during the growth period of rice plants. However, this increase varied over time from outside weather conditions, which resembled the salt stress in most natural conditions and thus met our primary purpose in developing salt-tolerant rice varieties and identifying genes/QTLs for salt tolerance. At maturity, a total of 126 BC_2F_4 plants showing significantly higher seed setting and yield than CY1 (the susceptible recurrent parent) were visually selected for measuring panicle number (PN), spikelet and filled grain numbers per panicle (SNP and FGN), and grain yield per plant (GYP, g). The number of selected salt-tolerant (ST) plants ranged from 13 from population CY1/SN265 to 50 from population CY1/X21 (Table 1).

The progeny testing of the selected 126 BC_2F_5 lines was conducted in the next normal season (May–Oct) under the same stress conditions in TAAS and normal non-stress conditions in the experimental farm of Institute of Crop Sciences, Chinese Academy of Agricultural Sciences. Under both experiments under salt stress and normal conditions, seeds of the 126 BC_2F_5 lines were sown in the seedling nurseries, and 30-day-old seedlings of each line were transplanted into a single-row plot with 11 plants in each plot at a spacing of 15 × 20 cm. The plots were arranged in a complete randomized block design with two replications for each of the selected ST lines. Four plots of CY1 were randomly inserted into the experiment under stress and non-stress conditions. The field arrangements and management of the experiments were the same for the stress and non-stress conditions except for the salt treatment, which was the same as that for the initial salt treatment in the previous season described above. Five representative plants in each plot were measured for heading date (days) and plant height (cm) in the field. At maturity, three representative plants from each plot were harvested and dried in an oven and then measured for PN, FGN, SNP, spikelet fertility (SF% = FGN/SNP × 100), 1000-grain weight (TGW, g), and GYP.

4.3. The Genotyping Experiment

Bulked fresh leaf tissues from each BC_2F_4 ILs were collected from the pooled leaf tissues of individual BC_2F_4 ILs from the seedling nursery. DNA was extracted using the CTAB method [53]. The parental polymorphism was surveyed using more than 600 common single sequence repeat (SSR) microsatellite markers (https://archive.gramene.org/markers/), and 210 evenly distributed polymorphic SSR markers were used for genotyping the 126 BC_2F_5 lines initially selected from the salt screening.

4.4. Statistical Analyses

ANOVA was used to summarize the variation of each trait among different ILs (G) between the stress treatments (S) and G × S interactions using the R software (http://www.r-project.org/). The R software was used for correlation analysis of different traits under the same or different treatments.

4.5. Identifying QTLs Affecting ST and Yield Traits under Non-Stress Normal Conditions

QTLs affecting ST (target trait) and non-target traits measured under non-stress conditions were identified using the data from replicated progeny testing described above and the strategy of selective introgression [24,48]. Specifically, for identifying ST QTLs, the replicated progeny testing confirmed that 78 of the ILs initially selected from four populations with *Xian* donors showed significantly ($p < 0.01$) higher GYP under salt stress, which were used for identifying ST QTLs. According to the population genetics theory [54,55], the confirmed ST ILs were expected to show significant over-introgression at loci associated with ST. Thus, χ^2 tests were performed to scan the whole genome to identify marker loci that showed significant ($p < 0.001$) deviation in genotypic frequencies in the ST ILs of a specific population from the whole genome donor introgression of the population, assuming that at most segregating loci in the genome, the donor alleles were not associated with ST and thus could be used as the random expectation for the χ^2 tests.

To identify QTLs affecting the seven yield traits, we first select ILs with significantly ($p < 0.01$) higher or lower trait values than CY1 for the seven yield traits under non-stress conditions of the progeny testing. Then, χ^2 tests were performed to scan the whole genome to identify marker loci that showed significant ($p < 0.001$) deviation in genotypic frequencies in the low- and high-trait value ILs. Because selection was also expected to result in strong non-random associations between or among loci contributing to the selected traits [56], we performed the multi-locus independence probability tests using the genotypic data of selected high- or low-trait value ILs to identify association groups (AGs), each consisting of a group of unlinked loci (markers) but perfectly associated loci in a set of high- (GYP, SNP, and PH) or low-trait (TGW) value ILs. The multi-locus independence probability was obtained by the formula, $P_{(AG)} = (p_i)^{rm} \bullet (1 - p_i)^{r(n-m)}$, where $p_{(AG)}$ is the probability of a group of r ($r \geq 2$) unlinked loci that are perfectly associated with one another in ILs selected for high- or low-trait value ILs, p_i is the expected frequencies of the donor introgression at rth locus in the selected ILs ($i = 1, 2, 3 \ldots r$), n is the number of ILs, m is the number of the ILs that have co-introgression of the donor alleles at the r unlinked loci, and $(n - m)$ is the number of ILs having no introgression at the r unlinked loci in the AG. Theoretically, an AG could be a group of loci acting epistatically in the positively regulating signaling pathway affecting the selected trait segregating in the ILs chosen [56]. Because very few ILs with significantly higher or lower trait values than CY1, we adopted double criteria to claim an AG associated with the selected traits, i.e., the donor introgression frequency, $F_{(donor)}$, at each of the loci in an AG was significantly higher than the whole population introgression ($p < 0.05$) based on the χ^2 test and $p_{(AG)} < 0.0001$ to minimize the false negative probability.

4.6. *Candidate Genes for the ST QTLs and Their Gene CDS Haplotype (gcHap) Diversity in Rice*

To identify candidate genes for detected ST QTLs, we searched all 35 QTL regions for possible candidate genes and found 25 harbor 43 cloned ST genes reported previously (Supplementary Table S4). To understand if and how the donor alleles at these candidate ST genes contributed to ST in the CY1 ILs, we performed a gcHap analysis on the 43 candidate ST genes in the 3000 core rice germplasm accessions [24,26] and the parents. Table S5 shows the gcHap diversity of the 43 ST genes.

5. Conclusions

Our results indicated that rice ST at the vegetative and reproductive stages had multiple complex phenological mechanisms controlled by large numbers of genes/QTLs. The Xian gene pool contains rich 'hidden' genetic variation for improving ST and yield potential, which could be efficiently exploited by selective introgression, including BC breeding, strong phenotypic selection and genome-wide characterization of donor introgression by marker genotyping and analyses. The developed ST ILs, plus their genetic information on the donor alleles for ST and yield traits, would provide a useful platform for developing superior ST and high-yield Geng varieties through breeding by design in the future.

Supplementary Materials: The following supporting information can be downloaded at: https://www.mdpi.com/article/10.3390/cimb45060305/s1. Table S1: Grain yield trait values of 126 BC_2F_4 plants selected from five Chaoyou 1 (CY1) BC_2F_4 populations under salt stress from tillering to maturity. Table S2: Performance of grain yield traits of 67 salt-tolerant BC_2F_5 introgression lines selected from four BC_2F_4 populations under non-stress (N) and salt stress (S) conditions. Table S3: Correlation between yield-related traits of the 113 ILs from four populations evaluated under normal (the lower triangle) and salt stress (the upper triangle) conditions of the replicated progeny testing. Table S4: Identification of 58 QTLs for salt tolerance (ST) in 35 genomic regions (bins) across the rice genome detected in the 78 ST ILs from four BC_2 CY1 populations. Table S5: QTLs affecting seven grain yield traits identified in low-(negative) or high (positive)-trait value ILs selected from four BC_2F_5 populations. Table S6: Summary of the negative selection for low-trait value ILs and positive selection for high-trait value introgression lines from the 113 initially selected for salt tolerance from four BC_2 populations and the summary of the QTL mapping results for salt tolerance and yield traits by selective introgression. Figure S1: Genomic distribution of 58 loci (on the right side of chromosomes, Table S5) affecting seven yield traits identified by negative selection in the low-trait value introgression lines from four BC_2F_4 populations derived from inter-subspecific crosses between a Geng (japonica) variety, Chaoyou 1 (CY1), and four Xian (indica) donors. At all these QTLs, the donor alleles were associated with reduced trait values. Figure S2: Haplotype networks of 21 ST genes as candidate genes for 15 ST QTLs, the frequency distributions of major alleles (gcHaps) at each of the ST genes in five major rice populations (Adm, Aus, Bas, Geng, and Xian) of the 3000 rice core collections (Wang et al. 2018), and the associations of major alleles (gcHaps) at each of the loci with salt tolerance or susceptibility detected in ST introgression lines from four BC_2 populations (Tables S4 and S7).

Author Contributions: Designed and conceived the research, L.M.; performed phenotyping and genotyping, L.M. and Y.S.; analyzed data, M.L., S.L., T.F., C.Z. (Chenyang Zhang), F.Z., H.L., L.L., C.Z. (Chaopu Zhang), W.Z. and E.L.; data curation Y.C.; wrote the paper, S.L., C.Z. (Chaopu Zhang), M.L. and L.M. All authors have read and agreed to the published version of the manuscript.

Funding: This research was funded by the National Natural Science Foundation of China (32101768, 32000471); Natural Science Foundation of Anhui Province (2108085MC9); The Science and Technology Major Project of Anhui Province (2021d06050002), the Improved Varieties Joint Research (Rice) Project of Anhui Province (the 14th five-year plan); Postgraduate Science Research Project of Anhui Province (YJS20210249); Postgraduate entrepreneurship and Innovation Program (SC2232, SC2233).

Institutional Review Board Statement: Not applicable.

Informed Consent Statement: Not applicable.

Data Availability Statement: The original contributions presented in the study are included in the article/Supplementary Materials; further inquiries can be directed to the corresponding author/s.

Conflicts of Interest: The authors declare no conflict of interest.

References

1. Ganie, S.A.; Molla, K.A.; Henry, R.J.; Bhat, K.; Mondal, T.K. Advances in understanding salt tolerance in rice. *Theor. Appl. Genet.* **2019**, *132*, 851–870. [CrossRef] [PubMed]
2. Qin, H.; Li, Y.; Huang, R. Advances and challenges in the breeding of salt-tolerant rice. *Int. J. Mol. Sci.* **2020**, *21*, 8385. [CrossRef] [PubMed]
3. Peltier, W.R.; Tushingham, A. Global sea level rise and the greenhouse effect: Might they be connected? *Science* **1989**, *244*, 806–810. [CrossRef] [PubMed]
4. Ismail, A.; Thomson, M.; Vergara, G.; Rahman, M.; Singh, R.; Gregorio, G.; Mackill, D. Designing resilient rice varieties for coastal deltas using modern breeding tools. In *Tropical Deltas and Coastal Zones: Food Production, Communities and Environment at the Land and Water Interface*; CABI: Wallingford, UK, 2010; pp. 154–165.
5. Quan, R.; Wang, J.; Yang, D.; Zhang, H.; Zhang, Z.; Huang, R. EIN3 and SOS2 synergistically modulate plant salt tolerance. *Sci. Rep.* **2017**, *7*, 44637. [CrossRef] [PubMed]
6. Xie, Y.; Xu, P.; Huang, J.; Ma, S.; Xie, X.; Tao, D.; Chen, L.; Liu, Y.-G. Interspecific hybrid sterility in rice is mediated by OgTPR1 at the S1 locus encoding a peptidase-like protein. *Mol. Plant* **2017**, *10*, 1137–1140. [CrossRef]
7. Pearson, G.A.; Ayers, A.; Eberhard, D. Relative salt tolerance of rice during germination and early seedling development. *Soil Sci.* **1966**, *102*, 151–156. [CrossRef]
8. Noor, A.U.Z.; Nurnabi Azad Jewel, G.; Haque, T.; Elias, S.M.; Biswas, S.; Rahman, M.S.; Seraj, Z.I. Validation of QTLs in Bangladeshi rice landrace Horkuch responsible for salt tolerance in seedling stage and maturation. *Acta Physiol. Plant.* **2019**, *41*, 173. [CrossRef]
9. Wing, R.A.; Purugganan, M.D.; Zhang, Q. The rice genome revolution: From an ancient grain to Green Super Rice. *Nat. Rev. Genet.* **2018**, *19*, 505–517. [CrossRef]
10. Li, Z.-K.; Xu, J.-L. Breeding for drought and salt tolerant rice (*Oryza sativa* L.): Progress and perspectives. In *Advances in Molecular Breeding Toward Drought and Salt Tolerant Crops*; Springer: Dordrecht, The Netherlands, 2007; pp. 531–564.
11. Mondal, S.; Septiningsih, E.M.; Singh, R.K.; Thomson, M.J. Mapping QTLs for Reproductive Stage Salinity Tolerance in Rice Using a Cross between Hasawi and BRRI dhan28. *Int. J. Mol. Sci.* **2022**, *23*, 11376. [CrossRef]
12. Batayeva, D.; Labaco, B.; Ye, C.; Li, X.; Usenbekov, B.; Rysbekova, A.; Dyuskalieva, G.; Vergara, G.; Reinke, R.; Leung, H. Genome-wide association study of seedling stage salinity tolerance in temperate *japonica* rice germplasm. *BMC Genet.* **2018**, *19*, 2. [CrossRef]

13. Kumar, P.; Choudhary, M.; Halder, T.; Prakash, N.R.; Singh, V.; Vineeth, V.T.; Sheoran, S.; Longmei, N.; Rakshit, S.; Siddique, K.H. Salinity stress tolerance and omics approaches: Revisiting the progress and achievements in major cereal crops. *Heredity* **2022**, *128*, 497–518. [CrossRef] [PubMed]
14. Singh, R.; Mishra, B.; Chauhan, M.; Yeo, A.; Flowers, S.; Flowers, T. Solution culture for screening rice varieties for sodicity tolerance. *J. Agric. Sci.* **2002**, *139*, 327–333. [CrossRef]
15. Jena, K.; Mackill, D. Molecular markers and their use in marker-assisted selection in rice. *Crop Sci.* **2008**, *48*, 1266–1276. [CrossRef]
16. Calapit-Palao, C.D.O. Identification of QTL for salinity tolerance at reproductive stage in rice (*Oryza sativa* L.). Master's Thesis, University of the Philippines Los Baños, Los Baños, Philippines, 2010; p. 116.
17. Singh, R.K.; Kota, S.; Flowers, T.J. Salt tolerance in rice: Seedling and reproductive stage QTL mapping come of age. *Theor. Appl. Genet.* **2021**, *134*, 3495–3533. [CrossRef]
18. Mohammadi, R.; Mendioro, M.S.; Diaz, G.Q.; Gregorio, G.B.; Singh, R.K. Mapping quantitative trait loci associated with yield and yield components under reproductive stage salinity stress in rice (*Oryza sativa* L.). *J. Genet.* **2013**, *92*, 433–443. [CrossRef]
19. Chen, C.; Norton, G.J.; Price, A.H. Genome-wide association mapping for salt tolerance of rice seedlings grown in hydroponic and soil systems using the Bengal and Assam Aus panel. *Front. Plant Sci.* **2020**, *11*, 576479. [CrossRef]
20. Takehisa, H.; Shimodate, T.; Fukuta, Y.; Ueda, T.; Yano, M.; Yamaya, T.; Kameya, T.; Sato, T. Identification of quantitative trait loci for plant growth of rice in paddy field flooded with salt water. *Field Crops Res.* **2004**, *89*, 85–95. [CrossRef]
21. Ammar, M.; Pandit, A.; Singh, R.; Sameena, S.; Chauhan, M.; Singh, A.; Sharma, P.; Gaikwad, K.; Sharma, T.; Mohapatra, T. Mapping of QTLs controlling Na^+, K^+ and Cl^- ion concentrations in salt tolerant *indica* rice variety CSR27. *J. Plant Biochem. Biotechnol.* **2009**, *18*, 139–150. [CrossRef]
22. Zhang, G.; Liu, Y.; Gui, R.; Wang, Z.; Li, Z.; Han, Y.; Guo, X.; Sun, J. Comparative multi-omics analysis of hypoxic germination tolerance in weedy rice embryos and coleoptiles. *Genomics* **2021**, *113*, 3337–3348. [CrossRef]
23. Murray, M.; Thompson, W. Protocol of DNA isolation. *Nucl. Acids Res.* **1980**, *8*, 4321–4326. [CrossRef]
24. Shi, J.; Zhou, X.-G.; Yan, Z.; Tabien, R.E.; Wilson, L.T.; Wang, L. Hybrid rice outperforms inbred rice in resistance to sheath blight and narrow brown leaf spot. *Plant Dis.* **2021**, *105*, 2981–2989. [CrossRef] [PubMed]
25. Wang, W.; Mauleon, R.; Hu, Z.; Chebotarov, D.; Tai, S.; Wu, Z.; Li, M. Genomic variation in 3,010 diverse accessions of Asian cultivated rice. *Nature* **2018**, *557*, 43–49. [CrossRef] [PubMed]
26. Hartl, D.L.; Clark, A.G.; Clark, A.G. *Principles of Population Genetics*; Sinauer associates Sunderland: Sunderland, MA, USA, 1997; Volume 116.
27. Falconer, D.; Mackay, T. *Introduction to Quantitative Genetics*; Longman Group: Harlow, UK, 1996; Volume 12.
28. Li, Z.-K.; Fu, B.-Y.; Gao, Y.-M.; Xu, J.-L.; Ali, J.; Lafitte, H.; Jiang, Y.-Z.; Rey, J.D.; Vijayakumar, C.; Maghirang, R. Genome-wide introgression lines and their use in genetic and molecular dissection of complex phenotypes in rice (*Oryza sativa* L.). *Plant Mol. Biol.* **2005**, *59*, 33–52. [CrossRef] [PubMed]
29. Zhang, F.; Zhai, H.-Q.; Paterson, A.H.; Xu, J.-L.; Gao, Y.-M.; Zheng, T.-Q.; Wu, R.-L.; Fu, B.-Y.; Ali, J.; Li, Z.-K. Dissecting genetic networks underlying complex phenotypes: The theoretical framework. *PLoS ONE* **2011**, *6*, e14541. [CrossRef]
30. Zhang, Z.; Liu, H.; Sun, C.; Ma, Q.; Bu, H.; Chong, K.; Xu, Y. A C2H2 zinc-finger protein OsZFP213 interacts with OsMAPK3 to enhance salt tolerance in rice. *J. Plant Physiol.* **2018**, *229*, 100–110. [CrossRef]
31. Shabala, S.; Alnayef, M.; Bose, J.; Chen, Z.-H.; Venkataraman, G.; Zhou, M.; Shabala, L.; Yu, M. Revealing the role of the Calcineurin B-like protein-interacting protein kinase 9 (CIPK9) in Rice adaptive responses to salinity, osmotic stress, and K^+ deficiency. *Plants* **2021**, *10*, 1513. [CrossRef]
32. You, J.; Zong, W.; Li, X.; Ning, J.; Hu, H.; Li, X.; Xiao, J.; Xiong, L. The SNAC1-targeted gene OsSRO1c modulates stomatal closure and oxidative stress tolerance by regulating hydrogen peroxide in rice. *J. Exp. Bot.* **2013**, *64*, 569–583. [CrossRef]
33. Qingjie, G.; Qiuxiang, L. Expression of rice OsAPX1 gene and its salt tolerance in tobacco. *Mol. Plant Breed.* **2007**, *5*, 1–7.
34. Fang, T.; Dong, Y.; Li, Y.; Chen, D.; Chen, X. The role of rice chloroplast ascorbate peroxidase in drought and high salt stress. *Plant Physiol.* **2015**, *51*, 2207–2213. [CrossRef]
35. Lee, S.-K.; Kim, B.-G.; Kwon, T.-R.; Jeong, M.-J.; Park, S.-R.; Lee, J.-W.; Byun, M.-O.; Kwon, H.-B.; Matthews, B.F.; Hong, C.-B. Overexpression of the mitogen-activated protein kinase gene OsMAPK33 enhances sensitivity to salt stress in rice (*Oryza sativa* L.). *J. Biosci.* **2011**, *36*, 139–151. [CrossRef]
36. Liu, X.; Wang, J.; Yu, Y.; Kong, L.; Liu, Y.; Liu, Z.; Li, H.; Wei, P.; Liu, M.; Zhou, H. Identification and characterization of the rice pre-harvest sprouting mutants involved in molybdenum cofactor biosynthesis. *New Phytol.* **2019**, *222*, 275–285. [CrossRef] [PubMed]
37. Li, X.; Yu, B.; Wu, Q.; Min, Q.; Zeng, R.; Xie, Z.; Huang, J. OsMADS23 phosphorylated by SAPK9 confers drought and salt tolerance by regulating ABA biosynthesis in rice. *PLoS Genet.* **2021**, *17*, e1009699. [CrossRef] [PubMed]
38. Akbar, M.; Ponnamperuma, F. Saline soils of South and Southeast Asia as potential rice lands. In *Rice Research Strategies for the Future*; IRRI: Makati, Philippines, 1982; pp. 265–281.
39. Lafitte, H.; Li, Z.; Vijayakumar, C.; Gao, Y.; Shi, Y.; Xu, J.; Fu, B.; Yu, S.; Ali, A.; Domingo, J. Improvement of rice drought tolerance through backcross breeding: Evaluation of donors and selection in drought nurseries. *Field Crops Res.* **2006**, *97*, 77–86. [CrossRef]
40. Ali, A.; Xu, J.; Ismail, A.; Fu, B.; Vijaykumar, C.; Gao, Y.; Domingo, J.; Maghirang, R.; Yu, S.; Gregorio, G. Hidden diversity for abiotic and biotic stress tolerances in the primary gene pool of rice revealed by a large backcross breeding program. *Field Crops Res.* **2006**, *97*, 66–76. [CrossRef]

41. He, Y.; Zheng, T.; Hao, X.; Wang, L.; Gao, Y.; Hua, Z.; Zhai, H.; Xu, J.; Xu, Z.; Zhu, L. Yield performances of *japonica* introgression lines selected for drought tolerance in a BC breeding programme. *Plant Breed.* **2010**, *129*, 167–175. [CrossRef]
42. Meng, L.; Lin, X.; Wang, J.; Chen, K.; Cui, Y.; Xu, J.; Li, Z. Simultaneous improvement in cold tolerance and yield of temperate *japonica* rice (*Oryza sativa* L.) by introgression breeding. *Plant Breed.* **2013**, *132*, 604–612. [CrossRef]
43. Zhang, Q.; Chen, Q.; Wang, S.; Hong, Y.; Wang, Z. Rice and cold stress: Methods for its evaluation and summary of cold tolerance-related quantitative trait loci. *Rice* **2014**, *7*, 24. [CrossRef]
44. Liang, Y.; Meng, L.; Lin, X.; Cui, Y.; Pang, Y.; Xu, J.; Li, Z. QTL and QTL networks for cold tolerance at the reproductive stage detected using selective introgression in rice. *PLoS ONE* **2018**, *13*, e0200846. [CrossRef]
45. Cui, Y.; Zhang, W.; Lin, X.; Xu, S.; Xu, J.; Li, Z. Simultaneous improvement and genetic dissection of drought tolerance using selected breeding populations of rice. *Front. Plant Sci.* **2018**, *9*, 320. [CrossRef]
46. Jagadish, S.K.; Craufurd, P.Q.; Wheeler, T. High temperature stress and spikelet fertility in rice (*Oryza sativa* L.). *J. Exp. Bot.* **2007**, *58*, 1627–1635. [CrossRef]
47. Liu, C.; Chen, K.; Zhao, X.; Wang, X.; Shen, C.; Zhu, Y.; Dai, M.; Qiu, X.; Yang, R.; Xing, D.; et al. Identification of genes for salt tolerance and yield-related traits in rice plants grown hydroponically and under saline field conditions by genome-wide association study. *Rice* **2019**, *12*, 24. [CrossRef] [PubMed]
48. Khatun, S.; Flowers, T. Effects of salinity on seed set in rice. *Plant Cell Environ.* **1995**, *18*, 61–67. [CrossRef]
49. Fan, X.; Jiang, H.; Meng, L.; Chen, J. Gene mapping, cloning and association analysis for salt tolerance in rice. *Inter Natl. J. Mol. Sci.* **2021**, *22*, 11674. [CrossRef] [PubMed]
50. Khan, M.A.; Abdullah, Z. Salinity–sodicity induced changes in reproductive physiology of rice (*Oryza sativa* L.) under dense soil conditions. *Environ. Exp. Bot.* **2003**, *49*, 145–157. [CrossRef]
51. Parida, A.K.; Das, A.B. Salt tolerance and salinity effects on plants: A review. *Ecotoxicol. Environ. Saf.* **2005**, *60*, 324–349. [CrossRef] [PubMed]
52. Jaiswal, S.; Gautam, R.; Singh, R.; Krishnamurthy, S.L.; Ali, S.; Sakthivel, K.; Iquebal, M.A.; Rai, A. Harmonizing technological advances in phenomics and genomics for enhanced salt tolerance in rice from a practical perspective. *Rice* **2019**, *12*, 89. [CrossRef]
53. Weber, D. Adaptive mechanisms of halophytes in desert regions. In *Salinity and Water Stress*; Springer: Dordrecht, The Netherlands, 2009; pp. 179–185.
54. Li, Z.-K.; Zhang, F. Rice breeding in the post-genomics era: From concept to practice. *Curr. Opin. Plant Biol.* **2013**, *16*, 261–269. [CrossRef]
55. Kim, J.H.; Jang, C.S. E3 ligase, the Oryza sativa salt-induced RING finger protein 4 (OsSIRP4), negatively regulates salt stress responses via degradation of the OsPEX11-1 protein. *Plant Mol. Biol.* **2021**, *105*, 231–245. [CrossRef]
56. Ren, Z.H.; Gao, J.P.; Li, L.G. A rice quantitative trait locus for salt tolerance encodes a sodium transporter. *Nat Genet.* **2005**, *37*, 1141–1146. [CrossRef]

current issues in molecular biology

MDPI

Article

Identification of Multiple Genetic Loci Related to Low-Temperature Tolerance during Germination in Maize (*Zea maize* L.) through a Genome-Wide Association Study

Tao Yu [1,2,3,*,†], Jianguo Zhang [1,2,3,*,†], Jingsheng Cao [1,2,3], Shujun Li [1,2,3], Quan Cai [1,2,3], Xin Li [1,2], Sinan Li [1,2], Yunlong Li [1,2], Changan He [2,4] and Xuena Ma [1,2]

1 Maize Research Institute of Heilongjiang Academy of Agricultural Sciences, Harbin 150086, China; caoj.s@163.com (J.C.); cq6539@163.com (Q.C.); maize_lee@163.com (X.L.); xuena19920207@163.com (X.M.)
2 Key Laboratory of Biology and Genetics Improvement of Maize in Northern Northeast Region, Ministry of Agriculture and Rural Affairs, Harbin 150086, China
3 Key Laboratory of Germplasm Resources Creation and Utilization of Maize, Harbin 150086, China
4 Keshan Branch of Heilongjiang Academy of Agricultural Sciences, Qiqihaer 161000, China
* Correspondence: yutaoweiwei@163.com (T.Y.); zhangjianguo72@163.com (J.Z.)
† These authors contributed equally to this work.

Abstract: Low-temperature stress during the germination stage is an important abiotic stress that affects the growth and development of northern spring maize and seriously restricts maize yield and quality. Although some quantitative trait locis (QTLs) related to low-temperature tolerance in maize have been detected, only a few can be commonly detected, and the QTL intervals are large, indicating that low-temperature tolerance is a complex trait that requires more in-depth research. In this study, 296 excellent inbred lines from domestic and foreign origins (America and Europe) were used as the study materials, and a low-coverage resequencing method was employed for genome sequencing. Five phenotypic traits related to low-temperature tolerance were used to assess the genetic diversity of maize through a genome-wide association study (GWAS). A total of 14 SNPs significantly associated with low-temperature tolerance were detected ($-\log10(P) > 4$), and an SNP consistently linked to low-temperature tolerance in the field and indoors during germination was utilized as a marker. This SNP, 14,070, was located on chromosome 5 at position 2,205,723, which explained 4.84–9.68% of the phenotypic variation. The aim of this study was to enrich the genetic theory of low-temperature tolerance in maize and provide support for the innovation of low-temperature tolerance resources and the breeding of new varieties.

Keywords: abiotic stress; QTLs; low-temperature tolerance; genome-wide association study (GWAS)

Citation: Yu, T.; Zhang, J.; Cao, J.; Li, S.; Cai, Q.; Li, X.; Li, S.; Li, Y.; He, C.; Ma, X. Identification of Multiple Genetic Loci Related to Low-Temperature Tolerance during Germination in Maize (*Zea maize* L.) through a Genome-Wide Association Study. *Curr. Issues Mol. Biol.* **2023**, *45*, 9634–9655. https://doi.org/10.3390/cimb45120602

Academic Editor: Angel Llamas

Received: 25 October 2023
Revised: 13 November 2023
Accepted: 21 November 2023
Published: 29 November 2023

1. Introduction

The northern spring corn area is an important corn production area and commercial grain base in China, located at the northern end of China's golden corn belt. However, due to the special geographical location and environmental conditions, low temperatures in spring are an important source of non-biotic stress that affects the seedling quality in this area, which seriously restricts the yield and quality of the corn produced. The low-temperature tolerance of maize belongs to a quantitative trait controlled by multiple genes. In recent years, with the development of molecular biology, scholars have carried out quantitative trait locis (QTL) analyses on its low-temperature tolerance, locating maize's low-temperature tolerance on chromosomes 1–10. One QTL was located in an interval on chromosome 6, which was associated with three low-temperature tolerance traits and could explain 18.1–32.8% of the phenotypic variation [1]; twenty-six QTLs, associated with seed vigor, were detected under low temperatures during maize's germination stage on chromosomes 2, 3, 5, and 9, alongside five meta-QTLs [2]. In the 176 IBM Syn10 doubled-haploid population from the B73 × Mo17 cross, there were thirteen QTLs associated with a low-temperature germination ability, three B73 upregulated genes, and five Mo17 upregulated genes found by combining the RNA-Seq technology and QTL analysis [3]. A recombinant inbred line population (IBM Syn4 RIL) from a B73 and Mo17 cross was used to identify QTLs and investigate the genetic architecture under low-temperature conditions at a young seedling stage: two QTLs (bin 1.02 and bin 5.05) with a high additive impact were detected, which were associated with cold tolerance [4]. A total of 406 recombinant inbred lines from a multi-parent, advanced-generation, intercross population were used and, as a result, many cold tolerance-related traits were recorded: the 858 SNPs were found that were significantly associated with all traits, which indicated that most QTLs are related to chlorophyll and Fv/Fm; the authors also located most of the QTLs in specific regions, particularly bin 10.04 [5]. An F_2 population was constructed from the cross of IB030 and Mo17 to map QTLs associated with cold tolerance via QTL-seq and transcriptomic integrative analyses, and two positively regulated genes (*ZmbZIP113* and *ZmTSAH1*) that control the low-temperature germination ability were identified [6]. Scholars performed QTL mapping on an IBM (intermated B73 × Mo17) Syn10 doubled-haploid (DH) population, and twenty-eight QTLs that significantly correlated with low-temperature germination were detected, and these QTLs explained 5.4–13.34% of the phenotypic variation. In addition, six QTL clusters were produced by fourteen overlapping QTLs on every chromosome, except for chromosomes 8 and 10 [7]. The identification of molecular marker loci associated with QTLs or genes can contribute to the study of the cold-tolerance mechanism of maize and could be further used for breeding cold-tolerant inbred lines or hybrids. The QTLs controlling low-temperature tolerance during the germination stage are distributed on chromosomes 1–10, and there are few QTLs that have been consistently identified using different methods and materials, with large intervals. At the same time, there is more than one main QTL interval, so it is necessary to continue to mine consistent, main QTLs and identify candidate genes.

In recent years, the construction of reference genomes, such as B73, MO17, W22, PH207, and CML247, has enabled the widespread application of high-throughput single-nucleotide sequence markers, greatly improving the accuracy and depth of maize's whole genome sequencing and marker development. Relying on the progress of whole-genome sequencing technology and the development of whole-genome association analysis models and methods, and due to the higher level of genetic diversity in the mapping populations, GWAS has been used to analyze the variations of maize seedling and germination traits under low-temperature conditions. This is because GWAS offers increased mapping resolution and accuracy. A total of 338 cross experiments showed that some QTLs for four seedling cold-tolerance traits were detected using GWAS; thirty-two significant loci and thirty-six candidate genes related to stress tolerance were identified, suggesting that heterosis may be related to maize's cold tolerance [8]. To identify and analyze cold-tolerance traits in 306 dent inbred lines and 206 European flint inbred lines from temperate regions, indirect cold-tolerance traits such as days from sowing to germination, relative chlorophyll content, and quantum yield of photosystem II were studied. Using the GWAS technology, 49,585 SNPs were used for genotyping, and associations between SNPs and cold-tolerance genes were located in both types. A total of 275 significant associated markers were found, and some candidate genes were consistent with current studies and previous reports [9]. A GWAS of 125 maize inbred lines was studied using 10 low-temperature tolerance traits during the seedling stage and the germination stage; finally, 43 SNPs were identified as being associated with low-temperature tolerance [10]. A study conducted a GWAS on 375 inbred lines grown outdoors and in an artificial climate chamber and identified 19 markers associated with low-temperature tolerance. These markers explained 5.7% to 52.5% of the phenotypic variation in the chlorophyll fluorescence parameters during the seedling stage. The candidate genes identified near the markers were related to ethylene signaling, brassinosteroid, and lignin synthesis [11]. A study employed two cold-tolerant inbred lines, 220 and P9-10, and two susceptible lines, Y1518 and PH4CV, to generate three $F_{2:3}$ populations to detect QTLs associated with the low-temperature germination ability of seeds. Forty-three QTLs were detected, explaining 0.62% to 39.44% of the phenotypic variation. Among them, 17 QTLs explained more than 10% of the phenotypic variation, with 16 inheriting the favorable alleles from the tolerant lines. After constructing a linkage map, three meta-QTLs were identified, including at least two initial QTLs from different populations. mQTL1-1 includes seven initial QTLs for germination and seedling traits, with three explaining more than 30% of the phenotypic variation [12]. GWAS was used to conduct a germination test on 282 inbred lines and 17 loci associated with cold tolerance were identified [13]. GWAS and QTL mapping were performed on two populations; a total of four associated SNPs and twelve QTLs related to cold tolerance were identified, and the results showed that the *Zm00001d002729* gene was a potential factor, with its overexpression being able to improve the cold tolerance of crops [14]. Using GWAS, a total of 30 SNPs were identified that were related to low-temperature tolerance during seed germination, and fourteen candidate genes were found to be directly related to these SNPs; in a further study of the linkage between these candidate genes and low-temperature tolerance, ten differentially expressed genes were identified via RNA-seq analysis [15]. Fifteen significant SNPs related to seed germination were identified via GWAS under cold stress in 300 inbred lines; among them, three genomic loci were repeatedly associated with multiple traits. In further candidate gene association analysis, *Zm00001d010458*, *Zm00001d050021*, *Zm00001d010454*, and *Zm00001d010459* were identified as cold-tolerance germination-related candidate genes [16]. A total of 187 significant SNPs were identified via GWAS in 836 maize inbred lines, and there were 159 QTLs for emergence and traits related to early growth [17]. Many of the QTL and GWAS analyses have been widely used to express large variations in cold tolerance of maize, and these cited results open up new possibilities for improving cold tolerance and understanding the molecular and genetic mechanism of cold tolerance in maize. In addition, QTL mapping and GWAS can be applied as resources for conducting marker-assisted selection of cold-tolerant varieties, and we can use genomic selection technology to predict cold-tolerant varieties in large maize populations [18].

In this study, a population of 296 excellent inbred lines of maize from China and abroad was used as the study material, and their genotypes were analyzed via genome resequencing. The germination stage was then subjected to low-temperature tolerance identification in the field and laboratory, and indicators such as germination rate and germination index were detected. The TASSEL 5.0 method was used for GWAS to identify associated SNPs, aiming to provide theoretical support and material resources for the gene mining and breeding of low-temperature tolerance in maize.

2. Materials and Methods

2.1. Plant Materials

We selected 296 representative inbred lines of maize (*Zea maize* L.) from both domestic and international sources, including 232 domestic lines, 36 US lines, and 28 European lines (see Appendix A). The seeds were provided by the maize research institute of Heilongjiang Academy of Agricultural Sciences, and the seed germination rate was above 95%.

2.2. Identification of Low-Temperature Tolerance during Germination in the Field

The experiment was conducted in the period of 2017–2019 at the experimental field of Heilongjiang Academy of Agricultural Sciences. The soil in the field was calcic soil, which is neutral, flat, and uniform. The experiment was conducted in two stages. In the first stage, seeds were sown as soon as the soil temperature at 5–10 cm depth reached and remained above 5 °C, while in the second stage, seeds were sown when the soil temperature at 5–10 cm depth remained stable at or above 10 °C. After sowing, timely irrigation was carried out. A randomized block design with two rows, each 5 m in length, with 20 cm between plants and 65 cm between rows, was employed with single-seed sowing, and three replicates were used. Daily records of soil temperature, maximum and minimum temperatures in the field, and daily average temperature were noted during the experiment.

Natural low-temperature treatment was applied to the seeded plots, and the number of seedlings that germinated was recorded accurately every day. After the cessation of seedling germination, the field seedling germination rate was calculated, and the relative seedling germination rate and relative seedling germination index were determined as follows:

$$\text{germination rate (\%)} = (\text{number of germination seeds/total number of seeds}) \times 100$$

$$\text{relative germination rate (\%)} = (\text{germination rate of early sowing treatment/ germination rate of appropriate sowing treatment}) \times 100$$

$$\text{germination index} = \sum Gt/Dt \text{ (Gt represents the number of germination seeds at time t, and Dt represents the corresponding days)}$$

$$\text{relative germination index (\%)} = (\text{germination index of early sowing treatment/ germination index of appropriate sowing treatment}) \times 100$$

2.3. *Identification of Low-Temperature Tolerance during Germination in the Laboratory*

Fifty plump seeds of each inbred line were selected, surface-sterilized with 0.5% sodium hypochlorite solution for 5 min, and then rinsed three times with sterile water. The sterilized seeds were transferred onto a culture dish lined with filter paper and covered with 3 cm thick vermiculite that was kept moist; these seeds were then allowed to germinate in a low-temperature incubator under dark conditions. Two low-temperature treatment stages were set up; the first included germination at 5 °C for 7 days, followed by 15 °C for 7 days and then 25 °C for another 7 days, whereas the control was germinated at 25 °C for 21 days. The germination of seedlings with germ breaking through the vermiculite was defined as germination, and the number of emerged seedlings was recorded daily. The experiment was carried out in three replicates. The germination rate was calculated, and the relative germination rate and relative germination index were determined to be 2.2.

2.4. *Phenotypic Analysis*

Data organization and analysis were performed using Microsoft Office Excel 2016 and R version 3.6.2 [19]. Basic statistical quantities were calculated using Microsoft Office Excel. ANOVA was performed using the aov function in the R language with a random blocking model [20]. Correlation analysis was conducted using the cor function in the R language.

2.5. *Analysis of Genotype*

2.5.1. Analysis of SNPs

The modified CTAB method [21] was used to extract the genomic DNA of 296 maize inbred lines, DNA quality was detected using a NanoDrop 2000 spectrophotometer (Thermo Fisher Scientific Inc., Kanagawa, Japan) and 0.80% agarose gel electrophoresis, and qualified DNA samples were used for SNP typing. In this study, the genotype of the association analysis population was analyzed by means of genotyping by sequencing (GBS). A combination of MseI, NlaIII, and EcoRI endonucleases was used to cleave the genomic DNA of the maize inbred lines, ligate the linker, construct the library, and sequence. After obtaining the original resequencing results, mutation detection was carried out using GATK (Genome Analysis Toolkit). Clean reads were compared to the reference sequence RefGen_v4 B73 using Bowtie2, and the resulting sam file was labeled, sorted, and removed through Picard to obtain a bam file for GATK. The indel around the bam file was re-aligned using GATK, and then SNP/INDEL analysis was performed using GATK's HaplotyeCaller command. After merging all obtained vcf files, the SNP genotype data of the 296 maize inbred lines were finally obtained.

2.5.2. Analysis of Population Structure

Population structure analysis was performed using the LEA software package v3.3.2 in R [22]. First, the TASSEL 5.0 software [23] was used to remove SNP markers with rare allele frequencies (minor allele frequencies, MAFs) of less than 0.05, and the remaining SNP markers were exported in the ped format. The ped files were converted to geno- and lfmm-formatted genotyping data using the ped2geno and ped2lfmm functions of the LEA software package v3.3.2. Then, the snmf function of the software package was used to calculate the population structure. The number of subpopulations was set from 1 to 10, and each subpopulation was repeated 10 times. The cross-entropy criterion for subpopulation allocation was calculated using the cross-validation method built into the snmf function, and the appropriate number of subpopulations was selected based on this criterion. The Q-matrix was determined based on the maximum genetic similarity of each inbred line.

2.5.3. Analysis of LD

In the analysis of linkage disequilibrium (LD), the TASSEL 5.0 software [23] was first used to remove SNP markers with a minor allele frequency (MAF) of less than 0.05, and the markers were divided into 10 categories according to the 10 chromosomes and arranged based on their physical position from smallest to largest using B73 RefGen_v4 as a reference. Then, a sliding window approach was used to calculate the LD between these SNP markers, with each window consisting of 100 SNPs and sliding by 1 SNP at a time. The LD between markers was measured using r^2 [24]. After obtaining the LD between pairs of SNP markers, an LD decay plot was generated as a function of the physical distance between the markers.

2.6. Analysis of Genome-Wide Association

Genome-wide association analysis was mainly performed using TASSEL 5.0 [23]. Based on the analysis of 296 maize inbred lines, high-quality SNP markers were selected for subsequent analysis by removing SNP markers with minor allele frequencies of less than 0.05 using the TASSEL software [23]. The kinship matrix was calculated using the TASSEL software to estimate the relatedness among the 296 maize inbred lines. The first 10 principal components were calculated using the TASSEL software's PCA function as population structure parameters. The low-temperature tolerance indices of the 296 domestic and foreign elite maize inbred lines were used, together with SNP genotypes, population structures, and relatedness, to perform genome-wide association analysis using a mixed linear model in the TASSEL software. False positives resulting from multiple comparisons in the genome-wide association analysis results were controlled using the Benjamini and Hochberg method for controlling the false discovery rate, and the false discovery rate was set to 0.10 [25].

3. Results

3.1. Phenotypic Analysis of Low-Temperature Tolerance during Germination in the Field

A variance analysis was performed based on the relative germination index and field relative germination rate of 296 maize inbred lines (Table 1). The results show that there were highly significant differences in genotype, environment, and the interaction between genotype and environment for the relative germination index, with all results reaching a significance level of 0.001. For the field germination rate, there were also highly significant differences in genotype and environment, and both reached a significance level of 0.001. Overall, the relative germination index and field germination rate indicate significant differences in low-temperature tolerance among different maize inbred lines.

The phenotypic analysis results of relative germination rate and relative germination index of the 296 inbred lines under natural field conditions are shown in Table 2. The minimum, maximum, and average values of the field average germination rate for the inbred lines are 28.00%, 100.00%, and 66.62%, respectively. The number of inbred lines that fall on the right side of the mean is higher than those on the left side of the mean. In comparison, the relative seedling germination index in 2017 was similar to that in 2018, which was 86.80% and 83.15%, respectively. Overall, the distribution of the field-averaged relative seedling germination rate, the 2017 relative seedling germination index, and the

2018 relative seedling germination index vary widely, and the low-temperature tolerance variation in the inbred lines is relatively high, with a generally normal distribution.

Table 1. The ANOVA analysis of traits related to low-temperature tolerance in the field from germination to the seedling stage.

Traits	Source of Variation	Degrees of Freedom	Sum of Squares	Mean Squares	*F*-Value	*p*-Value
	error	565	72,604	129	-	-
Relative germination index	Genotype	292	191,859	657	87.98	$<2 \times 10^{-10}$ ***
	Environment	1	4243	4243	568.15	$<2 \times 10^{-10}$ ***
	Block	1	4	4	0.52	0.47
	Genotype × environment	281	195,408	695	93.12	$<2 \times 10^{-10}$ ***
	Error	574	4287	7	-	-
Relative germination rate	Genotype	293	113,856	389	1.01	0.47
	Environment	1	12,858	12,858	33.33	$<2 \times 10^{-8}$ ***
	Error	285	109,946	386	-	-

Note: '***' indicates significance at the 0.001 level.

Table 2. The low-temperature tolerance phenotype statistics from germination to the seedling stage.

Traits	Number	Min.	Max.	Mean	Median	SD	Kurtosis	Skewness
Relative germination rate	293	28.00	100.00	66.62	67.00	14.10	−0.39	−0.25
Relative germination index in 2017	291	20.69	100.00	86.80	92.68	16.65	2.78	−1.66
Relative germination index in 2018	284	4.08	100.00	83.15	89.70	20.02	0.85	−1.17

Under suitable sowing and early sowing conditions, the average seedling germination rate of the 296 inbred lines was 86.87% and 62.48%, respectively. Low-temperature stress significantly reduced the germination rate of each inbred line. There were significant differences among the 296 inbred lines in their relative germination index, which could reduce genotypic differences among the inbred lines and better reflect the differences in their cold tolerance.

3.2. Phenotypic Analysis of Low-Temperature Tolerance during Germination in the Laboratory

The results of the variance analysis of indoor relative germination rate are shown in Table 3. The differences between genotypes, environments, and the interaction between genotype and environment were highly significant, reaching a significance level of 0.001. The differences between genotypes and blocks were also highly significant, reaching a significance level of 0.001. Overall, the indoor germination rate indicates significant differences in cold tolerance among different inbred lines.

Table 3. The ANOVA analysis of traits related to low-temperature tolerance in the laboratory during the germination stage.

Traits	Source of Variation	Degrees of Freedom	Sum of Squares	Mean Squares	*F*-Value	*p*-Value
Relative germination rate	Genotype	287	478,817	1668	12.98	$<2 \times 10^{-10}$ ***
	Environment	1	70,893	70,893	551.68	$<2 \times 10^{-10}$ ***
	Block	1	2	2	0.012	0.91
	Genotype × environment	280	98,028	350	2.72	$<2 \times 10^{-10}$ ***
	Error	565	72,604	129	-	-

Note: '***' indicates significance at the 0.001 level.

The phenotypic analysis results of the indoor relative germination rate of the inbred lines are shown in Table 4. From the table, it can be seen that the average relative germination rates in 2018 and 2019 were 79.51% and 84.60%, respectively. Overall, the distribution range of the indoor relative germination rates in 2018 and 2019 was relatively large, indicating a high variation in cold tolerance among different inbred lines.

Table 4. The low-temperature tolerance phenotype statistics during the germination stage.

Traits	Number	Min	Max	Mean	Median	SD	Kurtosis	Skewness
Relative germination index in 2018	281	0.00	100.00	79.51	76.00	25.25	0.09	−0.10
Relative germination index in 2019	288	0.00	100.00	84.60	92.00	20.09	4.68	−2.12

3.3. Correlation Analysis of Low-Temperature Tolerance during Germination between Field and Indoor

The correlation analysis showed that the indoor relative germination rate in 2018 was significantly correlated with the indoor relative germination rate in 2019 and the relative germination rate in the field, with correlation coefficients of 0.67 and 0.18, respectively, and both reached a significant level of 0.001 (Table 5). The indoor relative germination rate in 2019 was significantly correlated with the relative germination rate in the field, with a correlation coefficient of 0.20, which reached a significant level of 0.001. The relative germination rate in the field was significantly correlated with the field relative germination indices in 2017 and 2018, with correlation coefficients of 0.50 and 0.49, respectively, and both reached a significant level of 0.001. Among the significantly correlated indicators, the correlation coefficient between the indoor relative germination rate in 2018 and that in 2019 was the highest, reaching 0.67, while the correlation coefficient between the indoor relative germination rate in 2018 and the relative germination rate in the field was the lowest, at 0.18.

Table 5. The correlation analysis of low-temperature tolerance during the germination stage.

Traits	Correlation Coefficient				
	Indoor Relative Germination Rate (2018)	Indoor Relative Germination Rate (2019)	Field Relative Germination Rate	Field Relative Germination Index (2017)	Field Relative Germination Index (2018)
Indoor relative germination rate (2018)	1	0.67 ***	0.18 ***	0.07	0.02
Indoor relative germination rate (2019)	0.67 ***	1	0.20 ***	0.02	0.11
Field-relative germination rate	0.18 ***	0.20 ***	1	0.50 ***	0.49 ***
Field-relative germination index (2017)	0.07	0.02	0.50 ***	1	−0.06
Field-relative germination index (2018)	0.02	0.11	0.49 ***	−0.06	1

Note: '***' indicates significance at the 0.001 level.

3.4. Analysis of Genotype

3.4.1. Analysis of SNPs

A total of 24,042 high-quality SNP markers were identified across the entire maize genome (Table 6). The identified SNPs were distributed relatively evenly across the ten chromosomes, with the highest number found on chromosome 1 (3687 SNPs), followed by chromosome 2 (3217 SNPs). The lowest number of SNPs was found on chromosome 10 (2976 SNPs). Of the 24,042 SNPs identified, 98% had a minor allele frequency greater than 0.05, and 36% had a minor allele frequency greater than 0.1. Additionally, SNPs with different minor allele frequencies were distributed relatively evenly across the ten chromosomes.

Table 6. The allele frequency characteristics of SNP markers.

Chr.	SNP	Minor Allele Frequency (MAF)					
		>0.05	Percent (%)	>0.1	Percent (%)	>0.2	Percent (%)
1	3687	3599	98	1383	38	115	3
2	3217	3159	98	1174	36	89	3
3	2896	2833	98	930	32	91	3
4	2882	2802	97	947	33	95	3
5	2384	2334	98	838	35	94	4
6	1898	1868	98	715	38	90	5
7	1577	1532	97	580	37	57	4
8	1848	1793	97	635	34	62	3
9	2294	2264	99	854	37	111	5
10	1359	1313	97	529	39	36	3
Total	24,042	23,497	98	8585	36	840	3

3.4.2. Analysis of Population Structure

We used the snmf function with a cross-validation technique to calculate and select the appropriate number of subpopulations based on the standard, and we classified the individuals based on their maximum genetic similarity (Q-matrix). When the number of subpopulations is set from 1 to 10, the cross-entropy criterion for assigning subpopulations gradually decreases, but no obvious turning point is observed (Figure 1). When the number of subpopulations is varied, the clustering of the inbred lines is clearly distinguished (Figure 2). When the number of subpopulations is set to five, the 296 excellent inbred lines are divided into five subpopulations. Subpopulations A, B, C, D, and E include 21, 22, 178, 10, and 65 inbred lines, respectively.

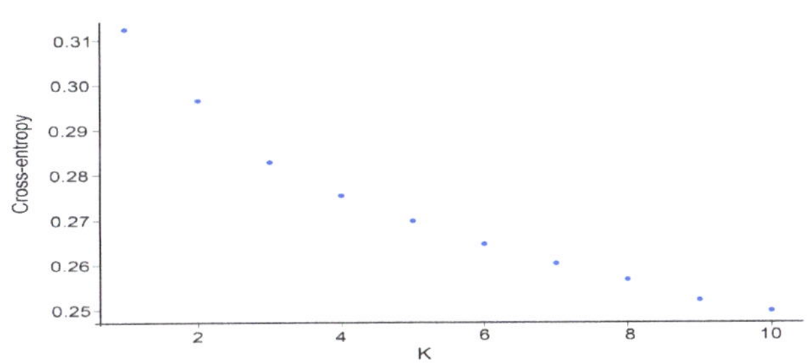

Figure 1. Variation in the clustering standard cross-entropy as the number of subgroups K increases.

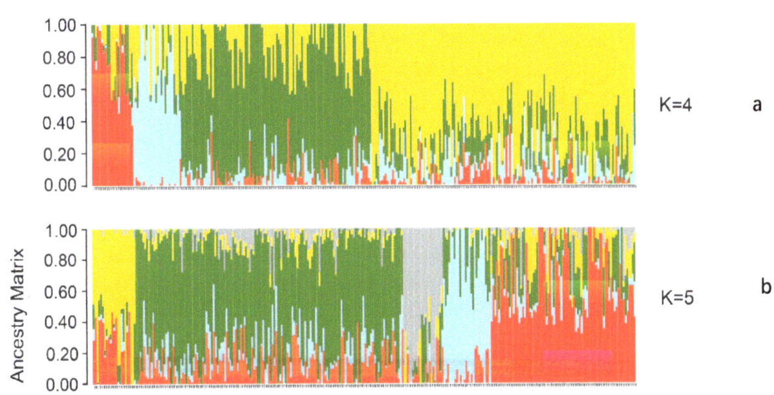

Figure 2. *Cont.*

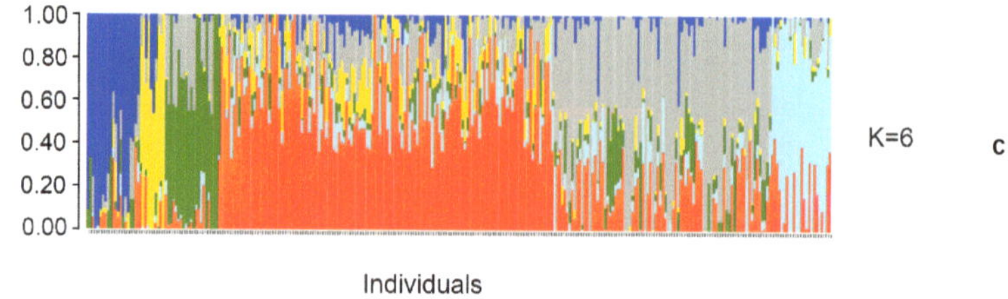

Figure 2. The population structure of subgroups from K = 4 to K = 6. Note: (**a**) K = 4; (**b**) K = 5; and (**c**) K = 6.

3.4.3. Analysis of LD

Using 23,497 SNP markers with an MAF greater than 0.05 for LD analysis, the LD r^2 between these SNP markers basically decreases with an increase in genetic distance between the markers, and all values are distributed between 0.000 and 1.000 (Table 7, Figure 3). The average value of LD between the SNP markers on chromosome 9 is the highest, at 0.122, while the average value on chromosome 7 is the lowest, at 0.052 (Table 7).

Table 7. LD and LD attenuation of maize chromosomes.

Chr.	LD Decay ($r^2 < 0.2$)	LD Decay ($r^2 < 0.1$)	Min	Max	Mean	Median	SD	Kurtosis	Skewness
1	120	410	0.000	1.000	0.062	0.008	0.152	13.871	3.605
2	320	1000	0.000	1.000	0.088	0.009	0.201	6.198	2.69
3	130	620	0.000	1.000	0.067	0.007	0.169	10.411	3.264
4	100	500	0.000	1.000	0.061	0.008	0.152	13.999	3.648
5	270	990	0.000	1.000	0.082	0.008	0.187	7.887	2.903
6	85	350	0.000	1.000	0.064	0.008	0.16	11.833	3.435
7	90	440	0.000	1.000	0.052	0.008	0.141	17.75	4.069
8	220	820	0.000	1.000	0.068	0.008	0.169	11.335	3.378
9	340	1990	0.000	1.000	0.122	0.011	0.24	2.903	2.069
10	70	340	0.000	1.000	0.07	0.01	0.165	11.326	3.351

Note: chr.1 to chr.10 represent chromosome 1 to chromosome 10 in maize, respectively.

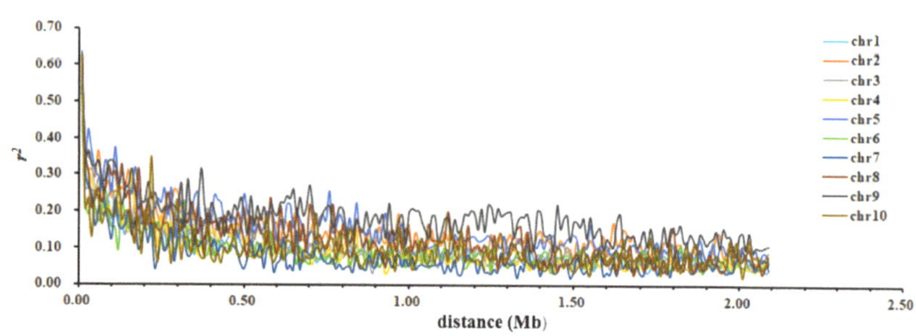

Figure 3. The attenuation in maize chromosome LD with an increase in SNP distance.

3.5. Analysis of Genome-Wide Association

3.5.1. Genome-Wide Association Analysis of Low-Temperature Tolerance in the Field

Using the MLM model in the TASSEL software at a significance threshold of $p < 1 \times 10^{-4}$, seven SNP markers associated with low-temperature tolerance were detected based on the average relative germination rate and the relative germination indices in the field in 2017 and 2018. These markers are located on chromosomes 5, 6, 7, and 10, and their phenotypic contributions range from 5.03% to 9.68% (Table 8, Figure 4). Among them, five significantly associated SNP markers were identified based on the relative germination index in 2018, including marker.17002, marker.17003, marker.17009, and marker.17105 located on chromosome 6 and marker.19874 located on chromosome 7, which explained 6.55%, 6.55%, 5.86%, 8.46%, and 7.30% of the phenotypic variation, respectively. No associated SNP loci were identified for the year 2017, which might be due to the lack of effective low-temperature stress between early sowing and sowing at the optimum time in that year.

Table 8. SNPs of maize with significant correlation with low-temperature tolerance.

Traits	Marker	Chr.	Physical Position	$-Lg\ (p)$	Contribution (%)
Field-relative germination rate	marker.14070	5	2,205,723	6.86	9.68
Relative germination index (2018)	marker.17002	6	64,236,775	4.56	6.55
Relative germination index (2018)	marker.17003	6	64,236,781	4.56	6.55
Relative germination index (2018)	marker.17009	6	64,298,566	4.15	5.86
Relative germination index (2018)	marker.17105	6	72,142,751	5.69	8.46
Relative germination index (2018)	marker.19874	7	180,326,388	5.01	7.30
Field-relative germination rate	marker.536	10	63,529,769	4.06	5.03

Note: The physical position of SNP markers was determined in reference to B73 RefGen-v4.

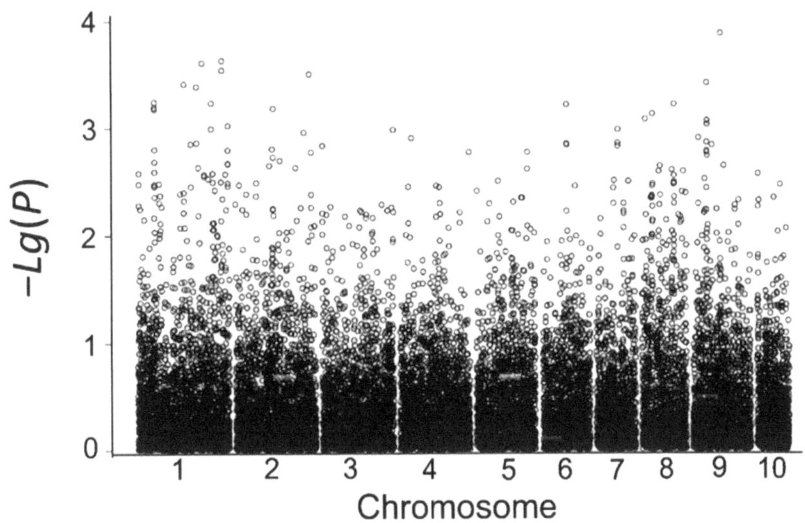

Figure 4. Correlation analysis based on the relative germination index in 2018.

3.5.2. Genome-Wide Association Analysis of Low-Temperature Tolerance in the Laboratory

Using the relative germination rate in the laboratory during the germination stage as the indicator, a total of 14 SNP loci associated with low-temperature tolerance during germination were detected; these SNP markers are located on chromosomes 1, 3, 4, 5, and 10, explaining 4.84% to 9.68% of the phenotypic variance. Among them, six significantly associated SNP markers were identified using the relative germination rate in 2018 (Table 9, Figure 5), and eight significantly associated SNP markers were identified using the relative germination rate in 2019 (Table 9, Figure 6).

Table 9. SNPs of maize with significant correlations with low-temperature tolerance.

Traits	Marker	Chr.	Physical Position	$-Lg\ (p)$	Contribution (%)
Relative germination rate (2019)	marker.1723	1	31,809,859	5.01	6.54
Relative germination rate (2018)	marker.1724	1	31,809,902	4.07	5.46
Relative germination rate (2019)	marker.1726	1	31,897,277	4.18	5.31
Relative germination rate (2019)	marker.1729	1	31,954,983	4.95	6.41
Relative germination rate (2018)	marker.8339	3	6,292,001	4.16	5.61
Relative germination rate (2019)	marker.8339	3	6,292,001	5.42	7.12
Relative germination rate (2018)	marker.8340	3	6,292,053	5.01	6.87
Relative germination rate (2019)	marker.8340	3	6,292,053	4.87	6.34
Relative germination rate (2019)	marker.12816	4	140,575,088	4.54	5.87
Relative germination rate (2018)	marker.14070	5	2,205,723	5.17	6.26
Relative germination rate (2018)	marker.190	10	22,696,941	4.13	5.56
Relative germination rate (2019)	marker.753	10	90,874,322	4.75	6.17
Relative germination rate (2019)	marker.14070	5	2,205,723	4.00	4.84
Relative germination rate (2018)	marker.843	10	100,622,715	4.52	6.10

Note: The physical position of SNP markers was determined in reference to B73 RefGen-v4.

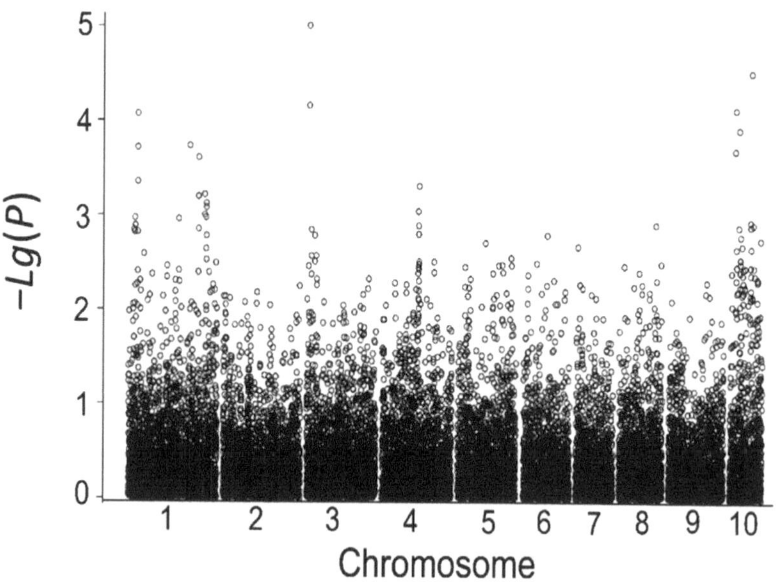

Figure 5. Correlation analysis based on the indoor relative germination index in 2018.

3.5.3. Consistency Analysis of SNP Markers Associated with Low-Temperature Tolerance

Eight significantly associated SNP markers were identified using the indoor relative germination rate in 2019, mainly distributed on chromosomes 1, 3, 4, and 10. When using the indoor relative germination rate in 2018 and the field relative germination index in 2018, five significantly associated SNP markers were identified for each. No significantly associated SNP markers were identified based on the germination index in 2017. Overall, significantly associated SNP markers were distributed on chromosomes 1, 3, 4, 5, 6, 7, and 10, with most markers on chromosome 1 (up to nine), and no significantly associated SNP markers were found on chromosomes 2, 8, and 9. The $-Lg(p)$ values of significantly associated SNP markers ranged from 4.00 to 6.86, with an average of 4.65. The phenotypic variation explained by a single SNP marker ranged from 4.84% to 9.68%, with an average of 6.13%.

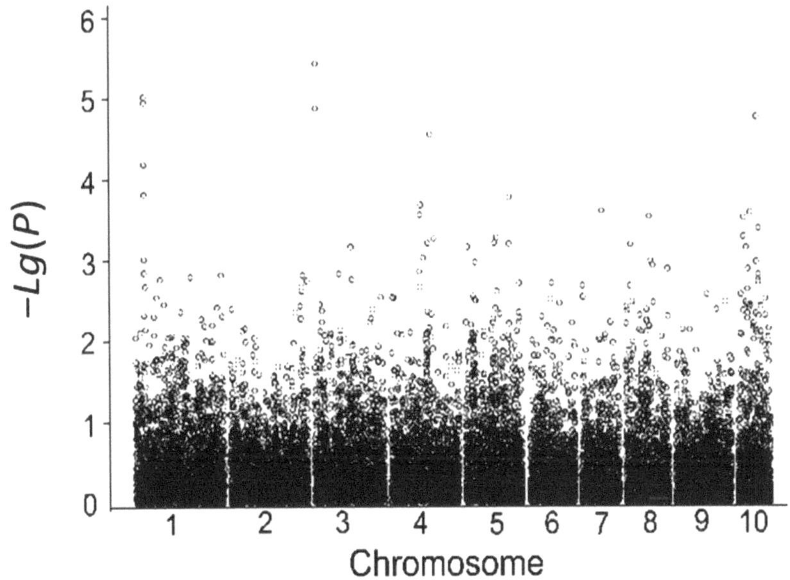

Figure 6. Correlation analysis based on the indoor relative germination index in 2019.

Significantly associated SNP markers also showed clustered distribution. Using the indoor relative germination rates in 2018 and 2019, four significantly associated SNP markers (marker.1723, marker.1724, marker.1726, and marker.1729) were identified in the interval of 31,809,859–31,954,983 on chromosome 1, with an average distance of 36.28 Kb between markers. Using the indoor relative germination rates in 2018 and 2019, a significantly associated SNP marker (marker.8339) was identified on chromosome 3 at position 6,292,001, explaining up to 5.61% of the phenotypic variation. Using the indoor relative germination rate in 2018 and the field relative germination rate in 2019, a significantly associated SNP marker (marker.8340) was identified on chromosome 3 at position 6,292,053, explaining up to 6.87% of the phenotypic variation. Using the indoor relative germination rate in 2019 and the field relative germination rate in 2019, a significantly associated SNP marker (marker.14070) was identified on chromosome 5 at position 2,205,723, explaining up to 9.68% of the phenotypic variation.

4. Discussion

In recent years, with the rapid development of sequencing technology and statistical algorithms, GWAS has become one of the most effective methods for identifying genetic variants associated with important agronomic traits in crops [26–29]. Compared to the traditional linkage analysis, GWAS can use natural populations as materials directly; it can detect more QTLs than traditional QTL mapping by using biparental populations because it uses a larger number of molecular markers and datasets from hundreds of maize inbred lines, which have a rich allelic diversity [30,31]. Moreover, GWAS can analyze multiple phenotypic traits in multiple environments and across multiple time points at the same time. Its high-throughput sequencing and high precision have greatly improved the efficiency of crop breeding [32,33]. Currently, GWAS has greatly advanced genetic research on maize functional genomics [34]; many agronomic traits such as flowering time, leaf angle, leaf size, and disease resistance have been identified in maize. For example, using 368 maize inbred lines and approximately 1 million SNPs, a GWAS analysis successfully detected 74 loci associated with seed oil content and fatty acid composition in maize [35,36]. The US-NAM population was used to detect maize flowering time variants, and a total of 90 flowering time regions were identified in the whole genome via GWAS; among them, one third of regions were associated with the environmental sensitivity of maize flowering time [37]. In another study, 513 inbred lines were used to identify 678 SNPs associated with 17 agronomic traits via GWAS, such as plant height, seed morphology, and flowering time; the results found that 54.3% of these SNPs were associated with at least two or more agronomic traits [38]. A total of 217 inbred lines were genotyped using the GBS technology, and 39 SNPs were identified to be significantly associated with fumonisin resistance in maize kernels based on GWAS analysis [39]. A panel of 143 elite lines were genotyped by using the MaizeSNP50 chip, combined with GWAS and transcriptome analysis; the results showed that 15 common quantitative trait nucleotides were associated with maize white spot, and SYN10137-PZA00131.14 was identified as a key genetic region for improving resistance to MWS; in this region, three candidate genes were identified [40].

Maize can grow in cool-temperate climates but is often exposed to cold temperatures in spring, which can affect seedling growth. Currently, although studies have shown that the growth and development of maize plants are closely related to low temperatures, the genetics of low-temperature tolerance in maize is not well understood. For example, low-temperature stress can increase the expression of related genes, resulting in the accumulation of folate in maize plants [41]. Cold stress can result in a series of physiological responses, such as the expression of osmotic stress-related genes, accumulation of ROS, activities of antioxidant enzymes, and levels of plant hormones and MDA production [42–45]; thus, plants need to stabilize cell membranes and biologically active proteins in order to survive under low-temperature conditions. However, low-temperature tolerance in maize is a complex trait because the identification and evaluation of low-temperature tolerance traits are complex and have not been standardized. Classic quantitative genetics studies have shown that low-temperature tolerance is controlled by multiple genes and is easily affected by environmental conditions. Quantitative genetic analyses of cold tolerance have shown that genotype, additive effects, growth stage, heterosis, and reciprocal and environmental factors are all involved in the expression of cold tolerance in maize [46]. Six maize lines were used to evaluate the expression of CAT, APX, SOD, and other genes; the results showed that there was heterosis for germination under cold stress, and non-additive genes were more important [47]. The studies cited above show that the genetics mechanisms of low-temperature tolerance in maize are very complicated.

Maize has rich genetic variability, a fast LD decay rate, and abundant information on SNP loci, so maize is an ideal model crop for GWAS analysis [48]. In this study, we used GWAS to identify the genetic loci associated with five traits related to low-temperature tolerance during germination. We identified 30 markers significantly associated with low-temperature tolerance, which were located on chromosomes 1, 2, 3, 4, 5, 6, 7, and 10. Two markers (marker.17002 and marker.17009) significantly associated with the relative germination index in the field in 2018 were located in bin 2.05. This interval has been mapped to several traits under different temperature conditions, including SPAD values [49], antioxidant activity under cold treatment, chlorophyll b, chlorophyll a + b, and Fv/Fm under different temperatures and sowing times [50]. Marker.7569, significantly associated with the relative germination index in the field in 2018, was located in bin 2.06 and was involved in the photosynthetic traits of the third leaf under 15 °C conditions, including CO_2 assimilation rate and ΦPSII [1]. Marker.19874, located in bin 2.08, was associated with hundred-grain weight under 14 °C/10 °C (day/night) conditions [51]. Some of these associated markers are consistent with previous studies on low-temperature tolerance, although the traits they are associated with may differ, which may be due to pleiotropy.

Eight significant SNPs related to relative germination rate were detected using the indoor relative germination rate in 2019, and five significant SNPs related to low-temperature tolerance were detected using both the indoor relative germination rate in 2018 and the field relative germination index in 2018. Overall, these significant SNPs related to low-temperature tolerance were distributed on chromosomes 1, 3, 4, 5, 6, 7, and 10, with most SNPs distributed on chromosome 1 (nine SNPs). Previous studies have also shown that

SNPs that are associated with seedling-related traits in maize under cold stress are concentrated on chromosomes 1, 2, 3, 5, 6, 8, and 10 [13,52,53]. The above research results further indicate that the cold tolerance of maize is a polygenic quantitative trait controlled by multiple genes. Using polygenic aggregation or multiple molecular markers for the genetic improvement of cold tolerance in maize is an effective strategy. No significant SNPs related to relative germination index were detected in 2017 using GWAS analysis, indicating that low-temperature tolerance in maize may be easily affected by environmental conditions, particularly climate conditions during the growing season. Some scholars have reported that the gene expression related to cold tolerance was affected by the environment of maize [54]. Under controlled conditions, the highest number of significant SNPs related to relative germination rate was detected using the indoor relative germination rate in 2019 (nine SNPs), further demonstrating that controlling low-temperature environmental conditions is important for identifying variations in low-temperature tolerance among different maize inbred lines. Therefore, future studies on identifying genes for low-temperature tolerance and improving maize varieties should focus more on phenotypic evaluations under artificial controlled low-temperature conditions.

Using the indoor relative germination rates in 2018 and 2019, four SNP markers significantly associated with low-temperature tolerance were identified in the range of 31,809,859–31,954,983 on chromosome 1, with an average distance of 36.28 Kb between markers. On chromosome 3, two SNP markers significantly associated with low-temperature tolerance were identified, namely marker.8339 and marker.8340. These markers can be directly applied or developed into easily detectable molecular markers for use in the marker-assisted selection for low-temperature tolerance in maize. In addition, some markers were found to be significantly associated with multiple traits, such as marker.1723 and marker.1724, which are only 43 bp apart and were associated with both indoor germination rates in 2018 and 2019. Marker.8339 and marker.8340 were associated with both indoor relative germination rates in 2018 and 2019, and marker.14070 was associated with both the field average relative germination rate and the 2019 indoor relative germination rate. The reason for the same markers being associated with multiple traits may be due to the strong correlation between traits, and it also indicates that the traits identified in this study are all effective for low temperature-tolerance identification. Another reason may be due to the pleiotropy of genes, where genes not only directly control the expression of a trait through the action of enzymes, but also affect many other traits through the modification of a particular trait. This requires further investigation into the function of candidate genes to avoid any negative effects of gene expression on maize breeding.

5. Conclusions

In the present study, GWAS was performed with 296 maize inbred lines, and a total of 14 SNPs significantly associated with low-temperature tolerance were detected. The SNP consistently linked to low-temperature tolerance in the field and indoors during germination was marker.14070, located on chromosome 5 at position 2,205,723, which explained 4.84–9.68% of the phenotypic variation.

Author Contributions: Conceptualization, T.Y., J.Z. and J.C.; data curation, J.Z., X.L. and Q.C.; formal analysis, T.Y., J.Z., J.C., S.L. (Shujun Li) and C.H.; funding acquisition, J.Z. and J.C.; investigation, T.Y., J.Z., C.H. and X.L.; methodology, J.Z., S.L. (Shujun Li), Y.L. and S.L. (Sinan Li); project administration, J.C.; resources, S.L. (Shujun Li) and Y.L.; software, T.Y., X.L. and C.H.; supervision, J.Z.; visualization, S.L. (Sinan Li) and Q.C.; writing—original draft preparation, T.Y.; writing—review and editing, S.L. (Sinan Li), J.C. and X.M. All authors have read and agreed to the published version of the manuscript.

Funding: This work was funded by the Innovation Project of Heilongjiang Academy of Agricultural Sciences (CX23JQ04, CX23ZD05); the Natural Science Foundation of Heilongjiang Province (LH2022C096); the National Key Research and Development Program of China (2021YFD1201001, 2021YFD1201001-2); and the Key Research and Development Program of Heilongjiang Province (JD22A010).

Institutional Review Board Statement: Not applicable.

Informed Consent Statement: Not applicable.

Data Availability Statement: The data used in this study are available within the text.

Conflicts of Interest: The authors declare no conflict of interest.

Appendix A

Table A1. The population distribution characteristics when the number of subgroups is 5.

NO.	Name	Ancestry Matrix					Subgroup
		A	B	C	D	E	
1	LX46	0.86	0	0.05	0	0.09	A
2	UH302	0.92	0	0	0.08	0	A
3	LX56	0.57	0.14	0	0	0.29	A
4	LX57	0.59	0.14	0	0	0.27	A
5	LX65	0.52	0.02	0.37	0.05	0.04	A
6	LX76	0.39	0.09	0.32	0	0.2	A
7	LX98	0.9	0.1	0	0	0	A
8	LX99	0.67	0.33	0	0	0	A
9	38P05f	0.77	0	0.07	0	0.16	A
10	38P05m	1	0	0	0	0	A
11	LX113	1	0	0	0	0	A
12	YA1M	0.9	0	0.04	0.03	0.03	A
13	LX147	0.85	0	0.06	0.03	0.06	A
14	LX148	0.88	0	0.05	0.03	0.04	A
15	LX151	0.74	0.09	0	0	0.17	A
16	LX156	0.67	0.33	0	0	0	A
17	LX162	0.37	0	0.37	0.16	0.1	A
18	LX171	0.53	0.03	0.06	0.2	0.18	A
19	UH303	0.91	0	0.05	0.04	0	A
20	LX194	0.45	0.15	0.27	0	0.13	A
21	688F	0.94	0.06	0	0	0	A
22	LX94	0	0.48	0.23	0.01	0.28	B
23	LX115	0.01	0.56	0.02	0	0.41	B
24	xy335f	0	0.97	0.03	0	0	B
25	LX132	0.01	0.44	0.15	0	0.4	B
26	LX137	0.04	0.47	0.11	0	0.38	D
27	LY88M	0	0.78	0.06	0	0.16	B
28	LX164	0	0.65	0.07	0	0.28	B
29	LX166	0	0.68	0	0	0.32	B
30	LX167	0	0.68	0.04	0.02	0.26	B
31	LX168	0	0.54	0.01	0	0.45	B
32	LX169	0	0.61	0.05	0	0.34	B
33	LX172	0.03	0.55	0	0	0.42	B
34	LX180	0	0.88	0.1	0.02	0	B
35	LX186	0	0.89	0.09	0.02	0	B
36	LX205	0	0.51	0.26	0	0.23	B
37	LX206	0.04	0.48	0.24	0.03	0.21	B
38	LY99M	0	1	0	0	0	B
39	MEIXI	0.03	0.43	0.17	0.04	0.33	B
40	101M	0.01	0.53	0	0	0.46	B
41	420F	0	1	0	0	0	B
42	738M	0	1	0	0	0	B
43	820F	0.05	0.55	0.33	0.07	0	B
44	LX1	0.05	0.04	0.49	0.09	0.33	C
45	LX2	0.03	0.05	0.62	0.16	0.14	C
46	LX3	0.01	0.06	0.64	0.2	0.09	C

Table A1. *Cont.*

NO.	Name	Ancestry Matrix					Subgroup
		A	B	C	D	E	
47	LX4	0	0.13	0.53	0.17	0.17	C
48	LX5	0.02	0.02	0.83	0.13	0	C
49	LX6	0	0.02	0.92	0.04	0.02	C
50	LX8	0.02	0.02	0.59	0.04	0.33	C
51	LX9	0	0	1	0	0	C
52	LX10	0.03	0.03	0.76	0.05	0.13	C
53	LuYuan92	0.02	0.03	0.72	0.08	0.15	C
54	LX12	0.25	0	0.45	0.22	0.08	C
55	LX13	0.02	0.05	0.53	0.05	0.35	C
56	LX14	0.02	0.05	0.59	0.04	0.3	C
57	LX15	0.03	0.09	0.69	0.05	0.14	C
58	LX17	0.02	0.03	0.69	0.04	0.22	C
59	LX18	0.1	0.03	0.44	0.3	0.13	C
60	H261	0	0.02	0.74	0.05	0.19	C
61	LX19	0.05	0.05	0.61	0.05	0.24	C
62	LX20	0	0.01	0.61	0.05	0.33	C
63	LX21	0.19	0.08	0.37	0.07	0.29	C
64	Q319	0.04	0.06	0.71	0.05	0.14	C
65	LX23	0.02	0.05	0.73	0.18	0.02	C
66	LX25	0.1	0.08	0.6	0.07	0.15	C
67	LX95	0	0.05	0.74	0.21	0	C
68	LX27	0	0	0.86	0	0.14	C
69	LX29	0.04	0.04	0.66	0.07	0.19	C
70	LX30	0	0.08	0.56	0.29	0.07	C
71	LX32	0	0	0.88	0.12	0	C
72	LX33	0	0.07	0.57	0.31	0.05	C
73	LX34	0	0.02	0.75	0.16	0.07	C
74	LX35	0.1	0.05	0.45	0.26	0.14	C
75	LX36	0.27	0.01	0.4	0.21	0.11	C
76	LX37	0.04	0.06	0.46	0.28	0.16	C
77	LX38	0	0	0.97	0	0.03	C
78	LX39	0	0	0.82	0.18	0	C
79	LX40	0.03	0	0.71	0	0.26	C
80	LX41	0	0	0.54	0.15	0.31	C
81	LX42	0.01	0.05	0.47	0.03	0.44	C
82	LX43	0	0	0.87	0.13	0	C
83	LX45	0.02	0.03	0.78	0.08	0.09	C
84	B317	0	0.01	0.84	0.08	0.07	C
85	LX48	0.01	0.02	0.77	0.11	0.09	C
86	B144	0.02	0.04	0.47	0.04	0.43	C
87	Si287	0	0	0.84	0.16	0	C
88	7-004	0	0	0.66	0	0.34	C
89	SD190	0.01	0.05	0.78	0.08	0.08	C
90	LX49	0.02	0.03	0.73	0.01	0.21	C
91	LX50	0	0	0.89	0	0.11	C
92	LX52	0.05	0	0.69	0.05	0.21	C
93	LX53	0	0	0.9	0.05	0.05	C
94	LX55	0.01	0.05	0.62	0.16	0.16	C
95	LX59	0.06	0.02	0.72	0.05	0.15	C
96	LX60	0.02	0.04	0.62	0.08	0.24	C
97	LX61	0.03	0.03	0.73	0.12	0.09	C
98	LX62	0	0	0.76	0.24	0	C
99	LX63	0	0.08	0.79	0.1	0.03	C
100	LX64	0	0.01	0.75	0.2	0.04	C
101	LX66	0.23	0	0.42	0.13	0.22	C

Table A1. *Cont.*

NO.	Name	Ancestry Matrix					Subgroup
		A	B	C	D	E	
102	LX67	0.05	0.07	0.53	0.08	0.27	C
103	LX68	0.02	0.04	0.33	0.29	0.32	C
104	LX69	0.02	0.01	0.55	0.29	0.13	C
105	LX71	0.02	0	0.87	0.04	0.07	C
106	LX74	0	0.21	0.58	0	0.21	C
107	LX75	0.04	0.18	0.41	0.05	0.32	C
108	LX82	0.03	0.01	0.48	0.22	0.26	C
109	LX83	0	0	0.74	0.22	0.04	C
110	LX84	0	0	0.8	0.19	0.01	C
111	LX85	0.03	0.06	0.49	0.02	0.4	C
112	LX86	0	0	0.93	0.07	0	C
113	LX87	0	0	0.95	0.05	0	C
114	LX90	0.02	0.04	0.47	0.06	0.41	C
115	LX92	0	0	0.74	0.26	0	C
116	LX93	0	0	0.66	0.3	0.04	C
117	LX100	0	0	0.53	0.29	0.18	C
118	LX102	0	0.03	0.86	0.09	0.02	C
119	LM33M	0.03	0.05	0.61	0.07	0.24	C
120	LX105	0.05	0	0.88	0.05	0.02	C
121	LX106	0.01	0.1	0.81	0.05	0.03	C
122	LX107	0.06	0	0.89	0.05	0	C
123	LX112	0.07	0.14	0.47	0.09	0.23	C
124	LX114	0.06	0.12	0.4	0.02	0.4	C
125	Huangzao4	0	0	0.86	0.14	0	C
126	Longxi53	0.04	0.04	0.41	0.13	0.38	C
127	706	0.08	0.05	0.71	0.11	0.05	C
128	LX117	0.01	0.01	0.96	0.02	0	C
129	LX118	0.03	0.1	0.72	0.06	0.09	C
130	LX119	0	0.03	0.78	0.14	0.05	C
131	K10	0	0.09	0.47	0.35	0.09	C
132	Chang3	0.02	0.03	0.6	0.13	0.22	C
133	Zhong7490-92	0.01	0.04	0.76	0.18	0.01	C
134	78599	0.03	0	0.91	0.06	0	C
135	478	0	0	0.89	0	0.11	C
136	He344	0	0.01	0.75	0.18	0.06	C
137	Longkang11	0	0	1	0	0	C
138	MO17	0.04	0.17	0.56	0.03	0.2	C
139	4F1	0	0	1	0	0	C
140	330	0	0	0.9	0.1	0	C
141	5003	0	0	0.98	0.02	0	C
142	B73	0.03	0.01	0.96	0	0	C
143	LX120	0.07	0.03	0.66	0.19	0.05	C
144	LX121	0.02	0.02	0.87	0.07	0.02	C
145	LX122	0	0.04	0.69	0.27	0	C
146	LX123	0.03	0	0.83	0.11	0.03	C
147	LX124	0	0	0.99	0.01	0	C
148	06S021	0	0	0.97	0	0.03	C
149	06S032	0	0	1	0	0	C
150	06S034	0.13	0.01	0.78	0.05	0.03	C
151	06S052	0	0	0.53	0.01	0.46	C
152	06S060	0	0.01	0.84	0	0.15	C
153	06S068	0	0	0.92	0	0.08	C
154	06S075	0	0	0.77	0.14	0.09	C
155	06S095	0	0.04	0.71	0.1	0.15	C
156	R117	0.01	0	0.91	0	0.08	C

Table A1. *Cont.*

NO.	Name	Ancestry Matrix					Subgroup
		A	B	C	D	E	
157	FLAF	0.13	0.01	0.48	0.22	0.16	C
158	YAM	0.07	0.08	0.66	0.09	0.1	C
159	LX125	0	0.05	0.81	0.14	0	C
160	Jidan27♂	0.01	0.01	0.49	0.04	0.45	C
161	LX126	0	0	0.59	0.01	0.4	C
162	LX127	0	0.03	0.75	0.04	0.18	C
163	LX128	0	0.02	0.83	0.13	0.02	C
164	LX129	0	0	0.98	0.02	0	C
165	LX130	0.01	0	0.67	0.01	0.31	C
166	LX131	0.03	0.13	0.61	0.08	0.15	C
167	Zheng58	0.01	0	0.87	0	0.12	C
168	Chang7-2	0	0.01	0.8	0.14	0.05	C
169	HY6M	0.18	0	0.49	0.19	0.14	C
170	HY6F	0.04	0.03	0.77	0.06	0.1	C
171	LX134	0.26	0	0.64	0.08	0.02	C
172	LX140	0	0	0.87	0	0.13	C
173	LX144	0.03	0.09	0.51	0.06	0.31	C
174	Co117-2	0.1	0.04	0.53	0.19	0.14	C
175	Co220	0.09	0.02	0.47	0.3	0.12	C
176	Co228	0.15	0.03	0.4	0.3	0.12	C
177	Co266	0.09	0.04	0.44	0.31	0.12	C
178	Co274	0.07	0.09	0.55	0.15	0.14	C
179	Co285	0.07	0.02	0.51	0.08	0.32	C
180	Co358	0.13	0.02	0.48	0.21	0.16	C
181	Co372	0.07	0.02	0.49	0.27	0.15	C
182	Co373	0.1	0.06	0.47	0.28	0.09	C
183	Co380	0.08	0.03	0.46	0.19	0.24	C
184	LY88F	0	0	0.8	0.1	0.1	C
185	LX149	0	0	0.94	0.06	0	C
186	LX150	0	0	0.81	0.08	0.11	C
187	LX152	0.02	0	0.93	0	0.05	C
188	LX153	0.06	0.02	0.71	0.09	0.12	C
189	LX154	0.01	0	0.87	0.12	0	C
190	LX155	0	0.02	0.83	0.15	0	C
191	H127RE	0.04	0.04	0.64	0.12	0.16	C
192	LX163	0.05	0.07	0.82	0	0.06	C
193	DY1	0	0.01	0.88	0.11	0	C
194	DY10	0.01	0.03	0.53	0	0.43	C
195	DY21	0.07	0.12	0.44	0.03	0.34	C
196	DY53	0.01	0	0.86	0.13	0	C
197	DY71	0.02	0.02	0.8	0.16	0	C
198	DY97	0	0.01	0.63	0	0.36	C
199	DY99	0.03	0.03	0.5	0.22	0.22	C
200	Dy13-17	0.14	0.03	0.46	0.2	0.17	C
201	LX176	0.03	0.02	0.95	0	0	C
202	LX177	0	0.06	0.9	0.04	0	C
203	LX178	0.03	0.06	0.57	0.06	0.28	C
204	LX181	0	0.07	0.59	0.06	0.28	C
205	LX184	0.01	0.03	0.67	0.01	0.28	C
206	LX185	0.04	0.22	0.48	0.04	0.22	C
207	LX188	0.03	0.26	0.37	0.02	0.32	C
208	LX190	0.09	0.17	0.4	0.02	0.32	C
209	LX193	0	0.02	0.95	0.03	0	C
210	LX195	0	0.01	0.91	0.08	0	C
211	LX196	0.07	0.01	0.77	0.03	0.12	C
212	LX197	0	0.01	0.75	0.01	0.23	C

Table A1. *Cont.*

NO.	Name	Ancestry Matrix					Subgroup
		A	B	C	D	E	
213	LX198	0	0.01	0.77	0.01	0.21	C
214	LX199	0.02	0	0.76	0	0.22	C
215	LX201	0.04	0.02	0.49	0.02	0.43	C
216	LX202	0.05	0	0.44	0.08	0.43	C
217	LX209	0.13	0.06	0.67	0.08	0.06	C
218	LX210	0.13	0.13	0.54	0.06	0.14	C
219	698F	0.04	0.04	0.68	0.01	0.23	C
220	820M	0.13	0.14	0.37	0.05	0.31	C
221	YA2M	0.27	0.09	0.47	0.06	0.11	C
222	LX26	0	0.03	0.17	0.69	0.11	D
223	UH004	0	0	0	1	0	D
224	FLAM	0	0.04	0.24	0.58	0.14	D
225	YAF	0	0	0	0.97	0.03	D
226	Co374	0.21	0.03	0.2	0.45	0.11	D
227	YA1F	0	0.02	0	0.94	0.04	D
228	LX159	0.36	0	0	0.64	0	D
229	LX160	0.47	0	0	0.53	0	D
230	LX161	0.21	0.1	0.06	0.52	0.11	D
231	UH009	0	0.01	0	0.93	0.06	D
232	Ji871	0	0	0.39	0.1	0.51	E
233	LX11	0.04	0.03	0.42	0.08	0.43	E
234	LX16	0	0.27	0.25	0.02	0.46	E
235	LX22	0	0	0.12	0.03	0.85	E
236	LX24	0	0.22	0	0	0.78	E
237	LX31	0.03	0.03	0.41	0.04	0.49	E
238	LX51	0.01	0	0.36	0.05	0.58	E
239	LX54	0	0	0.21	0	0.79	E
240	LX72	0.34	0.03	0.09	0	0.54	E
241	LX73	0.05	0.26	0.25	0.07	0.37	E
242	LX77	0.16	0.11	0.26	0.03	0.44	E
243	LX78	0.14	0.18	0.06	0	0.62	E
244	LX79	0.15	0.19	0.04	0	0.62	E
245	LX80	0.06	0.11	0.21	0.01	0.61	E
246	LX88	0.19	0.11	0.28	0	0.42	E
247	LX89	0.03	0.02	0.34	0.02	0.59	E
248	LX91	0	0	0.09	0.11	0.8	E
249	LX101	0.06	0.25	0.27	0.04	0.38	E
250	LX104	0	0	0.13	0.01	0.86	E
251	LX108	0	0.1	0	0	0.9	E
252	LX109	0.02	0.17	0.24	0.02	0.55	E
253	LX110	0.13	0.01	0.03	0.04	0.79	E
254	LX111	0.02	0.05	0.33	0.05	0.55	E
255	LX116	0	0.02	0.01	0	0.97	E
256	Lv28	0	0	0	0	1	E
257	LX124	0.38	0.01	0.21	0	0.4	E
258	xy335m	0	0.09	0.02	0	0.89	E
259	LX133	0	0.41	0.16	0.01	0.42	E
260	LX135	0	0.33	0.09	0.01	0.57	E
261	LX136	0	0.43	0.12	0.01	0.44	E
262	LX138	0	0.34	0.23	0.02	0.41	E
263	LX139	0.06	0.23	0.23	0.03	0.45	E
264	LX141	0.02	0.08	0.37	0.06	0.47	E
265	LX142	0	0	0	0	1	E
266	LX143	0	0	0	0	1	E
267	LX145	0	0	0	0.02	0.98	E
268	LX146	0.01	0.02	0.14	0.01	0.82	E

Table A1. *Cont.*

NO.	Name	Ancestry Matrix					Subgroup
		A	B	C	D	E	
269	Co371	0.07	0.09	0.28	0.17	0.39	E
270	LX165	0.3	0.04	0.3	0.01	0.35	E
271	LX170	0.02	0.41	0	0	0.57	E
272	DY7	0	0	0	0	1	E
273	DY29	0.01	0	0	0.01	0.98	E
274	DY36	0.05	0.03	0.14	0.02	0.76	E
275	DY49	0	0.12	0.2	0	0.68	E
276	LX173	0	0.07	0.09	0.02	0.82	E
277	LX174	0	0	0	0	1	E
278	LX175	0	0	0.04	0	0.96	E
279	LX179	0	0.09	0.34	0.05	0.52	E
280	LX182	0.42	0.09	0	0.02	0.47	E
281	LX183	0.06	0.1	0.24	0.2	0.4	E
282	LX187	0.01	0.03	0.36	0.07	0.53	E
283	LX191	0.04	0.23	0.27	0.04	0.42	E
284	LX192	0.18	0.26	0.24	0.01	0.31	E
285	LX200	0	0	0.45	0.04	0.51	E
286	LX203	0	0	0	0	1	E
287	LX204	0	0	0	0	1	E
288	LX207	0	0	0	0	1	E
289	LX208	0	0.07	0.18	0.02	0.73	E
290	252M	0.01	0.06	0.21	0.01	0.71	E
291	335MG	0	0.04	0	0	0.96	E
292	688M	0	0.04	0	0	0.96	E
293	XZD276	0	0.1	0	0	0.9	E
294	XZD170	0	0.08	0	0	0.92	E
295	XZD171	0.02	0.13	0.31	0.02	0.52	E
296	YA2F	0	0.33	0	0	0.67	E

References

1. Fracheboud, Y.; Ribaut, J.M.; Vaigas, M.; Messmer, R.; Stamp, P. Identification of quantitative trait loci for cold tolerance of photosynthesis in maize. *J. Exp. Bot.* **2002**, *376*, 1967–1977. [CrossRef]
2. Shi, Y.; Li, G.H.; Tian, Z.Q.; Wang, Z.Y.; Wang, X.B.; Zhu, Y.G.; Chen, Y.H.; Guo, S.L.; Qi, J.S.; Zhang, X.; et al. Genetic dissection of seed vigour traits in maize (*Zea mays* L.) under low-temperature conditions. *J. Genet.* **2016**, *95*, 1017–1022. [CrossRef] [PubMed]
3. Han, Q.; Zhu, Q.; Shen, Y.; Lee, M.; Lübberstedt, T.; Zhao, G. QTL Mapping Low-Temperature Germination Ability in the Maize IBM Syn10 DH Population. *Plants* **2022**, *14*, 214. [CrossRef] [PubMed]
4. Goering, R.; Larsen, S.; Tan, J.; Whelan, J.; Makarevitch, I. QTL mapping of seedling tolerance to exposure to low temperature in the maize IBM RIL population. *PLoS ONE* **2021**, *16*, e0254437. [CrossRef]
5. Yi, Q.; Malvar, R.A.; Álvarez-Iglesias, L.; Ordás, B.; Revilla, P. Dissecting the genetics of cold tolerance in a multiparental maize population. *Theor. Appl. Genet.* **2020**, *133*, 503–516. [CrossRef] [PubMed]
6. He, R.Y.; Zheng, J.J.; Chen, Y.; Pan, Z.Y.; Yang, T.; Zhou, Y.; Li, X.F.; Nan, X.; Li, Y.Z.; Cheng, M.J.; et al. QTL-seq and transcriptomic integrative analyses reveal two positively regulated genes that control the low-temperature germination ability of MTP-maize introgression lines. *Theor. Appl. Genet.* **2023**, *136*, 116. [CrossRef] [PubMed]
7. Xu, Q.; Wang, X.; Wang, Y.; Zhang, H.; Zhang, H.; Di, H.; Zhang, L.; Dong, L.; Zeng, X.; Liu, X.; et al. Combined QTL mapping and RNA-Seq pro-filing reveal candidate genes related to low-temperature tolerance in maize. *Mol. Breed.* **2022**, *42*, 33. [CrossRef] [PubMed]
8. Yan, J.; Wu, Y.; Li, W.; Qin, X.; Yue, B. Genetic mapping with testcrossing associations and $F_{2:3}$ populations reveals the importance of heterosis in chilling tolerance at maize seedling stage. *Sci. Rep.* **2017**, *7*, 427–428. [CrossRef]
9. Revilla, P.; Rodríguez, V.M.; Ordás, A.; Rincent, R.; Charcosset, A.; Giauffret, C. Association mapping for cold tolerance in two large maize inbred panels. *BMC Plant Biol.* **2016**, *16*, 127. [CrossRef]
10. Huang, J.; Zhang, J.; Li, W.; Hu, W.; Duan, L.; Feng, Y.; Qiu, F.Z.; Yue, B. Genome-wide association analysis of ten chilling tolerance indices at the germination and seedling stages in maize. *J. Integr. Plant Biol.* **2013**, *55*, 10. [CrossRef]
11. Strigens, A.; Grieder, C.; Haussmann, B.I.G.; Melchinger, A.E. Genetic variation among inbred lines and testcrosses of maize for early growth parameters and their relationship to final dry matter yield. *Crop Sci.* **2012**, *52*, 1084. [CrossRef]

12. Li, X.H.; Wang, G.H.; Fu, J.J.; Li, L.; Jia, G.Y.; Ren, L.S.; Lubberstedt, T.; Wang, G.Y.; Wang, J.H.; Gu, R.L. QTL mapping in three connected populations reveals a set of consensus genomic regions for low temperature germination ability in *Zea mays* L. *Front. Plant Sci.* **2018**, *9*, 65–72. [CrossRef]
13. Hu, G.H.; Li, Z.; Lu, Y.C.; Li, C.X.; Gong, S.C.; Yan, S.Q.; Li, G.L.; Wang, M.Q.; Ren, H.L.; Guan, H.T.; et al. Genome-wide association study identified multiple genetic loci on chilling resistance during germination in maize. *Sci. Rep.* **2017**, *7*, 10840. [CrossRef] [PubMed]
14. Jin, Y.; Zhang, Z.; Xi, Y.; Yang, Z.; Xiao, Z.; Guan, S.; Qu, J.; Wang, P.; Zhao, R. Identification and Functional Verification of Cold Tolerance Genes in Spring Maize Seedlings Based on a Genome-Wide Association Study and Quantitative Trait Locus Mapping. *Front. Plant Sci.* **2021**, *12*, 776972. [CrossRef]
15. Zhang, H.; Zhang, J.; Xu, Q.; Wang, D.; Di, H.; Huang, J.; Yang, X.; Wang, Z.; Zhang, L.; Dong, L.; et al. Identification of candidate tolerance genes to low-temperature during maize germination by GWAS and RNA-seqapproaches. *BMC Plant Biol.* **2020**, *20*, 333. [CrossRef] [PubMed]
16. Zhang, Y.; Liu, P.; Wang, C.; Zhang, N.; Zhu, Y.; Zou, C.; Yuan, G.; Yang, C.; Gao, S.; Pan, G.; et al. Genome-wide association study uncovers new genetic loci and candidate genes underlying seed chilling-germination in maize. *PeerJ* **2021**, *9*, e11707. [CrossRef]
17. Yi, Q.; Álvarez-Iglesias, L.; Malvar, R.A.; Romay, M.C.; Revilla, P.A. worldwide maize panel revealed new genetic variation for cold tolerance. *Theor. Appl. Genet.* **2021**, *4*, 1083–1094. [CrossRef]
18. Zhou, X.; Muhammad, I.; Lan, H.; Xia, C. Recent Advances in the Analysis of Cold Tolerance in Maize. *Front. Plant Sci.* **2022**, *13*, 866034. [CrossRef]
19. R Core Team. R: A language and environment for statistical computing. *Computing* **2011**, *1*, 12–21.
20. Kabacoff, R.I. *R in Action: Data Analysis and Graphics with R*; Simon and Schuster: New York, NY, USA, 2015.
21. Murray, M.; Thompson, W.F. Rapid isolation of high molecular weight plant DNA. *Nucleic Acids Res.* **1980**, *8*, 4321–4326. [CrossRef]
22. Frichot, E.F.O. Lea: An R package for landscape and ecological association studies. *Methods Ecol. Evol.* **2015**, *6*, 925–929. [CrossRef]
23. Bradbury, P.J.; Zhang, Z.; Kroon, D.E.; Casstevens, T.M.; Buckler, E. TASSEL: Software for association mapping of complex traits in diverse samples. *Bioinformatics* **2007**, *23*, 2633–2635. [CrossRef]
24. Gaut, B.S.; Long, A.D. The lowdown on linkage disequilibrium. *Plant Cell* **2003**, *15*, 1502–1506. [CrossRef] [PubMed]
25. Benjamini, Y.; Hochberg, Y. Controlling the false discovery rate: A practical and powerful approach to multiple testing. *J. R. Stat. Soc.* **1995**, *57*, 289–300. [CrossRef]
26. Wang, Q.; Tang, J.; Han, B.; Huang, X. Advances in genome-wide association studies of complex traits in rice. *Theor. Appl. Genet.* **2020**, *3*, 1415–1425. [CrossRef]
27. Alseekh, S.; Kostova, D.; Bulut, M.; Fernie, A.R. Genome-wide association studies: Assessing trait characteristics in model and crop plants. *Cell Mol. Life Sci.* **2021**, *8*, 5743–5754. [CrossRef] [PubMed]
28. Saini, D.K.; Chopra, Y.; Singh, J.; Sandhu, K.S.; Kumar, A.; Bazzer, S.; Srivastava, P. Comprehensive evaluation of mapping complex traits in wheat using genome-wide association studies. *Mol. Breed.* **2021**, *42*, 1. [CrossRef] [PubMed]
29. Xiao, Y.; Liu, H.; Wu, L.; Warburton, M.; Yan, J. Genome-wide Association Studies in Maize: Praise and Stargaze. *Mol. Plant* **2017**, *10*, 359–374. [CrossRef] [PubMed]
30. Weng, J.F.; Xie, C.X.; Hao, Z.F.; Wang, J.J.; Liu, C.L.; Li, M.S.; Zhang, D.G.; Bai, L.; Zhang, S.H.; Li, X.H. Genome-wide association study identifies candidate genes that affect plant height in Chinese elite maize (*Zea mays* L.) inbred lines. *PLoS ONE* **2011**, *6*, e29229. [CrossRef]
31. Huang, X.H.; Zhao, Y.; Wei, X.H.; Li, C.Y.; Wang, A.H.; Zhao, Q.; Li, W.J.; Guo, Y.L.; Deng, L.W.; Zhu, C.R.; et al. Genome-wide association study of flowering time and grain yield traits in a worldwide collection of rice germplasm. *Nat. Genet.* **2012**, *44*, 32–39. [CrossRef]
32. Xiao, Q.; Bai, X.; Zhang, C.; He, Y. Advanced high-throughput plant phenotyping techniques for genome-wide association studies: A review. *J. Adv. Res.* **2021**, *35*, 215–230. [CrossRef]
33. Gupta, P.K.; Kulwal, P.L.; Jaiswal, V. Association mapping in plants in the post-GWAS genomics era. *Adv. Genet.* **2019**, *104*, 75–154.
34. Wang, B.; Lin, Z.; Li, X.; Zhao, Y.; Zhao, B.; Wu, G.; Ma, X.; Wang, H.; Xie, Y.; Li, Q.; et al. Genome-wide selection and genetic improvement during modern maize breeding. *Nat. Genet.* **2020**, *52*, 565–571. [CrossRef]
35. Li, H.; Peng, Z.; Yang, X.; Wang, W.; Fu, J.; Wang, J.; Han, Y.; Chai, Y.; Guo, T.; Yang, N.; et al. Genome-wide association study dissects the genetic architecture of oil biosynthesis in maize kernels. *Nat. Genet.* **2013**, *45*, 43. [CrossRef]
36. Jia, G.; Huang, X.; Zhi, H.; Zhao, Y.; Zhao, Q.; Li, W.J.; Chai, Y.; Yang, L.F.; Liu, K.Y.; Lu, H.Y.; et al. A haplotype map of genomic variations and genome-wide association studies of agronomic traits in foxtail millet (*Setaria italica*). *Nat. Genet.* **2013**, *45*, 957–961. [CrossRef]
37. Li, Y.X.; Li, C.; Bradbury, P.J.; Liu, X.; Lu, F.; Romay, C.M.; Glaubitz, J.C.; Wu, X.; Peng, B.; Shi, Y.; et al. Identification of genetic variants associated with maize flowering time using an extremely large multi-genetic background population. *Plant J.* **2016**, *86*, 391–402. [CrossRef]
38. Yang, N.; Lu, Y.; Yang, X.; Huang, J.; Zhou, Y.; Ali, F.; Wen, W.; Liu, J.; Li, J.; Yan, J. Genome wide association studies using a new nonparametric model reveal the genetic architecture of 17 agronomic traits in an enlarged maize association panel. *PLoS Genet.* **2014**, *10*, e1004573. [CrossRef] [PubMed]
39. Samayoa, L.F.; Cao, A.; Santiago, R.; Malvar, R.A.; Butrón, A. Genome-wide association analysis for fumonisin content in maize kernels. *BMC Plant Biol.* **2019**, *19*, 166. [CrossRef] [PubMed]

40. Wang, D.; He, Y.; Nie, L.; Guo, S.; Tu, L.; Guo, X.; Wang, A.; Liu, P.; Zhu, Y.; Wu, X.; et al. Integrated IBD Analysis, GWAS Analysis and Transcriptome Analysis to Identify the Candidate Genes for White Spot Disease in Maize. *Int. J. Mol. Sci.* **2023**, *24*, 10005. [CrossRef]
41. Xiang, N.; Hu, J.G.; Wen, T.X.; Brennan, M.A.; Brennan, C.S.; Guo, X.B. Effects of temperature stress on the accumulation of ascorbic acid and folates in sweet corn (*Zea mays* L.) seedlings. *J. Sci. Food Agric.* **2020**, *100*, 1694–1701. [CrossRef] [PubMed]
42. Ritonga, F.N.; Chen, S. Physiological and Molecular Mechanism Involved in Cold Stress Tolerance in Plants. *Plants* **2020**, *9*, 560. [CrossRef] [PubMed]
43. Farhangi-Abriz, S.; Torabian, S. Antioxidant enzyme and osmotic adjustment changes in bean seedlings as affected by biochar under salt stress. *Ecotoxicol. Environ. Saf.* **2017**, *137*, 64–70. [CrossRef] [PubMed]
44. Nadarajah, K.K. ROS Homeostasis in Abiotic Stress Tolerance in Plants. *Int. J. Mol. Sci.* **2020**, *21*, 5208. [CrossRef] [PubMed]
45. Qin, P.; Zhang, G.; Hu, B.; Wu, J.; Chen, W.; Ren, Z.; Liu, Y.; Xie, J.; Yuan, H.; Tu, B.; et al. Leaf-derived ABA regulates rice seed development via a transporter-mediated and temperature-sensitive mechanism. *Sci. Adv.* **2021**, *7*, eabc8873. [CrossRef] [PubMed]
46. Ma, Y. emical Responses' and QTL/Genes' Identification. *Plants* **2022**, *11*, 2082. [CrossRef] [PubMed]
47. Costa Silva Neta, I.; Vilela de Resende Von Pinho, É.; de Abreu, V.M.; Rezende Vilela, D.; Santos, M.C.; Oliveira Dos Santos, H.; Diniz Cabral Ferreira, R.A.; Garcia Von Pinho, R.; Coelho de Castro Vasconcellos, R. Gene expression and genetic control to cold tolerance during maize seed germination. *BMC Plant Biol.* **2020**, *20*, 188. [CrossRef]
48. Liu, M.; Tan, X.; Yang, Y.; Liu, P.; Zhang, X.; Zhang, Y.; Wang, L.; Hu, Y.; Ma, L.; Li, Z.; et al. Analysis of the genetic architecture of maize kernel size traits by combined linkage and association mapping. *Plant Biotechnol. J.* **2020**, *18*, 207–221. [CrossRef]
49. Peraza, O.G.; Leipner, J.; Reimer, R.; Nguyen, H.T.; Fracheboud, Y. Temperature at night affects the genetic control of acclimation to cold in maize seedlings. *Maydica* **2012**, *56*, 367–377.
50. Jompuk, C. Mapping of quantitative trait loci associated with chilling tolerance in maize (*Zea mays* L.) seedlings grown under field conditions. *J. Exp. Bot.* **2005**, *56*, 1153–1163. [CrossRef]
51. Allam, M.; Revilla, P.; Djemel, A.; Tracy, W.F.; Ordás, B. Identification of QTLs involved in cold tolerance in sweetfield corn. *Euphytica* **2016**, *208*, 353–365. [CrossRef]
52. Zhou, Y.; Lu, Q.; Ma, J.; Wang, D.; Li, X.; Di, H.; Zhang, L.; Hu, X.; Dong, L.; Liu, X.; et al. Using a high density bin map to analyze quantitative trait locis of germination ability of maize at low temperatures. *Front. Plant Sci.* **2022**, *22*, 978941. [CrossRef] [PubMed]
53. He, R.Y.; Yang, T.; Zheng, J.J.; Pan, Z.Y.; Chen, Y.; Zhou, Y.; Li, X.F.; Li, Y.Z.; Iqbal, M.Z.; Yang, C.Y.; et al. QTL Mapping and a Transcriptome Integrative Analysis Uncover the Candidate Genes That Control the Cold Tolerance of Maize Introgression Lines at the Seedling Stage. *Int. J. Mol. Sci.* **2023**, *30*, 2629. [CrossRef] [PubMed]
54. Presterl, T.; Ouzunova, M.; Schmidt, W.; Möller, E.M.; Röber, F.K.; Knaak, C.; Ernst, K.; Westhoff, P.; Geiger, H.H. Quantitative trait loci for early plant vigour of maize grown in chilly environments. *Theor. Appl. Genet.* **2007**, *114*, 1059–1070. [CrossRef] [PubMed]

current issues in molecular biology

MDPI

Article

Enhancing Yield and Improving Grain Quality in Japonica Rice: Targeted EHD1 Editing via CRISPR-Cas9 in Low-Latitude Adaptation

Jian Song [1,†], Liqun Tang [1,†], Honghuan Fan [1], Xiaozheng Xu [2], Xinlu Peng [2], Yongtao Cui [1] and Jianjun Wang [1,*]

1 Institute of Crops and Nuclear Technology Utilization, Zhejiang Academy of Agricultural Sciences, Hangzhou 310021, China; song521125@163.com (J.S.); liquntang2013@126.com (L.T.); xixi615@163.com (H.F.); cuiyongtao20@163.com (Y.C.)

2 College of Advanced Agriculture Sciences, Zhejiang A&F University, Hangzhou 311300, China; 19550178539@163.com (X.X.); 15057155749@163.com (X.P.)

* Correspondence: wangjj4197@163.com or wangjj@zaas.ac.cn

† These authors contributed equally to this work.

Abstract: The "*Indica* to *Japonica*" initiative in China focuses on adapting *Japonica* rice varieties from the northeast to the unique photoperiod and temperature conditions of lower latitudes. While breeders can select varieties for their adaptability, the sensitivity to light and temperature often complicates and prolongs the process. Addressing the challenge of cultivating high-yield, superior-quality *Japonica* rice over expanded latitudinal ranges swiftly, in the face of these sensitivities, is critical. Our approach harnesses the CRISPR-Cas9 technology to edit the *EHD1* gene in the premium northeastern *Japonica* cultivars Jiyuanxiang 1 and Yinongxiang 12, which are distinguished by their exceptional grain quality—increased head rice rates, gel consistency, and reduced chalkiness and amylose content. Field trials showed that these new *ehd1* mutants not only surpass the wild types in yield when grown at low latitudes but also retain the desirable traits of their progenitors. Additionally, we found that disabling *Ehd1* boosts the activity of *Hd3a* and *RFT1*, postponing flowering by approximately one month in the *ehd1* mutants. This research presents a viable strategy for the accelerated breeding of elite northeastern *Japonica* rice by integrating genomic insights with gene-editing techniques suitable for low-latitude cultivation.

Keywords: northeastern *Japonica*; heading date; CRISPR/Cas9; *Ehd1*; low latitude

Citation: Song, J.; Tang, L.; Fan, H.; Xu, X.; Peng, X.; Cui, Y.; Wang, J. Enhancing Yield and Improving Grain Quality in Japonica Rice: Targeted EHD1 Editing via CRISPR-Cas9 in Low-Latitude Adaptation. *Curr. Issues Mol. Biol.* **2024**, *46*, 3741–3751. https://doi.org/10.3390/cimb46040233

Academic Editor: Vijai Bhadauria

Received: 25 March 2024
Revised: 15 April 2024
Accepted: 19 April 2024
Published: 22 April 2024

1. Introduction

The cultivation of high-yield crop varieties has significantly enhanced agricultural performance, particularly in adapting to diverse latitudes [1]. The photoperiod is a critical factor in this adaptation, as it significantly influences crop yields. In China, *Oryza sativa* L., predominantly a short-day (SD) crop, has evolved into two main subspecies: *indica* and *japonica* [2]. Traditionally, *indica* is cultivated in southern China's lower-latitude areas, while *Japonica* is grown in the mid-latitude regions of the northeastern plain and along the Yangtze River [3]. Northeastern *japonica* rice is increasingly preferred for its superior grain quality. However, attempts to breed northeastern *Japonica*-like rice at lower latitudes have faced challenges due to photosensitivity or temperature sensitivity [4–6]. Selecting the optimal flowering time for southern China's lower-latitude areas could enable the expansion of *Japonica* rice from north to south. However, traditional breeding methods are time-consuming [7]. Recent molecular genetic research has uncovered beneficial photoperiod-associated alleles in elite cultivars [8,9], paving the way for advanced gene-editing techniques, such as the clustered regularly interspaced short palindromic repeats (CRISPR)/CRISPR-associated 9 (Cas9) system [10,11].

Photoperiod conditions significantly promote rice heading [8]. Rice has evolved two distinct flowering pathways: a long-day (LD) flowering suppression pathway and a flowering-promotion pathway [8,12]. Under both LD and SD conditions, *Ehd1*, a B-type response regulator, plays a crucial role in integrating these signals. *Ehd1* enhances heading by elevating the expression of *Hd3a* and *RFT1* under SD conditions [13,14]. Various genes modulate the expression of *Ehd1*. For instance, *RID1/Ehd2/OsID1*, *SE5*, *OsMADS51*, and others positively influence *Ehd1* [15,16], while *Heading Date 1* (*Hd1*), *Ghd7*, *COL4*, and *DTH8/Ghd8/Hd5* have a negative impact [8,17–20]. *Hd1* is unique, suppressing heading under LD and promoting it under SD [17]. The roles of other key regulatory genes, like *Ghd7*, *Hd2*, and *DTH8*, are necessary for *Hd1* to reduce heading under LD [9,21]. *SE5* is involved in phytochrome chromophore biosynthesis [16]. Mutations in *se5* result in early heading under both SD and LD conditions. The *AP2* genes, containing the *miR172* target site, act as repressors of *Ehd1* [22]. Loss of *Ehd1* function leads to extended flowering times and exhibits elite agronomic traits at low latitudes, as demonstrated in the Longdao16 and Longdao24 rice varieties [3]. Therefore, *Ehd1* presents a potential target for genome editing to facilitate the adaptation of *Japonica* rice from north to south.

Addressing the pressing need to rapidly breed a diverse array of modern varieties, our study introduces a method using the CRISPR-Cas9 technique. We applied this technique to edit the *EHD1* gene in elite northeastern *Japonica* varieties, including Jiyuanxiang 1 and Yinongxiang 12, which are known for their superior grain quality (higher head rice rates; gel consistency; and lower chalky rice rates, chalkiness degree, and amylose content). The objectives include achieving suitable flowering times and elite plant architecture for modern rice production at low latitudes while preserving the superior grain quality traits of the progenitors. This endeavor represents an integration of gene editing with existing *Japonica* rice landraces and breeding strategies in northeastern China, aiming to enhance the adaptation of *Japonica* rice and breed new, modern varieties suitable for low-latitude regions.

2. Materials and Methods

2.1. Plant Materials and Growth Conditions

The transgenic plants used in this study were of the Yinongxiang and Jiyuanxiang varieties. All rice plants were cultivated in paddy fields under natural conditions at locations in Hangzhou and LingShui, managed by the Zhejiang Academy of Agricultural Sciences, China.

2.2. Agronomic Trait Measurements

Upon reaching maturity, primary agronomic traits such as plant height, tiller count, leaf length, leaf breadth, spikelet length, number of primary and secondary branches, and seed-setting rate were evaluated. For each trait, data were collected from three biological replicates.

2.3. Plasmid Construction

The CRISPR/Cas9 system was utilized to target the *EHD1* gene, with two specific target sites designed. gRNA1EHD1 and gRNA2EHD1 were initially digested and then assembled into an intermediate vector [23]. This intermediate vector was subsequently incorporated into the pC1300-Cas9 binary vector. Details of the target sequences can be found in Supplementary Table S3. The binary vector loaded with two sgRNAs was employed for genetic transformation using the Agrobacterium-mediated transformation method (strain EHA105), as outlined in previous studies [24].

2.4. Detection of Mutations

Genomic DNA was extracted from approximately 100 mg of leaf tissue from the transgenic rice plants by using the CTAB method. PCR amplification of the fragments surrounding the target sites was performed by using KOD FX DNA polymerase (Toyobo Co., Ltd., Osaka, Japan). The amplified DNA fragments were then sequenced using the Sanger method (Applied Biosystems, Foster City, CA, USA) and analyzed by the degenerate sequence decoding method [25,26]. To identify transgene-free plants, PCR amplification was conducted by using specific primers for the hygromycin phosphotransferase gene.

2.5. RNA Extraction and RT-PCR

Total RNA was extracted from leaf tissues by using TRIzol reagent (Thermo Fisher Scientific, Waltham, MA, USA). First-strand cDNA synthesis was performed by using 1 μg of RNA and the ReverTra Ace qPCR RT Master Mix with gDNA Remover (Toyobo Co., Ltd., Osaka, Japan). Real-time PCR was conducted on each cDNA sample in duplicate. The quantitative real-time PCR experiments were performed by using the Power SYBR Green PCR Master Mix kit (Applied Biosystems, Foster City, CA, USA) under specified conditions. The rice UBIQUITIN gene served as an internal control, and relative expression ratios were calculated by using the $2^{-\Delta\Delta CT}$ method. Gene-specific primers are listed in Table S3.

2.6. Photoperiod and Temperature Treatments

ycas-1, *jcas-2*, and WT plants were sown on 15 June 2022 at the Institute of Crops and Nuclear Technology Utilization, Zhejiang Academy of Agricultural Sciences. Fifty-eight-day-old plants underwent photoperiod and temperature treatments in growth chambers. Plants were treated for 7 days and sampled randomly every 3 days over a 10-day period, with each sampling involving three biological replicates.

2.7. Measurement of Amylose, Gel Consistency, and Alkali Spreading Value

Amylose content and gel consistency in mature grains of WT, *ycas-1*, and *jcas-2* plants were measured by using the concanavalin A-based method (Sigma-Aldrich, St. Louis, MO, USA). Gel consistency was determined by the alkali digestion test (Sigma-Aldrich, St. Louis, MO, USA), which involved evaluating starchy endosperm by using a standard seven-point numerical spreading scale. The alkali spreading value (ASV) classified the gelatinization temperature (GT) of rice grains into four groups: high (1–2), high–intermediate (3), intermediate (4–5), and low (6–7)

2.8. Statistical Analysis

Data were expressed as mean values ± standard deviations (SDs) based on three biological replicates. Statistical significance was determined by using Student's unpaired *t*-test and Duncan's multiple range test. Significance levels were set at 5% (indicated by a single asterisk, *) and 1% (indicated by a double asterisk, **) for the *t*-test. For Duncan's multiple range test, significance levels were also set at 5% and 1%, and these are indicated in the respective figures and tables with different letters to denote statistically significant differences among the groups.

3. Results

3.1. Early Flowering of Northeastern Japonica Rice at Low Latitudes

A study involving 119 high-grain-quality northeastern *Japonica* rice varieties in Hangzhou, a low-latitude region, revealed that these varieties flower 30–50 days earlier than the local cultivars, Zhe 08A and Zhejing 99 (Figure S1, Table S1). This early flowering, however, limits their direct application in low-latitude regions. The findings suggest the need for improvement in genes associated with the long-day flowering suppression pathway.

3.2. Both ehd1-Ji Yuan Xiang 1 and ehd1-Yi Nong Xiang 12 Display Delayed Heading Date and Good Field Performance

Among the varieties studied, Ji Yuan Xiang 1 and Yi Nong Xiang 12 were selected for modification of their flowering time. We employed the CRISPR/Cas9 system to target the *EHD1* gene in these varieties, building upon previous research for the construction of the CRISPR/Cas9 vector. The gRNA1 and gRNA2 targeting sites were strategically designed within the third exon of the *EHD1* gene (Figure 1A). Initial steps involved cloning the target sequences into SK-gRNA vectors, followed by assembling these gRNAs into a single intermediate vector to create a combined SK–gRNA1EHD1–gRNA2EHD1 structure. Subsequently, the sgRNAs were incorporated into the pC1300-Cas9 expression vector for genetic transformation. This resulted in 21 and 38 positive transgenic plants for Yinongxiang 12 and Jiyuanxiang 1, respectively, in the T0 generation (Table S2). The sequencing of the target regions in these plants revealed mutation rates of 0% and 47.6% at the gRNA1 and gRNA2 targeting sites in Yinongxiang 12, and 7.9% and 34.2% in Jiyuanxiang 1, respectively. These results demonstrate that the CRISPR/Cas9-mediated editing efficiency significantly varies between the two rice varieties Jiyuanxiang 1 and Yinongxiang 12, highlighting the influence of distinct genetic backgrounds on gene-editing outcomes. This underscores the necessity for tailored gene-editing strategies when targeting specific traits within different agricultural varieties.

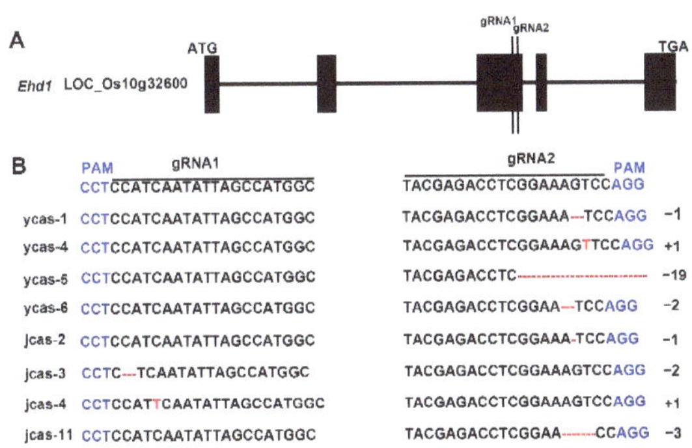

Figure 1. Illustration of *ehd1* mutant alleles in Jiyuanxiang 1 and Yinongxiang 12 varieties. (**A**) Targeted *EHD1* locus for CRISPR-Cas9 editing; (**B**) alignment of mutational sequences in the newly created *ehd1* alleles within the backgrounds of Jiyuanxiang 1 and Yinongxiang 12, compared with the wild type (WT) sequence. Protospacer Adjacent Motifs (PAMs) are marked in blue for clarity.

To assess the phenotypic consequences of the new *ehd1* mutations and eliminate selective marker genes and T-DNA elements, T0 progeny plants underwent DNA sequencing and were self-pollinated to produce T1 progeny. T2 progeny, either homozygous for *ehd1*/*ehd1* or free of T-DNA, were subjected to phenotypic screening and genotyping (Figure S2). The sequencing analysis identified deletions or insertions at the predicted editing sites in *ycas-1*, *ycas-4*, *ycas-5*, *ycas-6*, *jcas-2*, *jcas-3*, *jcas-4*, and *jcas-11*, potentially leading to altered protein functions (Figure 1B).

To investigate the impact of these allelic variations on yield, several agronomic traits were measured (Table 1). For instance, the average plant height of the mutants ranged from 102.7 cm to 111.7 cm (Figure 2A,D), showing a noticeable increase compared with the WT. Furthermore, the grain number per panicle in all new *ehd1* alleles demonstrated varying degrees of increase relative to the WT (Figure 2C,F). The number of tillers per plant in the mutant lines also saw a significant increase compared with the WT. Notably, the heading date in *ehd1* mutants was delayed by about 30 days (Figure 2B,E). In yield tests, the average yield of *ycas-1* and *jcas-2* across thirty plants each surpassed that of the WT (Yinongxiang 12 and Jiyuan xiang 1) (Table 1). The next phase involves applying these new *ehd1* alleles in breeding and fertilizer testing to maximize yield potential.

Table 1. Major agronomic traits of the *ehd1* mutant lines.

Rice Line	PH (cm)	PNPP	PL (cm)	PB	SB	SPP	SSR (%)
Yinongxiang 12	78.3 ± 2.1 a	7.3 ± 0.5 a	16.7 ± 0.3 a	5 ± 1.7 a	20.7 ± 1.5 a	76.3 ± 1.5 a	88.6 ± 1.9
ycas-1	102.7 ± 1.5 b	17.3 ± 0.6 b	23.2 ± 1.0 b	11.3 ± 0.6 b	40.3 ± 2.1 b	146.3 ± 10.2 b	88.3 ± 6.7
ycas-4	102.3 ± 1.2 b	19.3 ± 1.5 b	23.8 ± 0.3 b	12 ± 1 b	42.6 ± 4.0 b	148.7 ± 9.8 b	91.6 ± 3.2
ycas-5	103.7 ± 1.5 b	16.3 ± 1.5 b	23.7 ± 0.3 b	12 ± 1.7 b	44.3 ± 3.1 b	145.3 ± 6.0 b	87.5 ± 2.1
Jiyuanxiang 1	83.3 ± 1.2 a	10.3 ± 1.5 a	17.6 ± 0.3 a	10.0 ± 1.7 a	9.7 ± 1.2 a	99.3 ± 1.5 a	83.9 ± 3.5
jcas-2	109.3 ± 0.6 b	18.7 ± 3.1 b	27 ± 0.1 b	14.3 ± 1.5 b	25.3 ± 2.5 b	227.7 ± 14.3 b	92.0 ± 1.6
jcas-3	110.7 ± 1.5 b	16.7 ± 1.5 b	27.1 ± 0.3 b	14.0 ± 2.6 b	26.3 ± 1.2 b	229.3 ± 2.5 b	89.7 ± 1.7
jcas-4	111.7 ± 2.1 b	19.3 ± 1.5 b	26.6 ± 0.4 b	14.6 ± 2.1 b	22.6 ± 2.9 b	223.3 ± 5.9 b	91.2 ± 1.1

Note: The table presents average values for plant height (PH), panicle number per plant (PNPP), panicle length (PL), primary branches (PB), secondary branches (SB), spikelets per panicle (SPP), and seed-setting rate (SSR) with standard deviations (±SDs; $n = 10$). Statistical analysis of agronomic trait variations was performed by using Duncan's multiple range test ($p < 0.05$). The letters 'a' and 'b' indicate a highly significant difference between the two.

Figure 2. Characteristics of *ehd1* mutants derived from CRISPR-Cas9 editing with Jiyuanxiang 1 and Yinongxiang 12. (**A**) Physical characteristics of various *ehd1* mutants in the Yinongxiang 12 background, developed through gene editing. Scale bar represents 20 cm. (**B**) Extended basic vegetative growth periods observed in *ycas9-1*, *ycas9-4*, and *ycas9-5*. (**C**) Comparative analysis of panicle length between the wild type (Jiyuanxiang 1) and *jcas9-2*, *jcas9-3*, and *jcas9-4*. (**D**) Physical characteristics of different *ehd1* mutants with the Jiyuanxiang 1 background, developed through gene editing. Scale bar represents 20 cm. (**E**) Prolonged basic vegetative growth periods seen in *jcas9-2*, *jcas9-3*, and *jcas9-4*. (**F**) Panicle length comparison between the wild type (Jiyuanxiang 1) and *jcas9-2*, *jcas9-3*, and *jcas9-4*. Error bars represent means ± SDs ($n = 3$). Statistical analysis of agronomic trait variations was performed by using Duncan's multiple range test ($p < 0.05$). The letters 'a' and 'b' indicate a highly significant difference.

3.3. *Temperature Sensitivity of Modified ehd1 Variants*

To investigate the effects of *ehd1* mutations on photoperiod and temperature responses, experiments with *ycas-1* and *jcas-2* mutants were conducted. The results show that variations in photoperiods and temperatures impacted the mutants differently from the WT (Figures 3 and 4A,B). Both *ycas-1* and *jcas-2*, along with the WT, demonstrated temperature sensitivity; notably, *ycas-1* and *jcas-2* exhibited extended basic vegetative growth periods at 23 °C compared with 38 °C. Similarly, under photoperiod treatments, these mutants showed marginally longer basic vegetative growth periods under 11 h light conditions than under 14 h light conditions (Figures 3 and 4A,B).

Figure 3. Temperature sensitivity in *ehd1* mutants of Jiyuanxiang 1 and Yinongxiang 12. (**A**) Trait of Yi nongxiang under 38 °C, 14 h treatment. (**B**) Trait of ycas-1 under 38 °C, 14 h treatment. (**C**) Trait of Ji yuanxiang under 38 °C, 14 h treatment. (**D**) Trait of *jcas-2* under 38 °C, 14 h treatment. (**E**) Trait of Yi nongxiang under 38 °C, 11 h treatment. (**F**) Trait of *ycas-1* under 38 °C, 11 h treatment. (**G**) Trait of Ji yuanxiang under 38 °C, 11 h treatment. (**H**) Trait of *jcas-2* under 38 °C, 11 h treatment. (**I**) Trait of Yi nongxiang under 23 °C, 14 h treatment. (**J**) Trait of *ycas-1* under 23 °C, 14 h treatment. (**K**) Trait of Ji yuanxiang under 23 °C, 14 h treatment. (**L**) Trait of *jcas-2* under 23 °C, 14 h treatment. (**M**) Trait of Yi nongxiang under 23 °C, 11 h treatment. (**N**) Trait of *ycas-1* under 23 °C, 11 h treatment. (**O**) Trait of Ji yuanxiang under 23 °C, 11 h treatment. (**P**) Trait of *jcas-2* under 23 °C, 11 h treatment. Bar = 20 cm.

Additionally, the mRNA expression levels of *Hd3a* and *RFT1* were monitored in *ycas-1* and *jcas-2* under various photoperiod and temperature conditions. Initially, the expression levels of these genes in the mutants were akin to those in the WT. However, a significant increase was observed after 3 to 6 days of treatment (Figure 4C–F). This pattern suggests a negative correlation between the heading date and the expression levels of *Hd3a* and *RFT1*. This finding supports the hypothesis that the transcriptional activation of these genes largely depends on the presence of functional *Ehd1*. The observed photoperiod and temperature responses indicate that a loss of *Ehd1* function leads to the inactivation of *Hd3a* and *RFT1* expression, thereby resulting in a phenotype characterized by delayed flowering.

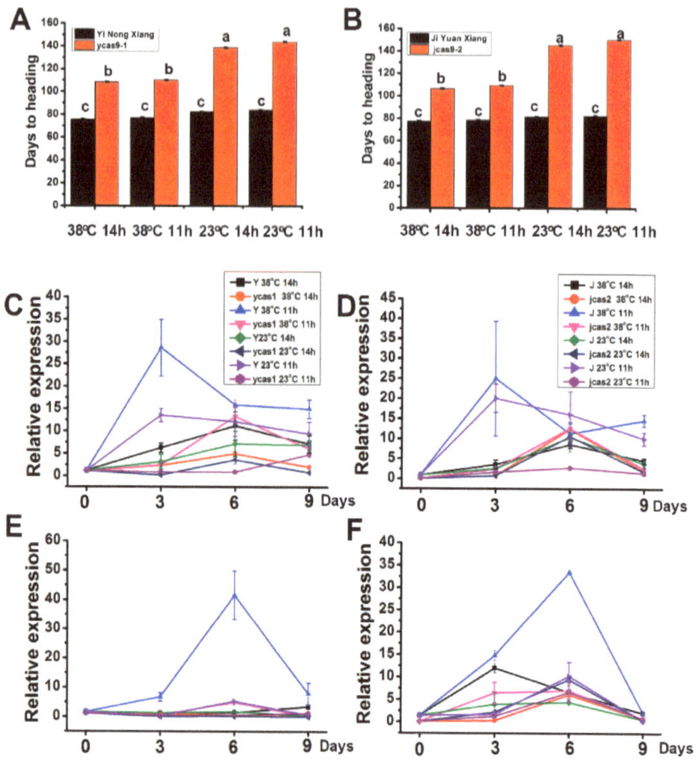

Figure 4. Heading dates and expression patterns of *Hd3a* and *RFT1* in the WT. Analysis of WT and mutant lines under different temperature and photoperiod conditions: artificial high temperature and long day (AHTLD), artificial high temperature and short day (AHTSD), artificial low temperature and long day (ALTLD), and artificial low temperature and short day (ALTSD). The study involved fifty-eight-day-old plants. (**A**,**B**) Heading day comparisons between Jiyuanxiang and *jcas9-2*, and Yinongxiang and *ycas9-1*, under varied light and temperature treatments (AHTLD, AHTSD, ALTLD, and ALTSD); (**C**–**F**) Expression patterns of *Hd3a* and *RFT1* in the presence and absence of *Ehd1* (Jiyuanxiang vs. *jcas9-2*; Yinongxiang vs. *ycas9-1*). Error bars represent means ± SDs (n = 3). Statistical analysis of agronomic trait variations was performed by using Duncan's multiple range test ($p < 0.05$). The letters 'a', 'b', and 'c' indicate a highly significant difference.

3.4. High Grain Yield and Quality of ehd1 Mutants in Low-Latitude Regions

The superior grain quality of *japonica* rice has significantly boosted its popularity in China. However, extended periods of basic vegetative growth can negatively impact both grain yield and quality. This study evaluated the grain quality of the ycas-1 and *jcas-2* variants by analyzing several grain quality characteristics, such as chalkiness degree, chalky rice rate, amylose content, gel consistency, and alkali spreading value. These characteristics were compared across Ji Yuan Xiang 1, Yi Nong Xiang 12, ycas-1, and *jcas-2* (Table 2). The results indicate that ycas-1 and *jcas-2*, cultivated in Hangzhou, showed a lower chalky rice rate and chalkiness degree than Ji Yuan Xiang 1 and Yi Nong Xiang 12, with no significant differences in other traits (*t*-test, $p > 0.1$). The reduced chalky rice rate and chalkiness degree in the *ehd1* mutant lines, *ycas-1* and *jcas-2,* might be attributed to the lower temperatures experienced during their grain-filling stage. Notably, the *ycas-1* and *jcas-2* lines demonstrated commendable grain quality when cultivated in Hangzhou.

Table 2. Comparative analysis of rice quality characteristics in *ehd1* mutant lines.

	Brown Rice Rate (%)	Milled Rice Rate (%)	Head Rice Rate (%)	Grain Length (mm)	Chalkiness Grain Rate (%)	Chalkiness (%)	Amylose (%)	Gel Consistency (mm)	Alkali Spreading Value
Yi nongxiang	81.41 ± 0.47	72.78 ± 0.72	69.85 ± 1.41	5.97 ± 0.05	52.67 ± 3.51 **	12.07 ± 1.66 **	17.46 ± 0.24 **	55.67 ± 5.13	6.60 ± 1.10
ycas9-1	82.99 ± 0.279	72.59 ± 0.29	67.51 ± 0.75	6.21 ± 0.01	44.00 ± 2.65	7.70 ± 0.25	15.51 ± 0.44	58.00 ± 3.60	6.50 ± 0.00
Ji yuanxiang	82.08 ± 0.44	74.38 ± 0.62	73.77 ± 0.78	4.63 ± 0.02	62.33 ± 7.64 **	20.03 ± 2.60 **	14.57 ± 0.31	62.67 ± 0.58	6.50 ± 0.00
jcas9-2	83.46 ± 0.31	74.40 ± 0.41	73.36 ± 0.43	4.67 ± 0.02	32.67 ± 1.53	3.70 ± 0.44	14.65 ± 0.16	62.67 ± 3.06	6.70 ± 0.17

Data represent the means ± SDs of three biological replicates (Student's *t*-test: ** $p < 0.01$).

4. Discussion

4.1. Advancements in Genome Editing for Rice Breeding

This study highlights the pivotal role of genome editing in rice gene research, breeding, and domestication. The CRISPR/Cas9 system, consisting of Cas9 protein and a single-stranded guide RNA (sgRNA), is now foundational in this field [11,27]. This technique, including methods like Prime genome editing and CRISPR/Cas9-based base editing, has seen successful application across diverse crops, like rice, sorghum, wheat, tomato, and maize [11,28–33]. Rice, with its small genome size, high transformation efficiency, extensive quality reference sequences, and varied genomic haplotypes, serves as an ideal model for genetic studies and breeding [8]. Numerous genes have been targeted for improvement in rice by using CRISPR/Cas9, including those associated with disease resistance, grain quality, and yield [11,31,34].

4.2. Tailoring Northeastern Japonica Rice for Low-Latitude Cultivation

Photoperiodic gene adaptation in rice is crucial for its suitability across different latitudes and directly impacts grain yield [9]. Single-nucleotide polymorphisms (SNPs) in these genes vary among varieties at different latitudes [35]. Studies have focused on latitude adaptation in rice through enhanced expression or mutations in flowering time genes [27]. For instance, the allele of *Heading date 1* (*Hd1*) in *indica* varieties has been backcrossed into a *Japonica* variety, facilitating the integration of *indica* traits into *Japonica* rice in southern China. Mutations in the *Ehd1* gene, particularly in its third exon, have shown promising results in delaying the heading date and enhancing suitability for low-latitude growth [3,27].

The adaptation of photoperiodic genes is crucial for the suitability of rice cultivation across different latitudes, which directly impacts grain yield. In our study, we focused on the adaptation of rice to low latitudes through modifications in the Ehd1 gene, which is instrumental in delaying flowering time. These genetic modifications are particularly important for northeastern *Japonica* rice varieties, which, despite their superior grain quality, face challenges in breeding due to their sensitivity to light and temperature.

The transformation of Jiyuanxiang 1 and Yinongxiang 12, varieties known for their high-quality grain, through CRISPR-Cas9-mediated editing demonstrates the potential to overcome these environmental sensitivities. By introducing mutations in the *Ehd1* gene, we delayed the flowering period to a time that avoids the peak temperatures of August, which typically range from an average low of 27 °C to a high of 35 °C. Instead, the edited varieties now flower in late August, with grain filling occurring in September, under milder temperature conditions of 22 °C to 29 °C. This strategic delay in the flowering time circumvents the high-temperature stress during the crucial grain-filling period, markedly improving grain quality. This evidence showcases the power of genome editing in mitigating adverse temperature effects on crop yield and quality and underscores the importance of precise genetic modifications for enhancing crop resilience and productivity in response to climate variability.

Furthermore, the observed differences in gene-editing efficiency between gRNA1 and gRNA2 could be attributed to variations in target site accessibility, sgRNA design, or genomic context, which may influence the CRISPR/Cas9 system's ability to induce precise edits. Understanding these nuances can guide improvements in sgRNA design and editing strategies, enhancing the precision and efficiency of genome editing applications in rice and other crops.

In summary, the modification of the *EHD1* gene in northeastern *japonica* rice varieties has proven effective for producing high-quality *Japonica* rice adapted to low latitudes. With the flowering times of the *ycas-1* and *jcas-2* variants closely aligning with local main varieties in Hangzhou, future efforts will focus on the broader application of these genetically edited varieties in rice breeding programs for low-latitude regions, thereby meeting the demands for high-quality rice production. This approach could serve as a blueprint for similar advancements in other crop varieties, highlighting the transformative impact of precision breeding technologies in modern agriculture.

Supplementary Materials: The following supporting information can be downloaded at: https://www.mdpi.com/article/10.3390/cimb46040233/s1, Figure S1: Frequency distribution of heading date of the elite northeastern plain japonica varieties and two control japonica varieties. Analysis performed using GraphPad Prism software (GraphPad Software, San Diego, CA, USA); Figure S2: Identification of hygromycin in different edited plants, Hygromycin detection performed using ELISA Kit (Sigma-Aldrich, St. Louis, MO, USA); Table S1: The heading date of elite northeastern plain japonica varieties in the Yangtze River region; Table S2: The number of plants detected in the two Yi Nong Xiang 12 PAM sites and Ji Yuan Xiang 1 PAM sites; Table S3: Primer sequences used in this study, Primers synthesized by Integrated DNA Technologies (IDT, Coralville, IA, USA).

Author Contributions: J.S., Y.C. and J.W. conceived the project and wrote the manuscript. L.T., X.X. and X.P. cloned the CRISPR-Cas9 plasmid, identified mutants, and performed qRT-PCR. H.F. performed phenotype observations. All authors have read and agreed to the published version of the manuscript.

Funding: This research was funded by National Natural Science Foundation of China, under grant No. 32101791; Zhejiang Provincial Natural Science Foundation, grant number LQ20C130007; and the "14th Five-Year Plan" Major Special Projects for Breeding New Rice Varieties of Zhejiang Province, grant number 2021C02063-1.

Institutional Review Board Statement: Not applicable.

Informed Consent Statement: Not applicable.

Data Availability Statement: The original data of this present study are available from the corresponding authors.

Acknowledgments: The authors thank the National Mid-term Genebank for Rice of China National Rice Research Institute for providing the rice germplasm collection.

Conflicts of Interest: The authors declare no conflicts of interest.

References

1. Eshed, Y.A.-O.; Lippman, Z.A.-O. Revolutions in agriculture chart a course for targeted breeding of old and new crops. *Science* **2019**, *366*, eaax0025. [CrossRef] [PubMed]
2. Huang, X.; Kurata, N.; Wei, X.; Wang, Z.-X.; Wang, A.; Zhao, Q.; Zhao, Y.; Liu, K.; Lu, H.; Li, W. A map of rice genome variation reveals the origin of cultivated rice. *Nature* **2012**, *490*, 497–501. [CrossRef] [PubMed]
3. Wu, M.; Liu, H.; Lin, Y.; Chen, J.; Fu, Y.; Luo, J.; Zhang, Z.; Liang, K.; Chen, S.; Wang, F. In-Frame and Frame-Shift Editing of the Ehd1 Gene to Develop Japonica Rice With Prolonged Basic Vegetative Growth Periods. *Front. Plant Sci.* **2020**, *11*, 307. [CrossRef] [PubMed]
4. Saito, H.; Yuan, Q.; Okumoto, Y.; Doi, K.; Yoshimura, A.; Inoue, H.; Teraishi, M.; Tsukiyama, T.; Tanisaka, T. Multiple alleles at Early flowering 1 locus making variation in the basic vegetative growth period in rice (*Oryza sativa* L.). *Theor. Appl. Genet.* **2009**, *119*, 315–323. [CrossRef] [PubMed]
5. Wei, X.J.; Jiang, L.; Xu, J.F.; Liu, X.; Liu, S.J.; Zhai, H.Q.; Wan, J.M. The distribution of japonica rice cultivars in the lower region of the Yangtze River valley is determined by its photoperiod-sensitivity and heading date genotypes. *J. Integr. Plant Biol.* **2009**, *51*, 922–932. [CrossRef] [PubMed]

6. Yuan, Q.; Saito, H.; Okumoto, Y.; Inoue, H.; Nishida, H.; Tsukiyama, T.; Teraishi, M.; Tanisaka, T. Identification of a novel gene ef7 conferring an extremely long basic vegetative growth phase in rice. *Theor. Appl. Genet.* **2009**, *119*, 675–684. [CrossRef] [PubMed]
7. Qiu, L.; Wu, Q.; Wang, X.; Han, J.; Zhuang, G.; Wang, H.; Shang, Z.; Tian, W.; Chen, Z.; Lin, Z. Forecasting rice latitude adaptation through a daylength-sensing-based environment adaptation simulator. *Nat. Food* **2021**, *2*, 348–362. [CrossRef]
8. Chen, R.; Deng, Y.; Ding, Y.; Guo, J.; Qiu, J.; Wang, B.; Wang, C.; Xie, Y.; Zhang, Z.; Chen, J.; et al. Rice functional genomics: Decades' efforts and roads ahead. *Sci. China Life Sci.* **2022**, *65*, 33–92. [CrossRef] [PubMed]
9. Zong, W.; Ren, D.; Huang, M.; Sun, K.; Feng, J.; Zhao, J.; Xiao, D.; Xie, W.; Liu, S.; Zhang, H. Strong photoperiod sensitivity is controlled by cooperation and competition among Hd1, Ghd7 and DTH8 in rice heading. *New Phytol.* **2021**, *229*, 1635–1649. [CrossRef] [PubMed]
10. Rothan, C.; Diouf, I.; Causse, M. Trait discovery and editing in tomato. *Plant J.* **2019**, *97*, 73–90. [CrossRef]
11. Cui, Y.; Hu, X.; Liang, G.; Feng, A.; Wang, F.; Ruan, S.; Dong, G.; Shen, L.; Zhang, B.; Chen, D. Production of novel beneficial alleles of a rice yield-related QTL by CRISPR/Cas9. *Plant Biotechnol. J.* **2020**, *18*, 1987. [CrossRef]
12. Cui, Y.; Zhu, M.; Song, J.; Fan, H.; Xu, X.; Wu, J.; Guo, L.; Wang, J. Expression dynamics of phytochrome genes for the shade-avoidance response in densely direct-seeding rice. *Front. Plant Sci.* **2022**, *13*, 1105882. [CrossRef] [PubMed]
13. Cho, L.-H.; Yoon, J.; Pasriga, R.; An, G. Homodimerization of Ehd1 is required to induce flowering in rice. *Plant Physiol.* **2016**, *170*, 2159–2171. [CrossRef] [PubMed]
14. Zhao, J.; Chen, H.; Ren, D.; Tang, H.; Qiu, R.; Feng, J.; Long, Y.; Niu, B.; Chen, D.; Zhong, T. Genetic interactions between diverged alleles of Early heading date 1 (Ehd1) and Heading date 3a (Hd3a)/RICE FLOWERING LOCUS T1 (RFT 1) control differential heading and contribute to regional adaptation in rice (*Oryza sativa*). *New Phytol.* **2015**, *208*, 936–948. [CrossRef] [PubMed]
15. Kim, S.L.; Lee, S.; Kim, H.J.; Nam, H.G.; An, G. OsMADS51 is a short-day flowering promoter that functions upstream of Ehd1, OsMADS14, and Hd3a. *Plant Physiol.* **2007**, *145*, 1484–1494. [CrossRef] [PubMed]
16. Andrés, F.; Galbraith, D.W.; Talón, M.; Domingo, C. Analysis of PHOTOPERIOD SENSITIVITY5 sheds light on the role of phytochromes in photoperiodic flowering in rice. *Plant Physiol.* **2009**, *151*, 681–690. [CrossRef] [PubMed]
17. Yano, M.; Katayose, Y.; Ashikari, M.; Yamanouchi, U.; Monna, L.; Fuse, T.; Baba, T.; Yamamoto, K.; Umehara, Y.; Nagamura, Y.; et al. Hd1, a Major Photoperiod Sensitivity Quantitative Trait Locus in Rice, Is Closely Related to the Arabidopsis Flowering Time Gene CONSTANS. *Plant Cell* **2000**, *12*, 2473–2483. [CrossRef] [PubMed]
18. Xue, W.; Xing, Y.; Weng, X.; Zhao, Y.; Tang, W.; Wang, L.; Zhou, H.; Yu, S.; Xu, C.; Li, X. Natural variation in Ghd7 is an important regulator of heading date and yield potential in rice. *Nat. Genet.* **2008**, *40*, 761–767. [CrossRef] [PubMed]
19. Lee, Y.S.; Jeong, D.H.; Lee, D.Y.; Yi, J.; Ryu, C.H.; Kim, S.L.; Jeong, H.J.; Choi, S.C.; Jin, P.; Yang, J. OsCOL4 is a constitutive flowering repressor upstream of Ehd1 and downstream of OsphyB. *Plant J.* **2010**, *63*, 18–30. [CrossRef] [PubMed]
20. Wei, X.; Xu, J.; Guo, H.; Jiang, L.; Chen, S.; Yu, C.; Zhou, Z.; Hu, P.; Zhai, H.; Wan, J. DTH8 suppresses flowering in rice, influencing plant height and yield potential simultaneously. *Plant Physiol.* **2010**, *153*, 1747–1758. [CrossRef] [PubMed]
21. Zhang, Z.; Hu, W.; Shen, G.; Liu, H.; Hu, Y.; Zhou, X.; Liu, T.; Xing, Y. Alternative functions of Hd1 in repressing or promoting heading are determined by Ghd7 status under long-day conditions. *Sci. Rep.* **2017**, *7*, 5388. [CrossRef] [PubMed]
22. Lee, Y.S.; Lee, D.Y.; Cho, L.H.; An, G. Rice miR172 induces flowering by suppressing OsIDS1 and SNB, two AP2 genes that negatively regulate expression of Ehd1 and florigens. *Rice* **2014**, *7*, 31. [CrossRef] [PubMed]
23. Shen, L.; Hua, Y.; Fu, Y.; Li, J.; Liu, Q.; Jiao, X.; Xin, G.; Wang, J.; Wang, X.; Yan, C.; et al. Erratum to: Rapid generation of genetic diversity by multiplex CRISPR/Cas9 genome editing in rice. *Sci. China Life Sci.* **2019**, *62*, 1270. [CrossRef] [PubMed]
24. Hiei, Y.; Komari, T.; Kubo, T. Transformation of rice mediated by Agrobacterium tumefaciens. *Plant Mol. Biol.* **1997**, *35*, 205–218. [CrossRef] [PubMed]
25. Liu, W.; Xie, X.; Ma, X.; Li, J.; Chen, J.; Liu, Y.G. DSDecode: A Web-Based Tool for Decoding of Sequencing Chromatograms for Genotyping of Targeted Mutations. *Mol. Plant* **2015**, *8*, 1431–1433. [CrossRef] [PubMed]
26. Ma, X.; Chen, L.; Zhu, Q.; Chen, Y.; Liu, Y.G. Rapid Decoding of Sequence-Specific Nuclease-Induced Heterozygous and Biallelic Mutations by Direct Sequencing of PCR Products. *Mol. Plant* **2015**, *8*, 1285–1287. [CrossRef] [PubMed]
27. Cui, Y.; Zhu, M.; Xu, Z.; Xu, Q. Assessment of the effect of ten heading time genes on reproductive transition and yield components in rice using a CRISPR/Cas9 system. *Theor. Appl. Genet.* **2019**, *132*, 1887–1896. [CrossRef] [PubMed]
28. Anzalone, A.V.; Koblan, L.W.; Liu, D.A.-O. Genome editing with CRISPR-Cas nucleases, base editors, transposases and prime editors. *Nat. Biotechnol.* **2020**, *38*, 824–844. [CrossRef] [PubMed]
29. Son, S.; Park, S.R. Challenges Facing CRISPR/Cas9-Based Genome Editing in Plants. *Front. Plant Sci.* **2022**, *13*, 902413. [CrossRef] [PubMed]
30. Tomlinson, L.; Yang, Y.; Emenecker, R.; Smoker, M.; Taylor, J.; Perkins, S.; Smith, J.; MacLean, D.; Olszewski, N.E.; Jones, J.D. Using CRISPR/Cas9 genome editing in tomato to create a gibberellin-responsive dominant dwarf DELLA allele. *Plant Biotechnol. J.* **2019**, *17*, 132–140. [CrossRef] [PubMed]
31. Macovei, A.; Sevilla, N.R.; Cantos, C.; Jonson, G.B.; Slamet-Loedin, I.; Cermak, T.; Voytas, D.F.; Choi, I.R.; Chadha-Mohanty, P. Novel alleles of rice eIF4G generated by CRISPR/Cas9-targeted mutagenesis confer resistance to Rice tungro spherical virus. *Plant Biotechnol. J.* **2018**, *16*, 1918–1927. [CrossRef] [PubMed]
32. Li, A.; Jia, S.; Yobi, A.; Ge, Z.; Sato, S.J.; Zhang, C.; Angelovici, R.; Clemente, T.E.; Holding, D.R. Editing of an alpha-kafirin gene family increases, digestibility and protein quality in sorghum. *Plant Physiol.* **2018**, *177*, 1425–1438. [CrossRef]

33. Liu, X.; Wu, S.; Xu, J.; Sui, C.; Wei, J. Application of CRISPR/Cas9 in plant biology. *Acta Pharm. Sin. B* **2017**, *7*, 292–302. [CrossRef]
34. Zhang, J.; Zhang, H.; Botella, J.R.; Zhu, J.K. Generation of new glutinous rice by CRISPR/Cas9-targeted mutagenesis of the Waxy gene in elite rice varieties. *J. Integr. Plant Biol.* **2018**, *60*, 369. [CrossRef] [PubMed]
35. Leng, Y.; Gao, Y.; Chen, L.; Yang, Y.; Huang, L.; Dai, L.; Ren, D.; Xu, Q.; Zhang, Y.; Ponce, K. Using Heading date 1 preponderant alleles from indica cultivars to breed high-yield, high-quality japonica rice varieties for cultivation in south China. *Plant Biotechnol. J.* **2020**, *18*, 119–128. [CrossRef] [PubMed]

current issues in molecular biology

MDPI

Article

miRNA Sequencing Analysis in Maize Roots Treated with Neutral and Alkaline Salts

Ziqi Chen †, Yang Liu †, Qi Wang, Jianbo Fei, Xiangguo Liu, Chuang Zhang * and Yuejia Yin *

Institute of Agricultural Biotechnology/Jilin Provincial Key Laboratory of Agricultural Biotechnology, Jilin Academy of Agricultural Sciences (Northeast Agricultural Research Center of China), Changchun 130033, China; chenziqi@cjaas.com (Z.C.); liuyang_20045894@163.com (Y.L.); wang-edu@foxmail.com (Q.W.); 13041382812@163.com (J.F.); lxgyyj@cjaas.com (X.L.)
* Correspondence: zhangchuang@cjaas.com (C.Z.); yinyuejia@cjaas.com (Y.Y.)
† These authors contributed equally to this work.

Abstract: Soil salinization/alkalization is a complex environmental factor that includes not only neutral salt NaCl but also other components like Na_2CO_3. miRNAs, as small molecules that regulate gene expression post-transcriptionally, are involved in plant responses to abiotic stress. In this study, maize seedling roots were treated for 5 h with 100 mM NaCl, 50 mM Na_2CO_3, and H_2O, respectively. Sequencing analysis of differentially expressed miRNAs under these conditions revealed that the Na_2CO_3 treatment group had the most differentially expressed miRNAs. Cluster analysis indicated their main involvement in the regulation of ion transport, binding, metabolism, and phenylpropanoid and flavonoid biosynthesis pathways. The unique differentially expressed miRNAs in the NaCl treatment group were related to the sulfur metabolism pathway. This indicates a significant difference in the response patterns of maize to different treatment groups. This study provides theoretical evidence and genetic resources for further analysis of the molecular mechanisms behind maize's salt–alkali tolerance.

Keywords: miRNA sequencing analysis; maize root; neutral salt; alkaline salts

Citation: Chen, Z.; Liu, Y.; Wang, Q.; Fei, J.; Liu, X.; Zhang, C.; Yin, Y. miRNA Sequencing Analysis in Maize Roots Treated with Neutral and Alkaline Salts. *Curr. Issues Mol. Biol.* **2024**, *46*, 8874–8889. https://doi.org/10.3390/cimb46080524

Academic Editor: Hany A. El-Shemy

Received: 3 July 2024
Revised: 4 August 2024
Accepted: 12 August 2024
Published: 15 August 2024

1. Introduction

Saline–alkali soils arise due to the progressive accumulation of salts within the soil matrix, detrimentally impacting agricultural productivity. Data from UNESCO and the Food and Agriculture Organization indicate that globally, saline–alkali soils encompass approximately 954 million hectares, with China accounting for 99.13 million hectares of this total [1]. In China, the genesis of such soils is predominantly linked to the accretion of carbonates, which elevates the soil's alkalinity to levels that are inhospitable for plant life, particularly in regions where salinity is acute. Given that saline–alkaline soils contain both neutral and basic salts, requiring plants to cope with various ionic stresses and pH changes, studying the comprehensive effects of salt and alkali on plants is of great significance for understanding the physiological response mechanisms of plants in saline–alkaline environments and for improving the salt–alkali tolerance of crops.

Unlike NaCl stress, Na_2CO_3 stress in plants arises from a complex interaction of sodium ion stress, high pH stress, and carbonate stress, with intricate mechanisms of action [2]. Previous studies on salinity–alkalinity stress focused mainly on neutral salts such as NaCl, identifying osmotic stress, ionic stress, and oxidative stress as the primary effects on plants [3]. Additionally, Na_2CO_3 stress incorporates a high pH effect, which is known to directly interfere with nutrient absorption, organic acid balance, and ion equilibrium, especially impacting pH stability at the cellular and whole-plant levels [4]. Owing to its triple stress effects of Na^+, high pH, and ${CO_3}^{2-}$, Na_2CO_3 poses greater risks to plants. The effects of Na^+ and ${CO_3}^{2-}$ stress on plants are primarily due to their toxic effects and the resulting nutritional imbalance. High concentrations of Na^+ disrupt the ionic balance within cells, replacing potassium ions (K^+) and interfering with normal physiological functions, including enzyme activity and water balance [5]. Furthermore, the accumulation of Na^+ increases the osmotic pressure of the soil solution, making it difficult for plant roots to absorb water, thus leading to water stress [6]. On the other hand, $CO_3{}^{2-}$ raises the pH of the soil, making it alkaline. High pH levels can lead to a decrease in the solubility of many essential elements (such as iron, manganese, and zinc), making it difficult for plants to absorb these necessary micronutrients [7]. Additionally, high concentrations of carbonate ions can interfere with plant metabolic activities, affecting normal root respiration and nutrient uptake. Therefore, Na^+ and $CO_3{}^{2-}$ stress primarily affects plants through ionic toxicity and nutritional imbalance, leading to inhibited growth and impaired physiological functions. Research by Parida and Das [8] indicates that Na_2CO_3 stress disrupts the dynamic balance of reactive oxygen species, causing oxidative damage to enzymes and photosynthetic apparatus, thereby suppressing the energy expenditure for CO_2 absorption. Badger and Price [9] found that Na_2CO_3 stress inhibits the activity of key enzymes in the Calvin cycle, reducing carbon fixation efficiency and photosynthesis. Miller et al. [10] observed enhanced lipid peroxidation under Na_2CO_3 stress, leading to significant electrolyte leakage and malondialdehyde (MDA) content, suggesting that the plasma membrane is the initial site of damage. Studies have shown that the concentration of Na^+ and K^+ on the surface of *Puccinellia tenuiflora* leaves increases with Na_2CO_3 concentration, as the plant expels salt ions through stomatal openings or excretes waxy substances, thereby achieving salinity–alkalinity tolerance [11,12]. Furthermore, Puccinellia tenuiflora's adaptive regulatory mechanism to high pH stress caused by alkalinity suggests that the plant adjusts the pH primarily through the accumulation and secretion of organic acids, such as citric acid, from the roots. Under $NaHCO_3$ stress, an increase in citrate content in leaves and roots was observed, along with the induction of vacuolar Na^+/H^+ antiporters, indicating their crucial role in pH regulation [13,14]. These studies indicate that Na_2CO_3 stress triggers a series of physiological and molecular changes in halophytes to combat alkaline salt stress. However, glycophytes like maize might have distinct mechanisms in responding to Na_2CO_3 stress compared to halophytes.

Plant stress responses involve complex and precise regulatory processes, and miRNAs, as a class of small RNAs crucial for post-transcriptional regulation, play an extremely important role in plant growth, development, and response to abiotic and biotic stresses by directly targeting functional genes or indirectly regulating downstream functional genes through targeting transcription factors. Over the years, with the widespread application of advanced high-throughput sequencing technologies, studies have reported that miRNAs in plants exhibit high conservation and play roles in plant growth and development, various treatment responses, and are considered important regulators in the post-transcriptional regulation of plant salt stress responses [15,16]. Xu et al. [17] propose future directions for the study of plant miRNAs under salt stress, suggesting that manipulating microRNAs could enhance the salinity resistance of crops. Chen et al. [18] employed a combination of PacBio isoform sequencing (Iso-Seq) and BGISEQ short-read RNA-seq to investigate the role of miRNAs in the salt stress response of the mangrove plant, Buch.-Ham. Through KEGG analysis, eighteen miRNA targets were associated with 'environmental information processing' and linked to plant hormone signal transduction (ko04075), the MAPK signaling pathway-plant (ko04016), and ABC transporters (ko02010).

Currently, most studies on miRNA responses to salt stress focus on neutral salts, while research on alkaline salts is relatively scarce, and the relevant molecular mechanisms remain unclear. This study aims to reveal the expression patterns of miRNAs under different salt stress conditions by comparing the differentially expressed miRNAs in maize roots under neutral salt and alkaline salt conditions. We will employ high-throughput sequencing technology to systematically analyze the miRNA expression profiles in maize roots under neutral and alkaline salt conditions, identifying significantly differentially expressed miRNAs. Furthermore, we will investigate the target genes and regulatory networks of these differentially expressed miRNAs under salt stress conditions to elucidate their mechanisms of action in plant salt stress responses. Additionally, this research aims to provide new insights into the post-transcriptional regulatory mechanisms of plants under salt stress and to offer theoretical support for enhancing crop resilience in saline–alkaline soils, facilitating the breeding of salt–alkaline tolerant crop varieties. By achieving these objectives, we hope to deepen our understanding of the miRNA regulatory mechanisms in maize under neutral and alkaline salt stress, thereby providing scientific evidence and technical support for improving crop growth and tolerance in saline–alkaline soils.

2. Materials and Methods

2.1. *Maize Seedling Stress Treatment*

B73 maize seeds were dark-incubated at 28 °C for 3–4 days, then transplanted to a greenhouse (25 °C, 16 h light/8 h dark) for further growth. Hydroponic culture was conducted using a 1/4 strength MS liquid medium. After 8–10 days of greenhouse cultivation, uniformly developed three-leaf stage maize plants were selected. The roots were treated with 100 mM NaCl, 50 mM Na_2CO_3, and H_2O for 5 h. Root tissues from each stress treatment were collected, flash-frozen in liquid nitrogen, and stored at −80 °C for subsequent analysis.

2.2. *miRNA Sequencing Analysis*

Fifteen samples from each group were randomly selected and divided into three sets, constituting three biological replicates for an initial comprehensive miRNA sequencing. The library construction and sequencing work was entrusted to LC Bio Technology Co., Ltd. (Hangzhou, China). The sequencing and analysis methods can be briefly described as follows.

The raw sequencing reads were processed using the in-house software ACGT101-miR (v4.2) to eliminate adapter dimers, low complexity sequences, common RNA families (rRNA, tRNA, snRNA, snoRNA), and repeats. Following this, unique sequences ranging from 18 to 25 nucleotides were mapped to species-specific precursors in miRBase 22.1 via a BLAST search to identify known miRNAs and novel 3p- and 5p-derived miRNAs. Length variations at both the 3′ and 5′ ends, as well as one internal mismatch, were permitted during alignment. Unique sequences mapping to mature miRNAs in the hairpin arms of specific species were classified as known miRNAs. Sequences aligning to the opposite arm of known species-specific precursor hairpins, relative to the annotated mature miRNA-containing arm, were designated as novel 5p- or 3p-derived miRNA candidates. Remaining sequences were mapped to other selected species' precursors (excluding the specific species) in miRBase 22.1 via BLAST, and the mapped pre-miRNAs were subsequently BLASTed against the specific species' genomes to determine their genomic locations. These sequences were defined as known miRNAs. Unmapped sequences were BLASTed against the specific genomes, and hairpin RNA structures containing these sequences were predicted from the flanking 120 nucleotide sequences using RNAfold software (http://rna.tbi.univie.ac.at/cgi-bin/RNAWebSuite/RNAfold.cgi, accessed on 13 August 2024. Vienna, Austria). The criteria for secondary structure prediction included the following: (1) number of nucleotides in one bulge in the stem ($\leq$12), (2) number of base pairs in the stem region of the predicted hairpin ($\geq$16), (3) free energy cutoff (kcal/mol ≤ -15), (4) length of hairpin (up and down stems plus terminal loop $\geq$ 50), (5) length of hairpin loop ($\leq$200), (6) number of nucleotides in one bulge in the mature region ($\leq$4), (7) number of biased errors in one bulge in the mature region ($\leq$2), (8) number of biased bulges in the mature region ($\leq$2), (9) number of errors in the mature region ($\leq$4), (10) number of base pairs in the mature region of the predicted hairpin ($\geq$12), and (11) percentage of the mature sequence in the stem ($\geq$80). The differential expression of miRNAs, based on normalized deep sequencing counts, was analyzed using ANOVA. Significance thresholds were set at 0.01 and 0.05 for each test.

The raw date of sequencing file could be accessible with the following link: https://www.ncbi.nlm.nih.gov/sra/PRJNA1126005 (accessed on 1 July 2024).

2.3. RNA Extraction, Quantification and Quantitative Real Time PCR (qPCR)

Total RNA was extracted from tissue samples utilizing the EasyPure® miRNA Kit (TransGen, Beijing, China), adhering strictly to the guidelines provided by the manufacturer. To assess the concentration and integrity of the extracted RNA, a Bioanalyzer 2100 instrument from Agilent Technologies (Santa Clara, CA, USA) was employed. Following the evaluation, cDNA synthesis was carried out using the TransScript® One-Step gDNA Removal (TransGen, Beijing, China) and cDNA Synthesis SuperMix (TransGen, Beijing, China) in accordance with the manufacturer's protocol. The resulting cDNA served as a template for PCR amplification, utilizing gene-specific primer pairs as listed in Table S12. Primers designated for selected miRNA and U6 detection were sourced from Sangon Biotech (Shanghai) Co., Ltd. (Shanghai, China). Quantitative real-time PCR (qPCR) analyses were conducted using the ABI StepOne system (Applied Biosystems Inc., Norwalk, CT, USA) employing the TransStart® Tip Green qPCR SuperMix (+Dye I) kit (TransGen, Beijing, China) for the reactions. Within this experimental setup, U6 RNA was implemented as the endogenous reference for normalizing miRNA expression levels. The expression fold changes in the miRNAs were quantified employing the ΔΔCt method. For the calculation of the relative expression levels of these genes, the $2^{-\Delta\Delta CT}$ method was utilized, with normalization against U6. This approach facilitated a precise quantification of miRNA expression alterations within the tissue samples under investigation [19].

2.4. Computational Prediction of miRNA Targets

To elucidate the possible molecular roles of the differentially expressed candidate miRNAs, two computational target prediction tools, TargetScan and miranda, were employed to identify potential miRNA binding sites. These algorithms were selected for their robust capabilities in predicting miRNA–mRNA interactions based on sequence complementarity and other structural features. The predictions generated by both TargetScan and miranda were subsequently integrated to enhance the reliability of the results. Overlapping predictions between the two algorithms were calculated to identify consensus targets, which are more likely to represent true miRNA–target interactions due to the agreement of independent computational models. Furthermore, the functional annotation of these abundant miRNAs and their predicted targets was conducted using in-house Perl scripts. These scripts were utilized to annotate the Gene Ontology (GO) terms and Kyoto Encyclopedia of Genes and Genomes (KEGG) pathways associated with the miRNA targets.

2.5. Ion Content Determination

Sample preparation: A volume of 200 μL of the sample was transferred into a 15 mL polypropylene tube with a screw cap, followed by the addition of 3.8 mL of diluent solution and subsequent vortexing [20].

Determination of sulfate in plants was conducted referencing the paper by Kurmanbayeva A et al. [21].

The methodology for assessing sodium (Na^+) and potassium (K^+) concentrations was delineated by Xu et al. [22].

Determination of other ion contents referenced the paper by AbdElgawad H et al. [23].

3. Results

3.1. Phenotypic Identification of Maize Roots under Neutral Salt and Alkaline Stress

B73 maize specimens grown hydroponically in a greenhouse were divided into three groups at the three-leaf stage with uniform growth. One group's roots were soaked in water for 5 h, while the roots of the other two groups were exposed to 100 mM NaCl and 50 mM Na_2CO_3 for the same period. No significant difference was observed between the roots treated with water and NaCl; however, those treated with Na_2CO_3 exhibited a notable yellowing (Figure 1). This indicates that maize roots under Na_2CO_3 stress may activate distinct metabolic pathways and regulatory networks compared to those treated with water and NaCl.

Figure 1. Phenotype of B73 maize roots treated under different conditions for 5 h, scale bar = 1 cm.

3.2. miRNA Sequencing Results and Data Analysis in Maize Roots under Different Salt Stress Conditions

miRNAs are well known as major post-transcriptional regulators that respond to salt stress in various plants. Roots, being in direct contact with saline soils, are the primary organs affected by salt stress. Therefore, studying the impact of salt stress on maize roots is of great importance.

To identify the differential expression of miRNAs in maize roots under different salt stress conditions, we performed miRNA sequencing analysis on the roots after 5 h of treatment, with three biological replicates for the control (water), NaCl-treated, and Na_2CO_3-treated groups. The analysis resulted in 11.91, 12.18, 9.92, 10.54, 14.75, 13.18, 15.61, 13.57, and 19.54 million reads, respectively. The sequences were aligned and filtered against mRNA, RFam (including rRNA, tRNA, snRNA, snoRNA, etc.), and the Repbase database, resulting in the filtered data referred to as 'Valid data'. The 'Valid data' was subsequently analyzed for miRNA identification and prediction (refer to Supplementary Table S1). The majority of the identified miRNAs were 20–24 nucleotides long, representing 95.74% of the total, consistent with the reported length of plant miRNAs (refer to Supplementary Table S2). A total of 325 conserved miRNAs in maize were identified from the miRBase database. In this study, we identified 164 conserved miRNAs and 276 newly predicted miRNAs across 50 families, with 34 families containing multiple miRNA members (refer to Supplementary Tables S3 and S4). The results indicate significant changes in miRNA expression levels in maize roots under various salt stress conditions. Nearly half of the conserved miRNAs showed altered expression, highlighting the extensive impact of salinity and alkalinity on transcriptional regulation in maize.

3.3. Differential miRNA Expression Analysis

To investigate the differential expression of miRNAs under different salt stress conditions, particularly carbonate stress, we analyzed miRNA sequencing results. A comparative analysis was performed on miRNA expression levels in the roots of maize subjected to water treatment, 100 mM NaCl treatment, and 50 mM Na_2CO_3 treatment. Pearson correlation analysis revealed a strong correlation among the sample groups, confirming the validity of the sampling (refer to Supplementary Figure S1A). The analysis of differentially expressed miRNAs in different comparison groups revealed that three miRNAs showed a differential expression between the NaCl vs. H_2O group and the Na_2CO_3 vs. H_2O group. There were 29 miRNAs that showed a differential expression between the NaCl vs. H_2O group and the Na_2CO_3 vs. NaCl group. A total of 48 miRNAs showed a differential expression between the Na_2CO_3 vs. H_2O group and the Na_2CO_3 vs. NaCl group. Additionally, two miRNAs showed differential expression in all three comparison groups (refer to Supplementary Figure S1B).

Significant differences were observed among the groups as follows: in the NaCl vs. H_2O comparison, forty-one miRNAs exhibited altered expression, with thirty-seven upregulated and four downregulated (refer to Supplementary Table S5); in the Na_2CO_3 vs. H_2O comparison, seventy-four miRNAs showed changes in expression, with twenty-seven upregulated and forty-seven downregulated (refer to Supplementary Table S6); and in the Na_2CO_3 vs. NaCl treatment group comparison, one hundred and fifty miRNAs displayed altered expression, with twenty-one upregulated and one hundred and twenty-nine downregulated (refer to Supplementary Table S7). Setting the screening criteria at $p \leq 0.05$ and a fold change ≥ 2, we identified that in the NaCl vs. H_2O comparison, 22 miRNAs exhibited altered expression, all of which were upregulated (refer to Supplementary Table S5); in the Na_2CO_3 vs. H_2O comparison, 53 differentially expressed miRNAs were identified, with 22 upregulated and 31 downregulated (refer to Supplementary Table S6). In conclusion, the number of differentially expressed miRNAs in the Na_2CO_3 treatment group significantly surpasses that in the NaCl treatment group, suggesting a notably greater impact of carbonate salt on plants compared to neutral salt. Additionally, the clustering analysis of differentially expressed miRNAs reveals notable distinctions between the Na_2CO_3 treatment group and the other treatment groups (Figure 2).

3.4. Prediction and Bioinformatics Analysis of Differentially Expressed miRNA Target Genes

To explore the regulatory role of miRNAs in maize roots under different salt stress conditions, targeted prediction, GO enrichment, and KEGG pathway analyses were performed on miRNAs uniquely expressed in distinct comparative groups identified through sequencing.

The 22 miRNAs exclusively expressed in the NaCl vs. H_2O group regulate 122 target genes. The GO enrichment analysis indicated that these genes are primarily associated with biological processes such as response to oxidative stress and transmembrane transport, cellular components such as integral components of the membrane, and molecular functions including transferase activity and metal ion binding (Figure 3). Remarkable KEGG pathway enrichment was observed in Monobactam biosynthesis and Selenocompound metabolism, among other pathways (Figure 4). Among these twenty-two differentially expressed miRNAs, two conserved miRNAs, *zma-miR395a-3p_L-1* and *zma-MIR395o-p5* from the MIR395 family, regulate seven genes, including *Zm00001d028164*, *Zm00001d033981*, and *Zm00001d013296*, which are involved in functions such as Sulfate transporter 2.2, ATP sulfurylase1, and ATP sulfurylase 1 chloroplastic. Their gene functions and enrichment in the GO and KEGG pathways are primarily associated with sulfate synthesis, metabolism, assimilation, and transport pathways, including GO:0000103 (sulfate assimilation), GO:0008272 (sulfate transport), and ko00920 (Sulfur metabolism). Additionally, other non-conserved miRNAs regulate genes associated with sulfate metabolism. For instance, *Zm00001d040796*, regulated by *PC-5p-19607_184* and annotated as Ubiquitin-like-specific protease ESD4, is enriched in GO:0008234 (cysteine-type peptidase activity), and *Zm00001d028818*, regulated by *PC-3p-45471_81* and annotated as Calpain-type cysteine protease DEK1, is enriched in GO:0008234 (cysteine-type peptidase activity) (refer to Supplementary Table S8). These findings indicate that NaCl treatment influences sulfate metabolism in maize roots.

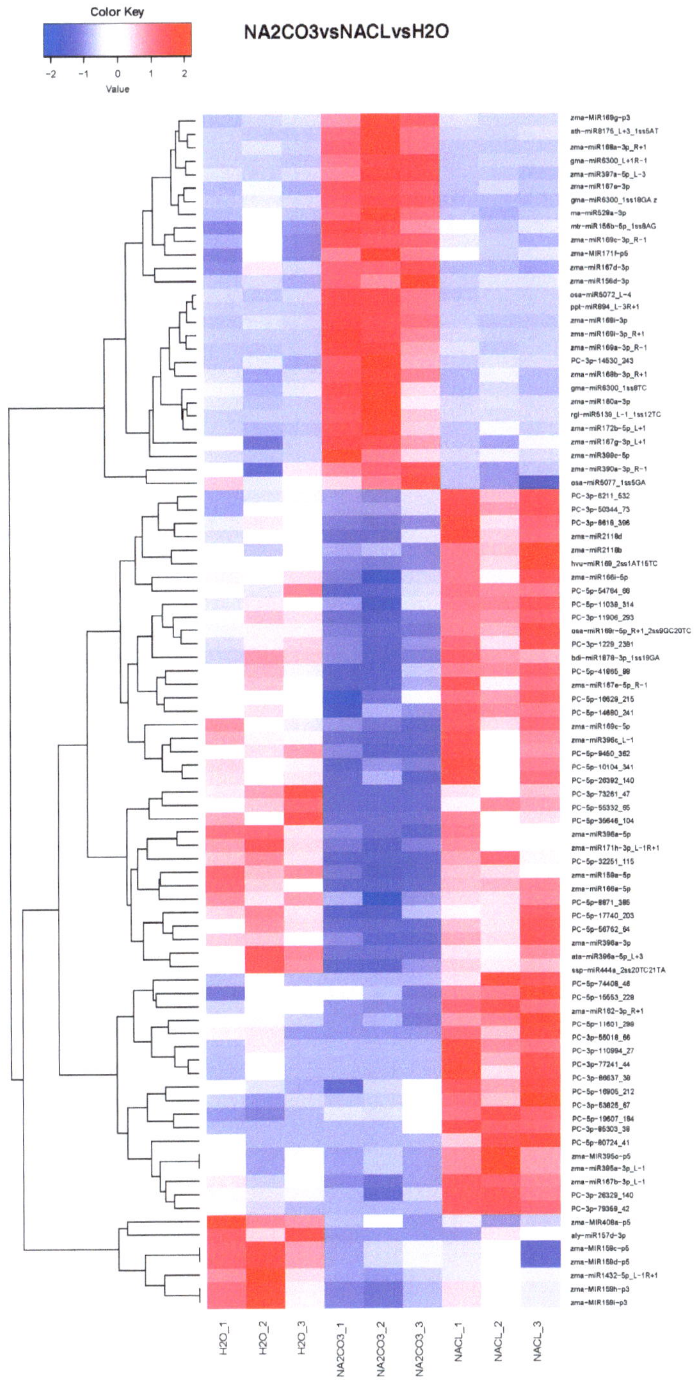

Figure 2. Cluster analysis of differentially expressed miRNAs across various treatment groups.

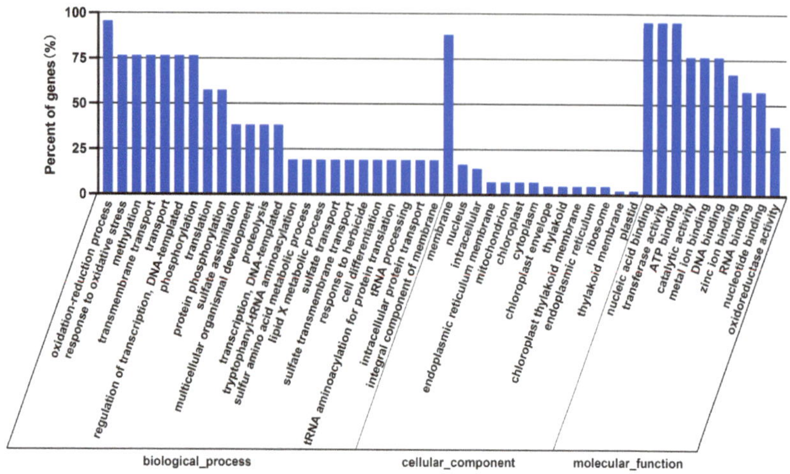

Figure 3. GO functional enrichment analysis of differential miRNA target genes in the NaCl vs. H_2O comparison group.

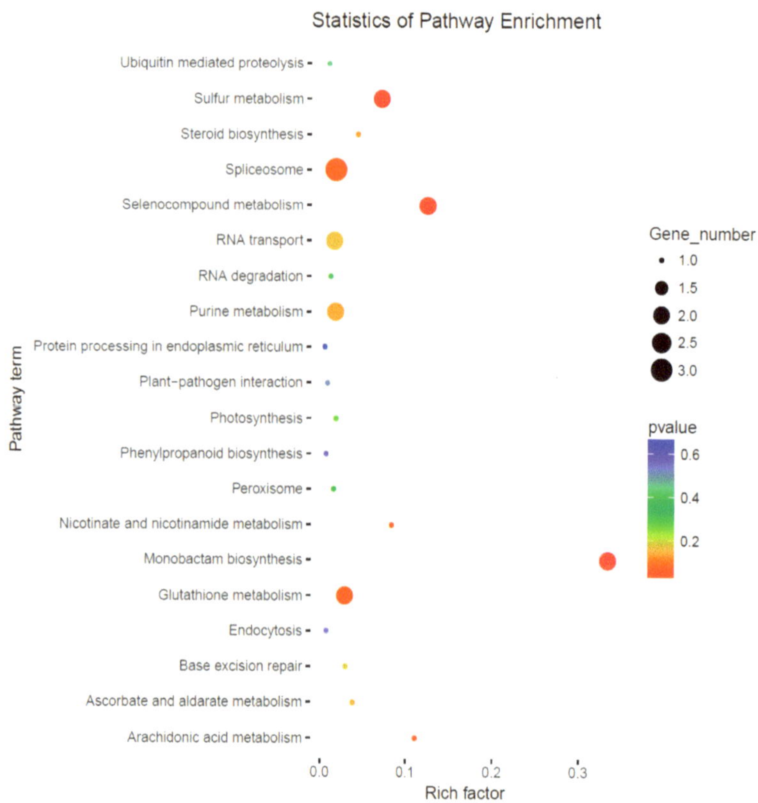

Figure 4. KEGG pathway enrichment analysis of differential miRNA target genes in the NaCl vs. H_2O comparison group.

In the Na_2CO_3 vs. H_2O comparison group, 53 uniquely expressed miRNAs regulate 379 target genes. The GO enrichment analysis reveals that these genes are primarily associated with biological processes such as the oxidation–reduction process, exocytosis, heme biosynthetic process, calcium ion transport, and lignin catabolic process; cellular components such as the extracellular region and exocyst; and molecular functions including copper ion binding and oxidoreductase activity (Figure 5). A remarkable KEGG pathway enrichment is observed in Phenylpropanoid biosynthesis, Flavonoid biosynthesis, Stilbenoid, diarylheptanoid, and gingerol biosynthesis pathways (Figure 6). Out of these 53 differentially expressed miRNAs, 32 are conserved miRNAs, belonging to the MIR156, MIR159, MIR167, MIR168, MIR169, MIR171, MIR396, MIR397, and MIR528 families. The 238 genes they regulate are involved in various functions, including *Zm00001d030062*, regulated by *zma-miR156d-3p* and annotated as LRR protein 4; *Zm00001d029550*, regulated by *zma-miR159a-5p* and annotated as Diacylglycerol kinase 1; *Zm00001d024744*, regulated by *zma-miR167g-3p_L+1* and annotated as PTI1-like tyrosine-protein kinase; *Zm00001d047683*, regulated by *zma-miR168b-3p_R+1* and annotated as NAD(P)-binding Rossmann-fold superfamily protein; *Zm00001d007181*, regulated by *zma-miR169i-3p* and annotated as Calcium-binding protein; *Zm00001d041592*, regulated by *zma-MIR171f-p5* and annotated as modifier of snc1; *Zm00001d033876*, regulated by *zma-miR396a-5p* and annotated as Growth-regulating factor 2; *Zm00001d017762*, regulated by *zma-miR397a-5p_L-3* and annotated as abscisic acid 8′-hydroxylase1; *Zm00001d020764*, regulated by *zma-miR528a-3p* and annotated as carbonic anhydrase5; and *Zm00001d020764*, regulated by *zma-miR528a-3p* and annotated as carbonic anhydrase5 (Supplementary Table S9).

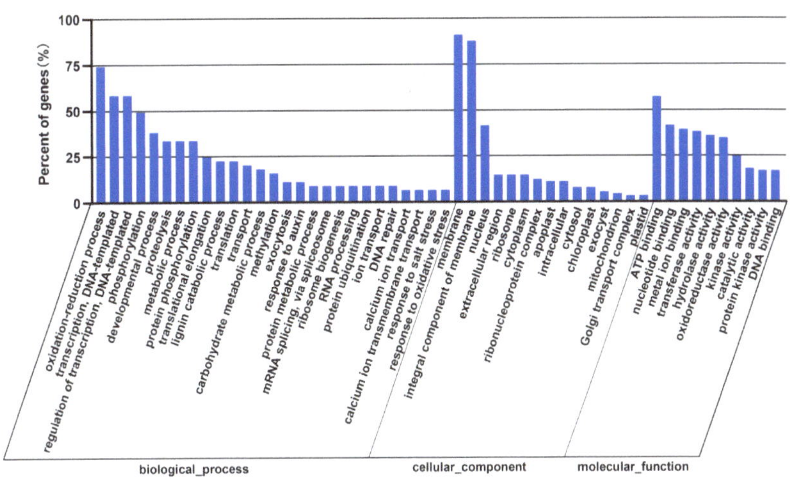

Figure 5. GO functional enrichment analysis of differentially expressed miRNA target genes in the Na_2CO_3 vs. H_2O comparison group.

3.5. Analysis of Differentially Expressed miRNAs Shared by Different Comparison Groups

To gain further insights into the potential mechanisms of miRNA responses to different salt stress conditions in maize roots, we analyzed the overlapping differentially expressed miRNAs between the NaCl vs. H_2O and Na_2CO_3 vs. H_2O comparison groups. We identified 36 differentially expressed miRNAs exhibiting consistent expression patterns in both groups, with 21 upregulated and 15 downregulated miRNAs (Supplementary Table S10). The upregulated conserved miRNAs belong to the MIR395, MIR172, MIR171, MIR169, MIR168, MIR162, and MIR156 families. The target genes of these miRNAs, including *Zm00001d040583*, *Zm00001d024477*, *Zm00001d009858*, *Zm00001d034635*, *Zm00001d039971*, *Zm00001d036593*, *Zm00001d014568*, etc., are enriched in the GO pathways, with processes such as metal ion binding, iron ion binding, flavonoid biosynthetic process, response to hormone, cellular response to DNA damage stimulus, DNA repair, and auxin-activated signaling pathway. In contrast, the downregulated, conserved miRNAs are primarily distributed among the MIR408, MIR396, MIR171, and MIR159 families. The target genes of these miRNAs, including *Zm00001d051265*, *Zm00001d019670*, *Zm00001d038107*, *Zm00001d007638*, etc., are enriched in the GO pathways via factors such as the integral component of the membrane, the regulation of the gibberellic acid-mediated signaling pathway, and the lipid metabolic process.

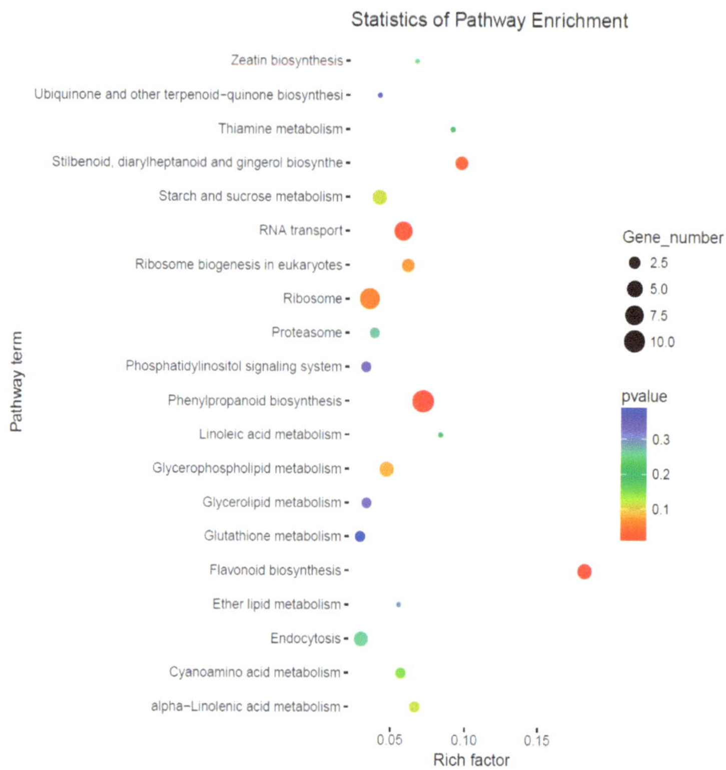

Figure 6. KEGG pathway enrichment analysis of differentially expressed miRNA target genes in the Na_2CO_3 vs. H_2O comparison group.

Apart from the miRNAs with consistent expression patterns, a subset of differentially expressed miRNAs showed contrasting expression patterns between the comparison groups (refer to Supplementary Table S11). A total of 58 miRNAs were found to be upregulated in the NaCl vs. H_2O group but downregulated in the Na_2CO_3 vs. H_2O group. These predominantly conserved miRNAs are members of the MIR162, MIR166, MIR167, MIR168, MIR169, MIR395, and MIR396 families. The target genes of these miRNAs, such as *Zm00001d018977*, *Zm00001d027412*, *Zm00001d018977*, *Zm00001d034493*, *Zm00001d026590*, *Zm00001d027874*, *Zm00001d028164*, *Zm00001d033981*, *Zm00001d015410*, etc., are enriched in GO pathways via factors including DNA binding, ATP binding, integral component of membrane, riboflavin biosynthetic process, response to hormone, CCAAT-binding factor complex, sulfate assimilation, metal ion binding, and more.

Conversely, 11 miRNAs were found to be downregulated in the NaCl vs. H_2O group but upregulated in the Na_2CO_3 vs. H_2O group. These predominantly conserved miRNAs belong to the MIR156, MIR167, MIR168, MIR169, MIR397, MIR528, and MIR2118 families. Their target genes, such as *Zm00001d004553*, *Zm00001d047587*, *Zm00001d024744*,

Zm00001d028837, *Zm00001d006682*, *Zm00001d015592*, *Zm00001d024477*, *Zm00001d033855*, *Zm00001d009532*, *Zm00001d020764*, etc., are associated with GO pathways via factors such as metal ion binding, glucose-6-phosphate dehydrogenase activity, peptidyl-tyrosine phosphorylation, nucleic acid binding, anaphase-promoting complex, oxidation-reduction process, integral component of membrane, auxin-activated signaling pathway, carbonate dehydratase activity, and more.

3.6. qPCR Validation of Sequencing Results

We selected 10 miRNAs from Supplementary Tables S10 and S11 for qPCR validation experiments. The qPCR validation results matched the sequencing data, confirming the reliability of the sequencing results (refer to Supplementary Figure S2).

3.7. The Changes in the Content of Different Ions in Maize Roots under Neutral and Alkaline Salt Stress

Excessive salt in the soil inhibits the absorption, translocation, and distribution of nutrients in plants, leading to an imbalance in plant ionic composition, thereby affecting plant physiological traits. Therefore, combining phenotypic observations and data analysis, we aim to investigate the changes in physiological indicators for each group under various salinity and alkalinity stress conditions, particularly the alterations in cations and anions, which could offer theoretical insights into the internal regulatory network shifts in maize under different salt–alkaline stresses. The testing of physiological and biochemical indicators in roots such as sodium, potassium, sulfate, calcium, and iron ions is proposed. We found that the content of various ions in different treatment groups was different. Compared with the normal water treatment group, except for Na^+ and $SO_4{}^{2-}$, the content of other ions showed a downward trend in the NaCl and Na_2CO_3 treatment groups. However, the degree of reduction was different, with Na^+, Ca^{2+}, $SO_4{}^{2-}$, and the K^+/Na^+ ratio being higher in the NaCl treatment group than in the Na_2CO_3 treatment group, except for K^+ content. In contrast, Fe^{2+} and Mg^{2+} showed an opposite trend, with these two metal ions having higher content in the Na_2CO_3 treatment group. The content of $SO_4{}^{2-}$ showed an upward trend in the NaCl treatment group and a downward trend in the Na_2CO_3 treatment group (Figure 7). These results suggest that plants mobilize different metabolic networks in response to neutral and alkaline salt stress.

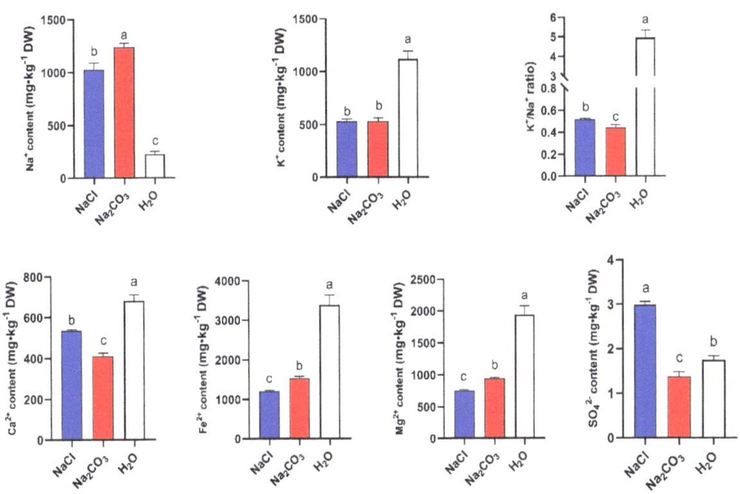

Figure 7. Alterations in ion content in maize roots under neutral and alkaline salts stresses, with different lowercase letters indicating statistical significance.

4. Discussion

Differential expression patterns of the same miRNA between salt-sensitive and salt-tolerant plants under salt stress indicate that the changes in miRNA expression may serve as a response mechanism to varying salt concentrations. Zhou et al. [24] employed a quantitative real-time polymerase chain reaction (qRT-PCR) to select five target genes belonging to the miRNA family, which are known to play a role in plant responses to salt stress. The study of plant miRNAs gained significant momentum from 2002 onwards, coinciding with the discovery of plant proteins involved in miRNA biogenesis. Satendra K et al. [25] provided a concise historical account highlighting the importance of miRNA discovery.

To identify the potential interactions between miRNAs and mRNAs in response to salt stress, a co-expression analysis was conducted using the same samples. The comparative expression analysis during germination, under both threshold (273 mM) and optimal (43 mM) NaCl treatments, revealed 13 differentially expressed miRNAs along with 23 corresponding target mRNAs. Furthermore, a comparison between forty-three mM NaCl and non-salt-stress conditions identified one differentially expressed miRNA and its corresponding target mRNA [26]. These findings furnish fundamental data for further investigation into the molecular mechanisms involved in the germination of salt-stressed seeds, while also serving as a reference for advancing salt tolerance during plant germination. In this study, we analyzed differentially expressed miRNAs in corn under NaCl stress and also under alkaline salt Na_2CO_3 stress, identifying 41 and 74 different miRNAs under these two stress conditions, respectively. We also found that under NaCl stress, a larger number of differentially expressed miRNAs were upregulated, accounting for over 90% of the total, whereas only 36% of the miRNAs were upregulated under Na_2CO_3 stress. Analyzing the number and expression patterns of different miRNAs in the two comparison groups also demonstrates that the factors mobilized by plants in response to Na_2CO_3 stress are much greater than those to NaCl stress, involving a more complex metabolic regulatory network mechanism.

In this study, we found that members of the MIR156, MIR169, MIR168, MIR162, MIR395, and MIR396 families were enriched in different treatment groups, with both similar and different expression patterns. This suggests that these miRNA members from the families may be involved in plant responses to different salt–alkali treatments, but their metabolic regulatory networks and mechanisms of action may be more complex. Wan et al. [27] found that the Osa-miR168-OsAGO1 module is involved in regulating rice salt stress tolerance. Researchers used STTM technology to silence miR168, which enhanced rice salt stress tolerance, with mutant plants showing significantly more roots and longer roots and stems under salt stress compared to wild-type plants. In this study, *zma-miR168a-5p* was downregulated under NaCl stress and upregulated under Na_2CO_3 stress. Its target gene, *Zm00001d006682*, is indeterminate domain 8, a member of the IDD family. IDD family genes encode transcription factors with zinc finger proteins and play a broad role in plant growth and development [28–31]. In Arabidopsis, Welch et al. [32] found that IDD transcription factor members JKD and MGP are part of the SCR-SHR complex. JKD and MGP are specifically expressed in root stem cells and are regulated by the *SCR* and *SHR* genes. Additionally, *SGR5* is expressed in the endodermis of the inflorescence stem, participating in the early gravitropic response of Arabidopsis stems [33]. It is evident that IDD genes not only regulate plant flowering but also participate in the regulation of plant growth and development. *zma-miR396c_L-1* was upregulated under NaCl stress and downregulated under Na_2CO_3 stress, regulating the target gene *Zm00001d018260*, which is Growth-regulating factor 6 (*GRF6*). At the transcriptional level, GRFs are typically regulated by the microRNA miR396 [34], and some studies have revealed other functions of GRFs in plant biology, such as flowering, seed, and root development; growth control under stress conditions; and the regulation of plant lifespan [35–41]. Yuan et al. [42] found that the Osa-miR396c-GRF module plays a significant role in regulating the salt stress tolerance of creeping bentgrass. The overexpression of *Osa-miR396c* in creeping bentgrass resulted in significantly reduced tiller number and length, and narrower and shorter leaves. Under salt stress, mutant plants showed increased leaf water content and chlorophyll content, reduced electrolyte leakage, and notably, an elevated expression of SALT OVERLY SENSITIVE1 (*OsSOS1*), leading to a decrease in the relative Na^+ content and an increase in the K^+/Na^+ ratio, demonstrating enhanced salt tolerance.

Members of the MIR408, MIR159, MIR171, and MIR172 families exhibited the same expression patterns in the two comparison groups, suggesting that these miRNA members may play a necessary role in plant responses to different salt–alkali treatments. In this study, *zma-miR159a-5p* was found to be downregulated in both comparison groups, regulating the target gene *Zm00001d029550*, which is diacylglycerol kinase 4. Diacylglycerol kinase (DGK) catalyzes the phosphorylation of lipids to produce diacylglycerol phosphate (PA), playing a significant role in biological activities. Previous studies have found that DGK plays an important role in plant responses to stress, such as in Arabidopsis, where Li et al. [43] discovered that *DGK5* and its catalytic product PA bind to *ABA2* and inhibit its activity, thereby affecting ABA synthesis under stress conditions and the plant's ability to cope with stress. Another differentially expressed miRNA, *zma-miR172e_L-1*, was upregulated in both comparison groups, regulating the target gene *Zm00001d035512*, which is AP2-EREBP-transcription factor 81. miR172, a highly conserved member of the miRNA family, is induced by various stresses such as low temperature, salt, and drought, and can also participate in plant responses to various environmental stresses by regulating AP2-type transcription factors [44–47]. Overexpressing *miRNA172c* or knocking out GmNNC1 in soybean can enhance its salt tolerance; conversely, interfering with *miR172* expression or overexpressing *GmNNC1* increases soybean's salt sensitivity [48]. Additionally, overexpressing soybean *miR172c* in Arabidopsis can significantly enhance its tolerance to drought and salt stress [49]. Although there is a relatively clear understanding of the mechanisms by which miR172 mediates plant responses to stress in Arabidopsis and soybean, research on other plants, especially important crops like corn, remains at the level of genome-wide expression analysis and has not yet involved the analysis of specific molecular mechanisms and the construction of regulatory networks.

Furthermore, members of the MIR166, MIR167, MIR397, MIR528, and MIR2118 families exhibited opposite expression patterns in the two comparison groups, indicating that these family members played unique roles under different salt and alkali stresses. *zma-miR167a-5p* was upregulated under NaCl stress and downregulated under Na_2CO_3 stress, regulating the target gene *Zm00001d026590*, which is ARF-transcription factor 30. Liu et al. [50] discovered that Zma-miR167 could reduce the transcription of *ZmARF3* and *ZmARF30*, resulting in decreased transcription levels of *ZmPAO1* and producing less hydrogen peroxide (H_2O_2), thereby enhancing maize's resistance to maize chlorotic mottle virus (MCMV). *zma-miR528a-3p* was downregulated under NaCl stress and upregulated under Na_2CO_3 stress, regulating the target gene *Zm00001d020764*, which is carbonic anhydrase5. Carbonic anhydrases (CAs) are a class of zinc-containing metalloenzymes that play an important role in various physiological processes such as cellular pH regulation, carbon dioxide transport, electrolyte balance, and cell homeostasis, which are crucial for cell survival and proliferation. The carbonyl sulfide (COS) absorbed by plants can be converted into CO_2 and Hydrogen Sulfide (H_2S) by carbonic anhydrase (CA) in cells [51]. High concentrations of H_2S tend to bind with ferrous ions, thereby modifying the structure of proteins containing ferrous ions (such as cytochrome oxidase, hemoglobin, myoglobin, etc.), which inhibits their activity and function, thus exhibiting cytotoxic effects [52–55].

In summary, miRNAs play a crucial role in plants' response to alkaline salt stress, but the current research on this topic is not sufficient. The specific effects of alkaline salt stress on plant miRNA expression and its regulatory mechanisms require further in-depth investigation. Our research aims to provide a theoretical basis for comprehensively understanding how crops respond to complex alkaline salt stress and for improving saline–alkali soils to expand crop cultivation areas. Through this research, we hope to provide a scientific basis for agricultural production, helping farmers to increase crop yields under saline–alkali conditions.

5. Conclusions

This study reveals the impact of soil salinization, a complex environmental factor, on plant gene expression by analyzing the differential expression of miRNAs in the roots of maize seedlings after treatment with 100 mM NaCl, 50 mM Na_2CO_3, and H_2O for 5 h. The results show that the Na_2CO_3 treatment group exhibited the highest number of differentially expressed miRNAs. The cluster analysis indicates that these differentially expressed miRNAs are primarily involved in the regulation of ion transport, binding, metabolism, phenylpropanoid, and flavonoid biosynthesis pathways. In contrast, the unique, differentially expressed miRNAs in the NaCl treatment group are associated with the sulfur metabolism pathways. This suggests that there are significant differences in the response patterns of maize to different saline–alkaline treatments. This study provides a theoretical basis and genetic resources for further analysis of the molecular mechanisms underlying salt–alkali tolerance in maize.

Supplementary Materials: The following supporting information can be downloaded at: https://www.mdpi.com/article/10.3390/cimb46080524/s1.

Author Contributions: Y.Y. and C.Z. designed the research; Z.C. and X.L. performed experiments; Z.C., Y.L., Q.W. and J.F. performed data analysis; Z.C. and Y.L. wrote the manuscript. Y.Y., C.Z., and X.L. reviewed the manuscript. All authors have read and agreed to the published version of the manuscript.

Funding: This work was supported by Jilin Scientific and Technological Development Program, grant number 20220101330JC; Jilin Agricultural Science and Technology Innovation Project, grant number CXGC2024ZD001 and CXGC2023SJ104; Germplasm Creation and Improvement of High-Starch High-Yield Corn in the Northeast Region, grant number NK202307040101.

Institutional Review Board Statement: Not applicable.

Informed Consent Statement: Not applicable.

Data Availability Statement: All relevant data are provided as figures or table within the paper.

Acknowledgments: We thank our contributors for their dedication and compliance through the many stages of this research as well as the editors and anonymous reviewers whose comments helped to greatly improve this paper.

Conflicts of Interest: The authors declare that they have no conflicts of interest or competing interests.

References

1. Zhang, L.M.; Liu, X.G.; Qu, X.N.; Yu, Y.; Han, S.P.; Dou, Y.; Xu, Y.Y.; Jing, H.C.; Hao, D.Y. Early transcriptomic adaptation to Na_2CO_3 stress altered the expression of a quarter of the total genes in the maize genome and exhibited shared and distinctive profiles with NaCl and high pH stresses. *J. Integr. Plant Biol.* **2013**, *55*, 1147–1165. [CrossRef] [PubMed]
2. Zhang, X.; Wei, L.; Wang, Z.; Wang, T. Physiological and molecular features of *Puccinellia tenuiflora* tolerating salt and alkaline-salt stress. *J. Integr. Plant Biol.* **2013**, *55*, 262–276. [CrossRef] [PubMed]
3. Yang, C.; Chong, J.; Li, C.; Kim, C.; Shi, D.; Wang, D. Osmotic adjustment and ion balance traits of an alkali resistant halophyte *Kochia sieversiana* during adaptation to salt and alkali conditions. *Plant Soil* **2007**, *294*, 263–276. [CrossRef]
4. Zhang, H.; Han, B.; Wang, T.; Chen, S.; Li, H.Y.; Zhang, Y.H.; Dai, S.J. Mechanisms of plant salt response: Insights from proteomics. *J. Proteome Res.* **2012**, *11*, 49–67. [CrossRef]
5. Munns, R.; Tester, M. Mechanisms of salinity tolerance. *Annu. Rev. Plant Biol.* **2008**, *59*, 651–681. [CrossRef]
6. Zhu, J.K. Plant salt tolerance. *Trends Plant Sci.* **2001**, *6*, 66–71. [CrossRef] [PubMed]
7. Rengasamy, P. Soil processes affecting crop production in salt-affected soils. *Funct. Plant Biol.* **2010**, *37*, 613–620. [CrossRef]
8. Parida, A.K.; Das, A.B. Salt tolerance and salinity effects on plants: A review. *Ecotoxicol. Environ. Saf.* **2005**, *60*, 324–349. [CrossRef]
9. Badger, M.R.; Price, G.D. The role of carbonic anhydrase in photosynthesis. *Annu. Rev. Plant Biol.* **1994**, *45*, 369–392. [CrossRef]
10. Miller, G.A.D.; Suzuki, N.; Ciftci-Yilmaz, S.; Mittler, R.O.N. Reactive oxygen species homeostasis and signalling during drought and salinity stresses. *Plant Cell Environ.* **2010**, *33*, 453–467. [CrossRef]
11. Sun, G.R.; Peng, Y.Z.; Shao, H.B.; Chu, L.Y.; Zhao, X.N.; Min, H.Y. Does *Puccinellia tenuiflora* have the ability of salt exudation? *Colloids Surf. B Biointerfaces* **2005**, *46*, 197–203.
12. Yu, J.; Chen, S.; Wang, T.; Sun, G.; Dai, S. Comparative proteomic analysis of *Puccinellia tenuiflora* leaves under Na_2CO_3 stress. *Int. J. Mol. Sci.* **2013**, *14*, 1740–1762. [CrossRef]

13. Kobayashi, S.; Abe, N.; Yoshida, K.T.; Liu, S.; Takano, T. Molecular cloning and characterization of plasma membrane-and vacuolar-type Na^+/H^+ antiporters of an alkaline-salt-tolerant monocot, *Puccinellia tenuiflora*. *J. Plant Res.* **2012**, *125*, 587–594. [CrossRef]
14. Zhong, N.Q.; Han, L.B.; Wu, X.M.; Wang, L.L.; Wang, F.; Ma, Y.H.; Xia, G.X. Ectopic Expression of a Bacterium NhaD-type Na^+/H^+ Antiporter Leads to Increased Tolerance to Combined Salt/Alkali Stresses. *J. Integr. Plant Biol.* **2012**, *54*, 412–421. [CrossRef]
15. Ma, Z.; Jiang, J.; Hu, Z.; Lyu, T.; Yang, Y.; Jiang, J.; Cao, J. Over-expression of miR158 causes pollen abortion in *Brassica campestris* ssp. *chinensis*. *Plant Mol. Biol.* **2016**, *93*, 313–326. [CrossRef]
16. Chen, Z.; Huo, Q.; Yang, H.; Jian, H.; Qu, C.; Lu, K.; Li, J. Joint RNA-Seq and miRNA profiling analyses to reveal molecular mechanisms in regulating thickness of pod canopy in *Brassica napus*. *Genes* **2019**, *10*, 591. [CrossRef]
17. Xu, T.; Zhang, L.; Yang, Z.; Wei, Y.; Dong, T. Identification and functional characterization of plant miRNA under salt stress shed light on salinity resistance improvement through miRNA manipulation in crops. *Front. Plant Sci.* **2021**, *12*, 665439. [CrossRef]
18. Chen, B.; Ding, Z.; Zhou, X.; Wang, Y.; Huang, F.; Sun, J.; Chen, J.; Han, W. Integrated full-length transcriptome and microRNA sequencing approaches provide insights into salt tolerance in mangrove (*Sonneratia apetala* Buch.-Ham.). *Front. Genet.* **2022**, *13*, 932832. [CrossRef]
19. Liu, M.; Li, M.; Ruan, J.; Jia, J.; Ge, C.; Cao, W. Analysis of microRNA Expression Profiles in Broiler Muscle Tissues by Feeding Different Levels of Guanidinoacetic Acid. *Curr. Issues Mol. Biol.* **2024**, *46*, 3713–3728. [CrossRef] [PubMed]
20. Kim, Y.-h.; Ra, W.-j.; Cho, S.; Choi, S.; Soh, B.; Joo, Y.; Lee, K.-W. Method Validation for Determination of Thalliumby Inductively Coupled Plasma Mass Spectrometry and Monitoring of Various Foods in South Korea. *Molecules* **2021**, *26*, 6729. [CrossRef] [PubMed]
21. Kurmanbayeva, A.; Brychkova, G.; Bekturova, A.; Khozin, I.; Sagi, M. Determination of Total Sulfur, Sulfate, Sulfite, Thiosulfate, and Sulfolipids in Plants. *Methods Mol. Biol.* **2017**, *1670*, 255–265.
22. Xu, J.; Li, H.D.; Chen, L.Q.; Wang, Y.; Liu, L.L.; He, L.; Wu, W.H. A protein kinase, interacting with two calcineurin B-like proteins, regulates K^+ transporter AKT1 in Arabidopsis. *Cell* **2006**, *125*, 1347–1360. [CrossRef] [PubMed]
23. AbdElgawad, H.; Peshev, D.; Zinta, G.; Van den Ende, W.; Janssens, I.A.; Asard, H. Climate extreme effects on the chemical composition of temperate grassland species under ambient and elevated CO_2: A comparison of fructan and non-fructan accumulators. *PLoS ONE* **2014**, *9*, e92044. [CrossRef] [PubMed]
24. Zhou, M.; Li, J.; Zhang, X.; Jia, Y. MicroRNA-mediated regulation of plant responses to abiotic stress. *J. Integr. Plant Biol.* **2012**, *54*, 885–897.
25. Satendra, K.M.; Surekha, A.; Sailaja, B.; Sheshu, M.M.; Voleti, S.R. MicroRNAs and their role in salt stress response in plants. *Genes Genom.* **2013**, *35*, 377–388.
26. Zhang, H.; Liu, X.; Yang, X.; Wu, H.; Zhu, J.; Zhang, H. MicroRNA–mRNA Integrated Analysis Reveals Roles for MicroRNAs in A Typical Halophyte, *Reaumuria soongorica*, During Seed Germination Under Salt Stress. *Plants* **2020**, *9*, 803.
27. Wan, J.; Meng, S.; Wang, Q.; Zhao, J.; Qiu, X.; Wang, L.; Li, J.; Lin, Y.; Mu, L.; Dang, K.; et al. Suppression of microRNA168 enhances salt tolerance in rice (*Oryza sativa* L.). *BMC Plant Biol.* **2022**, *22*, 563. [CrossRef]
28. Frankel, A.D.; Pabo, C.O. Fingering too many proteins. *Cell* **1988**, *56*, 675. [CrossRef] [PubMed]
29. Bland, J.M.; Altman, D.G. Multiple significance tests: The Bonferroni method. *BMJ* **1995**, *310*, 170. [CrossRef] [PubMed]
30. Takatsuji, H. Zinc-finger transcription factors in plants. *Cell. Mol. Life Sci.* **1998**, *54*, 582–596. [CrossRef]
31. Merk, H.L.; Yarnes, S.C.; Van Deynze, A. Trait diversity and potential for selection indices based on variation among regionally adapted processing tomato germplasm. *J. Am. Soc. Hort. Sci.* **2012**, *137*, 427–437. [CrossRef]
32. Welch, D.; Hassan, H.; Blilou, I.; Immink, R.; Heidstra, R. Arabidopsis JACKDAW and MAGPIE zinc finger proteins delimit asymmetric cell division and stabilize tissue boundaries by restricting SHORT-ROOT action. *Genes Dev.* **2007**, *21*, 2196–2204. [CrossRef] [PubMed]
33. Kim, J.Y.; Ryu, J.Y.; Baek, K.; Park, C.M. High temperature attenuates the gravitropism of inflorescence stems by inducing *SHOOT GRAVITROPISM 5* alternative splicing in Arabidopsis. *New Phytol.* **2016**, *209*, 265–279. [CrossRef]
34. Omidbakhshfard, M.A.; Proost, S.; Fujikura, U.; Mueller-Roeber, B. Growth-regulating factors (GRFs): A small transcription factor family with important functions in plant biology. *Mol. Plant* **2015**, *8*, 998–1010. [CrossRef] [PubMed]
35. Kim, J.H.; Choi, D.; Kende, H. The AtGRF family of putative transcription factors is involved in leaf and cotyledon growth in Arabidopsis. *Plant J.* **2003**, *36*, 94–104. [CrossRef] [PubMed]
36. Hewezi, T.; Baum, T.J. Complex feedback regulations govern the expression of miRNA396 and its GRF target genes. *Plant Signal. Behav.* **2012**, *7*, 749–751. [CrossRef]
37. Bao, M.; Bian, H.; Zha, Y.; Li, F.; Sun, Y.; Bai, B.; Chen, Z.; Wang, J.; Zhu, M.; Han, N. miR396a-mediated basic helix-loop-helix transcription factor *bHLH74* repression acts as a regulator for root growth in Arabidopsis seedlings. *Plant Cell Physiol.* **2014**, *55*, 1343–1353. [CrossRef]
38. Debernardi, J.M.; Mecchia, M.A.; Vercruyssen, L.; Smaczniak, C.; Kaufmann, K.; Inzé, D.; Rodriguez, R.E.; Palatnik, J.F. Post-transcriptional control of GRF transcription factors by microRNA miR396 and GIF co-activator affects leaf size and longevity. *Plant J.* **2014**, *79*, 413–426. [CrossRef]
39. Liu, H.; Guo, S.; Xu, Y.; Li, C.; Zhang, Z.; Zhang, D.; Xu, S.; Zhang, C.; Chong, K. OsmiR396d-regulated OsGRFs function in floral organogenesis in rice through binding to their targets *OsJMJ706* and *OsCR4*. *Plant Physiol.* **2014**, *165*, 160–174. [CrossRef]

40. Pajoro, A.; Madrigal, P.; Muiño, J.M.; Matus, J.T.; Jin, J.; Mecchia, M.A.; Debernardi, J.M.; Palatnik, J.F.; Balazadeh, S.; Arif, M.; et al. Dynamics of chromatin accessibility and gene regulation by MADS-domain transcription factors in flower development. *Genome Biol.* **2014**, *15*, R41. [CrossRef]
41. Liang, G.; He, H.; Li, Y.; Yu, D. Molecular mechanism of microRNA396 mediating pistil development in Arabidopsis. *Plant Physiol.* **2015**, *170*, 2490–2504. [CrossRef] [PubMed]
42. Yuan, S.; Zhao, J.; Li, Z.; Hu, Q.; Yuan, N.; Zhou, M.; Xia, X.; Noorai, R.; Saski, C.; Li, S.; et al. MicroRNA396-mediated alteration in plant development and salinity stress response in creeping bentgrass. *Horticulturae* **2019**, *6*, 48. [CrossRef] [PubMed]
43. Li, J.; Yao, S.; Kim, S.C.; Wang, X. Lipid phosphorylation by a diacylglycerol kinase suppresses ABA biosynthesis to regulate plant stress responses. *Mol. Plant* **2024**, *17*, 342–358. [CrossRef] [PubMed]
44. Zhou, X.; Wang, G.; Sutoh, K.; Zhu, J.K.; Zhang, W. Identification of cold-inducible microRNAs in plants by transcriptome analysis. Biochim. Biophys. *Acta Gene Regul. Mech.* **2008**, *1779*, 780–788.
45. Frazier, T.P.; Sun, G.L.; Burklew, C.E.; Zhang, B.H. Salt and drought stresses induce the aberrant expression of microRNA genes in tobacco. *Mol. Biotechnol.* **2011**, *49*, 159–165. [CrossRef] [PubMed]
46. Yang, J.W.; Zhang, N.; Ma, C.Y.; Qu, Y.; Si, H.J.; Wang, D. Prediction and verification of microRNAs related to proline accumulation under drought stress in potato. *Comput. Biol. Chem.* **2013**, *46*, 48–54. [CrossRef] [PubMed]
47. Candar-Cakir, B.; Arican, E.; Zhang, B.H. Small RNA and degradome deep sequencing reveals drought-and tissue-specific microRNAs and their important roles in drought-sensitive and drought-tolerant tomato genotypes. *Plant Biotechnol. J.* **2016**, *14*, 1727–1746. [CrossRef] [PubMed]
48. Sahito, Z.A.; Wang, L.; Sun, Z.; Yan, Q.; Zhang, X.; Jiang, Q.; Ullah, I.; Tong, Y.; Li, X. The miR172c-NNC1 module modulates root plastic development in response to salt in soybean. *BMC Plant Biol.* **2017**, *17*, 229. [CrossRef] [PubMed]
49. Zou, Y.; Wang, Y.; Wang, L.; Yang, L.; Wang, R.; Li, X. miR172b controls the transition to autotrophic development inhibited by ABA in Arabidopsis. *PLoS ONE* **2013**, *8*, e64770. [CrossRef]
50. Liu, X.; Liu, S.; Chen, X.; Prasanna, B.M.; Ni, Z.; Li, X.; He, Y.; Fan, Z.; Zhou, T. Maize miR167-ARF3/30-polyamine oxidase 1 module-regulated H_2O_2 production confers resistance to maize chlorotic mottle virus. *Plant Physiol.* **2022**, *189*, 1065–1082. [CrossRef]
51. Yamasaki, H.; Cohen, M.F. Biological consilience of hydrogen sulfide and nitric oxide in plants: Gases of primordial earth linking plant, microbial and animal physiologies. *Nitric Oxide* **2016**, *55/56*, 91–100. [CrossRef]
52. Hancock, J.T.; Whiteman, M. Hydrogen sulfide signaling: Interactions with nitric oxide and reactive oxygen species. *Ann. N. Y. Acad. Sci.* **2016**, *1365*, 5–14. [CrossRef]
53. Li, Z.G.; Min, X.; Zhou, Z.H. Hydrogen Sulfide: A Signal Molecule in Plant Cross-Adaptation. *Front. Plant Sci.* **2016**, *7*, 1621. [CrossRef]
54. Corpas, F.J. Hydrogen sulfide: A new warrior against abiotic stress. *Trends Plant Sci.* **2019**, *24*, 983–988. [CrossRef]
55. Hancock, J.H. Hydrogen sulfide and environmental stresses. *Environ. Exp. Bot.* **2019**, *161*, 50–56. [CrossRef]

current issues in molecular biology

MDPI

Article

Functional Validation of Different Alternative Splicing Variants of the *Chrysanthemum lavandulifolium ClNUM1* Gene in Tobacco

Wenxin Zhang, Hai Wang, Yuning Guo, Xueying Hao, Yanxi Li, Wenting He, Xiang Zhao, Shiyi Cai and Xuebin Song *

College of Landscape Architecture and Forestry, Qingdao Agricultural University, Qingdao 266109, China; 20212210009@stu.qau.edu.cn (W.Z.); wh672831@126.com (H.W.); 20212110007@stu.qau.edu.cn (Y.G.); 20212210002@stu.qau.edu.cn (X.H.); lyx33091@126.com (Y.L.); qauylhwt@163.com (W.H.); crazyz18@163.com (X.Z.); qaucaishiyi@163.com (S.C.)

* Correspondence: xuebinsong@sina.cn

Abstract: The Asteraceae are widely distributed throughout the world, with diverse functions and large genomes. Many of these genes remain undiscovered and unstudied. In this study, we discovered a new gene *ClNUM1* in *Chrysanthemum lavandulifolium* and studied its function. In this study, bioinformatics, RT-qPCR, paraffin sectioning, and tobacco transgenics were utilized to bioinformatically analyze and functionally study the three variable splice variants of the unknown gene *ClNUM1* cloned from *C. lavandulifolium*. The results showed that *ClNUM1.1* and *ClNUM1.2* had selective 3′ splicing and selective 5′ splicing, and *ClNUM1.3* had selective 5′ splicing. When the corresponding transgenic tobacco plants were subjected to abiotic stress treatment, in the tobacco seedlings, the *ClNUM1.1* gene and the *ClNUM1.2* gene enhanced salt and low-temperature tolerance and the *ClNUM1.3* gene enhanced low-temperature tolerance; in mature tobacco plants, the *ClNUM1.1* gene was able to enhance salt and low-temperature tolerance, and the *ClNUM1.2* and *ClNUM1.3* genes were able to enhance low-temperature tolerance. In summary, there are differences in the functions of the different splice variants and the different seedling stages of transgenic tobacco, but all of them enhanced the resistance of tobacco to a certain extent. The analysis and functional characterization of the *ClNUM1* gene provided new potential genes and research directions for abiotic resistance breeding in *Chrysanthemum*.

Keywords: *C. lavandulifolium*; *ClNUM1*; alternative splicing; abiotic stress

Citation: Zhang, W.; Wang, H.; Guo, Y.; Hao, X.; Li, Y.; He, W.; Zhao, X.; Cai, S.; Song, X. Functional Validation of Different Alternative Splicing Variants of the *Chrysanthemum lavandulifolium ClNUM1* Gene in Tobacco. *Curr. Issues Mol. Biol.* **2024**, *46*, 5242–5256. https://doi.org/10.3390/cimb46060314

Academic Editor: Vijai Bhadauria

Received: 8 April 2024
Revised: 10 May 2024
Accepted: 17 May 2024
Published: 25 May 2024

1. Introduction

Chrysanthemums are traditional Chinese flowers and among the most important ornamental flowers in the world, with high ornamental and application value [1]; however, at the same time, *Chrysanthemums* are very susceptible to abiotic stresses, which greatly limits their application, and their genetic background is relatively complex, so progress in *Chrysanthemum* resistance research has been relatively slow [2]. *C. lavandulifolium* is a diploid plant of the genus *Chrysanthemum* in the family Asteraceae with a relatively simple genetic background and is also one of the important parents of modern *Chrysanthemum* species, making it a model plant for the study of plants in the family Asteraceae. Therefore, genetics studies in *C. lavandulifolium* are important for resistance breeding in *Chrysanthemum* [3].

Studies on the resistance of Chrysanthemum plants have been carried out in many aspects, such as heat resistance and insect resistance, and the main bases are NAC, MYB, and WRKY transcription factors family [4–12]. For example, the *SND1* and *NST1* genes in the NAC family regulate secondary cell wall and lignin synthesis, increasing the broad-spectrum resistance of *Leucanthemella linearis* [13]. *Chrysanthemum* × *morifolium*

CmWRKY15-1 enhanced resistance to chrysanthemum white rust by regulating the expression of *CmNPR1* [14]. *Chrysanthemum indicum* var.aromaticum *CiMYB32* responds to drought stress [15]. In the process of studying the *NST1* gene of *C. lavandulifolium*, we accidentally cloned the unknown gene *ClNUM1* and its alternative splicing and the gene has a conserved structural domain of Nuclease-associated DNA-binding domain 3 (NUMOD3).

NUMOD3 is a homing endonucleases and related proteins. Sitbon et al. identified four new short conserved sequence structural domains in homing endonucleases and related proteins [16]. NUMODs were named the new structural domains nuclear-associated modular DNA-binding structural domains. These domains are modular and occur in various combinations. One structural domain consists of a motif whose structure has been described as a new sequence-specific DNA-binding helix. Sequence similarity suggests that the other two structural domains are novel helix–turn–helix DNA-binding domains. Four families of homing endonucleases are known—HNH, GIY-YIG, His-Cys box, and LAGLIDADG—each characterized by and named after a short-conserved sequence motif in the structural domain of its nuclease [17]. Among them, NUMOD3 is found in single-stranded and tandem repeats of GIY-YIG and HNH proteins and includes the DNA-binding domain of the I-TevI homologous endonuclease [18,19]. It forms a unique extended structure that wraps around the DNA. This region binds to DNA in a sequence-specific manner, helically inserting and twisting the DNA groove [20].

Alternative splicing events, including alternative 5′ and 3′ splice site selection, exon skipping, and intron retention [21], are widespread mechanisms in eukaryotes [22]. In alternative splicing, a single gene produces different protein variants through different splice site combinations [23–25]. Selective splicing is involved in the regulation of various plant biological activities, including plant responses to biotic and abiotic stresses [26,27], regulation of plant flowering [28], and regulation of the plant biological clock [29]. Chaudhary et al. proposed that under stress conditions, plants buffer the level of normal protein synthesis by alternative splicing, reduce the translation of a large portion of the transcriptome, and produce protein isoforms that are adapted to the protein isoforms needed under stress [30]. Splice variants have been studied in plants such as maize, rice, citrus, and *Arabidopsis thaliana* [31–34], but there are still few studies on *Chrysanthemum*.

Since the conserved structural domain of these three splice variants is NUMOD3, we named these three splice variants *ClNUM1.1*, *ClNUM1.2*, and *ClNUM1.3*. The nucleotide sequences of the three variable splices were compared with the homologous gene *CsNST1* in the NCBI database, and the homology of the nucleotide sequences was found to be as high as 83%. Afterward, the three splice variants of this gene were transferred into tobacco, and the biological functions of the different variants in the model plant tobacco were identified via RT-qPCR, morphometric indexes, physiological indexes, and paraffin sections, to study the function of *ClNUM1* in plants under abiotic stress so as to explore potential functional genes for resistance breeding of *Chrysanthemum*.

2. Materials and Methods

2.1. Plant Materials, Vectors and Strains

The *C. lavandulifolium* histocultures and wild-type big-leaf tobacco (*Nicotiana tabacum* var. *macrophylla*), vector Super35S::GFP, *E. coli* strain DH5α, and Agrobacterium strain GV3101 used in this study were obtained from the group of Associate Prof. Xuebin Song at Qingdao Agricultural University.

2.2. Homologous Gene Cloning

Cynara scolymus is a plant in the family Asteraceae and is closely related to *C. lavandulifolium*. The coding region of the *CsNST1* gene found in the NCBI database (accession number LOC112517473, NCBI) was used as a template for primer design, and the primer sequences were as follows (Table 1):

Table 1. The primer sequence for *CsNST1* gene.

	***CsNST1* Primer Sequence**
NST1-F	5′-ATGCTGTCTTCTACTTTGTAAGCTC-3′
NST1-R	5′-TTAATACCCTGTTTCAGAAAATGGATG-3′

Then, the gene was cloned using *C. lavandulifolium* as a template according to the instructions of 2 × Phanta Flash Master Mix (Dye Plus) (Vazyme Biotech, Nanjing, China), and the obtained products were subjected to agarose gel electrophoresis migration experiments and then recovered by gel recovery using a FastPure Gel DNA Extraction Mini Kit (Vazyme Biotech, Nanjing, China), and the obtained products were sent to the company for sequencing to obtain different splice variants of *ClNUM1* gene: *ClNUM1.1*, *ClNUM1.2* and *ClNUM1.3*.

2.3. *Stress Treatment of Chrysanthemum lavandulifolium*

We transplanted *C. lavandulifolium* seedlings of 5 cm in height and good growth condition from sterile planting bottles (1/2 MS + 30 g/L sucrose + 6 g/LAGAR) and planted them in an artificial climate chamber at 25 °C, 16 h of light, 8 h of darkness, 3000 lux of light and 70% air humidity. Different treatments were applied to *C. lavandulifolium* after one week of incubation. The plants exposed to salt treatment were watered with 200 mmol/L NaCl, the drought treatment simulates drought conditions by not watering for a specified period of time, and the plants exposed to low-temperature treatment were placed in an incubator at 4 °C and sampled at 0 h, 1 h, 4 h, 8 h, 12 h, 24 h, and 36 h. We carried out at least three replicates per treatment.

2.4. *RNA Extraction and RT-qPCR*

We carried out RT-qPCR using wild-type *C. lavandulifolium* under different treatments to verify the expression patterns of the three transcripts under stress conditions. A certain amount of *C. lavandulifolium* and tobacco tissues was quick-frozen in liquid nitrogen and then quickly ground into powder and transferred to RNase-free 2 mL centrifuge tubes. Then, total RNA was extracted according to the instructions for the FastPure Plant Total RNA Isolation Kit (Vazyme Biotech, Nanjing, China). cDNA synthesis was performed according to the instructions for the HiScript III RT SuperMix for qPCR (+gDNA wiper) kit (Vazyme Biotech, Nanjing, China). RT-qPCR was performed on a StepONE Plus system (Applied Biosystems, Waltham, MA, USA) using SYBR qPCR Master Mix (Vazyme Biotech, Nanjing, China) and specific primers. The data obtained were analyzed by the $2^{-\Delta\Delta Ct}$ method, and the internal reference gene was *Actin7* (Table 2).

Table 2. The primer sequence for *Actin7* gene.

	***Actin7* Primer Sequence**
Actin7-F	5′-TCCGGCTATGTATGTTGCTATTC-3′
Actin7-R	5′-AATCTTCATCAAGGGATCGGTAAG-3′

2.5. *Agrobacterium Transformation Method*

Three splice variants were recombined with the overexpression vector Super35S::GFP using the homologous recombination kit from Vazyme (ClonExpress® Ultra One Step Cloning Kit). The recombinant plasmids were transferred into the receptor cells of Agrobacterium tumefaciens GV3101. The transformed A. tumefaciens colonies were selected on LB-agar plates containing 50 mg L^{-1} kanamycin, 50 mg L^{-1} rifampicin, and 50 mg L^{-1} gentamicin. Positive colonies were picked for PCR amplification and identification, after which the colonies were subjected to the configuration of infiltration solution and transformed tobacco as described [35]. After screening and characterization, the transgenic tobaccos were obtained.

2.6. Transgenic Tobacco Resistance Screening and Management and Maintenance Methods

In this study, Super35S::GFP was used as the control group (CK), and Super35S::GFP, Super35S::*ClNUM1.1*, Super35S::*ClNUM1.2*, and Super35S::*ClNUM1.3* tobacco seeds were sown in screening medium (MS + 50 mg/L Hgy + 30 g/L sucrose + 6 g/L agar) and incubated at 25 °C with 16 h of light and 8 h of darkness and 3000 lux light intensity. After the growth of true leaves, tobacco seedlings with similar growth status were picked and transplanted to the stress treatment medium in an aseptic environment as well as to an artificial climate chamber for subsequent treatment experiments.

2.7. Abiotic Stress Treatments and Morphometric Measurements

2.7.1. Stress Treatment of Transgenic Tobacco at the Seedling Stage

Tobacco seedlings with the same growth status in the screening medium were transferred to MS, MS + 200 mmol/L NaCl, and MS + 200 μmol/L ABA(Abscisic-Acid) media, ABA is used to simulate drought stress [13]. The incubation conditions of the control, salt, and ABA treatments were at 25 °C, 16 h of light and 8 h of darkness, and 3000 lux light intensity, and that of the low-temperature treatment was at 4 °C, 16 h of light and 8 h of darkness, and 3000 lux light intensity. After 24 days of treatment in the medium, the tobacco plants were carefully removed, and the roots were carefully cleaned from the medium. After cleaning, the control, salt, ABA, and low-temperature treated tobacco plants were subjected to fresh weight and root length determination (Table S1).

2.7.2. Stress Treatment of Transgenic Tobacco at the Mature Stage

Tobacco seedlings with the same growth status in the screening medium were selected and transferred to a charcoal substrate for two to three weeks and were subjected to control, salt, drought, and low-temperature treatments when they reached maturity. The control treatments were watered every other week, the salt treatments were watered with 200 mmol/L NaCl every other week, the drought treatments were subjected to simulated drought conditions by artificially controlling the water content of the soil, and the low-temperature treatments were placed in a 4 °C incubator. The control, drought, and salt treatments were all incubated at 25 °C. After 15 days of treatment in the artificial climate chamber, the growth status of the tobacco was photographed, and the height of the stalks was measured. Afterward, the stress-treated tobacco was rehydrated, and the height of the stalks was measured and photographed after seven days (Table S2).

2.8. Measurement of Physiological Parameters

To better illustrate the function of the splice variants, the chlorophyll content was determined in control and treated tobacco seedlings grown in an artificial climate chamber. Leaf samples were rinsed with distilled water and blotted dry with filter paper. The sides of the main veins of the leaves were cut into filaments <1 mm in width, and a 0.2 g sample was weighed out. The leaves were then immersed in 25 mL of 95% ethanol and left in the dark for 3 days. Absorbance was measured at wavelengths of 470 mm, 649 mm, and 665 mm using a UV spectrophotometer (Hitachi; Tokyo, Japan), and, finally, the total chlorophyll content was calculated.

2.9. Paraffin Sectioning

We selected prime-aged tobacco with the same growth state and cut a 1.5 cm long part from 1/3 of the stem of the transgenic tobacco. The selected stalks were treated in FAA fixative (70% alcohol/formalin/acetic acid 18:1:1) for 24 h. The material was soaked in a mixture of hydrogen peroxide and glacial acetic acid (1:1) for 48 h to soften the material and then dehydrated with ethanol and embedded in paraffin wax. The samples were divided into 10 μm sections by a slicer. Afterward, fenugreek solid green staining was performed, and the thickness of the cell wall and the size of the cells were observed using a light microscope.

2.10. Statistical Analysis

All experiments were set up with three or more biological replicates. Data are expressed as the mean ± SD (standard deviation) and were analyzed by a *t*-test (* $p \leq 0.05$ and ** $p \leq 0.01$).

3. Results

3.1. Bioinformatics Analysis of the ClNUM1 Splice Variants

Three alternative splicing variants of *ClNUM1* were cloned from *C. lavandulifolium*, named *ClNUM1.1*, *ClNUM1.2*, and *ClNUM1.3* (Figure S1). *ClNUM1.1* was 1164 bp in total length and encoded 370 amino acids, while *ClNUM1.2* was 1176 bp and encoded 373 amino acids; *CLNUM1.3* was 1134 bp and encoded 362 amino acids (Figure 1a). *ClNUM1.1* and *ClNUM1.2* had both selective 3′ splicing and selective 5′ splicing, and *ClNUM1.3* had selective 5′ splicing (Figure 1b).

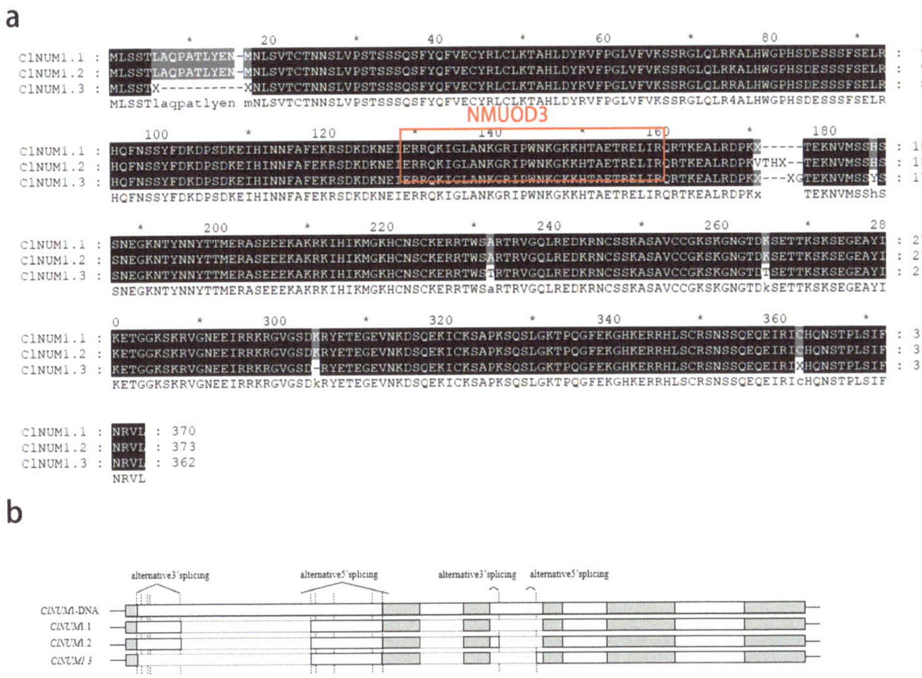

Figure 1. Bioinformatics analysis of three different splice variants. (**a**) Comparison of amino acid sequences. Black is the same part of the amino acid, gray is the different part of the amino acid. (* represents the length sequence number of the omitted amino acid.) (**b**) Schematic structure of *ClNUM1.1*, *ClNUM1.2*, and *ClNUM1.3*. The gray parts represent exons, and the white parts represent introns.

However, three variable splices of this gene were found in the cloning results. Alternative splicing generates multiple mRNA transcripts from a single gene by assembling exons differently using selective splice sites in precursor mRNAs.

Further analysis of the three variables found that the conserved structural domains of these three variable splices were different from those of the NST gene, and the three variable splices belonged to the conserved structural domain of NUMOD3 after comparison.

3.2. *ClNUM1.1*, *ClNUM1.2*, and *ClNUM1.3* in *C. lavandulifolium* Show Different Responses to Stress Treatments

To further investigate the response of *ClNUM1.1*, *ClNUM1.2*, and *ClNUM1.3* in *C. lavandulifolium* under abiotic stress conditions, we comparatively analyzed the expression levels of the *ClNUM1.1*, *ClNUM1.2*, and *ClNUM1.3* genes in wild-type *C. lavandulifolium* under different stress treatments (Figure 2).

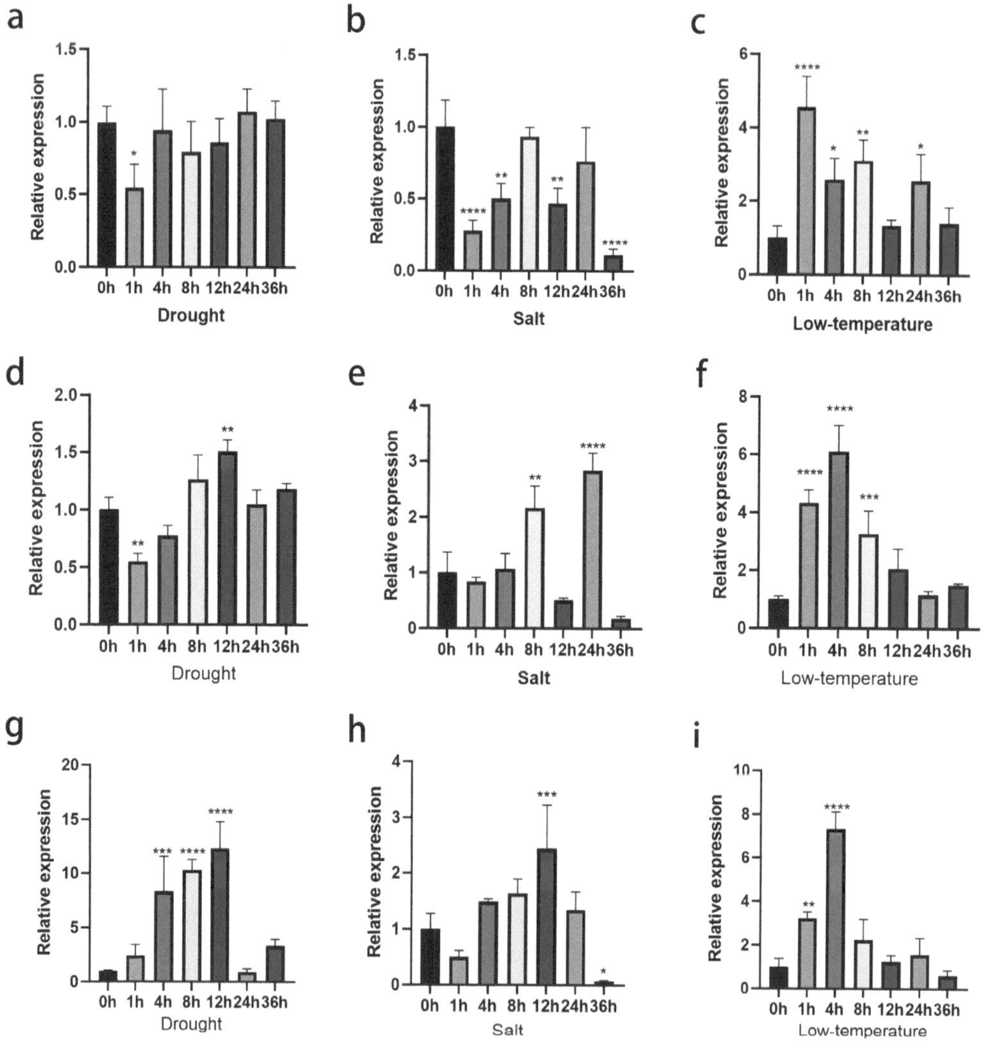

Figure 2. Expression levels of *ClNUM1.1*, *ClNUM1.2*, and *ClNUM1.3* in *C. lavandulifolium* under stress treatments. (**a**–**c**) Expression levels of *ClNUM1.1* in *C. lavandulifolium* under drought, salt, and low-temperature treatments. (**d**–**f**) Expression levels of *ClNUM1.2* in *C. lavandulifolium* under drought, salt, and low-temperature treatments. (**g**–**i**) Expression levels of *ClNUM1.3* in *C. lavandulifolium* under drought, salt, and low-temperature treatments. All experiments were set up with three or more biological replicates (independent sample *t*-test; * $p < 0.05$, ** $p < 0.01$, *** $p < 0.001$, **** $p < 0.0001$).

The expression of *ClNUM1.1* under drought and salt treatments was lower than that before the treatments, with a minimum decrease of 45.5% at 1 h in drought treatment and a minimum decrease of 88.7% at 36 h in salt treatment, while the expression of *ClNUM1.1* under low-temperature treatment showed an overall trend higher than that before the treatments and reached a peak at 1 h, with a 3.56-fold increase in expression. Under drought treatment, the expression of *ClNUM1.2* at 1 h and 12 h was significantly different from that before treatment, with a decrease of 45% and an increase of 50.7%, respectively. Under salt treatment, the expression increased significantly at 8 h and 24 h, with an increase of 1.16-fold and 1.84-fold, respectively, and the expression increased under low-temperature treatment compared with that before treatment, with increases of 3.34-fold, 5.13-fold, and 2.29-fold at 1 h, 4 h, and 8 h, respectively. The expression of *ClNUM1.3* under drought treatment was elevated compared with that before treatment, increasing 7.30-fold, 9.30-fold, and 11.33-fold at 4 h, 8 h, and 12 h. Under salt treatment the expression of *ClNUM1.3* was highest at 12 h, increasing 1.44-fold, and there was a significant increase in the expression of *ClNUM1.3* under low-temperature treatment at 1 h and 4 h, increasing 2.22-fold and 6.33-fold, respectively, compared with that before treatment.

These results clearly show that *ClNUM1.1*, *ClNUM1.2*, and *ClNUM1.3* in *C. lavandulifolium* all responded under the stress treatments, but the different splice variants responded differently to each treatment.

3.3. Phenotypic Analysis of Young ClNUM1 Transgenic Tobacco under Abiotic Stress

To investigate the function of the *ClNUM1.1*, *ClNUM1.2*, and *ClNUM1.3* splice variants, the present study was carried out to determine the root length, fresh weight, and chlorophyll content of the treated transgenic tobacco by abiotic stress treatment. After 24 days of treatment, we found that compared with before treatment, tobacco grew significantly under CK, NaCl, and low-temperature treatments, but the growth signs were not obvious under ABA conditions (Figure 3). The measurements showed that there was no significant difference in root length among the transgenic tobacco plants at 25 °C with 16 h of light and 8 h of darkness and 3000 lux, and the fresh weight of *ClNUM1.1* and *ClNUM1.3* was not significantly different from that of CK; the fresh weight of *ClNUM1.2* increased by 47.5% compared with that of CK; under ABA treatment, *ClNUM1.2* and *ClNUM1.3* showed an increase in root length and fresh weight compared to CK, their root length increased by 32.7% and 18.2%, respectively, and their fresh weight increased by 53.4%, showing a significant difference, while *ClNUM1.1* showed an increase in root length by 106.1% compared to CK, and its fresh weight was not significantly different; the root length and fresh weight of all three transgenic tobacco lines were increased under low-temperature treatment compared with CK, with a 30.6%, 35.2%, and 16.5% increase in root length and 251%, 183.8%, and 87.4% increase in fresh weight for *ClNUM1.1*, *ClNUM1.2*, and *ClNUM1.3*, respectively, showing significant differences; under salt treatment, the fresh weight of all three transgenic tobacco lines was increased and significantly different from CK, and the root length of *ClNUM1.1* and *ClNUM1.2* was increased by 10.8% and 39%, respectively, compared to that of CK, but the root length of *ClNUM1.3* showed little change compared to that of CK (Figure 4). We found that the growth of CK was inhibited under all the stress treatments, as well as the growth of the *ClNUM1.1*, *ClNUM1.2*, and *ClNUM1.3* transgenic tobacco lines under ABA treatment.

The experimental results showed that *ClNUM1.1* transgenic tobacco had a strong ability to tolerate salt and low temperature at the seedling stage; *ClNUM1.2* transgenic tobacco had a strong ability to tolerate salt and low temperature; and *ClNUM1.3* transgenic tobacco had a strong ability to tolerate low temperature.

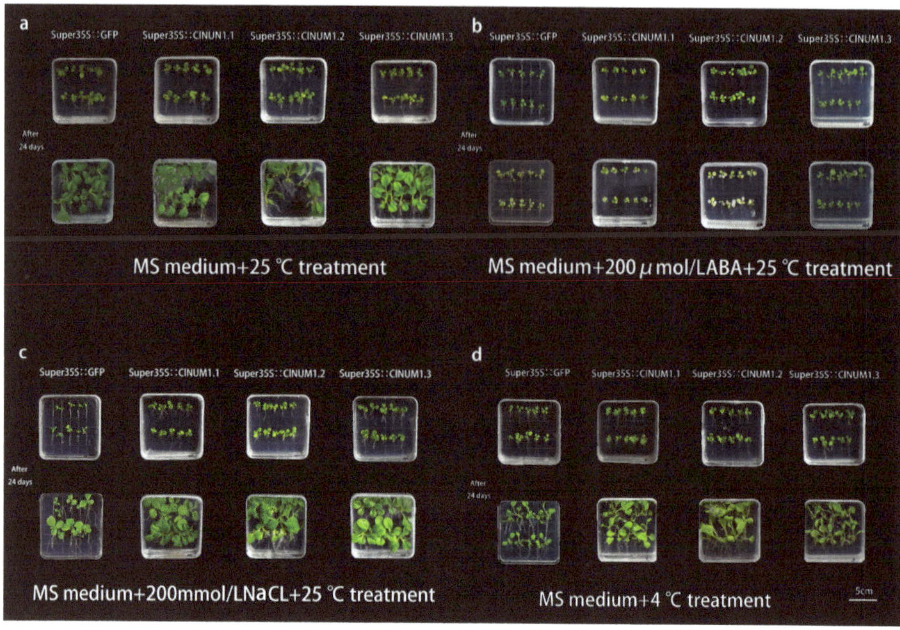

Figure 3. Growth status of tobacco under different treatments in the medium: (**a**) growth status under control treatment, (**b**) growth status under ABA treatment, (**c**) growth status under salt treatment, and (**d**) growth status under low-temperature treatment.

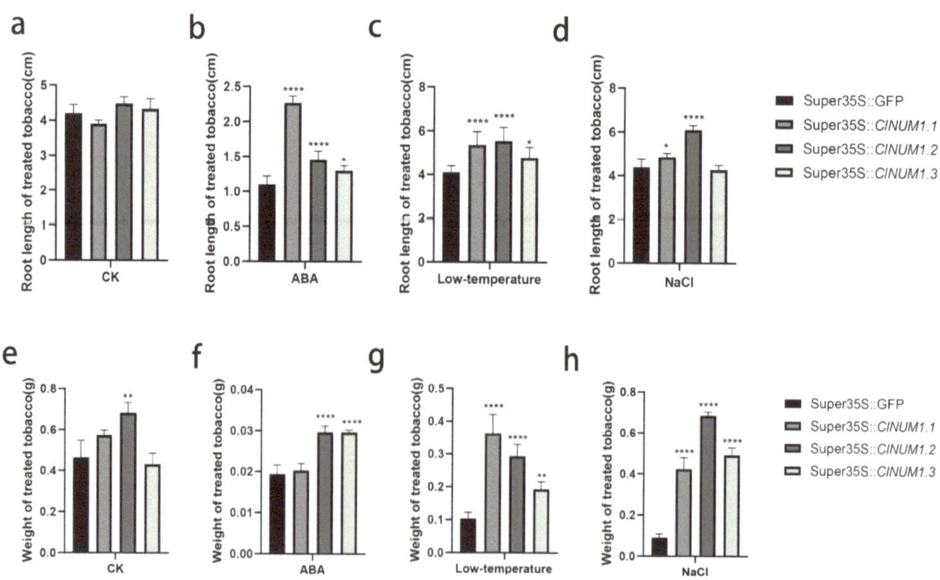

Figure 4. Root length and fresh weight of tobacco under different treatments after 24 days. (**a**–**d**) Root length of transgenic tobacco under control, drought, low-temperature, and salt treatments. (**e**–**h**) Fresh weight of transgenic tobacco under control, drought, low-temperature, and salt treatments All experiments were set up with three or more biological replicates (independent sample t-test; * $p < 0.05$, ** $p < 0.01$, **** $p < 0.0001$).

3.4. Phenotypic Study of ClNUM1 Transgenic Tobacco at Maturity under Abiotic Stress and Determination of Physiological Indexes

To further illustrate the function of the splice variants under abiotic stress treatments, we observed the growth status of mature transgenic tobacco seedlings under stress treatments and statistically compiled the changes in the stem height of tobacco before and after rehydration under the stress treatments, as well as the chlorophyll content of the transgenic tobacco after the treatments.

In this study, it was found that under the control treatment, transgenic tobacco did not show a significant difference in stem height change after treatment with the same growth rate, but after rewatering, *ClNUM1.1*, *ClNUM1.2*, and *ClNUM1.3* showed significantly slower growth rates compared to CK (Figure 5).

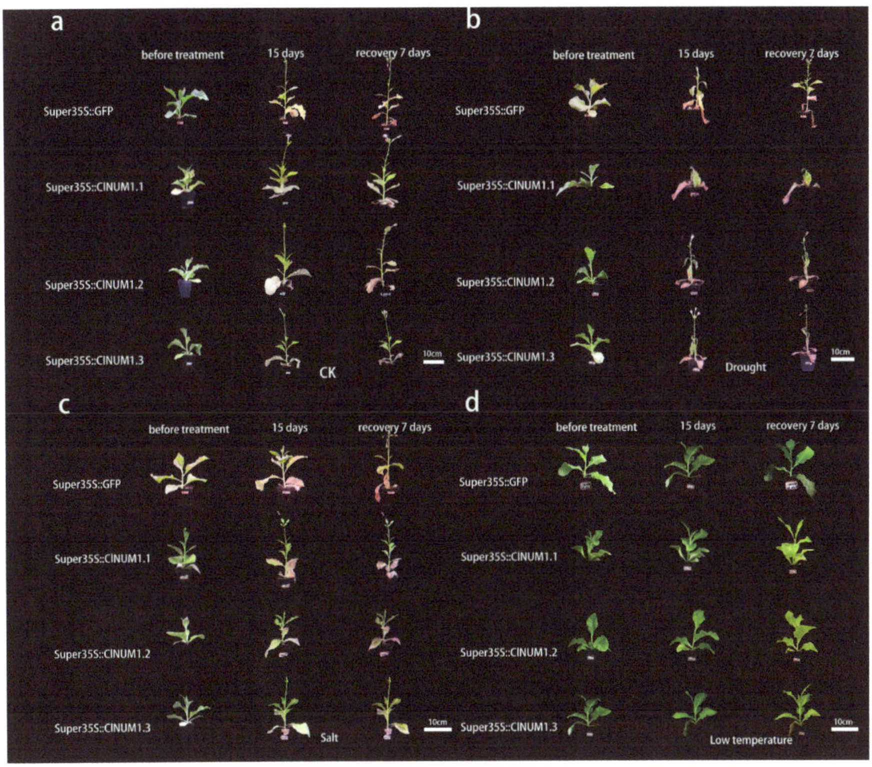

Figure 5. Growth states of mature tobacco seedlings under different treatments. (**a**) Growth state under the control treatment. (**b**) Growth state under drought treatment. (**c**) Growth state under salt treatment. (**d**) Growth state under low-temperature treatment.

Under drought treatment, the stem height of *ClNUM1.2* and *ClNUM1.3* increased by 1.5 cm and 9.1 cm compared with that of CK, but the growth recovery rate of *ClNUM1* transgenic tobacco was significantly slower than that of CK after rehydration; the stem height of *ClNUM1.1* exhibited a slower growth rate than that of CK after both treatment and rehydration. Under salt treatment, the stem height of *ClNUM1.1* increased by 0.27 cm and 0.05 cm after both treatment and rehydration compared with that of CK; the stem height of *ClNUM1.2* grew at a slower rate than that of CK after treatment but recovered very quickly after rehydration; and the average stem height of *ClNUM1.3* was 5.5 cm higher than that of CK after treatment but recovered very slowly after rehydration. The average stalk heights of *ClNUM1.1*, *ClNUM1.2*, and *ClNUM1.3* were 0.13 cm, 0.54 cm, and 1.1 cm higher than that of CK under the low-temperature treatment, and they recovered better than CK after rewatering (Figure 6a,b).

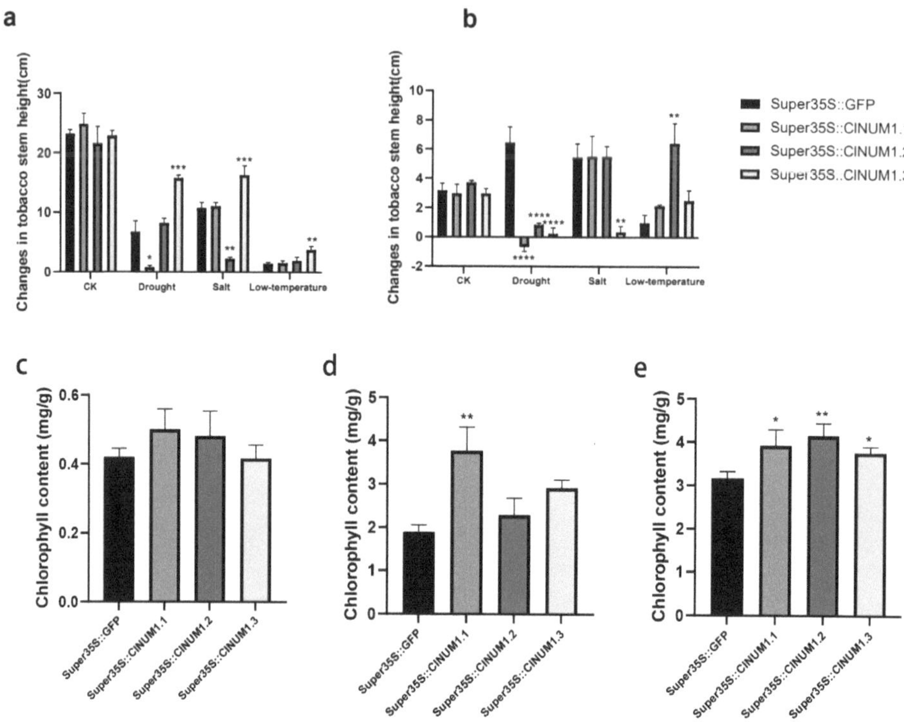

Figure 6. Changes in stem height of mature tobacco seedlings after different treatments and after rehydration and chlorophyll content after treatments. (**a**) Changes in stem height of tobacco after stress treatment. (**b**) Changes in stem height after 7 days of restoration to normal growth conditions under salt, drought, and low-temperature stresses. (**c**) Chlorophyll content after drought treatment. (**d**) Chlorophyll content after salt treatment. (**e**) Chlorophyll content after low-temperature treatment. All experiments were set up with three or more biological replicates (independent sample *t*-test; * $p < 0.05$, ** $p < 0.01$, *** $p < 0.001$, **** $p < 0.0001$).

Under drought treatment, the chlorophyll content in *ClNUM1* transgenic tobacco was not significantly different from that in CK; under salt treatment, the chlorophyll content of *ClNUM1.1* transgenic tobacco was twice as high as that of CK, with a significant difference, and the chlorophyll content of *ClNUM1.2* and *ClNUM1.3* transgenic tobacco was 1.2 and 1.5 times as high as that of CK, with an increase from CK but no significant difference; the chlorophyll contents of *ClNUM1.1*, *ClNUM1.2*, and *ClNUM1.3* transgenic tobacco were 1.24, 1.31, and 1.2 times higher than those of CK, respectively, under low-temperature treatment, all of which were significant differences (Figure 6c–e).

The test results showed that *ClNUM1.1* transgenic tobacco had strong salt and low-temperature tolerance at the mature seedling stage; *ClNUM1.2* transgenic tobacco had strong low-temperature tolerance; and *ClNUM1.3* transgenic tobacco had strong low-temperature tolerance.

3.5. Anatomical Analysis of Transgenic Tobacco Stalks

Paraffin sections were taken from one-third of 12-week-old transgenic tobacco stalks and stained with Senka solid green, and the structures of the epidermis, cortex, bast, xylem, and pith could be observed under a light microscope (Figure 7a–h). Both saffron and solid green stain the cell walls of plants; saffron stains the lignified and corky cell walls as well as the ducts of plants red, while solid green stains the cellulose cell walls as well as the sieve tubes of plants green. We measured cell wall thickness and cell size of twenty each in the xylem of transgenic tobacco stalks (Figure 7i,j). Through observation, we found that the cell wall thickness of *ClNUM1.1* and *ClNUM1.3* transgenic tobacco increased compared to that of CK by 1.08-fold and 1.16-fold, respectively, and the cell size also increased compared to that of CK by 1.06 and 1.11 times, respectively. The cell wall thickness and cell size of *ClNUM1.3* transgenic tobacco were significantly different; the cell wall thickness and cell size of *ClNUM1.2* transgenic tobacco were both reduced compared to those of CK with no significant differences.

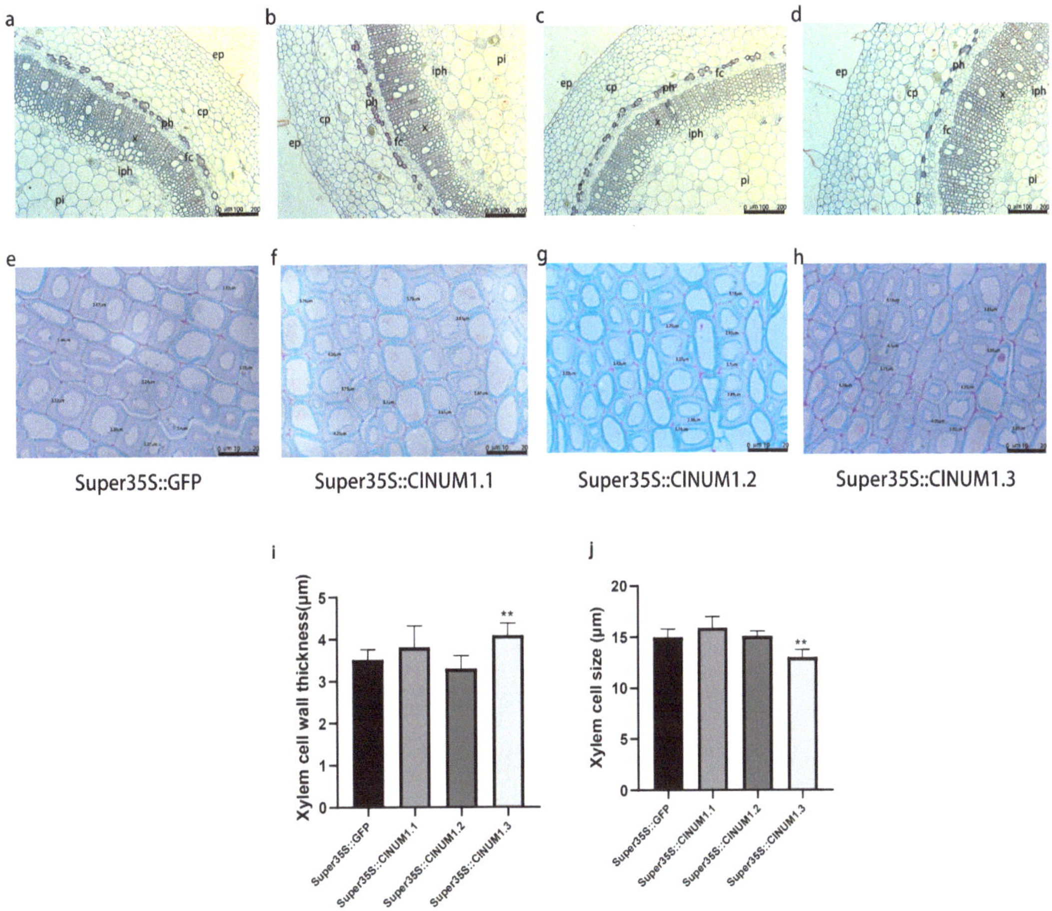

Figure 7. Paraffin sections of transgenic tobacco stems as well as xylem cell wall thickness and xylem cell size of different genotypes of tobacco. (**a**–**h**) Stem cross-sections of CK, *ClNUM1.1*, *ClNUM1.2*, and *ClNUM1.3* transgenic tobacco under normal growth conditions. (**i**) Xylem cell wall thickness of different genotypes of tobacco. (**j**) Xylem cell size of different genotypes of tobacco. Twenty cell sizes and cell wall thicknesses were measured for each transgenic plant. ep = epidermis, cp = cortical parenchyma, ph = phloem, iph = inner phloem, fc = osteocortex, x = xylem, pi = pith. (independent sample *t*-test; ** $p < 0.01$.)

4. Discussion

Alternative splicing is an important pathway for eukaryotes to generate significant regulatory and proteomic complexity. Alternative splicing has two main outcomes—proteome diversification and gene expression regulation [36]—and alternative splicing can increase transcriptome variability and complexity. It is considered to be one of the possible sources of large phenotypic differences between species. In the clipping type, alternative splicing of 5′ and 3′ splice sites is performed by selecting either a 5′ splice donor or a 3′ splice acceptor to retain or splice all or part of the exon sequence [37]. In this paper, the splicing variants we obtained were A5SS and A3SS, which enriches the chrysanthemum variable splicing types. In many studies, alternative splicing has been shown to play an important role in the growth and development of plants and animals, as well as in the response to adversity.

The *ClNUM1* gene was discovered by chance during the cloning of the homologous gene *CcNST1* in *C. lavandulifolium*, and a NUMOD3 conserved domain was found in this gene. Since the gene is a new and unknown gene, we named it *ClNUM1*. In addition, during the cloning process, we found three splicing variants of the *ClNUM1* gene, named *ClNUM1.1*, *ClNUM1.2*, and *ClNUM1.3*.

We characterized the functions of the different splice variants by analyzing morphological and physiological indices of seedling and mature lines of transgenic tobacco under abiotic stress treatments. It was found that the three splice variants enhanced the stress resistance of tobacco (seedling and maturity) and promoted plant growth to some extent, which was consistent with the function of the NAC transcription factor family [38,39].

In the seedling stress experiment, we found that the root length and fresh weight of *ClNUM1* transgenic tobacco were not positively correlated under stress conditions, and sometimes the root length increased but fresh weight decreased or vice versa. In the stress experiment at maturity, we found two interesting phenomena: one is that the growth potential of *ClNUM1* transgenic tobacco under stress conditions was significantly higher than that of CK after treatment, but the recovery rate of the growth potential after rewatering was slower than that of CK. The other is that the growth potential of *ClNUM1* transgenic tobacco after treatment was not significantly different from that of CK, but the recovery rate of the growth potential after rewatering significantly differed from that of CK. We speculate that this may be a response of the plant to resist external environmental changes.

Previous studies have found that NST can thicken the secondary cell wall of Populus and Arabidopsis [40,41], and the change of the secondary cell wall can enhance grapevine and Populus resistance to abiotic stress [42,43]. By observing the cross-section of transgenic tobacco, it was found that compared with CK, the secondary cell wall thickness of transgenic tobacco has changed to a certain extent and the resistance to abiotic stress has been enhanced. We speculated that this gene can regulate the change of plant cell wall thickness and thus regulate plant resistance; this is consistent with previous studies. Abiotic stress will affect the chlorophyll metabolism of plants and thus affect the chlorophyll content of plants [44]. The strength of plant photosynthesis can be reflected by measuring the chlorophyll content so as to evaluate the growth status of plants under abiotic stress. Under the treatment, the chlorophyll content of transgenic tobacco was higher than CK, indicating that the growth status of transgenic tobacco was better than CK, and the resistance was higher than CK.

We have made a preliminary validation for *ClNUM1* function in tobacco based on this experiment; however, how *ClNUM1* in *C. lavandulifolium* regulates plant stress resistance more precisely by affecting secondary cell walls remains to be further studied. Subsequent studies on genes related to secondary cell wall synthesis will reveal the regulatory network and resistance mechanism of *ClNUM1* in *C. lavandulifolium* under abiotic stress.

5. Conclusions

In this study, by analyzing the amino acid sequences of the three splice variants of the *ClNUM1* gene of *C. lavandulifolium*, we found that *ClNUM1.1* and *ClNUM1.2* had two selective splicing types and *ClNUM1.3* had selective 5′ splicing. The function of the splice variants of the unknown *C. lavandulifolium* gene *ClNUM1* was verified in tobacco. It was found that *ClNUM1.1* enhanced salt and low-temperature tolerance, *ClNUM1.2* enhanced salt and low-temperature tolerance, and *ClNUM1.3* enhanced low-temperature tolerance in tobacco. By observing the paraffin sections, we found that the thickness of the xylem cells was related to the growth rate of the plant, with thick cell walls in fast-growing plants and thin cell walls in slow-growing plants. However, the molecular mechanism underlying the role of the *ClNUM1* gene in secondary cell wall synthesis and the synthesis of secondary cell wall components such as lignin, cellulose, and polysaccharides is not clear, which is something we need to focus on in future studies. In this study, we discovered a new unknown gene, *ClNUM1*, in *C. lavandulifolium* and verified the function of the splice variants of this gene in tobacco, thus providing a new potential gene for resistance breeding in *Chrysanthemum* and enriching the diversity of resistance breeding in *Chrysanthemum*.

Supplementary Materials: The following supporting information can be downloaded at: https://www.mdpi.com/article/10.3390/cimb46060314/s1, Figure S1: The figure shows a comparison of the nucleotide sequences of *ClNUM1* with its three spliced variants. Table S1: Mean and SD of root length and fresh weight of *ClNUM1* transgenic tobacco at seedling stage after treatment. Table S2: Mean and SD of stem height change in mature *ClNUM1* transgenic tobacco after treatment and after recovery.

Author Contributions: W.Z., H.W., X.S., and Y.G. carried out the majority of the research and the data analysis and completed the draft preparation and revision. X.H., Y.L., W.H., S.C., and X.Z. completed the experimental design, the data analysis, and the writing of the manuscript. All authors have read and agreed to the published version of the manuscript.

Funding: This research was funded by the National Natural Science Foundation Youth Fund Project (32101580), the Survey and Collection Project of Herbal Plant Germplasm Resources at Shandong Provincial Forest and Grass Germplasm Resources Center (6602423134), and the Qingdao Agricultural University Doctoral Start-Up Fund (6631122019).

Institutional Review Board Statement: Not applicable.

Informed Consent Statement: Not applicable.

Data Availability Statement: The original contributions presented in the study are included in the article/Supplementary Materials, further inquiries can be directed to the corresponding author.

Conflicts of Interest: The authors declare no conflicts of interest.

References

1. Yu, L.; Dang, J.; Liu, X.; Jiang, H. Investigation and Research on the Applied Value of *Chrysanthemum*. *Contemp. Hortic.* **2022**, *13*, 21–23. [CrossRef]
2. Cheng, X. Interspecific Hybridization between *Chrysanthemum* Species and Research on Resistant Germplasm Innovation. Master's Thesis, University of Nanjing Agriculture, Nanjing, China, 2012.
3. Peng, Y. Salt-Tolerance Analysis of *Chrysanthemum lavandulifolium* and Transformation of *DlBADH* Promoters to Chrysanthemum. Master's Thesis, Beijing Forestry University, Beijing, China, 2011.
4. An, C.; Sheng, L.P.; Du, X.P.; Wang, Y.J.; Zhang, Y.; Song, A.P.; Jiang, J.F.; Guan, Z.Y.; Fang, W.M.; Chen, F.D.; et al. Overexpression of *CmMYB15* provides chrysanthemum resistance to aphids by regulating the biosynthesis of lignin. *Hortic. Res.* **2019**, *6*, 84. [CrossRef]
5. Huang, H.; Wang, Y.; Wang, S.L.; Wu, X.; Yang, K.; Niu, Y.J.; Dai, S.L. Transcriptome-wide survey and expression analysis of stress-responsive NAC genes in *Chrysanthemum lavandulifolium*. *Plant Sci.* **2012**, *193*, 18–27. [CrossRef]
6. Wang, K.; Wu, Y.H.; Tian, X.Q.; Bai, Z.Y.; Liang, Q.Y.; Liu, Q.L.; Pan, Y.Z.; Zhang, L.; Jiang, B.B. Overexpression of *DgWRKY4* Enhances Salt Tolerance in Chrysanthemum Seedlings. *Front. Plant Sci.* **2017**, *8*, 1592. [CrossRef]
7. Jaffar, M.A.; Song, A.P.; Faheem, M.; Chen, S.M.; Jiang, J.F.; Liu, C.; Fan, Q.Q.; Chen, F.D. Involvement of *CmWRKY10* in Drought Tolerance of *Chrysanthemum* through the ABA-Signaling Pathway. *Int. J. Mol. Sci.* **2016**, *17*, 693. [CrossRef] [PubMed]
8. Li, C.N.; Ng, C.K.Y.; Fan, L.M. MYB transcription factors, active players in abiotic stress signaling. *Environ. Exp. Bot.* **2015**, *114*, 80–91. [CrossRef]
9. Li, F.; Zhang, Y.; Tian, C.; Wang, X.H.; Zhou, L.J.; Jiang, J.F.; Wang, L.K.; Chen, F.D.; Chen, S.M. Molecular module of *CmMYB15-like-Cm4CL2* regulating lignin biosynthesis of chrysanthemum (*Chrysanthemum morifolium*) in response to aphid (*Macrosiphoniella sanborni*) feeding. *New Phytol.* **2023**, *237*, 1776–1793. [CrossRef] [PubMed]
10. Wang, K.; Zhong, M.; Wu, Y.H.; Bai, Z.Y.; Liang, Q.Y.; Liu, Q.L.; Pan, Y.Z.; Zhang, L.; Jiang, B.B.; Jia, Y.; et al. Overexpression of a chrysanthemum transcription factor gene *DgNAC1* improves the salinity tolerance in chrysanthemum. *Plant Cell Rep.* **2017**, *36*, 571–581. [CrossRef] [PubMed]

11. Wang, Y.J.; Sheng, L.P.; Zhang, H.R.; Du, X.P.; An, C.; Xia, X.L.; Chen, F.D.; Jiang, J.F.; Chen, S.M. *CmMYB19* Over-Expression Improves Aphid Tolerance in *Chrysanthemum* by Promoting Lignin Synthesis. *Int. J. Mol. Sci.* **2017**, *18*, 619. [CrossRef] [PubMed]
12. Zhou, L.J.; Geng, Z.Q.; Wang, Y.X.; Wang, Y.G.; Liu, S.H.; Chen, C.W.; Song, A.P.; Jiang, J.F.; Chen, S.M.; Chen, F.D. A novel transcription factor CmMYB012 inhibits flavone and anthocyanin biosynthesis in response to high temperatures in chrysanthemum. *Hortic. Res.* **2021**, *8*, 248. [CrossRef]
13. Wang, H.; Guo, Y.N.; Hao, X.Y.; Zhang, W.X.; Xu, Y.X.; He, W.T.; Li, Y.X.; Cai, S.Y.; Zhao, X.; Song, X.B. Alternative Splicing for Leucanthemella linearis NST1Contributes to Variable Abiotic Stress Resistance in Transgenic Tobacco. *Genes* **2023**, *14*, 1549. [CrossRef] [PubMed]
14. Gao, G. *Chrysanthemum CmWRKY15-1* Enhanced Resistance to *Chrysanthemum* White Rust by Regulating the Expression of CmNPR1. Master's Thesis, Shenyang Agricultural University, Shenyang, China, 2022.
15. Meng, R. Study on the Function of *Shennong Chrysanthemum CiMYB32* in Response to Drought Stress. Master's Thesis, Northeast Forestry University, Harbin, China, 2023.
16. Sitbon, E.; Pietrokovski, S. New types of conserved sequence domains in DNA-binding regions of homing endonucleases. *Trends Biochem. Sci.* **2003**, *28*, 473–477. [CrossRef]
17. Belfort, M.; Roberts, R.J. Homing endonucleases: Keeping the house in order. *Nucleic Acids Res.* **1997**, *25*, 3379–3388. [CrossRef]
18. Lagerback, P.; Carlson, K. Amino Acid Residues in the GIY-YIG Endonuclease II of Phage T4 Affecting Sequence Recognition and Binding as Well as Catalysis. *J. Bacteriol.* **2008**, *190*, 5533–5544. [CrossRef]
19. Andersson, C.E.; Lagerback, P.; Carlson, K. Structure of Bacteriophage T4 Endonuclease II Mutant E118A, a Tetrameric GIY-YIG Enzyme. *J. Mol. Biol.* **2010**, *379*, 1003–1016. [CrossRef]
20. Van Roey, P.; Waddling, C.A.; Fox, K.M.; Belfort, M.; Derbyshire, V. Intertwined structure of the DNA-binding domain of intron endonuclease I-TevI with its substrate. *EMBO J.* **2001**, *20*, 3631–3637. [CrossRef]
21. Raczynska, K.D.; Simpson, C.G.; Ciesiolka, A.; Szewc, L.; Lewandowska, D.; McNicol, J.; Szweykowska-Kulinska, Z.; Brown, J.W.S.; Jarmolowski, A. Involvement of the nuclear cap-binding protein complex in alternative splicing in Arabidopsis thaliana. *Nucleic Acids Res.* **2010**, *38*, 265–278. [CrossRef] [PubMed]
22. Yang, X.; Jia, Z.; Pu, Q.; Tian, Y.; Zhu, F.; Liu, Y. ABA Mediates Plant Development and Abiotic Stress via Alternative Splicing. *Int. J. Mol. Sci.* **2022**, *23*, 3796. [CrossRef]
23. James, A.B.; Syed, N.H.; Bordage, S.; Marshall, J.; Nimmo, G.A.; Jenkins, G.I.; Herzyk, P.; Brown, J.W.S.; Nimmo, H.G. Alternative Splicing Mediates Responses of the Arabidopsis Circadian Clock to Temperature Changes. *Plant Cell* **2012**, *24*, 961–981. [CrossRef] [PubMed]
24. Cui, Z.; Xu, Q.; Wang, X. Regulation of the circadian clock through pre-mRNA splicing in *Arabidopsis*. *J. Exp. Bot.* **2014**, *65*, 1973–1980. [CrossRef] [PubMed]
25. Fuchs, A.; Riegler, S.; Ayatollahi, Z.; Cavallari, N.; Giono, L.E.; Nimeth, B.A.; Mutanwad, K.V.; Schweighofer, A.; Lucyshyn, D.; Barta, A.; et al. Targeting alternative splicing by RNAi: From the differential impact on splice variants to triggering artificial pre-mRNA splicing. *Nucleic Acids Res.* **2021**, *49*, 1133–1151. [CrossRef] [PubMed]
26. Ren, Z.; Liu, Y.; Kang, D.; Fan, K.; Wang, C.; Wang, G.; Liu, Y. Two alternative splicing variants of maize *HKT1;1* confer salt tolerance in transgenic tobacco plants. *Plant Cell Tissue Organ Cult.* **2015**, *123*, 569–578. [CrossRef]
27. Lv, B.; Hu, K.; Tian, T.; Wei, K.; Zhang, F.; Jia, Y.; Tian, H.; Ding, Z. The pre-mRNA splicing factor RDM16 regulates root stem cell maintenance in *Arabidopsis*. *J. Integr. Plant Biol.* **2021**, *63*, 662–675. [CrossRef] [PubMed]
28. Lu, H.; Deng, Q.; Wu, M.; Wang, Z.; Wei, D.; Wang, H.; Xiang, H.; Zhang, H.; Tang, Q. Mechanisms of alternative splicing in regulating plant flowering: A review. *Chin. J. Biotechnol.* **2021**, *37*, 2991–3004. [CrossRef]
29. Simpson, C.G.; Fuller, J.; Calixto, C.P.G.; McNicol, J.; Booth, C.; Brown, J.W.S.; Staiger, D. Monitoring Alternative Splicing Changes in Arabidopsis Circadian Clock Genes. *Methods Mol. Biol.* **2016**, *1398*, 119–132. [CrossRef] [PubMed]
30. Chaudhary, S.; Jabre, I.; Reddy, A.S.N.; Staiger, D.; Syed, N.H. Perspective on Alternative Splicing and Proteome Complexity in Plants. *Trends Plant Sci.* **2019**, *24*, 496–506. [CrossRef]
31. Thatcher, S.R.; Danilevskaya, O.N.; Meng, X.; Beatty, M.; Zastrow-Hayes, G.; Harris, C.; Van Allen, B.; Habben, J.; Li, B.L. Genome-Wide Analysis of Alternative Splicing during Development and Drought Stress in Maize. *Plant Physiol.* **2016**, *170*, 586–599. [CrossRef] [PubMed]
32. Alhabsi, A.; Butt, H.; Kirschner, G.K.; Blilou, I.; Mahfouz, M.M.; Sunkar, R. SCR106 splicing factor modulates abiotic stress responses by maintaining RNA splicing in rice. *J. Exp. Bot.* **2023**, *75*, 802–818. [CrossRef] [PubMed]
33. Ye, L.-X.; Wu, Y.-M.; Zhang, J.-X.; Zhang, J.-X.; Zhou, H.; Zeng, R.-F.; Zheng, W.-X.; Qiu, M.-Q.; Zhou, J.-J.; Xie, Z.-Z.; et al. A bZIP transcription factor CiFD regulates drought- and low-temperature-induced flowering by alternative splicing in citrus. *J. Integr. Plant Biol.* **2023**, *65*, 674–691. [CrossRef]
34. Albaqami, M.; Laluk, K.; Reddy, A.S.N. The Arabidopsis splicing regulator SR45 confers salt tolerance in a splice isoform-dependent manner. *Plant Mol. Biol.* **2019**, *100*, 379–390. [CrossRef]
35. Horsch, R.B.; Fry, J.E.; Hoffmann, N.L.; Wallroth, M.; Eichholtz, D.; Rogers, S.G.; Fraley, R.T. A simple and general method for transferring genes into plants. *Science* **1985**, *227*, 1229–1231. [CrossRef] [PubMed]
36. Smith, C.C.R.; Tittes, S.; Mendieta, J.P.; Collier-zans, E.; Rowe, H.C.; Rieseberg, L.H.; Kane, N.C. Genetics of alternative splicing evolution during sunflower domestication. *Proc. Natl. Acad. Sci. USA* **2018**, *115*, 6768–6773. [CrossRef]

37. Iñiguez, L.P.; Ramírez, M.; Barbazuk, W.B.; Hernández, G. Identification and analysis of alternative splicing events in *Phaseolus vulgaris* and Glycine max. *BMC Genom.* **2017**, *18*, 650. [CrossRef]
38. Xu, Y.; Cheng, J.; Hu, H.; Yan, L.; Jia, J.; Wu, B. Genome-Wide Identification of NAC Family Genes in Oat and Functional Characterization of AsNAC109 in Abiotic Stress Tolerance. *Plants* **2024**, *13*, 1017. [CrossRef]
39. Chen, N.; Shao, Q.; Lu, Q.; Li, X.; Gao, Y.; Xiao, Q. Research progress on function of NAC transcription factors in tomato (*Solanum lycopersicum* L.). *Euphytica* **2023**, *219*, 22. [CrossRef]
40. Takata, N.; Awano, T.; Nakata, M.T.; Sano, Y.; Sakamoto, S.; Mitsuda, N.; Taniguchi, T. Populus NST/SND orthologs are key regulators of secondary cell wall formation in wood fibers, phloem fibers and xylem ray parenchyma cells. *Tree Physiol.* **2019**, *39*, 514–525. [CrossRef] [PubMed]
41. Zhang, Q.; Luo, F.; Zhong, Y.; He, J.; Li, L. Modulation of NAC transcription factor NST1 activity by XYLEM NAC DOMAIN1 regulates secondary cell wall formation in Arabidopsis. *J. Exp. Bot.* **2020**, *71*, 1449–1458. [CrossRef] [PubMed]
42. Guo, H.; Zhang, C.; Wang, Y.; Zhang, Y.; Zhang, Y.; Wang, Y.; Wang, C. Expression profiles of genes regulated by BplMYB46 in *Betula platyphylla*. *J. For. Res.* **2019**, *30*, 2267–2276. [CrossRef]
43. Malacarne, G.; Lagreze, J.; Martin, B.R.S.; Malnoy, M.; Moretto, M.; Moser, C.; Costa, L.D. Insights into the cell-wall dynamics in grapevine berries during ripening and in response to biotic and abiotic stress. *Plant Mol. Biol.* **2024**, *114*, 38. [CrossRef] [PubMed]
44. Li, X.; Zhang, W.; Niu, D.; Liu, X. Effects of abiotic stress on chlorophyll metabolism. *Plant Sci.* **2024**, *342*, 112030. [CrossRef]

current issues in molecular biology

MDPI

Article

Overexpression of *KvCHX* Enhances Salt Tolerance in *Arabidopsis thaliana* Seedlings

Yuqi Guo [1], Chengrong Zhu [1] and Zengyuan Tian [2,*]

[1] School of Life Sciences, Zhengzhou University, Zhengzhou 450001, China; guo_yuqi2021@126.com (Y.G.); 18856300358@163.com (C.Z.)

[2] School of Agricultural Sciences, Zhengzhou University, Zhengzhou 450001, China

* Correspondence: tianzengyuan@zzu.edu.cn

Abstract: The CHX (cation/H^+ exchanger) family plays an important role in the transmembrane transport of cation/H^+ in plants. The aim of this study was to identify and functionally analyze the *KvCHX* gene in the halophyte *Kosteletzkya virginica* to investigate its role in regulating the K^+/Na^+ ratio under salinity tolerance. Based on a partial gene sequence of EST from *K. virginica*, the full-length DNA sequence of the *KvCHX* gene was obtained using genome walking technology. Structural analysis and phylogenetic relationship analysis showed that the *KvCHX* gene was closely related to the *AtCHX17* gene. The *KvCHX* overexpression vector was successfully constructed and transformed into Arabidopsis via floral dipping. Arabidopsis seedlings overexpressing *KvCHX* showed an enhanced tolerance to salt stress compared with wild-type plants. Transgenic Arabidopsis seedlings grew better under K^+ deficiency than WT. The results showed that *KvCHX* could promote the uptake of K^+, increase the ratio of K^+/Na^+, and promote the growth of plants under K^+ deficiency and treatment with NaCl solution. *KvCHX* is involved in K^+ transport and improves plant salt tolerance by coordinating K^+ acquisition and homeostasis.

Keywords: *Kosteletzkya virginica*; *KvCHX*; transgenic arabidopsis; salt stress; K^+ transport

Citation: Guo, Y.; Zhu, C.; Tian, Z. Overexpression of *KvCHX* Enhances Salt Tolerance in *Arabidopsis thaliana* Seedlings. *Curr. Issues Mol. Biol.* **2023**, *45*, 9692–9708. https://doi.org/10.3390/cimb45120605

Academic Editor: Vijai Bhadauria

Received: 27 October 2023
Revised: 23 November 2023
Accepted: 27 November 2023
Published: 1 December 2023

1. Introduction

Plants are often affected by different environments, such as high salt levels, drought, and extreme temperatures [1]. Salt stress is the main abiotic factor affecting plant growth and development. Currently, at least 800 million hectares of land in the world are affected by salinization, which is more than 6% of the world's total land area [2]. Generally, soils with a soil solution electrical conductivity (EC) of 4 dS/m (deciSiemens per meter) or greater are referred to as saline soils [3]. Plants are classified into glycophytes and halophytes based on their ability to grow in highly saline soils. Halophytes have a high salinity tolerance and can survive and reproduce in environments with salt concentrations around or above 200 mm NaCl [4]. In contrast, at high salt concentrations [5], glycophytes are subjected to ionic, osmotic, and secondary stresses.

The major effect of salt stress is the accumulation of Na^+ and Cl^- ions in plant tissues exposed to saline soils [6]. The massive uptake of Na^+ and Cl^- results in a severe ionic imbalance. High Na^+ concentrations inhibit the uptake of K^+ ions, which are important elements for growth and development, and low K^+ concentrations lead to decreased productivity and even death [7]. To ensure survival in this adverse environment, halophytes have evolved their own mechanisms for salt tolerance. Some halophytes reduce harmful ion concentrations by removing harmful ions from roots and shoots through salt glands, and some halophytes are able to tolerate high concentrations of salt, while others adopt measures such as ion compartments, osmoregulation, and the synthesis of compatible solutes [3,4]. Therefore, the salinity tolerance of halophytes depends on the controlled uptake and fractionation of Na^+, K^+, and Cl^- and the synthesis of organic solutes [4].

Among them, K^+ is the most abundant cation in plant cells, accounting for about 10% of the dry weight of plants [8]. Potassium ions mediate many physiological reactions in plants, such as stomatal closure and opening, leaf movement, and plant tropism driven by swelling pressure caused by K^+ [9–11]. High concentrations of Na^+ are toxic to plants and have a negative effect on plant growth, but Na^+ is greater than K^+ in saline soils. Therefore, faced with competition from Na^+ [12], halophytes must choose to absorb K^+.

To regulate K^+ uptake and balance, plants have evolved transport systems containing a large number of transporters, such as HAK/KUP/KT (K^+/H^+ transporter), HKT/TRK (K^+/H^+ or K^+/Na^+ transporter), CPA (cation/H^+ anti-transporter), and Shaker channels. At present, some *CHX* genes in Arabidopsis have been revealed to have their functions through heterologous expression. They may play an important role in regulating membrane cation and pH homeostasis [13]. *AtCHX20* has been confirmed as a K^+ transporter for guard cells [14]. Similarly, in the latest research, the *GmSALT3* gene in soybeans has been confirmed to be an endometrial localization protein with transport ability and plays an important role in salt tolerance [15]. *Ksteletzkya virginica* is a dicotyledon halophyte (http://plants.usda.gov, accessed on 26 October 2023) native to the southeastern United States, which often grows in coastal soils containing 0.3–2.5% sodium salts (mainly NaCl) [16]. As a seaside mallow, *K. virginica* is a perennial dicotylote and halophyte native to brackish portions of coastal tidal marshes in the mid-Atlantic and southeastern United States and is considered an obligate wetland species. It extends north to Long Island, New York, south to Florida, and along the eastern Texas coastal plain [17]. Blanchard synonymized *K. Pentacarpos* (found in Eurasia) with *K. virginica*. With a lifespan of about 11 years and a high seed protein (32%) and oil (22%) content, *K. virginica* is attractive not only for animal feed and biomass production in coastal areas unsuitable for conventional agriculture [18], but also for biodiesel and ethanol production [19]. In 1992, *K. virginica* was introduced into Jiangsu Province, China. *K. virginica* can be grown in saline or arid lands and can be irrigated with salt or sea water. The salt physiological properties of *K. virginica* were investigated and the results showed its high salt tolerance [20]. However, the specific mechanism of salinity tolerance in *K. virginica* remains elusive. In a previous report, we found the expression of *K. virginica CHX* (*KvCHX*) was up-regulated in roots under salt stress and obtained the EST of the *KvCHX* gene [21]. Therefore, we propose a hypothesis that *KvCHX* may be involved in the salinity tolerance of *K. Virginia*. Previous studies have shown that comparing the AtCHX protein with fungal and bacterial CPA2 proteins reveals that the CHX protein has K^+, Na^+, and H^+ transport functions [22]. Moreover, after phylogenetic analysis of the plant CHX gene family, it was found that it has important genetic relationships with many ion transport genes [23]. Here, we verified the function of the *CHX* gene of *K. virgineca*, which is associated with its ability to maintain Na^+ and K^+ homeostasis in transgenic Arabidopsis seedlings.

In this study, a DNA sequence of *KvCHX* was obtained based on a *KvCHX* known fragment [21] using genome walking technology. Protein structure analysis showed that the 2400 bp CDS of *KvCHX* encodes a polypeptide containing 799 amino acids. The predicted KvCHX protein consists of an amino-terminal domain with 12 transmembrane spans (430 residues) and a hydrophilic domain at the carboxyl end, which is proposed to have regulatory roles. Subcellular localization analysis showed that KvCHX was localized to the cytoplasmic membrane. Phylogenetic relationship analysis implied that KvCHX was very similar to AtCHX17; both of them shared higher identity in amino-terminal hydrophobic domains than other members of the *AtCHX* gene family, especially TM5 and TM11 (91.30% and 90.00% identity, respectively). To investigate the role of *KvCHX* in K^+ transport and salinity tolerance, an overexpression vector 2×35S-KvCHX-Nos was constructed and transformed into *Arabidopsis thaliana* ecotype Columbia (WT) using floral dipping. The T_3 transgenic lines of Arabidopsis seedlings could promote the absorption of K^+, increase the K^+/Na^+ ratio, and contribute to the growth of plants under K^+ deficient conditions and treatment with NaCl solution. In summary, the results indicated that *KvCHX* enhances salt tolerance in transgenic Arabidopsis by regulating the K^+/Na^+ ratio under salt stress.

2. Materials and Methods

2.1. Plant Materials and Stress Treatments

A. thaliana ecotype Columbia (WT) was used in this study. Seeds of *K. virginica* were soaked overnight in fresh water and then sown in pots filled with vermiculite in a greenhouse at 26/20 °C with day/night of 15/9 h.

WT and T_3 transgenic Arabidopsis seeds were sterilized with sodium hypochlorite and placed in MS medium for 2 weeks. Seedlings (5 plants per treatment) were grown under hydroponic conditions in a greenhouse. For greenhouse cultivation, seedlings used in the transgenic experiment were transplanted into pots filled with vermiculite and soil (3:1) and watered with Hoagland solution. For hydroponic cultivation, the seedlings were transferred to nutrient solution containing 1 mM $Ca(NO_3)_2$, 1 mM KH_2PO_4, 1 mM KNO_3, 1 mM $MgSO_4$, 50 μM Na-Fe-EDTA, 50 μM H_3BO_3, 0.05 μM CoCl, 0.05 μM $CuSO_4$, 15 μM $ZnSO_4$, 50 μM $MnSO_4$, and 3 μM Na_2MoO_4 [24]. The environmental parameters in the growth chamber were as follows: light/dark cycle of 8/16 h, light intensity of 300 μmol $s^{-1}m^{-2}$par, temperature of 24 °C/22 °C, humidity of 70%. The nutrient solution was renewed once a week during the first part of the culture and twice a week during the experiment.

In the K^+ deficient treatment, plants (five plants per treatment) were selected at the rosette stage at the beginning of each experiment [25]. Plant roots were rinsed in 0.2 mM $CaSO_4$ for 5 min and then transferred to 1/2 K^+ nutrient solution. In 1/2 K^+ nutrient solution, KH_2PO_4 was replaced by 1 mM NaH_2PO_4. In the salt treatment, the basal nutrient solution was supplemented with 0 mM, 100 mM, 150 mM, and 200 mM NaCl, respectively.

2.2. Isolation and Sequence Analysis of KvCHX, Protein Structure, Phylogenetic Tree, and Multiple Alignment

Genomic DNA of *K. Virginica* was extracted from 10-day-old seedlings using the NuClean PlantGen DNA Kit (Biotech, Hefei, China) according to the manufacturer's instructions and examined via electrophoresis on 1.5% agarose gels.

To obtain the full-length DNA sequence of *KvCHX*, the 5′ sequence was extended three times, and the 3′ sequence once, via TAIL PCR, using the genome Walking toolkit (Takara D316) based on the known fragment of the *KvCHX* gene (FK816439). Gene-specific primers (GSPs) were designed each time based on the obtained sequences using genomic DNA as a template (Table 1). For each genome walking, three rounds of TAIL-PCR (Thermal Asymmetric Interlaced PCR) were performed using the GSPs and degenerate primers for the 5′-end flanking sequence of the known segment of the *CHX* gene.

Table 1. Primers used for the genome walking and sequencing of the *KvCHX* gene.

GSP1-5′-2	5′-CCCAAAACTTCAGGACCAGG-3′	GSP1-3′-2	5′-CCTGGTCCTGAAGTTTTGGG-3′
GSP1-5′-3	5′-GCTTCGTGATTGTCGTGTCC-3′	GSP1-3′-3	5′-CAATCTTTTCGTGGTGGGCC-3′
GSP2-5′-1	5′-GGGGAGAAGAGTTTGTTCCTG-3′	GSP3-5′-1	5′-CAACTATCTCGCCGATGACC-3′
GSP2-5′-2	5′-TACGGGCTAACACAGGGAAG-3′	GSP3-5′-2	5′-GCGAGGATACGAGTGAGTGC-3′
GSP2-5′-3	5′-AAGGTGTCCAGCACCGTTAG-3′	GSP3-5′-3	5′-GTACCGCTATCACCGACGAA-3′
RV-M	5′-GAGCGGATAACAATTTCACACAGG-3′	M13-47	5′-CGCCAGGGTTTTCCCAGTCACGAC-3′

Note: The first genome walking primers of 5′-end flanking sequence: GSP1-5′-1~GSP1-5′-3. The second genome walking primers of 5′-end flanking sequence: GSP2-5′-1~GSP2-5′-3. The third genome walking primers of 5′-end flanking sequence: GSP3-5′-1~GSP3-5′-3. The genome walking primers of 3′ -end flanking sequence: GSP1-3′-1~GSP1-3′-3. The primers for the sequencing of genome walking products: RV-M and M13-47.

This technique is well suited to cloning the flanking sequences of genes based on the partial known sequences of genes, especially when the species under study has not yet been fully sequenced. A total of three gene-specific primers (GSPs) close to the flanking sequences at the 5′ end of known fragments were designed and synthesized sequentially. The primer GSP1 is upstream of GSP2, and the primer GSP2 is upstream of GSP3. TAIL-PCR was performed with four arbitrary degenerate primers combined with specific primers (GSP3, GSP2, and GSP1). For each chromosome walk, TAIL-PCR, including three rounds of PCR, was performed using three specific primers designed according to known sequences, which were designated as GSP1-5′-1~GSP1-5′-3, GSP2-5′-1~GSP2-5′-3, and GSP3-5′-1~GSP3-5′-3, respectively, for 5′-end flanking cloning every time, and GSP1-3′-1~GSP1-3′-3 for 3′-end flanking cloning one time only. Firstly, four arbitrary degenerate primers (AP primers) in combination with the specific primer GSP3 were used for first TAIL-PCR. Three rounds of TAIL-PCR were carried out on a Biorad Thermal Cycler. The products from the first round of TAIL-PCR were used as templates for the secondary PCR, and the products from the second round of TAIL-PCR were used as templates for the third round of PCR, and were then isolated from agarose and cloned into the T vector for nucleotide sequencing. Secondly, the specific primer GSP2 was designed based on the above sequencing results, and the second chromosome walk was further performed on the basis of the first. A final chromosome walk was then performed using the GSP1 primer dependent on the results of the second sequencing to obtain the 5′-end flanking sequence. Each product from each of the third TAIL-PCR rounds was separated on an agarose gel and sequenced. The resulting target sequence was determined vis analyzing the overlapping amplified fragments of the three 5′-end flanking discrete PCR products. Finally, the 3′-end flanking sequences of the known segments of the *CHX* gene were isolated using TAIL-PCR with GSP1-3′. The volume of a single PCR was 50 μL, containing 2 μL of genomic DNA, 5 μL of 10 × LA PCR BufferII (Mg^+ plus), 8 μL of dNTP mixture (2.5 mM each), 0.5 μL TakaRa LA Taq (5 U/μL), 1uL of AP primer (100 pmol/μL), 1 μL of GSP primer (10 pmol/μL), and 32.5 μL of ddH_2O. Table 2 summarizes the thermal cycling conditions. PCR products were detected via 2% agarose gel electrophoresis and purified (AxyPrepTM DNA Gel Extraction kit) and cloned into PMD18-T vector (TaKaRa, Beijing, China) for sequencing. The full-length *KvCHX* gene sequence was obtained.

Table 2. The PCR programs used in TAIL-PCR reactions.

PCR Reaction	Cycle No.	Thermal Condition
The first round of PCR(GSP-5′-1/AP)	1	94 °C 1min, 98 °C 1min;
	5	94 °C (30 s), 60 °C (1 min), 72 °C (3 min)
	1	94 °C (30 s), 25 °C (3 min), 72 °C (3 min)
	15	94 °C (30 s), 60 °C (1 min), 72 °C (3 min); 94 °C (30 s), 60 °C (1 min), 72 °C (3 min); 94 °C(30 s), 44 °C (1 min), 72 °C (3 min)
	1	72 °C (10 min)
The second round of PCR(GSP-5′-2/AP)	15	94 °C (30 s), 60 °C (1 min), 72 °C (3 min); 94 °C (30 s), 60 °C (1 min), 72 °C (3 min); 94 °C (30 s), 44 °C (1 min), 72 °C (3 min)
	1	72 °C (10min)
The third round of PCR(GSP-5′-3/AP)	15	94 °C (30 s), 60 °C (1 min), 72 °C (3 min); 94 °C (30 s), 60 °C (1 min), 72 °C (3 min); 94 °C (30 s), 44 °C (1 min),72 °C (3 min)
	1	72 °C, 10 min

Gene structure prediction was analyzed using SoftBerry FGENESH (http://www.softberry.com/berry.phtml, accessed on 26 October 2023). Protein prediction was performed using the ExPASy server (http://web.expasy.org/protparam/, accessed on 26 October 2023). Signal sequence analysis and nuclear localization signal analysis were performed using the SignalP 4.1 server and NLS prediction, respectively. ProtComp 9.0 was used for subcellular localization analysis, and TMHMM Server 2.0 for investigating the transmembrane domain. SWISS-MODEL was used for protein tertiary structure prediction. The multiple sequence alignment was performed between *AtCHX*s and *KvCHX* using ClastalX (1.81) for homology analysis, and the phylogenetic tree was constructed using MEGA4.1 to compare the evolutionary relationships of *KvCHX* with *AtCHX*s.

2.3. Vector Construction and Transformation

The *KvCHX* gene was amplified via PCR from *K. virginica* genome DNA using the primers of *CHX*-F (5′-CACCGGCAGGGGAAGTGAAATC-3′) and *CHX*-R (5′-TGTTGTTTCTA CATATCCTTCG-3′). The transformation vector was constructed using Gateway® Technology with the ClonaseTM Kit (Invitrogen, Carlsbad, CA, USA) according to the manufacturer's protocol (www.invitrogen.com/gateway, accessed on 26 October 2023) [26]. The recombinant plasmid P2×35S::GFP-KvCHX was introduced into Agrobacterium tumefaciens strains GV3101 [27] using the liquid nitrogen freeze–thaw method [28]. Arabidopsis was transformed with A. tumefaciens transformation containing the P2×35S::GFP-CHX plasmid using floral dipping technique [29]. Transgenic Arabidopsis was grown in a greenhouse at 24/22 °C, 70% relative humidity, and a 16/8 h light/dark cycle. Transformed Arabidopsis seeds were harvested from transformed plants (T_0) and grown on MS medium containing 50 mg L^{-1} hygromycin. Hygromycin-resistant lines were selected for PCR identification. Homozygous T_3 progeny from the T_2 population were selected and further verified via Western blot and salt tolerance analyses [30].

2.4. Confirmation of Transgenic Arabidopsis Plant

Genomic DNA was extracted from 6-week-old seedling leaves of T_3 transgenic plants and wild-type control via CTAB. PCR was performed using primers GFP-FP (5′-ATGAGTAA AGGAGAAGAACTTTTCACTG GA-3′) and CHX-R2 (5′-CATTTGTGTTTGTATCGG CTGC-3), and the amplification was performed using the program: 94 °C for 2 min, followed by 35 cycles of program (94 °C for 30 s, 60 °C for 30 s, 72 °C for 2 min), and terminated with an extension at 72 °C for 5 min. The PCR products were detected via electrophoresis on 1.5% agarose gel.

The PCR amplification was carried out using genomic DNA extracted from 6-week-old seedling leaves of T_3 transgenic plants and wild-type control via CTAB. The primers used for PCR were GFP-FP (5′-ATGAGTAAAGGAGAAGAACTTTTCACTGGA-3′) and CHX-R2 (5′-CATTTGTGTTTGTATCGG CTGC-3′). The PCR program consisted of an initial denaturation at 94 °C for 2 min, followed by 35 cycles of denaturation at 94 °C for 30 s, annealing at 60 °C for 30 s, and extension at 72 °C for 2 min, and a final extension at 72 °C for 5 min. The PCR products were then analyzed via electrophoresis on a 1.5% agarose gel.

2.5. Western Blot Analysis of Transgenic Arabidopsis Plant

WT and transgenic Arabidopsis plants were grown under hydroponic conditions in basal nutrient solution. After 10 days of cultivation, the seedlings were transferred to nutrient solution for 10 days. Total protein was extracted from WT and transgenic Arabidopsis seedlings with a buffer consisting of 50 mM Tris/HCl (pH 8.0), 150 mM NaCl, 1 mM EDTA, and 0.2% (*w/v*) Triton X-100, 4% β-mercaptoethanol, 1 mM dithiothreitol (DTT), and 1% (*v/v*) protease inhibitor cocktail (BBI Life Science, Shanghai, China) and then used for protein quantification with BCA protein quantitative Kit (Boster, Pleasanton, CA, USA). The protein samples (200 μg amounts) were electrophoresed in 8% SDS-PAGE, and the gels were transferred to nitrocellulose membranes. The membranes were blocked with TBST buffer (10 mM Tris/HCl (pH 7.5), 150 mM NaCl, and 0.05% Tween-20) supplemented with 5% non-fat milk for 2 h and incubated with primary antibodies (Anti-GFP antibody, Abcam (Cambridge, UK), diluted at 1:200) in TBST buffer with 5% BSA overnight at 4 °C. Afterwards, the membranes were washed three times (10 min each) with TBST buffer and incubated with the secondary antibodies (Goat Anti-Mouse IgG H&L (HRP), Abcam, dilution at 1:1000) for 2 h. After being washed three times with TBST buffer, the membranes were incubated with a chromogenic agent Enhanced HRP-DAB Chromogenic Substrate Kit (Boster).

2.6. Subcellular Localization Assay

Fluorescence of KvCHX-GFP fusion protein in transgenic plants was observed using confocal laser-scanning microscopes. The roots and leaves of 4-week-old transgenic plants were selected, and the epidermis was removed for observation.

2.7. Salt Stress Treatments Involving the Transgenic Arabidopsis Plant

Seeds of transgenic and WT plants were sterilized with sodium hypochlorite and planted on MS medium for 2 weeks. Then, the seedlings were transferred into a greenhouse and fixed on polyvinyl chloride (PVC) plates folating on 1/2 Hogland nutrient solution supplemented with 0 mM, 100 mM, 150 mM, and 200 mM NaCl, respectively, for 7 days. The salt treatment nutrient solution was changed every 3 days, and after 7 days of treatment, the growth conditions of seedling, such as root length and leaf number, and physiological indexes were measured. The proline content of WT and transgenic plants was measured as described by [31], and the malondialdehyde (MDA) content was determined according to the method of Song [32].

2.8. K^+ Deficient Treatments of Transgenic Arabidopsis Plant

Seeds of transgenic Arabidopsis plants and WT were collected and sterilized with sodium hypochlorite, then planted in MS medium for germinating and cultivating for 10 days. Then, WT and transgenic Arabidopsis plants were transferred into Hogland nutrient solution with 1/2 K^+ (KNO_3) concentration for 20 days. The leaf number, branch number, fresh weight, and dry weight were determined.

2.9. Determination of Potassium and Sodium Ion Content

The second fully expanded rosette leaves and the mature zone of the roots (without the root tip) were collected and washed in 0.2 mM $CaSO_4$ for 5 min. To quantify the potassium and sodium ion content, tissues harvested from five plants per treatment were oven-dried at 100 °C for 2 h and at 85 °C for 18 h. Subsequently, the plant material was digested with HNO_3 at 80 °C and dissolved with ddH_2O for ion extraction. The concentrations of cation, sodium, and potassium ions in tissue samples were determined using a Thermo fisher ICP spectrometer. All measurements were performed in triplicate.

3. Results

3.1. Isolation and Sequence Analysis of KvCHX

BLASTX analysis of the 496 bp known fragment of *KvCHX* (FK816439) revealed high similarity between the *KvCHX* fragment and *CHXs* described in other species, such as cocoa, cherry, grape, tomato, corn, and Arabidopsis. To obtain the full-length *KvCHX* sequence, three rounds of TAIL-PCR were performed with template genomic DNA and the primers GSP1-5′-1~GSP1-5′-3, GSP1-3′-1~GSP1-3′-3 for 5′ flanking extension firstly and 3′ flanking extension, respectively. The PCR products were examined via electrophoresis on 2% agarose gel, yielding two fragments of approximately 1800 bp (Figure 1A) and 1400 bp (Figure 1D). The final round of TAIL-PCR products were cloned into the pMD18-T vector and sequenced. Sequence analysis (BioXM 2.6) showed that the overlapping regions of these fragments were the same as those of known *KvCHX* fragments. These three fragments include the known *KvCHX* fragment and the two sequenced fragments above, and we obtained a fragment of about 3440 bp via assembly. BLASTX analysis showed that the amino acid sequence of this fragment was significantly similar to that of the CHX proteins of cacao, prunus persica, and grape.

Via the second genome walking for the 5′ flanking-sequence, we obtained two fragments, of approximately 800 bp and 600 bp, respectively. Sequencing analysis showed that the 670 bp fragment was a specific product (Figure 1B, 3rd). The last genome walking for the 5′ flanking-sequence allowed us to obtain a fragment of approximately 611bp (Figure 1C, 3rd). Finally, we obtained a 4333 bp fragment by assembling these fragments (the 496 bp, 3440 bp, 670 bp, and 611 bp fragments). Sequence analysis suggested that it encodes a complete protein.

To further obtain the assembled 3440 bp upstream of the 5′ flanking-sequence, genome walking was performed twice based on the obtained 5′ flanking sequence. The second genome walking for the 5′ flanking sequence yielded two fragments, of approximately 800 bp and 600 bp, respectively. Sequencing analysis showed that the 670bp fragment was a specific product (Figure 1B, 3rd). The last genome walking of the 5′ flanking sequence allowed us to acquire a fragment of approximately 611 bp (Figure 1C, 3rd). Finally, we assembled these fragments (496 bp, 3440 bp, 670 bp, and 611 bp) together to obtain a fragment of 4333 bp. Sequence analysis showed that it encodes a complete protein.

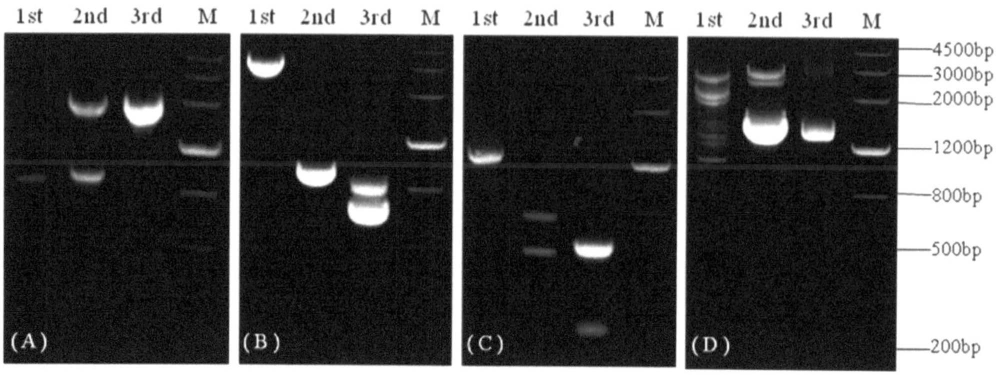

Figure 1. PCR products during the genome walking of *KvCHX*. (**A–D**) represent the genome walking products of the 3′-end sequence and the third, the second, and the first genome walking for the 5′-end sequence, respectively.

3.2. Analysis of KvCHX Gene Structure

The full-length open reading frame (ORF) of *KvCHX* was analyzed using SoftBerry FGENESH and BLASTX (Figure 2), and the results showed that the *KvCHX* gene had an ORF of 2562 bp (from 627 bp to 3189 bp). There were three exons (627 bp–827 bp, 913 bp–1914 bp, and 1993 bp–3189 bp, respectively) and two introns (828 bp–912 bp and 1915 bp–1992 bp, respectively). The CAAT Box and TATA Box were located at 50 bp and 100 bp, respectively. The TSS was located at 125 bp, and the Poly (A) tail of the mRNA was located at 3354 bp.

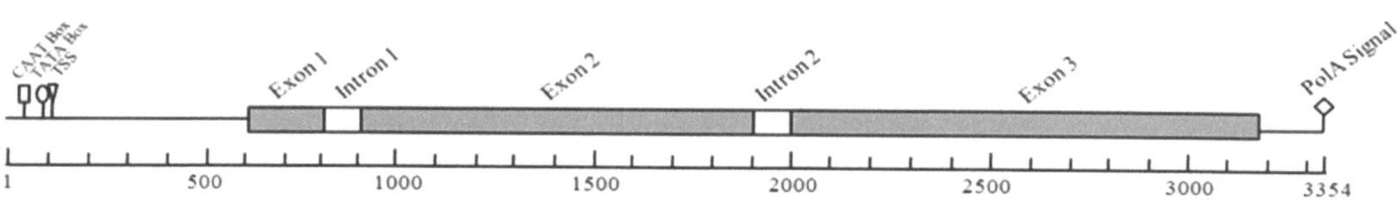

Figure 2. Structural sketch of the *KvCHX* gene.

3.3. Structure and Function Analysis of KvCHX Protein

The physical and chemical properties of the *KvCHX* protein were analyzed using the ProtParam program. The results showed that *KvCHX* encoded a polypeptide of 799 amino acids with a molecular weight of 86.6 kDa and a theoretical pI of 8.86. Its molecular formula was $C_{3934}H_{6355}N_{1023}O_{1097}S_{33}$. The *KvCHX* protein was predicted to be a hydrophobic basic lipid-soluble protein. Signal sequence analysis and nuclear localization signal analysis showed that the *KvCHX* protein had no signal peptide and nuclear localization signal. Transmembrane domain analysis revealed that *KvCHX* has 12 putative transmembrane domains in a highly hydrophobic N-terminal region (Figure 3). Subcellular localization analysis showed that the protein was located on the plasma membrane.

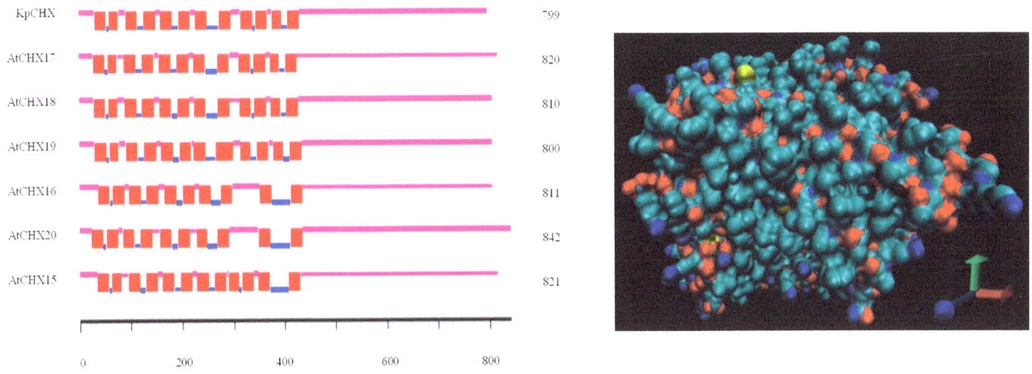

Figure 3. Transmembrane spans of the predicted amino acid sequences of KvCHX and AtCHX15~AtCHX20 (left) and tertiary structure prediction of the KvCHX protein from different perspectives using SWISS-MODEL. The protein visualization model is a single atom drawing method, where O is marked in red, N is marked in blue, C is marked in cyan, and S is marked in yellow (right).

3.4. Phylogenetic Analysis and Multiple Alignment of CHXs

To compare the evolutionary relationships, the putative *KvCHX* and *AtCHX* family were used to construct the phylogenetic tree using MEGA-X with the neighbor-joining (NJ) method and 1000 bootstrap replicates. The phylogenetic tree showed that a total of 28 members of the Arabidopsis *CHX* family could be divided into five subclades [21], among which *AtCHX15-AtCHX20*, *AtCHX21*, and *AtCHX23* belonged to the fourth class, and the *KvCHX* gene was closely related to the *AtCHX17* gene (Figure 4).

To further reveal the structure of the *KvCHX* sequence, the software ClastalX (1.81) was used to align the *KvHXK* sequence and *AtCHX17–19* members as the reference to find the conserved domain, so as to determine the similarity between the AtCHX protein sequence and *KvCHX*. As a result, similar to *AtCHX* family members *AtCHX*(17~19), the hydrophobic domain of *KvCHX* contains 12 conserved transmembrane domains, including 430 N-terminal amino acid residues, and a C-terminal hydrophilic domain located outside the plasma membrane, which was predicted to have a regulatory role (Figure 5). Multiple sequence analysis also showed that *KvCHX* and *AtCHX 17–19* had a high identity in the N-terminal hydrophobic transmembrane domain, and the identity level was 80.00%, 79.31% and 77.01%, respectively. In addition, among the 12 conserved transmembrane domains, 21 out of 23 amino acid residues in the TM5 domain, and 18 out of 23 amino acid residues in the TM11 domain, were completely identical, with 91.30% and 90% identity levels to homologous sequences, respectively. This indicated that TM5 and TM11 may play a greater direct role in restoring cation/H^+ transport.

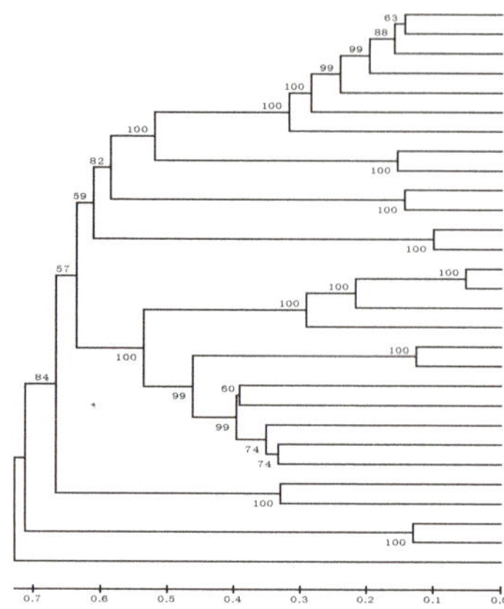

Figure 4. Phylogenetic relationship among *KvCHX* and 28 members of the *Arabidopsis thaliana* CHX family. *AtCHX1*, AY926465.1; *AtCHX2*, AEE36238.1; *AtCHX03*, AY926466.1; *AtCHX04*, DQ499020.1; *AtCHX05*, NP_172294.2; *AtCHX06a*, AEE28253.1; *AtCHX06b*, AEE28251.1; *AtCHX07*, AEC08087.1; *AtCHX08*, AY926468.1; *AtCHX09*; AY926469.1; *AtCHX10*, AEE77969.1; *AtCHX11*, AEE77968.1; *AtCHX12*, AEE77967.1; *AtCHX13*, EFH573351.1; *AtCHX14*, EFH65886.1; *AtCHX15*, AEC06246.1; *AtCHX16*, AEE34204.1; *AtCHX17*, EFH43951.1; *AtCHX18*, EFH44842.1; *AtCHX19*, ANM63400.1; *AtCHX20*, ANM63347.1; *AtCHX21*, ANM62431.1; *AtCHX23*, AEE27860.1; *AtCHX24*, EFH44758.1; *AtCHX25*, EFH42526.1; *AtCHX26*, AED90376.1; *AtCHX27*, AD90377.2; *AtCHX28*, AEE7888.1.

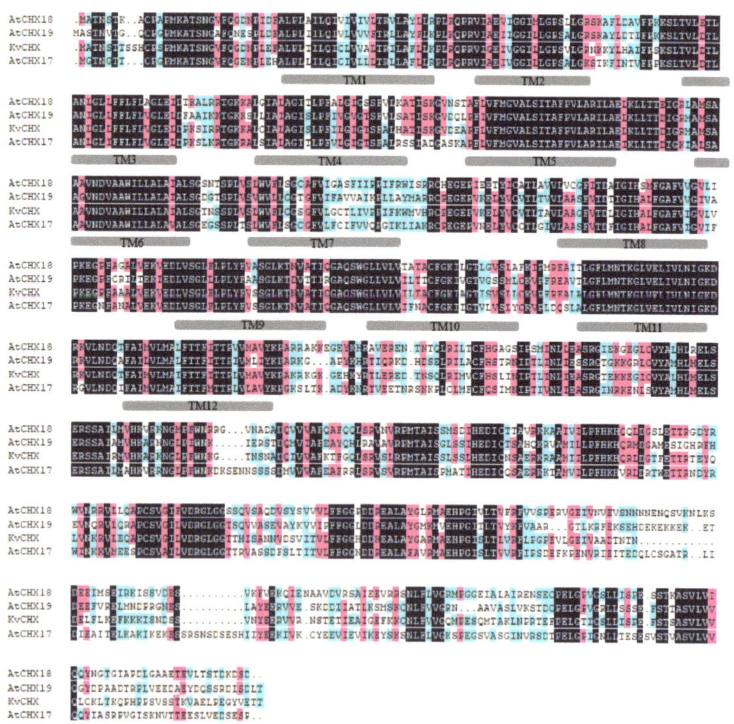

Figure 5. Multiple amino acid sequence alignments of *KvCHX* and *AtCHX17–19*. Identical amino acids are shaded in black. Similar amino acid residues are shaded in pink and light blue.

3.5. Expression and Subcellular Localization of Transgenic A. Thaliana

To investigate the function of the *KvHXK* gene under salt stress, the pMDC45 vector containing GFP-KvCHX was transformed into A. tumefaciens strain GV3101, and then successfully transferred into Arabidopsis plants via the floral dipping technique. To check the integration of the *KvCHX* gene in different transgenic lines, PCR analysis used to amplify the integrated fragment was performed using genomic DNA from leaves from WT and T_3 transgenic seedlings. The bands of the integrated fragment of 2978 bp predicted were displayed (Figure 6A). Our result showed that the *KvCHX* gene from 17 seedlings had been integrated into the genome of *A. thaliana*.

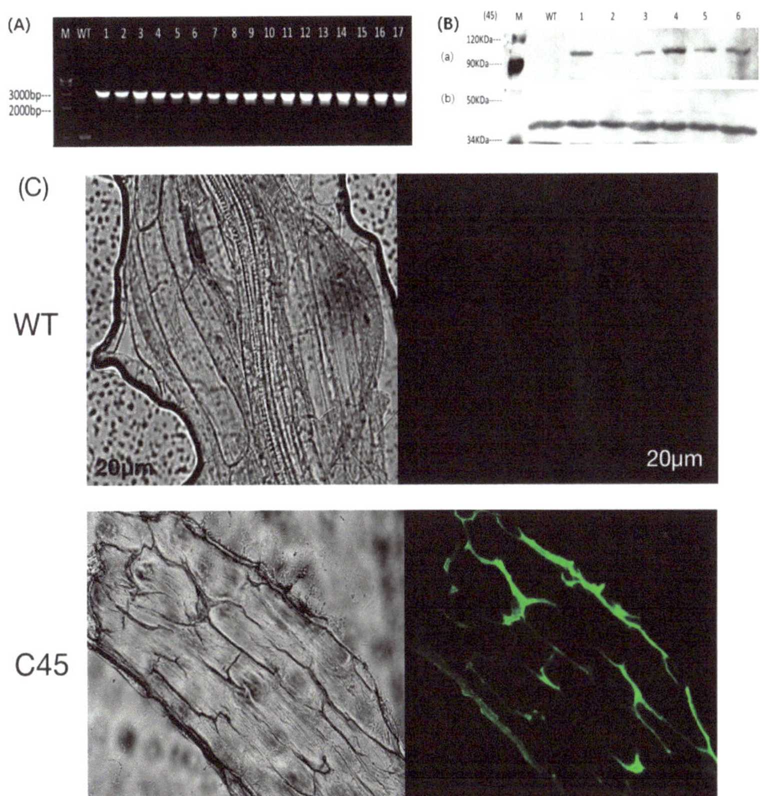

Figure 6. (**A**) PCR product of WT and transgenic Arabidopsis lines. M, 1 Kb Ladder; WT, wild type; 1–17, the transgenic lines. (**B**) Western blot analysis of transgenic Arabidopsis. M, protein marker; WT, wild type; 1–6, C45 transgenic lines. (a) KpCHX-GFP protein; (b) GAPDH as reference protein. (**C**) The GFP fluorescence observation of the root in WT and transgenic plant seedlings (C45) (GFP: at 488 nm excitation and emitted between 515 and 565 nm).

The GFP-KvCHX fusion protein in transgenic Arabidopsis seedlings was analyzed for the expression of the *KvCHX* gene. The KpCHX protein had a molecular weight of 86.6 KDa, and together with the GFP's molecular weight of 27 KDa, it was expected to form a 113 KDa fusion protein. Western blot analysis showed that there was one obvious protein band between 90 KDa and 120 KDa. The band close to 120 KDa was predicted as the target band (Figure 6B). Therefore, the KvCHX protein was successfully detected in transgenic plants under salt stress and 1/2 K^+ deficiency treatments.

To investigate the subcellular localization of *KvCHX*, the young root tip and leaf of transgenic seedlings (designated as C45) and WT were selected to observe the fluorescence. A green fluorescence signal was present in the plasma membrane (Figure 6C). The result showed that the KvCHX protein exclusively localized to the plasma membrane. This result was consistent with the *KvCHX* structure using the in silico approaches mentioned above.

3.6. Overexpression of KvCHX Enhances Tolerance to Salt Stress in Transgenic Arabidopsis

Through there was no significant difference in germination rate between WT and transgenic plants (C45) cultured in MS medium, there was a significant difference in phenotype. Under salt stress treatment, root elongation was severely retarded (Figure 7A,B), and leaves gradually lost greenness (Figure 7C,D). Compared with WT, transgenic plants showed particularly vigorous root development, and leaves remained green under a 150 mM NaCl concentration of salt stress. With increasing NaCl concentration, the root length, fresh weight, and dry weight of transgenic plants and WT all decreased to varying degrees (Figure 7B,E,F). It was noteworthy that the root length and fresh and dry weight of transgenic plants, however, were significantly higher than those of WT plants at the same NaCl concentration. In the same concentration of NaCl solution, transgenic plants had significantly more leaves than WT plants (Figure 7C,D). In addition, prolines are an important indicator that represents the extent to which a plant is tolerant to salt stress; our results suggested that the proline content of transgenic plants is higher than that of WT under salt stress, especially in 200 mM NaCl. As an important indicator of salt tolerance, transgenic plants had a higher proline content than wild-type plants under salt stress, especially under 200 mM NaCl stress (Figure 7G). In transgenic plants and WT, salt stress induced MDA production with increasing NaCl concentration, but in transgenic plants, it was significantly lower than in WT (Figure 7H) in the same concentration of NaCl solution. Thus, the results showed that *KvCHX* enhanced salt tolerance in transgenic Arabidopsis seedlings.

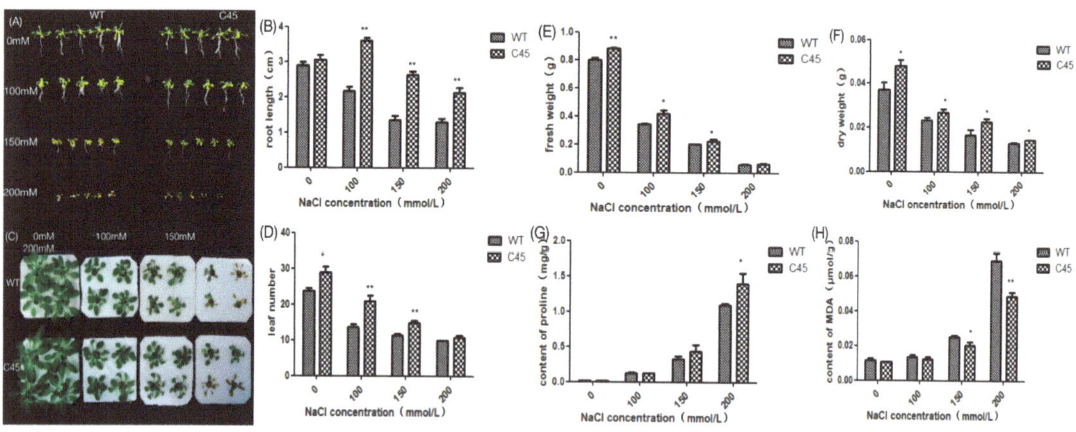

Figure 7. Effects of salt stress on WT and transgenic Arabidopsis lines. (**A**) Seedings of WT and C45 transgenic Arabidopsis in MS medium supplemented with NaCl. (**B**) Root length of (**A**). (**C**) Seedlings of transgenic Arabidopsis (C45) and WT were planted in basic nutrient solution to rosette growth stage, and then transferred into nutrient solution supplemented with 0 mM, 100 mM, 150 mM, and 200 mM NaCl for 7 days. (**D–H**) The leaf numbers, fresh and dry weight, MDA content, and proline content of (**C**), respectively. (Values are mean ± SD for three independent replicates, $n = 5$.) *, $p < 0.05$; **, $p < 0.01$ in two-tailed Student's t-tests.

3.7. KvCHX-Overexpressing Plants Grew Better Than WT under K^+ Deficient Conditions

After 20 days of culture in 1/2 K^+ nutrient solution, WT and transgenic plants (C45) showed a different appearance. The transgenic plants were significantly vigorous, while WT shriveled under K^+ deficient conditions (Figure 8). Moreover, the transgenic plants showed an increase in root length compared to the control plants (Figure 8A,B). The number of leaves and branches (Figure 8C), fresh weight (Figure 8D), and dry weight (Figure 8E) of the transgenic lines were higher than those of the WT lines. Transgenic plants grew better under K deficiency than WT.

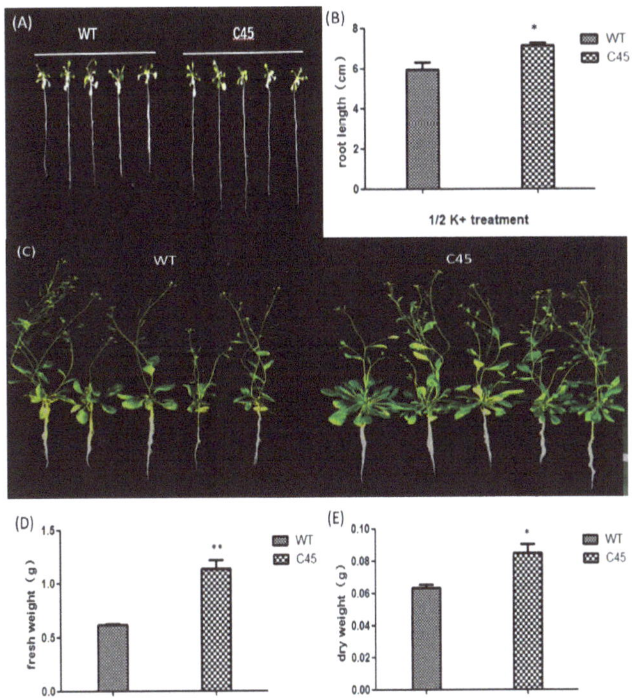

Figure 8. WT and transgenic Arabidopsis lines under deficient K^+ treatment. (**A**) Seedlings of WT and transgenic (C45) plants were grown in basal nutrient solution for 10 days, and then transferred into 1/2 K^+ nutrient solution for 20 days. (**B**) Root length of WT and transgenic plants. (**C**) Branch numbers of WT and transgenic plants. (**D**,**E**) Fresh weight and dry weight of transgenic lines and WT. (Values are mean ± SD for three independent replicates, n = 5.) *, $p < 0.05$; **, $p < 0.01$ in two-tailed Student's *t*-tests.

3.8. Determination of Ion Contents of Transgenic Arabidopsis Lines under Deficient K^+ Treatment and Treatment with NaCl

The root and leaf K^+/Na^+ ratios of WT and C45 transgenic plants grown in basic nutrition solution were almost the same (Figure 9A). However, under 1/2 K^+ treatment, the K^+/Na^+ ratio in the roots and leaves of C45 transgenic plants were both higher than that of WT (Figure 9B). In the roots, the K^+/Na^+ ratio of C45 transgenic plants was higher than that of WT by 28.69%, and that of C45 transgenic plants was higher than that of WT by 26.93% in the leaves. Similarly, the K^+/Na^+ ratio of C45 transgenic plants was higher in roots and leaves than that of WT. These results indicate that in the K^+-deficient environment or salt stress conditions, the *KpCHX* gene could promote the absorption of K^+ and the excretion of Na^+ in transgenic Arabidopsis, which could help maintain a high level of intracellular K^+/Na^+ ratio and promote plant growth.

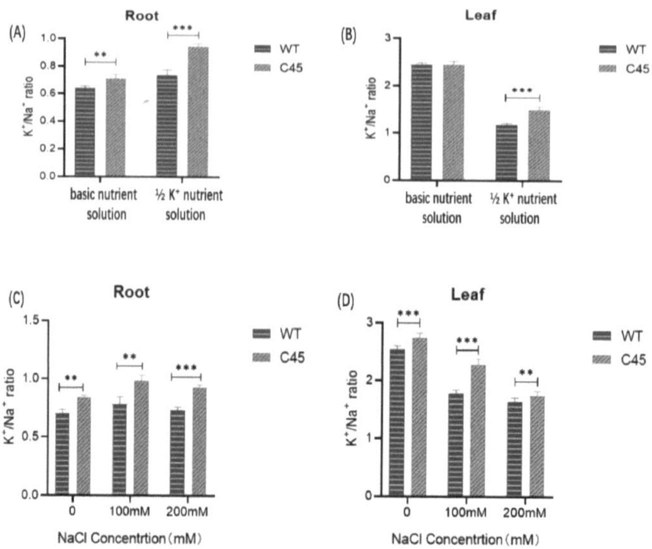

Figure 9. Comparison of the K^+/Na^+ ratios from roots and leaves between C45 transgenic plants and WT seedlings. (**A**) The K^+/Na^+ ratio from the roots of plants grown in basic nutrient solution and in 1/2 K^+ nutrient solution. (**B**) The K^+/Na^+ ratio from the leaves of plants grown in basic nutrient solution and in 1/2 K+ nutrient solution. (**C**) The K^+/Na^+ ratio from the roots of plants under salt stress. (**D**) The K^+/Na^+ ratio from the leaves of plants under salt stress. Error bars indicate mean ± SD (values are mean ± SD for three independent replicates, $n = 6$). **, $p < 0.01$; ***, $p < 0.001$ in two-tailed Student's t-tests.

4. Discussion

Soil salinity is an important stress that limits plant growth and yield [33]. High concentrations of Na^+ are toxic and have a negative effect on the growth of glycophyte plants. When plants grow in saline or alkaline soil environment, excessive sodium ions accumulate in the cytoplasm. A higher concentration of sodium ions is harmful for plant growth and disrupts enzymatic functions [34]. Halophytes have the ability to withstand salt stress and feature salt-tolerant genes to counter the adverse effects of salt stress [35,36]. In addition to the biosynthesis of osmoprotectants, activation of antioxidant enzymes, synthesis of polyamines and regulation of plant hormones, the survival strategies of halophytes also include other mechanisms under higher-saline soil [37].

K. viginica is a dicotyledonous halophyte that has been studied at the physiological level. However, the molecular mechanism of its tolerance to salt stress remains unknown. In this study, the *KvCHX* gene was isolated for the first time using the genome walking method. PCR and Western blot analysis showed that the *KvCHX* gene was integrated into the Arabidopsis genome, and the GFP-KvCHX fusion protein was expressed in transgenic plants. Via subcellular localization assay, KvCHX was localized on plasma membrane in the root. Transgenic *A. thaliana* showed better growth than WT under K^+ deficiency treatment and salt stress.

In this study, *KvCHX* was specifically investigated to clarify its function of tolerance to salt stress. Phylogenetic and multiple comparison analyses revealed that *KvCHX* shared the highest homology with the *AtCHX17* gene, and coded for a putative cation/H^+ transporter. Compared with WT, the overexpression of *KvCHX* promoted root development and increased the leaf number, fresh weight, and dry weight under salt tolerance, indicating that *KvCHX* enhanced the salt tolerance of transgenic Arabidopsis seedlings. With 1/2 K^+ nutrient solution or NaCl stress, the K^+/Na^+ ratio of transgenic plants was higher than that of WT seedlings, and the K^+/Na^+ ratio of leaves was higher than that of roots. The results showed that *KvCHX* had a role in regulating the Na^+ and K^+ balance with 1/2 K^+ nutrient solution and NaCl stress. The salt tolerance mechanism of the halophyte *K. virginica* may depend on the absorption and sequestration of Na^+, K^+ acquisition, and the maintenance of the K^+/Na^+ ratio to avoid the damage of NaCl toxicity.

Ion channels and transporters play an important role in plant salt tolerance. A number of ion transporters related functionally have been identified in plants via structural homology. Among them, members of the CPA gene superfamily are important transporters. They include the Na^+/H^+ exchanger (NHX), K^+ efflux antiporter (KEA), and cation/H^+ exchanger (CHX) families in plant genomes, with a conserved Na^+/H^+ exchanger domain [38–40]. The compartmentation of Na^+ in the vacuoles by K^+/Na^+-specific NHX1-type antiporters and the maintenance of low Na^+ in cytosol by HKT1-type transporters prevents or reduces the movement of Na^+ to aboveground parts of plants [41]. Studies have shown that halophytes require high K^+/Na^+ ratios for normal cellular function [42].

The CPA2 family contains two subfamilies. A member of the CPA2 family, from Saccharomyces cerevisiae, appeared to mediate intracellular K^+ flux. In fact, besides KHA from microorganisms, the *CHX* genes of Arabidopsis, soybean, rice, and other plants were identified as being involved in tolerance to salt stress, which has the function of cation/H^+ transport [38]. *CHX* is a large family with 28 members in Arabidopsis [22], among which *AtCHX13* is considered to be a plasma membrane K^+ transporter, playing a role in increasing plant K^+ uptake in a potassium-deficient environment [43]. *AtCHX16-20* have different regulatory effects on K^+ and pH homeostasis in different cellular compartments [44]. A variety of environmental stresses, including high salt, potassium deficiency, ABA, and acidic mediators, can up-regulate *AtCHX17* transcripts [24]. The *AtCHX17* knockout mutant lines accumulate less K^+ in response to salt stress and K^+ starvation than the wild type. CHX20 is mainly localized to the membrane of the endosomal system and not only maintains K^+ homeostasis, but also affects the pH under certain conditions [14]. *AtCHX21* and *AtCHX23* are involved in K^+ homeostasis in the female gametophyte, and *AtCHX21* is also involved in regulating xylem Na^+ concentration and Na^+ accumulation in leaves [45]. According to previous research on the model plant *Arabidopsis*, *AtCHX17-19*, which is highly homologous to the *KvCHX* gene, is located on the membrane system, which is highly similar to the localization of *KvCHX* in our study [14]. *GmSALT3* belongs to the *CHX* family and is predominately expressed in root phloem- and xylem-associated cells under both saline and non-saline conditions [46]. *GmSALT3* improves the ability of salt-tolerant near-isogenic soybean lines under saline stress through preventing excessive ROS accumulation in roots, and potentially modulating Ca^{2+} signaling, vesicle trafficking, and the formation of diffusion barriers [15]. GmSALT3 confers net shoot exclusion for both Na^+ and Cl^- and improves salt tolerance in soybeans. *GmCHX1* from salt-tolerant soybeans was shown to protect plants via Na^+ exclusion under salt stress [47]. *GmCHX20a* and *GmCHX1* might work complementally through a concerted effort to address both osmotic stress and ionic stress as a result of elevated salinity. An overexpression of *GsCHX19.3* in Arabidopsis improved plant tolerance to under saline–alkali stress by reducing the Na^+ concentration and increasing the K^+/Na^+ ratio [48]. *OsCHX14* is regulated by the JA signaling pathway, capable of transporting K^+, Rb^+, and Cs^+ in vivo, and plays an important role in K^+ homeostasis during flowering in rice [49].

Halophytes prevent Na^+ accumulation in the cytoplasm through a variety of mechanisms, including Na^+ extrusion and/or the intracellular compartmentalization of Na^+, along with the recirculation of Na^+ out of the shoot and the up-regulation of Na^+/H^+ antiporters in the plasma membrane [50,51]. Here, we demonstrate that the *KvCHX* gene participates in regulating K^+/Na^+ to reduce damage to *K. virginica* seedlings under salt stress. Our study shows that the *KvCHX* gene confers salt tolerance to *A. Thaliana* by taking up and translocating K^+ to withstand salt stress. The *KvCHX* gene may play a key role in selective K^+ accumulation and transport from roots to leaves at the cellular and whole-plant levels. These results provide genetic and biochemical evidence that the *KvCHX* protein plays a major role in the balance between K^+ influxes and possibly Na^+ modulation via regulating different transporters, directly or indirectly.

Improving salt tolerance is an important goal in plant breeding. Genes associated with halophyte survival strategies can be used to modify glycophytes through protecting and maintaining the function and structure of cellular components. In the current research, we report that the *CHX* gene from *K. virgineca*, a salt marsh halophyte, maintains Na^+ and K^+ homeostasis in transgenic Arabidopsis lines. *KvCHX* encodes a K^+, Na^+/H^+ exchanger, which not only mediates K^+ uptake, but also has a role in regulating cellular ion homeostasis through the expulsion of Na^+ as cation antiporters under NaCl stress. At present, GmSALT3, as a *CHX* family gene in soybeans, has been shown for the first time to improve plant salt tolerance by promoting the recycling of Cl^- in the phloem. Although our study reveals the potential mechanism by which the *KvCHX* gene enhances the salt stress ability of Arabidopsis, further research is still needed [14]. It is not clear which initial signals can directly initiate the upstream and downstream cascade involving K^+ activation and Na^+ expulsion, but it is conceivable that *KvCHX* or other related cation exchangers may have significant effects on the osmotic pressure, volume, and pH of plant cell compartments. In addition, *CHX* may also be involved in the polarized growth of plants, which is of great significance to follow-up research. Therefore, future work needs to focus on the exploration of molecular signals to reveal new insights into K^+/Na^+ regulation for plants in response to abiotic stress.

Author Contributions: All authors contributed to the study conception and design. Material preparation, data collection, and analysis were performed by Z.T. and C.Z. The first draft of the manuscript was written by Z.T. and Y.G. reviewed and revised the manuscript. Z.T. is the corresponding author and finalized the manuscript. All authors have read and agreed to the published version of the manuscript.

Funding: Scientific and Technological Project (231111111200) in Henan Province, China.

Institutional Review Board Statement: Not applicable.

Informed Consent Statement: Not applicable.

Data Availability Statement: Data are contained within the article.

Acknowledgments: We thank Denise M. Seliskar (Biotechnology Research, Delaware University) for providing the seeds of *K. virginica*.

Conflicts of Interest: The authors declare no conflict of interest.

References

1. De Leonardis, A.M.; Marone, D.; Mazzucotelli, E.; Neffar, F.; Rizza, F.; Di Fonzo, N.; Cattivelli, L.; Mastrangelo, A.M. Durum wheat genes up-regulated in the early phases of cold stress are modulated by drought in a developmental and genotype dependent manner. *Plant Sci.* **2007**, *172*, 1005–1016. [CrossRef]
2. Yan, K.; Shao, H.; Shao, C.; Chen, P.; Zhao, S.; Brestic, M.; Chen, X. Physiological adaptive mechanisms of plants grown in saline soil and implications for sustainable saline agriculture in coastal zone. *Acta Physiol. Plant.* **2013**, *35*, 2867–2878. [CrossRef]
3. Tang, X.; Mu, X.; Shao, H.; Wang, H.; Brestic, M. Global plant-responding mechanisms to salt stress: Physiological and molecular levels and implications in biotechnology. *Crit. Rev. Biotechnol.* **2015**, *35*, 425–437. [CrossRef] [PubMed]
4. Flowers, T.J.; Colmer, T.D. Salinity tolerance in halophytes. *New Phytol.* **2008**, *179*, 945–963. [CrossRef] [PubMed]
5. Zhu, J. Salt and drought stress signal transduction in plants. *Annu. Rev. Plant Biol.* **2002**, *53*, 247–273. [CrossRef]
6. Gupta, B.; Huang, B. Mechanism of salinity tolerance in plants: Physiological, biochemical, and molecular characterization. *Int. J. Genom.* **2014**, *2014*, 701596. [CrossRef]
7. James, R.A.; Blake, C.; Byrt, C.S.; Munns, R. Major genes for Na^+ exclusion, *Nax1* and *Nax2* (wheat *HKT1;4* and *HKT1;5*), decrease Na^+ accumulation in bread wheat leaves under saline and waterlogged conditions. *J. Exp. Bot.* **2011**, *62*, 2939–2947. [CrossRef]
8. Szczerba, M.W.; Britto, D.T.; Kronzucker, H.J. K^+ transport in plants: Physiology and molecular biology. *J. Plant Physiol.* **2009**, *166*, 447–466. [CrossRef]
9. Elumalai, R.P.; Nagpal, P.; Reed, J.W. A mutation in the Arabidopsis *KT2/KUP2* potassium transporter gene affects shoot cell expansion. *Plant Cell* **2002**, *14*, 119–131. [CrossRef]
10. Maathuis, F.J.M.; Sanders, D. Mechanisms of potassium absorption by higher plant roots. *Physiol. Plant.* **1996**, *96*, 158–168. [CrossRef]

11. Philippar, K.; Fuchs, I.; Luthen, H.; Hoth, S.; Bauer, C.S.; Haga, K.; Thiel, G.; Ljung, K.; Sandberg, G.; Bottger, M.; et al. Auxin-induced K^+ channel expression represents an essential step in coleoptile growth and gravitropism. *Proc. Natl. Acad. Sci. USA* **1999**, *96*, 12186–12191. [CrossRef] [PubMed]
12. Maathuis, F.; Amtmann, A. K^+ Nutrition and Na^+ Toxicity: The Basis of Cellular K^+/Na^+ Ratios. *Ann. Bot.* **1999**, *84*, 123–133. [CrossRef]
13. Sze, H.; Chanroj, S. Plant Endomembrane Dynamics: Studies of K^+/H^+ Antiporters Provide Insights on the Effects of pH and Ion Homeostasis. *Plant Physiol.* **2018**, *177*, 875–895. [CrossRef] [PubMed]
14. Padmanaban, S.; Chanroj, S.; Kwak, J.M.; Li, X.; Ward, J.M.; Sze, H. Participation of endomembrane cation/H^+ exchanger *AtCHX20* in osmoregulation of guard cells. *Plant Physiol.* **2007**, *144*, 82–93. [CrossRef] [PubMed]
15. Qu, Y.; Guan, R.; Bose, J.; Henderson, S.W.; Wege, S.; Qiu, L.; Gilliham, M. Soybean CHX-type ion transport protein GmSALT3 confers leaf Na^+ exclusion via a root derived mechanism, and Cl^- exclusion via a shoot derived process. *Plant Cell Environ.* **2021**, *44*, 856–869. [CrossRef] [PubMed]
16. Zhou, G.; Xia, Y.; Ma, B.L.; Feng, C.; Qin, P. Culture of Seashore Mallow under Different Salinity Levels Using Plastic Nutrient-Rich Matrices and Transplantation. *Agron. J.* **2010**, *102*, 395–402. [CrossRef]
17. Godfrey, R.K.; Wooten, J.W. *Aquatic and Wetland Plants of Southeastern United States. Vol. II: Dicotyledons*; University of Georgia Press: Athens, Greece, 1981.
18. Islam, M.N.; Wilson, C.A.; Watkins, T.R. Nutritional evaluation of seashore mallow seed, *Kosteletzkya virginica*. *J. Agric. Food Chem.* **1982**, *30*, 1195–1198. [CrossRef]
19. Moser, B.R.; Dien, B.S.; Seliskar, D.M.; Gallagher, J.L. Seashore mallow (*Kosteletzkya pentacarpos*) as a salt-tolerant feedstock for production of biodiesel and ethanol. *Renew. Energy* **2013**, *50*, 833–839. [CrossRef]
20. Wang, H.; Tang, X.; Wang, H.; Shao, H. Physiological responses of *Kosteletzkya virginica* to coastal wetland soil. *Sci. World J.* **2015**, *2015*, 354581. [CrossRef]
21. Guo, Y.; Tian, Z.; Qin, G.; Yan, D.; Zhang, J.; Zhou, W.; Qin, P. Gene expression of halophyte *Kosteletzkya virginica* seedlings under salt stress at early stage. *Genetica* **2009**, *137*, 189–199. [CrossRef]
22. Sze, H.; Padmanaban, S.; Cellier, F.; Honys, D.; Cheng, N.; Bock, K.W.; Conéjéro, G.; Li, X.; Twell, D.; Ward, J.M.; et al. Expression patterns of a novel *AtCHX* gene family highlight potential roles in osmotic adjustment and K^+ homeostasis in pollen development. *Plant Physiol.* **2004**, *136*, 2532–2547. [CrossRef] [PubMed]
23. Ramírez, J.; Ramírez, O.; Saldaña, C.; Coria, R.; Peña, A. A *Saccharomyces cerevisiae* mutant lacking a K+/H+ exchanger. *J. Bacteriol.* **1998**, *180*, 5860–5865. [CrossRef] [PubMed]
24. Cellier, F.; Conéjéro, G.; Ricaud, L.; Luu, D.T.; Lepetit, M.; Gosti, F.; Casse, F. Characterization of *AtCHX17*, a member of the cation/H^+ exchangers, CHX family, from *Arabidopsis thaliana* suggests a role in K^+ homeostasis. *Plant J.* **2004**, *39*, 834–846. [CrossRef] [PubMed]
25. Boyes, D.C.; Zayed, A.M.; Ascenzi, R.; McCaskill, A.J.; Hoffman, N.E.; Davis, K.R.; Gorlach, J. Growth stage-based phenotypic analysis of Arabidopsis: A model for high throughput functional genomics in plants. *Plant Cell* **2001**, *13*, 1499–1510. [CrossRef]
26. Karimi, M.; Inzé, D.; Depicker, A. GATEWAY vectors for Agrobacterium-mediated plant transformation. *Trends Plant Sci.* **2002**, *7*, 193–195. [CrossRef] [PubMed]
27. Curtis, M.D.; Grossniklaus, U. A gateway cloning vector set for high-throughput functional analysis of genes in planta. *Plant Physiol.* **2003**, *133*, 462–469. [CrossRef] [PubMed]
28. Hofgen, R.; Willmitzer, L. Storage of competent cells for *Agrobacterium transformation*. *Nucleic Acids Res.* **1988**, *16*, 9877. [CrossRef]
29. Clough, S.J.; Bent, A.F. Floral dip: A simplified method for Agrobacterium-mediated transformation of *Arabidopsis thaliana*. *Plant J.* **1998**, *16*, 735–743. [CrossRef]
30. Chen, S.; Tian, Z.Y.; Guo, Y.Q. Characterization of hexokinase gene family members in *Glycine max* and functional analysis of *GmHXK2* under salt stress. *Front. Genet.* **2023**, *23*, 1135290. [CrossRef]
31. Bates, L.S.; Waldren, R.P.; Teare, I.D. Rapid determination of free proline for water-stress studies. *Plant Soil* **1973**, *39*, 205–207. [CrossRef]
32. Song, S.; Chen, Y.; Chen, J.; Dai, X.; Zhang, W. Physiological mechanisms underlying *OsNAC5*-dependent tolerance of rice plants to abiotic stress. *Planta* **2011**, *234*, 331–345. [CrossRef] [PubMed]
33. Szabolcs, I. *Salt-Affected Soils*; CRC Press Inc.: Boca Raton, FL, USA, 1989.
34. Tester, M.; Davenport, R. Na^+ tolerance and Na^+ transport in higher plants. *Ann. Bot.* **2003**, *91*, 503–527. [CrossRef] [PubMed]
35. Asada, K. Production and scavenging of reactive oxygen species in chloroplasts and their functions. *Plant Physiol.* **2006**, *141*, 391–396. [CrossRef] [PubMed]
36. Yu, J.; Chen, S.; Zhao, Q.; Wang, T.; Yang, C.; Diaz, C.; Sun, G.; Dai, S. Physiological and proteomic analysis of salinity tolerance in *Puccinellia tenuiflora*. *J. Proteome Res.* **2011**, *10*, 3852–3870. [CrossRef] [PubMed]
37. Kumari, A.; Das, P.; Parida, A.K.; Agarwal, P.K. Proteomics, metabolomics, and ionomics perspectives of salinity tolerance in halophytes *Front. Plant Sci.* **2015**, *6*, 537.
38. Ye, C.; Yang, X.; Xia, X.; Yin, W. Comparative analysis of cation/proton antiporter superfamily in plants. *Gene* **2013**, *521*, 245–251. [CrossRef]
39. Brett, C.L.; Donowitz, M.; Rao, R. Evolutionary origins of eukaryotic sodium/proton exchangers. *Am. J. Physiol. Cell Physiol.* **2005**, *288*, C223–C239. [CrossRef]

40. Punta, M.; Coggill, P.C.; Eberhardt, R.Y.; Mistry, J.; Tate, J.; Boursnell, C.; Pang, N.; Forslund, K.; Ceric, G.; Clements, J.; et al. The Pfam protein families database. *Nucleic Acids Res.* **2012**, *40*, D290–D301. [CrossRef]
41. Zhu, J. Regulation of ion homeostasis under salt stress. *Curr. Opin. Plant Biol.* **2003**, *6*, 441–445. [CrossRef]
42. Bose, J.; Rodrigo-Moreno, A.; Lai, D.; Xie, Y.; Shen, W.; Shabala, S. Rapid regulation of the plasma membrane H^+-ATPase activity is essential to salinity tolerance in two halophyte species, *Atriplex lentiformis* and *Chenopodium quinoa*. *Ann. Bot.* **2015**, *115*, 481–494. [CrossRef]
43. Zhao, J.; Cheng, N.; Motes, C.M.; Blancaflor, E.B.; Moore, M.; Gonzales, N.; Padmanaban, S.; Sze, H.; Ward, J.M.; Hirschi, K.D. AtCHX13 Is a Plasma Membrane K^+ Transporter. *Plant Physiol.* **2008**, *148*, 796–807. [CrossRef] [PubMed]
44. Chanroj, S.; Lu, Y.; Padmanaban, S.; Nanatani, K.; Uozumi, N.; Rao, R.; Sze, H. Plant-specific cation/H^+ exchanger 17 and its homologs are endomembrane K^+ transporters with roles in protein sorting. *J. Biol. Chem.* **2011**, *286*, 33931–33941. [CrossRef] [PubMed]
45. Hall, D.; Evans, A.R.; Newbury, H.J.; Pritchard, J. Functional analysis of CHX, 21: A putative sodium transporter in Arabidopsis. *J. Exp. Bot.* **2006**, *57*, 1201–1210. [CrossRef] [PubMed]
46. Qu, Y.; Guan, R.; Yu, L.; Berkowitz, O.; David, R.; Whelan, J.; Ford, M.; Wege, S.; Qiu, L.; Gilliham, M. Enhanced reactive oxygen detoxification occurs in salt-stressed soybean roots expressing *GmSALT3*. *Physiol. Plant* **2022**, *174*, e13709. [CrossRef] [PubMed]
47. Jia, Q.; Li, M.W.; Zheng, C.; Xu, Y.; Sun, S.; Li, Z.; Wong, F.L.; Song, J.; Lin, W.; Li, Q.; et al. The soybean plasma membrane-localized cation/H^+ exchanger *GmCHX20a* plays a negative role under salt stress. *Physiol. Plant* **2021**, *171*, 714–727. [CrossRef] [PubMed]
48. Jia, B.; Sun, M.; DuanMu, H.; Ding, X.; Liu, B.; Zhu, Y.; Sun, X. GsCHX19.3, a member of cation/H^+ exchanger superfamily from wild soybean contributes to high salinity and carbonate alkaline tolerance. *Sci. Rep.* **2017**, *7*, 9423. [CrossRef]
49. Chen, Y.; Ma, J.; Miller, A.J.; Luo, B.; Wang, M.; Zhu, Z.; Ouwerkerk, P.B.F. OsCHX14 is Involved in the K^+ Homeostasis in Rice (*Oryza sativa*) Flowers. *Plant. Cell Physiol.* **2016**, *57*, 1530–1543. [CrossRef]
50. Hamada, A.; Shono, M.; Xia, T.; Ohta, M.; Hayashi, Y.; Tanaka, A.; Hayakawa, T. Isolation and characterization of a Na^+/H^+ antiporter gene from the halophyte *Atriplex gmelini*. *Plant Mol. Biol.* **2001**, *46*, 35–42. [CrossRef]
51. Shi, H.; Lee, B.H.; Wu, S.J.; Zhu, J. Overexpression of a plasma membrane Na^+/H^+ antiporter gene improves salt tolerance in *Arabidopsis thaliana*. *Nat. Biotechnol.* **2003**, *21*, 81–85. [CrossRef]

current issues in molecular biology

Article

Transcriptomic Approach for Investigation of *Solanum* spp. Resistance upon Early-Stage Broomrape Parasitism

Maria Gerakari [1], Vasiliki Kotsira [2,3], Aliki Kapazoglou [4], Spyros Tastsoglou [2,3], Anastasios Katsileros [1], Demosthenis Chachalis [5], Artemis G. Hatzigeorgiou [2,3,*] and Eleni Tani [1,*]

1 Laboratory of Plant Breeding and Biometry, Agricultural University of Athens, 11855 Athens, Greece; mgerakari@aua.gr (M.G.); katsileros@aua.gr (A.K.)
2 Hellenic Pasteur Institute, 11521 Athens, Greece; vasilikikotsira@hotmail.com (V.K.); spiros.tastsoglou@gmail.com (S.T.)
3 DIANA-Lab, Department of Computer Science and Biomedical Informatics, University of Thessaly, 35131 Lamia, Greece
4 Hellenic Agricultural Organization-Dimitra (ELGO-DIMITRA), Department of Vitis, Institute of Olive Tree, Subtropical Crops and Viticulture (IOSV), Sofokli Venizelou 1, Lykovrysi, 14123 Athens, Greece; kapazoglou@elgo.gr
5 Laboratory of Weed Science, Benaki Phytopathological Institute, 14561 Kifisia, Greece; d.chachalis@bpi.gr
* Correspondence: artemis.hatzigeorgiou@gmail.com (A.G.H.); etani@aua.gr (E.T.)

Abstract: Tomato (*Solanum lycopersicum*) is a major horticultural crop of high economic importance. *Phelipanche* and *Orobanche* genera (broomrapes) are parasitic weeds, constituting biotic stressors that impact tomato production. Developing varieties with tolerance to broomrapes has become imperative for sustainable agriculture. *Solanum pennellii*, a wild relative of cultivated tomato, has been utilized as breeding material for *S. lycopersicum*. In the present study, it is the first time that an in-depth analysis has been conducted for these two specific introgression lines (ILs), IL6-2 and IL6-3 (*S. lycopersicum* X *S. pennellii*), which were employed to identify genes and metabolic pathways associated with resistance against broomrape. Comparative transcriptomic analysis revealed a multitude of differentially expressed genes (DEGs) in roots, especially in the resistant genotype IL6-3, several of which were validated by quantitative PCR. DEG and pathway enrichment analysis (PEA) revealed diverse molecular mechanisms that can potentially be implicated in the host's defense response and the establishment of resistance. The identified DEGs were mostly up-regulated in response to broomrape parasitism and play crucial roles in various processes different from strigolactone regulation. Our findings indicate that, in addition to the essential role of strigolactone metabolism, multiple cellular processes may be involved in the tomato's response to broomrapes. The insights gained from this study will enhance our understanding and facilitate molecular breeding methods regarding broomrape parasitism. Moreover, they will assist in developing sustainable strategies and providing alternative solutions for weed management in tomatoes and other agronomically important crops.

Keywords: *Solanum* spp.; broomrape; parasitism; transcriptomics

Citation: Gerakari, M.; Kotsira, V.; Kapazoglou, A.; Tastsoglou, S.; Katsileros, A.; Chachalis, D.; Hatzigeorgiou, A.G.; Tani, E. Transcriptomic Approach for Investigation of *Solanum* spp. Resistance upon Early-Stage Broomrape Parasitism. *Curr. Issues Mol. Biol.* **2024**, *46*, 9047–9073. https://doi.org/10.3390/cimb46080535

Academic Editor: Dongchul Kang

Received: 3 July 2024
Revised: 28 July 2024
Accepted: 2 August 2024
Published: 18 August 2024

1. Introduction

Climate change impacts food security directly and indirectly. It imposes new limitations on resources essential to plant growth, affects crop–weed interactions, and exacerbates the negative effect of weed infestation on crop productivity, leading to severe losses in crop yield [1]. This can shift the balance of competition between host plants and parasitic weeds, potentially favoring the parasite over the host. Weeds comprise the major biotic factor limiting crop production worldwide [2]. Parasitic plants are scientifically intriguing and agriculturally important weeds that are spreading worldwide whereas at the same time, the means to control them remain limited [3]. Considering that parasitic weeds have evolved resistance towards some of the most widely applied herbicides, effective weed management has become an imperative for ensuring crop productivity [4]. The efficacy of classic herbicides has decreased due to their extensive use over the years [5], but they are not considered a sustainable agronomic practice.

The parasitic weed family *Orobanchaceae* (broomrapes) causes severe damage to economically important crops including legumes, sunflowers, and tomatoes [6]. They are widespread in Mediterranean areas, Asia, and Southern and Eastern Europe. Broomrape species attack dicotyledonous crops, depending entirely on their hosts for all nutritional requirements [7]. Due to their physiology, underground parasitism, achlorophyllous nature, and the hardly controlled seed bank, broomrapes cannot be effectively controlled by the usual management strategies designed for non-parasitic weeds [8]. Most of these strategies, such as preventive measures, cultivation practices, and biological and chemical control [9], fail to provide satisfactory control, as they are not feasible on a large scale, and often, they are not financially sustainable or environmentally safe [10,11]. One of these broomrape species namely *Phelipanche ramosa* is extremely widespread in the Mediterranean region and can parasitize many host crops of high nutritional and economic importance, including oilseed rape (*Brassica napus*), hemp (*Cannabis sativa*), tomato (*Solanum lycopersicum*), tobacco (*Nicotiana tabacum*), sunflower (*Helianthus annuus*), and melon (*Cucumis melo*) [12,13]. The recent state of the art regarding parasitic weeds–plant host interactions at the molecular level focuses on a newly discovered family of phytohormones, strigolactones (SLs), and their demonstrated involvement in host recognition and evolution of parasitic plants. Research by Cheng et al. [6] suggests that the development of low SL-producing lines in tomatoes may constitute a promising approach to combat the parasitic weed *Phelipanche ramosa*, as SLs are the main germination stimulants of this parasite. In this study, new findings have revealed that except SLs, there are also other metabolic pathways, and the respective underlying genes may be implicated in the complex interactions between host plants and parasitic weeds through the viewpoint of climate change.

Cultivated tomato is an important and very nutritional food resource for humans [14] and many commercial varieties are infected by broomrape species [15]. However, little is known about tomato defense mechanisms against broomrape parasitism, and despite extensive screening, no single immune strategy or strong resistance against broomrape has been reported [16]. Breeding for host resistance against holoparasitic weeds remains one of the most effective and sustainable management strategies to date, and considerable effort has been placed towards this goal for several crops such as peas and sunflower, over the last two decades [17]. Nevertheless, breeding for resistant varieties constitutes an ongoing struggle, as germplasm with broomrape-resistant traits remains scarce. Plant breeding programs based on molecular techniques seem to be the most promising and sustainable approach to effectively handle/manage the phenomenon of parasitism. A plethora of molecular and non-molecular techniques have been deployed in tomatoes aiming to increase the resistance of tomatoes to *Phelipanche* sp., such as the application of chemical or targeted mutagenesis and CRISPR/Cas9-mediated mutagenesis of specific targets as well as the use of appropriate rootstocks resistant to broomrape parasitism [18–21]. The vast genetic diversity found in the tomato's wild relatives is associated often with resilience to different biotic and abiotic stress factors and adaptation mechanisms to extreme environmental conditions [22,23]. *Solanum pennellii*, a wild relative of the cultivated tomato, which is endemic to the Andean regions in South America, has evolved to thrive in arid habitats and has been used as breeding material for the cultivated *S. lycopersicum*, due to its stress tolerance to extreme conditions and unusual morphology [24]. *S. pennellii* ILs have been widely used to map QTLs influencing yield and fruit quality and are involved in heterosis [25]. ILs are also utilized as valuable genetic resources to improve abiotic stress tolerance [26].

RNA-Seq is probably the most widely applied high-throughput method to study gene expression at the transcript/isoform level. It involves the sequencing, quantification, and comparative analysis of the entire set of transcripts in an organism across tissues, developmental stages, and different conditions to assess the up- or down-regulation of genes involved in the relevant cellular processes [27]. PEA, also known as functional enrichment analysis (or pathway analysis), is typically applied to identify known molecular processes that are significantly enriched with genes that are altered in case samples in comparison to a control, facilitating interpretation and subsequent hypothesis generation [28].

Previous works employed the same approach to study the interaction of tomatoes with biotic stress factors. For example, RNA-sequencing and tissue-specific expression analyses were conducted to analyze the compatible interaction of tomatoes with *Verticillium dahliae* and *Cuscuta campestris*, respectively [29,30]. These studies aimed at developing novel sustainable breeding tools towards improved crop adaptability to environmental stresses that have been occurring with ever-increasing frequency and intensity over the last decades. Herein we provide, for the first time, a comparative transcriptomic and pathway analysis among putative tomato genotypes resistant and susceptible to the holoparasitic weed *P. ramosa*. Our study reveals differentially expressed genes in the host roots upon infection and metabolic pathways which may play a crucial role in the tomato response to parasitic infection and the establishment of resistance.

In the present study, a commercial tomato hybrid of *S. lycopersicum*, "Formula", and two introgression lines (ILs) that were created by Eshed et al. [31] by crossing *S. lycopersicum* X *S. pennellii* (LA4060/IL6-2 and LA4062/IL6-3) were subjected to RNA-Seq and transcriptomic analysis to identify genes potentially implicated in tomato resistance against broomrape parasitism. In the present study, we did not use any synthetic bio-stimulants and we investigated the processes of parasitism and host response under natural conditions to simulate field environments and produce outcomes that could be practically useful to the farmers. IL6-2 and IL6-3 were selected based on previous work by Bai et al. [32] which showed resistance-associated QTLs residing in these lines. The analysis uncovered a large array of genes that were differentially expressed between parasitized and non-parasitized tomato plants in all three genotypes. Out of these, fourteen differentially expressed genes (DEGs) were further analyzed and validated using qPCR to reveal statistically significant differences in parasitized and non-parasitized samples within and between varieties. Selected genes were found to be implicated in specific aspects of host/parasite interactions such as the regulation of fatty acids and the role of several transcription factors which have been found to be involved in the regulation of plant secondary metabolism and their responses to environmental stressors [33–35]. PEA was also carried out revealing several key metabolic pathways to be involved in plants' response against broomrape across the different tomato genotypes.

2. Materials and Methods

2.1. Plant Material and Treatment

A well-known commercial tomato hybrid, 'Formula', with high susceptibility to broomrape parasitism as documented by the farmers and two introgression line (IL) populations (LA4060/IL6-2 and LA4062/IL6-3) developed by Eshed et al. [31] were selected for broomrape parasitism and molecular studies. These two specific ILs were selected based on the study of Bai et al. [32], as described. Each IL is homozygous for a single introgression from *S. pennellii* (LA0716) in the background of the cultivar (c.v) M82 (LA3475) of *S. lycopersicum* [31]. Fifty plants from each genotype were sown on disks under controlled temperature conditions (19–25 °C) in the greenhouse. When the fourth pair of true tomato leaves was visible (BBCH code 14, according to the tomato phenological growth stages key), forty plants were inoculated with *O. ramosa* seeds, while ten were maintained as non-parasitized control plants. Root infection observations began 30 days post-inoculation (Figure 1), and sampling was carried out at the flower spike development stage of broomrape. Tomato roots were collected both from control and parasitized plants, carefully removing the parasite from the roots of the host. Measurements of plant growth parameters (plant height and weight, soil plant analysis development (SPAD), and relative water content (RWC)) (Supplementary Tables S1, S2, S3, and S4, respectively) were recorded after roots' inoculation with broomrape seeds. A one-way analysis of variance (ANOVA) was performed with six levels consisting of inoculated and non-inoculated plants of three genotypes.

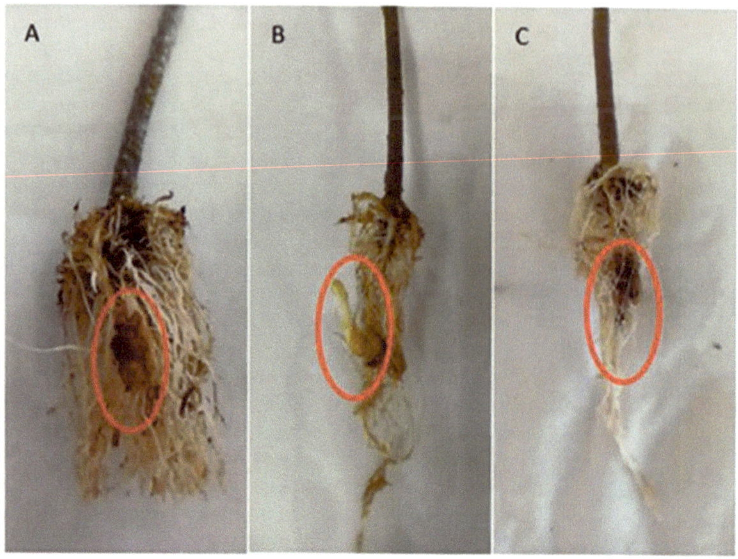

Figure 1. Parasitized tomato roots one month after inoculation with *O. ramosa* seeds. (**A**) Formula genotype parasitized with *O. ramosa* at the tubercule stage. (**B**) IL6-2 genotype parasitized with *O. ramosa* at the emergence stage. (**C**) IL6-3 genotype parasitized with *O. ramosa* at the early tubercule stage. Broomrape parasitism on tomato roots is presented in the red circles.

2.2. Total RNA Extraction and Transcriptome Sequencing

Total RNA was extracted from tomato roots using the NucleoSpin RNA kit (MACHEREY-NAGEL, Dueren, Germany) according to the manufacturer's instructions. Thermo Scientific NanoDrop ND-1000 Spectrophotometer UV/Vis was used to determine the concentration and quality of the RNA. Each sample's concentrations ranged between 85 ng/uL and 260 ng/uL, and six biological replicates (three control plants and three parasitized plants) for each genotype were sequenced. Samples were subjected to cDNA library preparation (poly-A selection) and sequenced (150 nt paired-end reads) on an Illumina NovaSeq 6000 system (Novogene Europe, Cambridge, UK). On average 13.9 million reads were obtained per library (Supplementary File S2).

2.3. Transcriptomic Analysis

2.3.1. Quality Control and Pre-Processing

Read quality was assessed with FastQCsuite [36]. In order to remove adapter sequences and trim low-quality reads from the raw sequencing data, the Cutadapt [37] tool was utilized, setting the minimum read length parameter at 30 and the quality threshold at 15.

2.3.2. Genomic Alignment and Quantification

Pre-processed reads were subjected to splice-aware genomic alignment and quantification using STAR (v2.7.10) [38] and RSEM (v1.3.1) [39], respectively. Regarding sequencing libraries from *S. lycopersicum* samples (i.e., Fc1-3 and Fp1-3), a STAR genome index (STAR—runModegenomeGenerate) and an RSEM reference index (rsem-prepare-reference) were constructed using the GCF_000188115.5_SL3.1_genomic.fna genome assembly and providing the respective gene models (GCF_000188115.5_SL3.1_genomic.gtf). Genomic alignment (STAR—alignReads), allowing known and novel splice junctions and translating alignments into transcript coordinates (—quantModeTranscriptomeSAM), and quantification (rsem-calculate-expression) was performed, achieving a median 93.9% mapping rate. For sequencing libraries derived from introgression lines (i.e., IL6-2c1-3, IL6-2p2, IL6-2p3, IL6-2p5, IL6-3c1, IL6-3c2, IL6-3c4, and IL6-3p1-3), the mapping strategy inspired by the Szymański et al. [40] study was followed. Briefly, a combined STAR genomic index corresponding to both the *S. lycopersicum* (GCF_000188115.5_SL3.1_genomic.fna) and the *S. pennellii* (GCF_001406875.1_SPENNV200_genomic.fna) genomes was created. After splice-aware genomic alignment with STAR (median value of a 94.6% mapping rate), the alignment SAM records of each library were processed to create three discrete subsets:

(i) Slyc_subset, which contains multiple alignments (i.e., "NH" SAM tag > 1) on both *S. lycopersicum* and *S. pennellii* presenting the highest alignment score ("AS" tag in SAM) within *S. lycopersicum* and multiple alignments solely within *S. lycopersicum*, as well as unique alignments in *S. lycopersicum* (a median value of 85.2% of transcriptome-aligned reads across all datasets).
(ii) Penn_subset, which contains multiple alignments (i.e., "NH" SAM tag > 1) on both *S. lycopersicum* and *S. pennellii* presenting the highest alignment score within *S. pennellii* and multiple alignments solely within *S. pennellii*, as well as unique alignments in *S. pennellii* (a median value of 6.6% of transcriptome-aligned reads across all datasets).
(iii) Ambi_subset, which contains multiple alignments on both *S. lycopersicum* and *S. pennellii* presenting equally high alignment scores (i.e., ties) in both species (a median value of 8.1% of transcriptome-aligned reads across all datasets). Entries on *S. pennellii* were retained for these ambiguous alignments for downstream analysis.

RSEM (rsem-calculate-expression), provided with a combined reference index containing gene models of both species, was used on each set separately to produce read count estimates. Supplementary Tables S5 and S6 present alignment and quantification metrics for all generated datasets.

2.4. Differential Expression Analysis

As previously described, the analysis proceeded following the strategy outlined by Szymański et al. [40]. The subset of alignments in both genomes (Ambi subset) was generated for *S. pennellii*. Differential expression analysis was performed using the programming language R (v4.2.2; R Core Team 2021) and the R package edgeR (version 3.40.2) [41]. Specifically, the differential expression analysis for each line was conducted twice: once for the genes aligned in the *S. pennellii* genome and again for the genes aligned in the *S. lycopersicum* genome. Low-abundance genes were filtered by employing the filterByExpr function with default arguments. The weighted trimmed mean of the M-values (TMM) normalization method and quasi-likelihood F-testing were applied to assess abundance change in the following pair-wise contrasts: (i) Fc vs. Fp (Formula); (ii) IL6-2c vs. IL6-2p (*S. pennellii* genes); (iii) IL6-2c vs. IL6-2p (*S. lycopersicum* genes); (iv) IL6-3c vs. IL6-3p (*S. pennellii* genes); and (v) IL6-3c vs. IL6-3p (*S. lycopersicum* genes). The false discovery rate (FDR) was controlled using the Benjamini–Hochberg procedure [42], and the statistical significance threshold was set at p adjusted to <0.05.

2.5. Functional Enrichment Analysis

The limma R package (version 3.54.2) [43] was utilized for pathway enrichment analysis, employing the kegga function with default parameters and the KEGG database. The differentially expressed genes were subjected to pathway analysis, applying two different approaches due to the absence of a standardized method in the literature. In the first approach, separate analyses were conducted for each of the species of each introgression line, having the genes originating from *S. lycopersicum* considered separately from the genes derived from *S. pennellii*. In the second approach, since the number of pathways and their names were the same for the two different species of each introgression line, genes from both were merged into a single gene set within each pathway, resulting in a single pathway analysis for each introgression line. Each of these approaches has its own advantages and limitations. In the first scenario, the study is analogous to examining two distinct organisms, potentially leading to information loss within the hybrid formula. In the second instance, the presence of duplicates in pathways is highly probable, due to the inability to identify corresponding genes between the two species owing to discrepancies in gene IDs and names within the annotation files, possibly impacting the subsequent statistical analysis. A pathway was deemed significant if its corresponding adjusted *p*-value (Benjamini–Hochberg correction) was below the threshold of 0.05.

2.6. Visualizations

Visualizations were generated using the following R packages for each task: volcano plots were generated using ggplot2 (version 3.4.4) [44], bar plots were created using both ggplot2 and ggpubr (version 0.6.0) [45], heatmaps were generated using limma (version 3.54.2), gplots (version 3.1.3.1) [46], and gridExtra (version 2.3) [47], and Venn diagrams were created using VennDiagram (version 1.7.3) [48].

2.7. Gene Expression Analysis Validation by qRT-PCR

Quantitative Reverse Transcription–PCR (qRT-PCR) was performed to verify the RNA-Seq results. Total RNA was extracted from the three genotypes (Formula, IL6-2, and IL6-3) and used for RNA-Seq analysis, as described above. First-strand cDNA was synthesized from 500 ng of total RNA using the PrimeScript™ 1st strand cDNA Synthesis kit (Takara Bio Inc. Lab Supplies P.GALANIS & CO, Athens, Greece). Gene-specific primers for selected DEGs were designed using the NCBI Primer-Blast tool (https://www.ncbi.nlm.nih.gov/tools/primer-blast/, accessed on 30 January 2023). The list of the DEGs selected for qPCR validation and their data after bioinformatic analysis are presented in Supplementary Table S7. Primers and GenBank accession numbers of genes are listed in Supplementary Table S8. qPCR was performed in a total volume of 20 μL using the PowerUpTM SYBRTM Green Master Mix (Applied Biosystems) and was conducted in a StepOnePlusTM Real-Time PCR System machine (Thermo Fisher Scientific, Waltham, MA, USA). PCR conditions were applied according to the manufacturers' instructions. Each reaction was performed in three technical triplicates along with the internal control reaction. Gene expression levels were determined with StepOnePlusTM Real-Time PCR manual 2-delta-delta-Ct method [49], and R programming language (v4.2.2; R Core Team 2021) software was applied for the statistical analysis. Three biological repeats were used for the experiment.

3. Results

3.1. Plant Growth

Measurements regarding height, weight, and soil plant analysis development (SPAD) were recorded on four different dates (Supplementary Tables S1–S3), while one RWC measurement was carried out too (Supplementary Table S4). The results did not indicate any significant differences for weight and RWC measurements. Regarding the comparison of plant height between non-inoculated and inoculated plants, on average, inoculated plants were taller. SPAD measurements presented differences only in the third week after inoculation when Formula appeared to be the most affected genotype with decreased SPAD values. Inoculated plants gave better SPAD values than non-inoculated plants for the three genotypes on all other dates. The above results agree with a previous report, demonstrating that host plants begin to be significantly affected approximately 40 days post-infection, highlighting the fact that early stages of parasitism do not elicit profound morphological/physiological differences across genotypes [50]. Furthermore, during the experimental procedures, visual observations were taking place regarding the percentage of parasitism in the infected plants of all genotypes. Forty plants were inoculated with broomrape seeds in total for the three genotypes studied. We observed 80% parasitism at Formula roots, while IL6-2 and IL6-3 had 45% and 35% parasitized roots, respectively.

3.2. Comparative Transcriptomic Analysis Reveals Differentially Regulated Genes upon Parasitism

During the sampling day, phenotypic observations on roots regarding the number of parasitized plants revealed that the IL6-3 line was less infected compared to the other two genotypes, while Formula had the most infected roots. These observations confirmed the initial hypothesis that the commercial hybrid is possibly more susceptible compared to the introgression lines. The transcriptomic analysis results unveiled a significant number of DEGs, displaying both up- and down-regulation across most of the tested genotypes (Figure 2 and Supplementary File S2). The analysis for each genotype involved comparing the parasitized samples with their respective controls. The examination of two introgression lines (ILs) was conducted separately for genes originating from *S. pennellii* and *S. lycopersicum*. Thus, for each line, two distinct comparisons were performed. The results are visually presented in the volcano plots below (Figure 3B,C for IL6-2 and Figure 3D,E for IL6-3). Differential expression analysis revealed a notable number of DEGs in all genotypes at the parasitized plants with a very clear distinction between up-regulated and down-regulated genes. Moreover, it should be emphasized that in both ILs, a considerable number of up- and down-regulated genes originated from the *S. pennellii* genome (i.e., 1475 and 1940 for IL6-2 and IL6-3, respectively). Additionally, as depicted in Figure 2, the number of significant genes for *S. lycopersicum* was 5428 and 5158 for IL6-2 and IL6-3, respectively. Heatmap depictions indicated differences in gene expression comparing parasitized and non-parasitized tomato roots of the three studied genotypes. In each heatmap, the top 50 DEGs are presented for each genotype (three control samples and three parasitized samples per genotype). In Formula's parasitized roots (Fp), 14 out of 50 DEGs show up-regulation, while the same 14 genes are down-regulated for the controls of Formula (Fc) (Figure 4A). In IL6-2, we observe that genes' up-regulation is more targeted in parasitized plants (IL6-2p) compared to controls (IL6-2c) as there are 4 out of 50 up-regulated genes (2 originated from *S. lycopersicum* and 2 from *S. pennellii*). The rest of the DEGs are down-regulated in parasitized plants, while in control samples, the same genes are up-regulated (Figure 4B). Regarding IL6-3, the vast majority of DEGs are down-regulated in parasitized plants (IL6-3p), while in control samples, the same genes are up-regulated (Figure 4C). All 50 genes in the two ILs originate from both *S. lycopersicum* and *S. pennellii*. From these results, we can assume that ILs' resistance mechanisms against broomrape parasitism are more targeted which is presented by the up-regulation of very few genes.

3.3. The 14 DEGs in Response to Broomrape Parasitism

Following transcriptomic analysis, we opted to select key candidate genes for validation and further study. Based on DEGs and the existing literature, fourteen genes were subjected to qPCR analysis. The genes' selection for validation was based on the bioinformatic analysis results of the RNA-Seq (Supplementary File S2, Supplementary Table S7) as well as on previous research studies. Table 1 elucidates the description of the studied genes in the NCBI and SolGenomics databases and the gene names in the current study.

Table 1. Gene descriptions, IDs according to NCBI and SolGenomics databases, and gene names used in this manuscript.

NCBI Gene Description	Gene Names in This Manuscript	GeneIDs	SolycIDS
omega-3 fatty acid desaturase, chloroplastic (FAD)	*OFA*	LOC107023114	Solyc06g051400
transcription factor bHLH35 (bHLH)	*BHLH35*	LOC107024383	Solyc07g018010
1-phosphatidylinositol-3-phosphate 5-kinase FAB1B-like (FAB)	*FAB1B*	LOC101246905	Solyc03g123570
F-box protein At1g78280 (F-box)	*F-BOX*	LOC107017080	Solyc04g074490
glutamate decarboxylase 4 (GAD)	*GLDE*	LOC107016348	Solyc04g025530

Table 1. *Cont.*

NCBI Gene Description	Gene Names in This Manuscript	GeneIDs	SolycIDS
MLO-like protein 9	*MLO*	LOC101254181	Solyc02g038806
phospholipase D gamma 1-like (PLD)	*PPDG*	LOC107007945	Solyc01g091910
zinc finger CCCH domain-containing protein 32 (ZFP)	*ZINC*	LOC101250699	Solyc06g008740
CaM2 calmodulin 2	*CaM2*	SlCaM2	Solyc10g081170
ATP-citrate synthase beta chain protein 2-like	*ACLA2*	LOC107007996	Solyc01g059880
alpha/beta-hydrolase DWARF14-like	*D14*	LOC101259838	Solyc02g092770
GABA-TP3 gamma-aminobutyrate transaminase subunit precursor isozyme 3	*GABA*	GABA-TP3	Solyc12g006450
phytoene synthase 1	*PSY*	LOC107014634	Solyc03g031860
zinc finger protein ZPR1-like	*ZPR1*	LOC101251441	Solyc02g069120

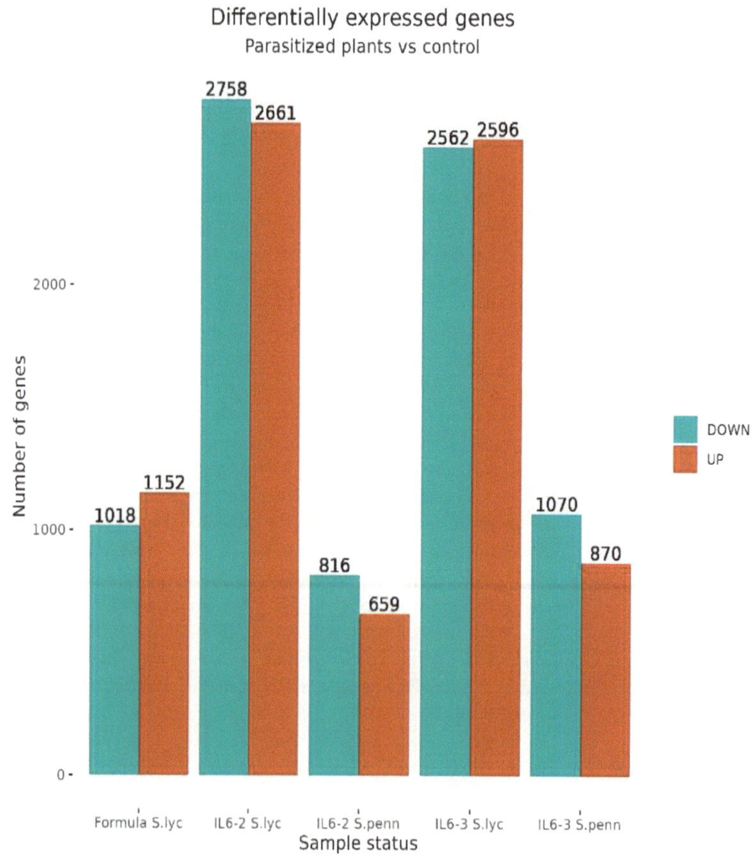

Figure 2. Bar plot illustrating numbers of significant DEGs (FDR < 0.05) for the Formula genotype and the two introgression lines, based on the results obtained from edgeR. Each line produced two sets of results since differential expression analysis was conducted once for genes originating from the *S. pennellii* genome and the second for genes originating from the *S. lycopersicum* genome. Down-regulated genes in the presence of *Phelipanche* are represented in the blue color, while up-regulated ones are shown in red.

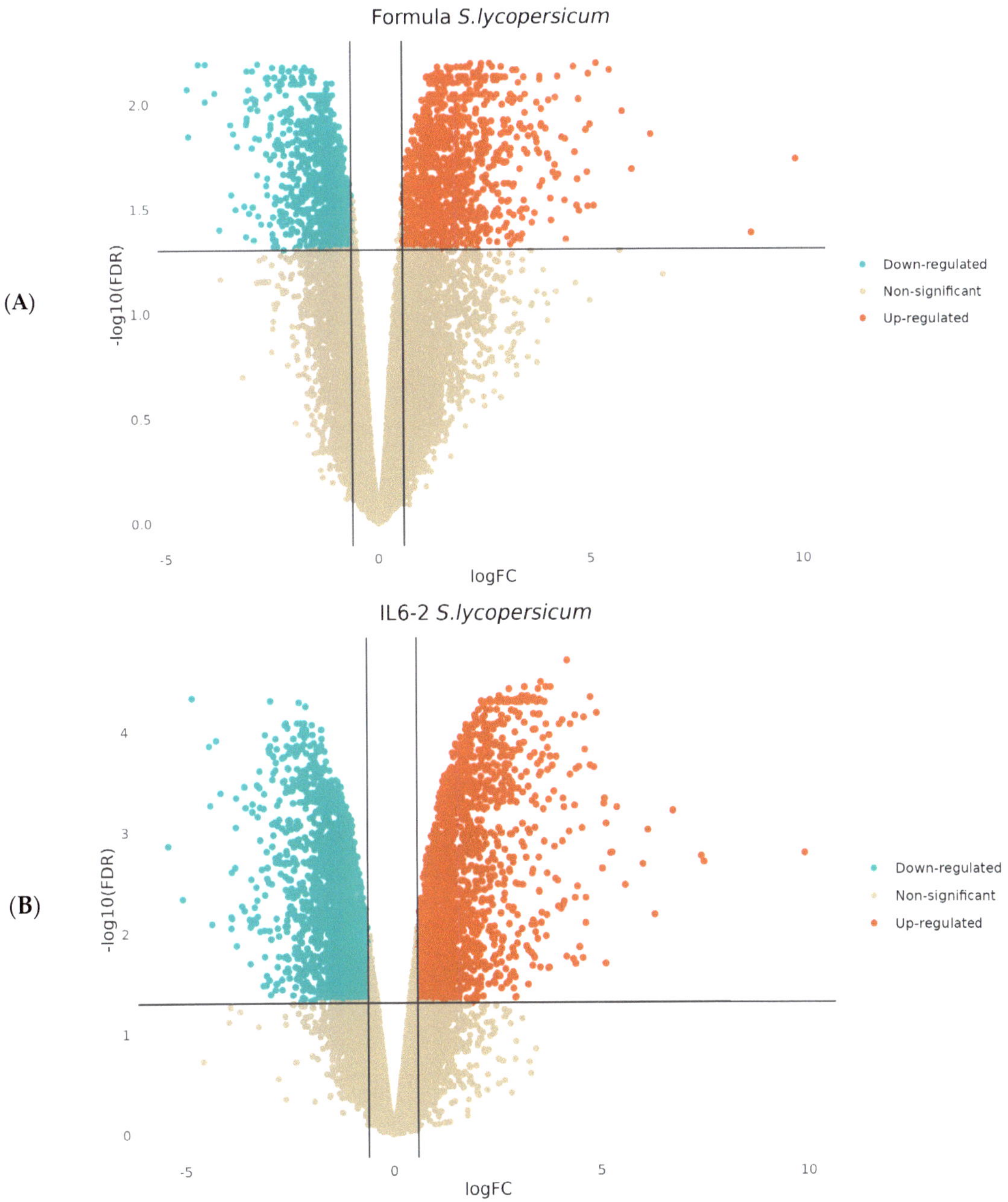

Figure 3. *Cont.*

(C)

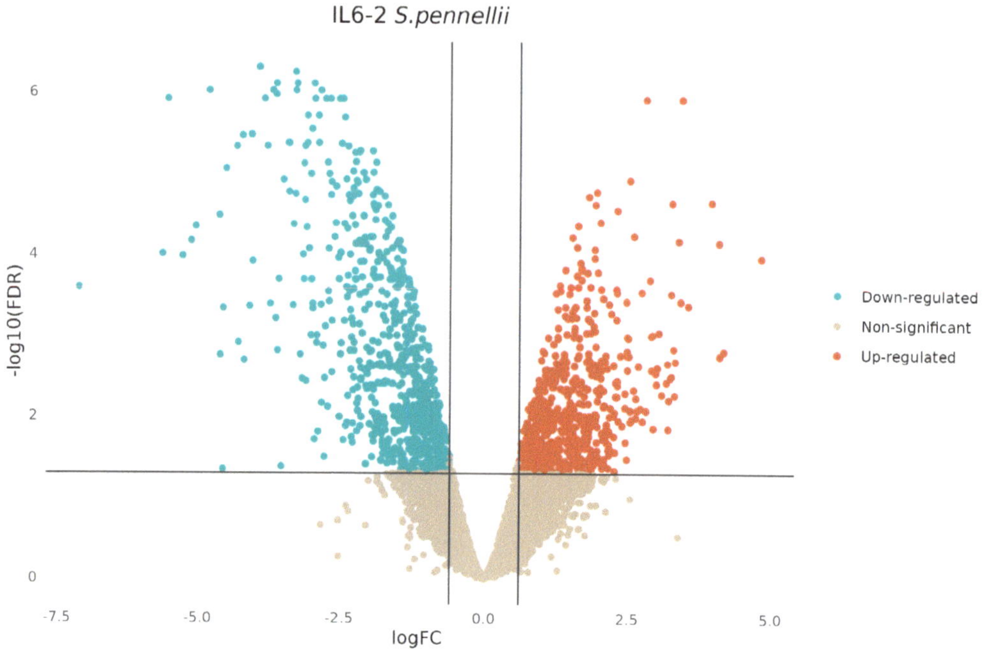

(D)

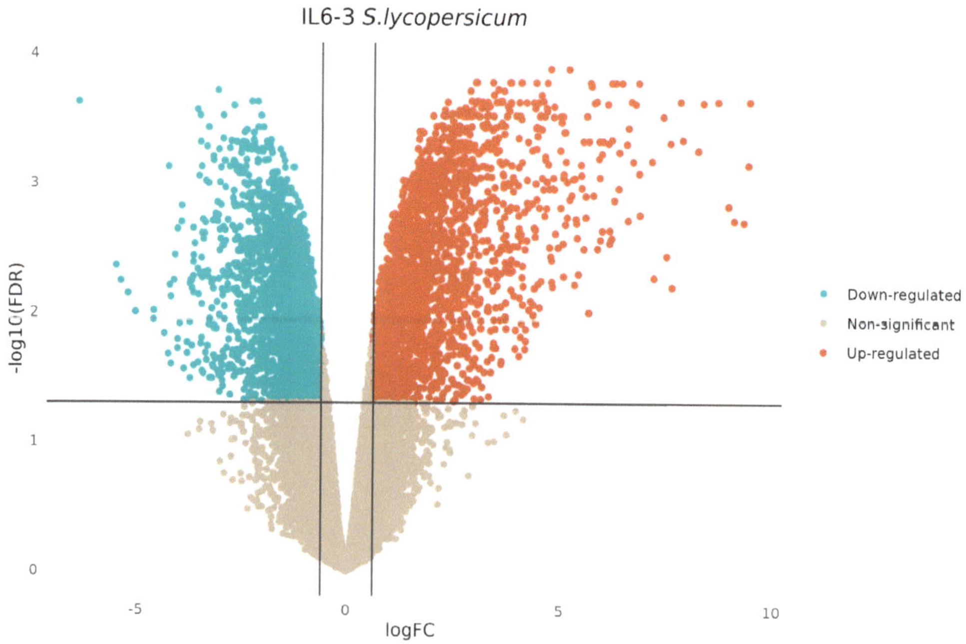

Figure 3. *Cont.*

(**E**)

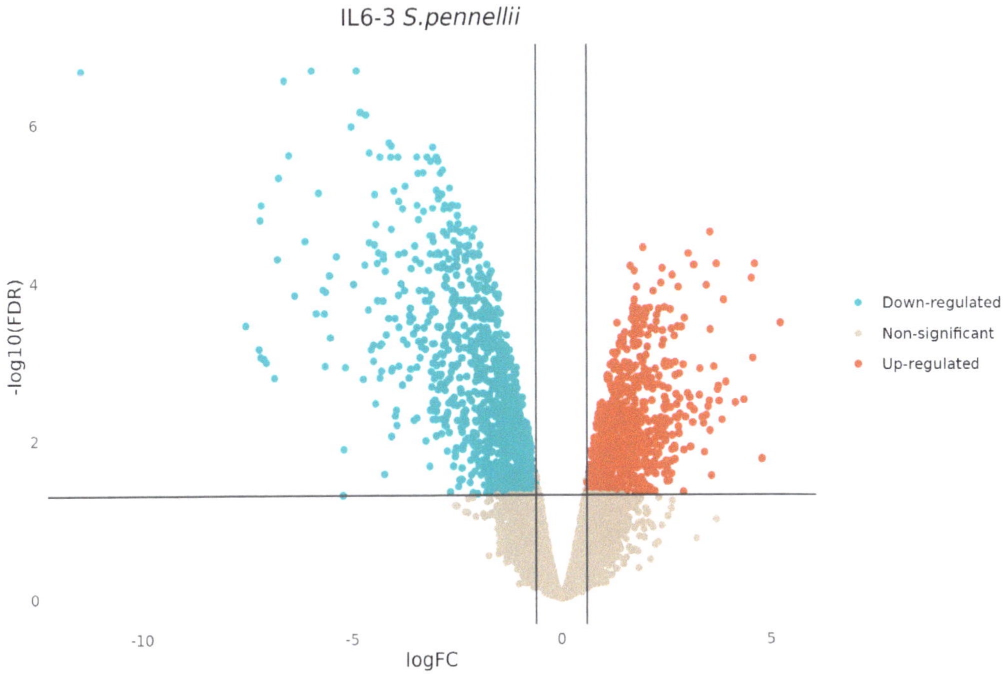

Figure 3. Volcano plots demonstrating results of differential expression analysis between parasitized and non-parasitized plant samples of (**A**) 'Formula' with *S. lycopersicum* annotated genome, (**B**) IL6-2 with *S. lycopersicum* annotated genome, (**C**) IL6-2 with *S. pennellii* annotated genome, (**D**) IL6-3 with *S. lycopersicum* annotated genome, and (**E**) IL6-3 with *S. pennellii* annotated genome. Genes are represented as points and are colored if $|\log_2(\text{fold change})| > 0.6$ and adjusted to $p < 0.05$. Up-regulated genes are highlighted in red, while down-regulated genes are highlighted in blue color.

3.4. qPCR Validation Highlights Prominent Changes upon Parasitism in IL6-3 Line

Relative expression levels were normalized with Actin (ACT) gene expression levels, and statistically significant differences were determined at a 95% confidence interval. All genes were expressed in the genotypes under study; however, statistically significant differences were observed for 9 out of 14 candidate genes, especially for the genotype IL6-3. With respect to *OFA*, *BHLH35*, *FAB1B*, *FBOX*, *GLDE*, *MLO*, *ZINC*, *PPDG*, and *CaM2* DEGs, the relative expression levels with statistically significant differences were observed in parasitized plants, mostly in IL6-3, validating the transcriptomic analysis results. In particular, *OFA*, *BHLH35*, *FAB1B*, *FBOX*, *GLDE*, and *MLO* DEGs displayed significant up-regulation in the parasitized plant of IL6-3 (Figure 5).

Specifically, in parasitized IL6-3, *OFA* and *BHLH35* exhibited a significant ~18- and 11-fold increase in gene expression, respectively, as compared to the control samples (Figure 5A,B). Similarly, *FAB1B* and *FBOX* displayed a significant but lower up-regulation of 3- and 5-fold, respectively, compared to their controls (Figure 5C,D). Finally, a 4- and 2-fold increase was evidenced for *GLDE* and *MLO*, respectively, in parasitized IL6-3 (Figure 5E,F).

Significant down-regulation for the parasitized plants of IL6-3 was also observed among the fourteen selected genes. In particular, *PPDG* and *ZINC* displayed a 2- and 3-fold decrease in parasitized IL6-3, respectively (Figure 6A,B). *CaM2* gene discriminated its relevant expression only in parasitized plants of Formula, where an 8-fold change was observed compared to the control plants (Figure 6C).

(A)

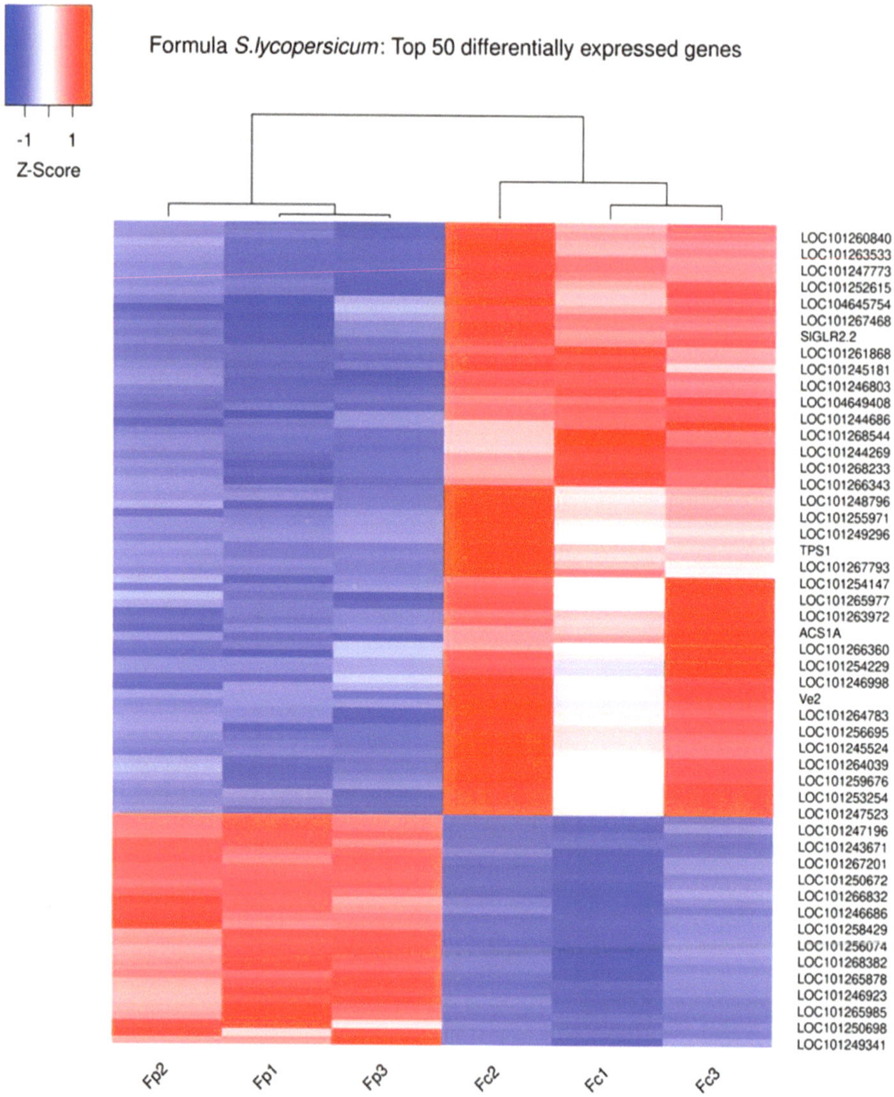

Figure 4. *Cont.*

(**B**)

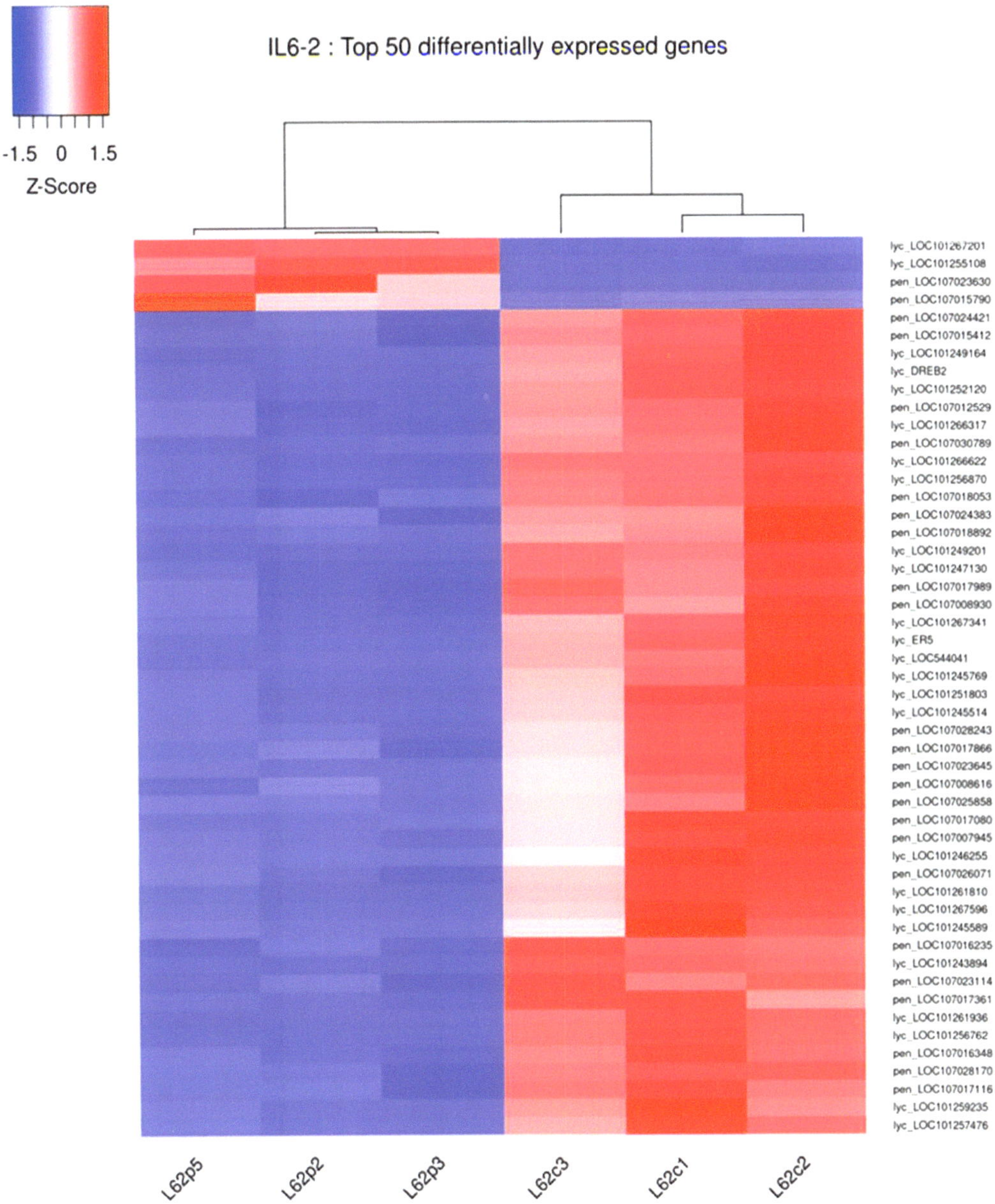

Figure 4. *Cont.*

(C)

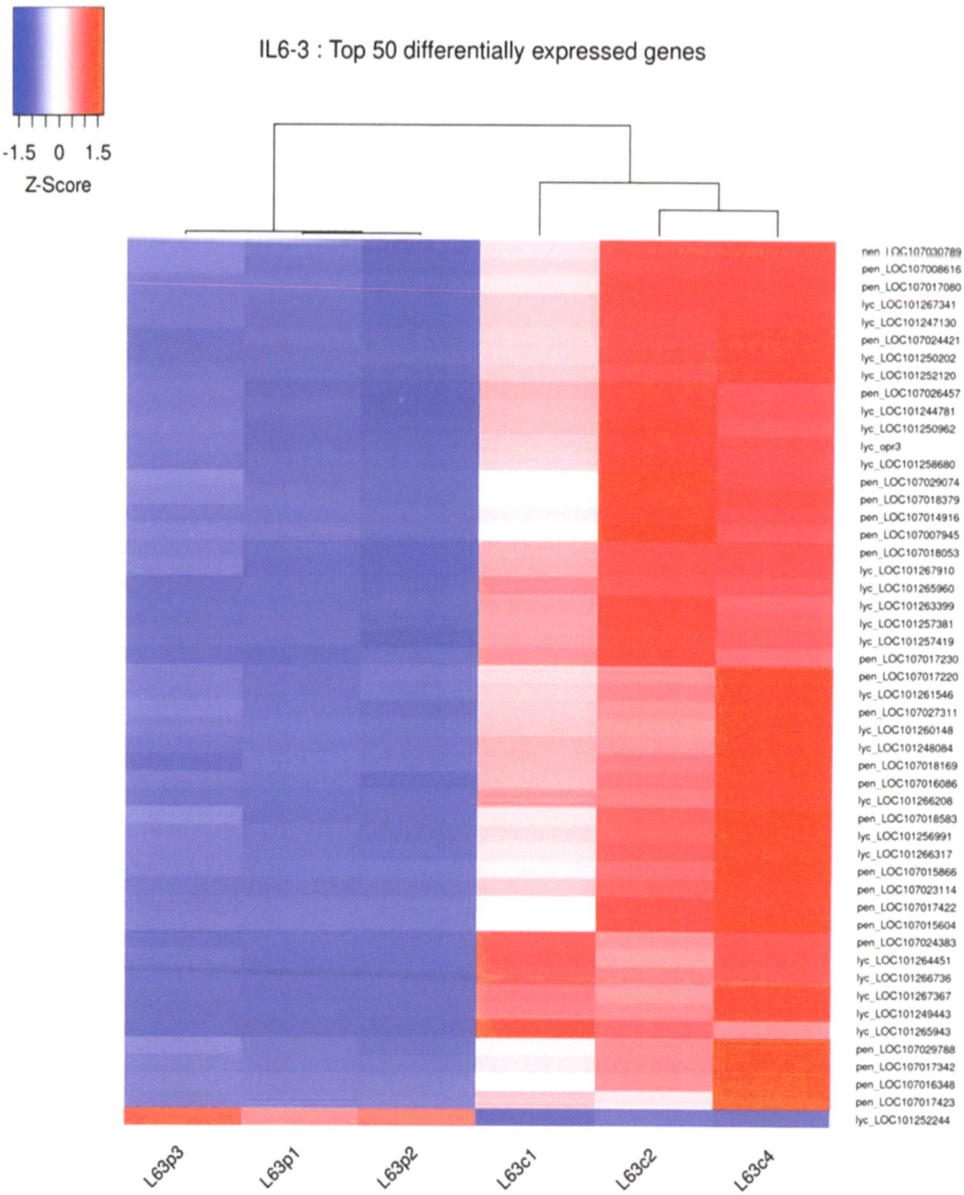

Figure 4. (**A**) Top 50 DEGs (up- and down-regulated) in Formula genotype. Fp1, Fp2, and Fp3 correspond to parasitized plants of Formula hybrid, while Fc1, Fc2, and Fc3 refer to Formula controls (i.e., non-infected roots). (**B**) Top 50 DEGs (up- and down-regulated) in IL6-2 genotype. IL6-2p2, IL6-2p3, and IL6-2p5 correspond to parasitized plants of IL6-2, while IL6-2c1, IL6-2c2, and IL6-2c3 refer to IL6-2 controls. (**C**) Top 50 DEGs (up- and down-regulated) in IL6-3 genotype. IL6-3p1, IL6-3p2, and IL6-3p3 correspond to parasitized plants of IL6-3, while IL6-3c1, IL6-3c2, and IL6-3c4 refer to IL6-3 controls. Color-coding represents z-score-scaled normalized read counts for each gene across samples.

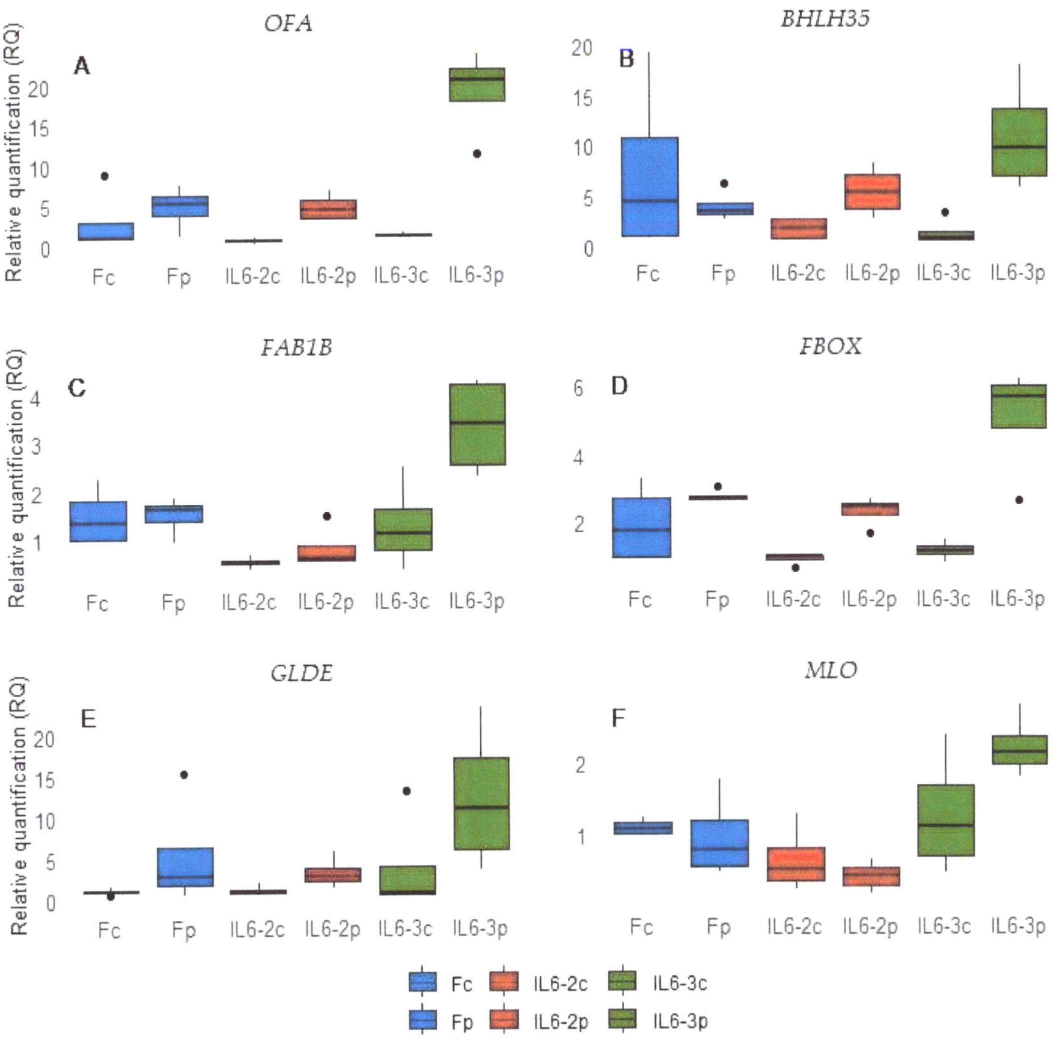

Figure 5. Relative expression levels per genotype and treatment (control vs. parasitized plants) of 6 genes denoted as DEGs in the RNA-Seq experiments, which appear as significantly up-regulated in IL6-3 parasitized plants via qPCR. Treatments: Formula control (Fc), Formula parasitized (Fp), introgression line 6-2 control (IL6-2c) and parasitized (IL6-2p), and introgression line 6-3 control (IL6-3c) and parasitized (IL6-3p). Statistically significant differences were determined at $p < 0.05$ comparing the relative expression between the control and the parasitized plants/genotype.

ACLA2, *D14*, *GABA*, *PSY*, and *ZPR1* genes were additionally selected for qPCR analysis based on previous publications regarding transcriptional responses to biotic stressors. They were expressed in all tested genotypes in both parasitic and control plants; however, no statistically significant differences were revealed in their relative gene expression (Supplementary Figure S1).

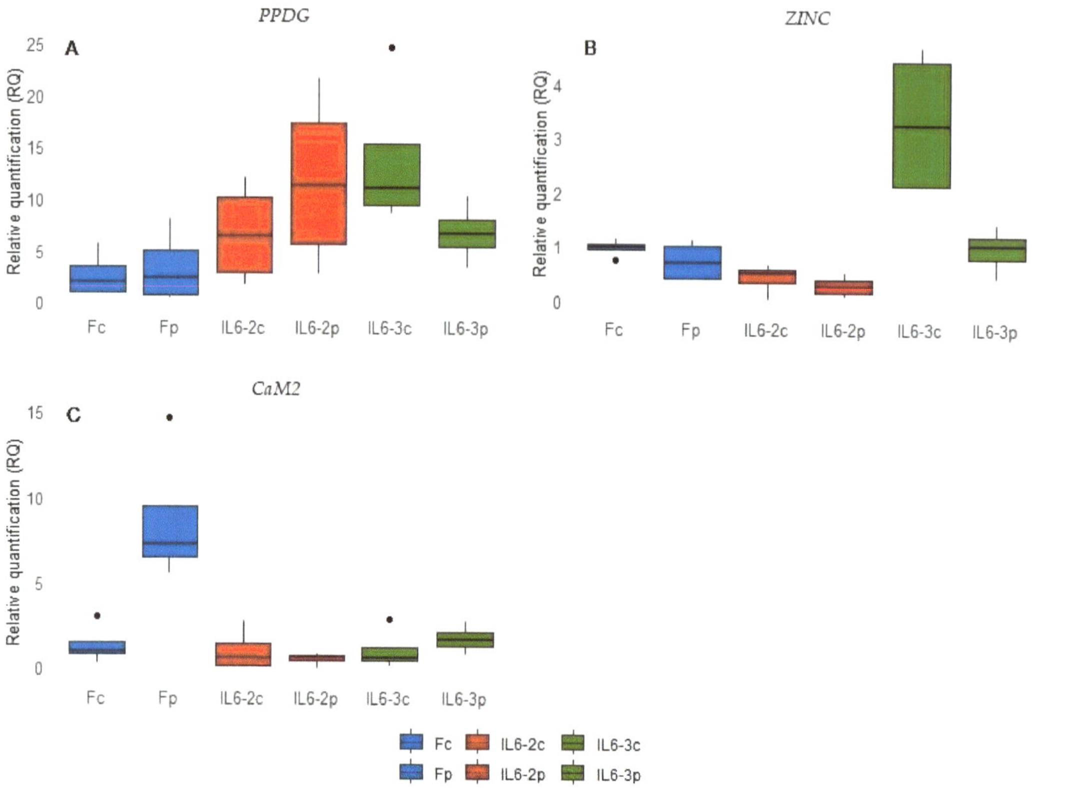

Figure 6. Relative expression levels per genotype and treatment (control vs. parasitized plants) of 3 genes denoted as DEGs in the RNA-Seq experiments, which appear as significantly up-regulated in IL6-3 control plants compared to parasitized ones ((**A**,**B**), *PPDG*, and *ZINC*, respectively) and up-regulated in Formula parasitized plants ((**C**), *CaM2*). Treatments: Formula control (Fc), Formula parasitized (Fp), introgression line 6-2 control (IL6-2c) and parasitized (IL6-2p), and introgression line 6-3 control (IL6-3c) and parasitized (IL6-3p). Statistically significant differences were determined at $p < 0.05$ comparing the relative expression between the control and the parasitized plants/genotype.

3.5. KEGG Enrichment Analysis of DEGs

PEA of the RNA-Seq findings was carried out with two distinct approaches. It is worth noting that approximately 11,000 genes in both species (out of ~31,000 total) are annotated with regard to biological pathways in the Kyoto Encyclopedia of Genes and Genomes (KEGG) resource [51], and thus used as background for testing. Regarding the first approach, we analyzed the genes of each species separately in each introgression line, and the top 20 significantly enriched pathways are presented in Figures 7 and 8. Specifically, the number of significantly enriched pathways for IL6-2 was 32 regarding *S. lycopersicum* and 36 regarding *S. pennellii*, respectively. For line IL6-3, the number of significantly enhanced pathways was 35 regarding both *S. lycopersicum* and *S. pennellii*. For the second approach, in which genes of the two species that belong to the same pathway were merged into a single set, the results can be found in Supplementary Figure S2. Additionally, a Venn diagram was generated to visualize the number of significantly enriched pathways found to be shared/unique among the two different species in each IL (Figure 9).

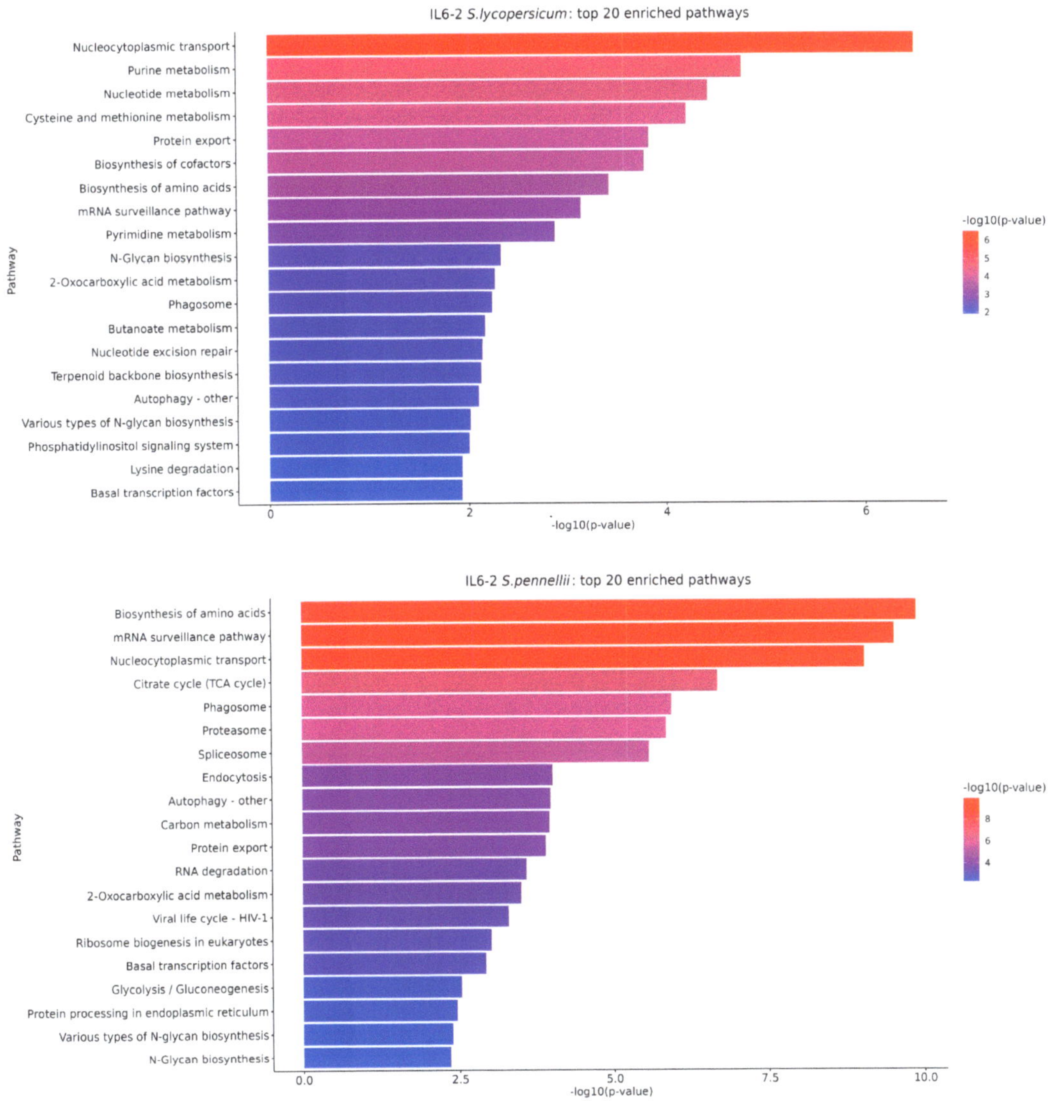

Figure 7. The top 20 significantly enriched pathways, obtained from the first approach pathway analysis for introgression line IL6-2, with $p < 0.05$.

Notably, out of the 14 genes subjected to qPCR validation, only 4 are currently included in pathway annotations provided by KEGG. Out of these, 3 pertain to metabolic and signaling pathways that were found to be significantly enriched by DEGs in our transcriptomic analyses (Table 2).

Table 2. Here, 4 out of 14 DEGs studied expressed in annotated pathways in KEGG database.

Gene Name	NCBI ID	SolycID	Pathway Name	Species	Significant Pathway Found
FAB1B	LOC101246905	Solyc03g123570	Inositol phosphate metabolism	*S. lycopersicum*	-

Table 2. *Cont.*

Gene Name	NCBI ID	SolycID	Pathway Name	Species	Significant Pathway Found
GLDE	LOC107016348	Solyc04g025530	Alanine, aspartate, and glutamate metabolism	*S. pennelli*	IL6-2_penn, IL6-3_penn
PPDG	LOC107007945	Solyc01g091910	Glycerophospholipid metabolism	*S. pennelli*	IL6-2_merged, IL6-3_merged
CaM2	SlCaM2	Solyc10g081170	MAPK signaling pathway–plant	*S. lycopersicum*	IL6-2_merged, IL6-3_merged

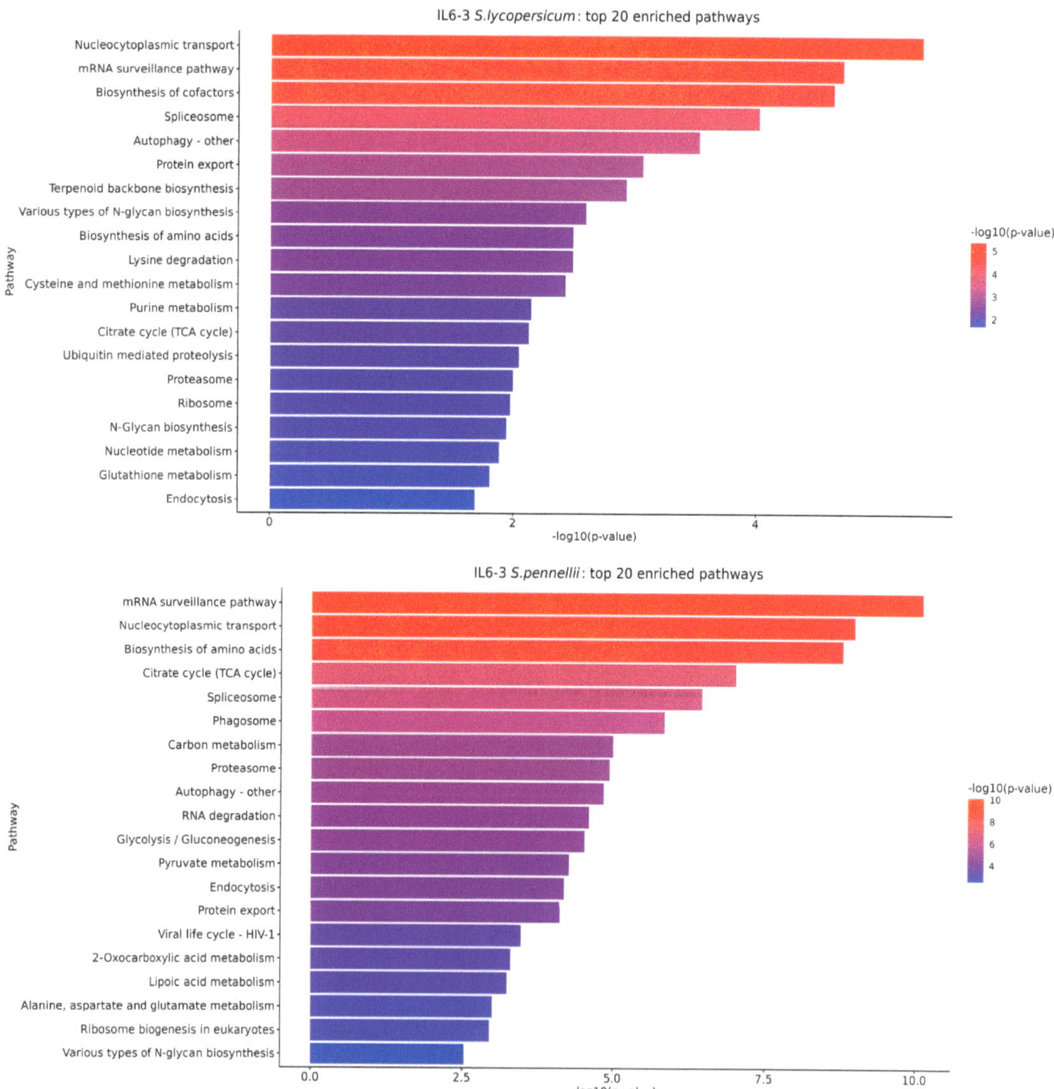

Figure 8. The top 20 significantly enriched pathways, obtained from the first approach pathway analysis for introgression line IL6-3, with $p < 0.05$.

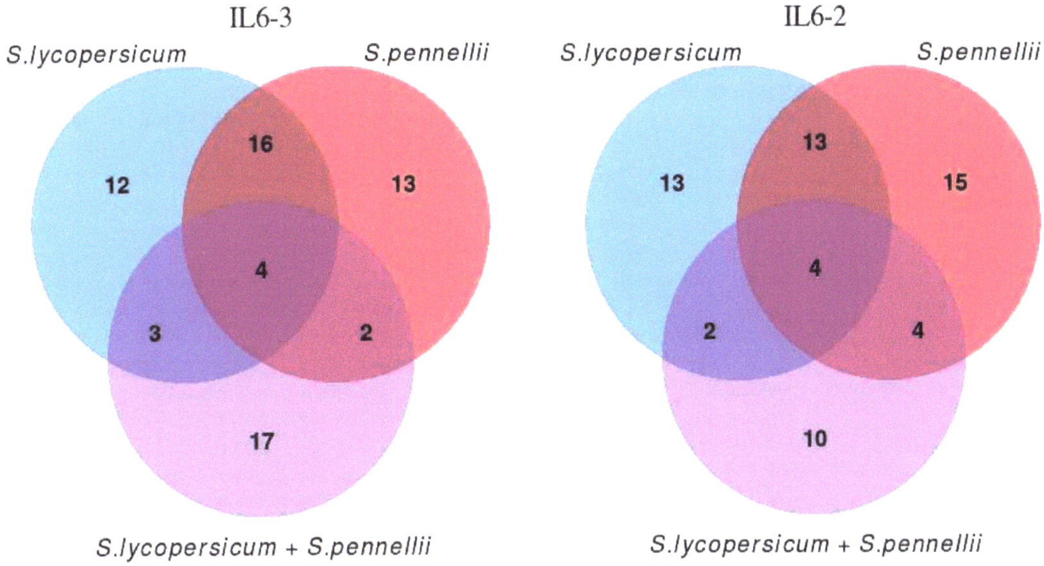

Figure 9. Venn diagram illustrating numbers of common pathways identified as significant across different genomic annotations.

4. Discussion

Managing broomrapes effectively requires a combination of strategies. The difficulties associated with their control, such as host specificity, seed bank persistence, underground growth, etc., are well documented (and reported in the international bibliography) [6,52]. However, the efficacy of these strategies in combating the parasite remains very constrained. Notably, timely prevention of root parasitism prior to attachment of the parasite to the root is of crucial importance [8]. The present study aimed to unravel significant differences in the expression of key genes during the early stage of parasitic infection, without utilizing stimulants, but instead by mimicking field conditions of parasitism. Differentially expressed key genes will be further investigated towards developing appropriate marker tools for the identification of genetic material with broomrape tolerance. Notably, several transcriptomic studies regarding tolerance to *Orobanche* have investigated the molecular response to infection at a single time point [53,54], which resembles our approach.

RNA-Seq analyses of host plants upon parasitism can provide a high-throughput view of the alterations induced as they interact with a parasite, including changes in gene expression, biochemical pathways, and host defense regulatory elements. Key parameters that are of particular interest include resistance genes, defense signaling pathways, hormonal regulatory components, and secondary metabolites implicated in the response to the parasite infection and development of resistance [55,56]. A plethora of transcriptomic analyses has been carried out in the past in tomato cultivars parasitized with broomrape species [56–58].

Our study is the first to perform an in-depth analysis of two specific ILs (IL6-2 and IL6-3) upon infection with the holoparasite and to unravel important DEGs originating from the *S. lycopersicum* and/or the *S. pennellii* genome. The aim was to validate several DEGs derived from transcriptomic analysis and relate them with significant pathways identified from each species (*S. lycopersicum* and *S. pennellii*). The overwhelming majority of DEGs revealed in this study are related to cellular and metabolic processes, responses to stimuli, and the regulation of protein synthesis. Prior studies on parasite–host interactions and crop resistance have focused on genes related to the metabolic pathway of strigolactones, a group of terpenoid lactones derived from carotenoids [3]. Fernández-Aparicio et al. [59]

carried out research in fava beans, demonstrating that low strigolactone exudation within fava bean germplasm contributes to broomrape resistance. A recent review discusses different germination stimulants and their potential role in host specificity [60]. While very promising, strigolactone's regulation is not the only approach to manipulate broomrape resistance in crops. A recent transcriptomic study has demonstrated the significant contribution of the salicylic acid (SA) pathway on sunflower resistance to *Phelipanche cumana* during parasitism [61].

Section [-15]Our transcriptomic analysis revealed DEGs between parasitized and control plants, the vast majority of which (a) were found up-regulated in response to broomrape parasitism and (b) have crucial roles in processes other than strigolactone regulation. Omega-3 fatty acid desaturases (*FADs*) are enzymes that catalyze the conversion of omega-3 fatty acids from their precursor omega-6 fatty acids. In plants, *FADs* are important for maintaining membrane fluidity and stability, being key factors contributing to tolerance to various environmental stresses [62,63]. The studied FAD family member (named *OFA* gene throughout the manuscript) exhibited a 20-fold increase in parasitized roots of IL6-3 compared to the control and remained at control levels in the other two genotypes under study. The results by Shi et al. [63] associate an increased expression of *CbFAD3* in tobacco to tolerance against multiple stressors. Regarding the *FAB* gene family, *FAB1A* and *FAB1B* genes in *Arabidopsis* are genes that are important for the maintenance of endomembrane homeostasis [64]. The *FAB1B* gene, known for its role in regulating plant growth and stress responses, encodes a phosphatidylinositol-3-phosphate 5-kinase, an enzyme involved in the regulation of membrane homeostasis, trafficking, and cellular signaling [64]. Both genes, *FAD* (*OFA*) and *FAB1B* were found overexpressed in the parasitized roots of IL6-3, indicating/suggesting potential functional roles under biotic stress conditions, such as broomrape infestation. This constitutes the first research work implicating a role for *FAB* genes in broomrape tolerance, while very recently published data link *FAD-like* genes to broomrape resistance [65]. These findings highlight (*OFA*) and *FAB1B* as promising candidates for further study towards a better understanding of the defense mechanisms of tomato cultivars against broomrape infestation.

The basic helix–loop–helix (BHLH) family of transcription factors is extensively studied and has been implicated in the regulation of plant secondary metabolism [66], associated with responses to environmental stressors [67]. Jiang et al. [68] have reported that heterologous overexpression of the *Anthuriuman draeanum BHLH35* gene (*AaBHLH35*) in *Arabidopsis* transgenic lines enhanced tolerance to cold and drought stresses, supporting the notion that *BHLH35* is an essential factor for abiotic stress tolerance. Considering the findings of studies linking increased *BHLH* expression with enhanced tolerance to *Phelipanche* species [67], we hypothesize that *BHLH35* could be another promising candidate gene for conferring partial resistance to *P. ramosa* in IL6-3.

The *F-box* gene family is one of the most abundant and pleiotropic families in plants [69,70], with members that can be either positive or negative regulators of hormone biosynthesis [71–73]. The involvement of the F-box protein *MAX2* in strigolactone biosynthesis has been demonstrated [74], and further structural and functional analysis of our *F-box* gene can elucidate its correlation to strigolactone biosynthesis and its specific role in broomrape tolerance.

The *MLO* gene encodes a negative regulator of defense responses in plants, which has been reported as a key regulator of resistance to powdery mildew in tomatoes [75]. In general *MLO* gene appears to be important in regulating plant defense responses to biotic stress factors [76].

The *glutamate decarboxylase 4* gene *GAD* (named *GLDE* gene throughout the manuscript) encodes the glutamate decarboxylase 4 (GAD4) protein. In plants, *GAD* genes are crucial in regulating GABA levels, which in turn affects plant growth and development, as well as the response to environmental stresses [77]. Calmodulin (CaM) is a calcium-binding protein that also controls various cellular processes, including signal transduction, gene expression, and stress responses. Plants employ the divalent cation calcium (Ca^{2+}) as a second messenger in relaying these endogenous (developmental) and exogenous (environmental) signals to appropriate cellular responses [78]. Very few studies have reported the implication of both genes to stress tolerance. Rajani et al. [79] demonstrated that maize plants expressing the deregulated *AtGAD1* exhibit severe chlorosis, retarded growth phenotype, high levels of GABA, and Ca^{2+}/CaM-independent *GAD* activity, indicating that both genes are involved in the response of plants to stress factors. These findings agree with the results of the present study, where *GLDE* and *CaM2* are found up-regulated in broomrape-infected plants of IL6-3 and Formula genotypes, respectively.

Phospholipase D (PLD) enzymes cleave phospholipids to generate the lipid second messenger, phosphatidic acid (PA). In *Arabidopsis thaliana*, *PLDγ1-like* (named *PPDG* gene throughout the manuscript) is expressed in various plant tissues and is involved in the regulation of plant responses to abiotic and biotic stresses [80]. Zinc finger CCCH domain-containing proteins (ZFPs) are a subfamily of zinc finger proteins that contain a conserved CCCH motif. Tomatoes feature several *ZFPs-encoding* genes, including the Zinc finger CCCH domain-containing protein 32 (SlZFP32), a transcription factor involved in regulating genes implicated in abiotic stress responses [81]. *PPDG* and *ZINC* genes are down-regulated in parasitized plants compared to the controls, especially in IL6-3, revealing an interesting pathway of potential tolerance for further study.

Additionally, three other genes *CaM2*, *PPDG*, and *ZINC* displayed down-regulation in parasitized IL6-3 roots, an expected outcome based on their gene ontology description. More specifically, based on the description of gene ontology data at the Sol genomics database, the *CaM2* gene (GO:0075135) is implicated in processes in which the symbiont organism (the host organism—tomato) stops, prevents, or reduces the frequency, rate, or extent of calcium or calmodulin-mediated signal transduction. Therefore, it could be suggested that *CaM2* should be suppressed in the tolerant host genotypes inoculated with broomrapes. Formulas' parasitized plants exhibited a marked increase in *CaM2* expression, as compared to control plants as well as to the other two ILs, which supports the notion of Formula's susceptibility to the parasite.

Moreover, the *PPDG* gene studied in this work has been found to be a negative regulator in various biotic and abiotic stressors. More specifically, according to the gene ontology description, *PPDG* presents negative regulation of the response to biotic stimulus (GO: 0002832) like stringolactones (SLs), a major mechanism involved in broomrape germination. Furthermore, it provokes negative regulation of filamentous growth of a population of unicellular organisms in response to a chemical stimulus (GO: 1900438) (filamentous growth also inoculates the hosts' roots, and the hosts use similar resistant mechanisms to avoid parasitism like in broomrapes). It also appears as a negative regulator of defense response (GO:0031348). These are only a few examples of *PPDG*'s negative regulation of plant tolerance to biotic stressors. These results agree with the findings of this study, where *PPDG* is down-regulated in IL6-3 parasitized plants, further supporting our suggestion that this specific introgression line is the most tolerant of the three genotypes tested.

Regarding the *ZINC* gene, and based on the gene ontology data, it codes for a host cell extracellular matrix binding protein, interacting selectively and non-covalently with the extracellular matrix of a host cell (GO:0046810). In our study, *ZINC* displayed up-regulation in control roots (non-parasitized with broomrape) in IL6-3, compared to all the other genotypes. It might serve as a repressor of transcription whose down-regulation in parasitized IL6-3 roots allows for downstream stress-responsive genes to be transcribed. In conclusion, the three down-regulated genes found in this research are all in accordance with the gene ontology data of the Solgenomics database, which is the most specific, updated, and trustworthy database for Solanaceae plants. Based on these findings, these genes could be further studied to demonstrate a putative role in tomato resistance against *Phelipanche*. Our results are in line with existing reports regarding the involvement of these genes in abiotic/biotic stressors and imply possible roles in broomrape resistance. The differential gene expression could possibly be due to the *S. pennellii* genome or to the complementary interaction of both genomes. Future studies assessing the precise genomic content of these lines, as well as their epigenomic activity under parasitism, could shed more light on the exact mechanisms. Partial resistance against a different broomrape species, *Phelipanche aegyptiaca*, was also found by Bai et al. [75] in these ILs.

PEA identifies biological terms that are overrepresented in a group of genes more than would be expected by chance and ranks these terms by significance [82]. Dai et al. [83] carried out a comparative transcriptome analysis between the cultivated and wild tomato genotypes under abiotic stressors (i.e., salt and drought stress). Their enrichment results using the KEGG resource revealed numerous implicated pathways of interest, some of which are also found through our analysis, including the biosynthesis of amino acids, carbon metabolism, glycolysis/glyconeogenesis, and the ribosome pathway (Figures 6 and 7). A de novo comparative transcriptome analysis of DEGs in the scion of homografted and heterografted tomato seedlings carried out by Wang et al. [84] presented the 10 most significantly enriched KEGG pathways among DEGs detected between hetero- and homografted tomato seedlings. Grafting might cause abiotic stress effects to the seedlings, and their analysis pinpointed pathways like our study, such as the MARK-signaling pathway plant, plant hormone signal transduction (Supplementary Figure S2), the biosynthesis of amino acids, protein processing in the endoplasmic reticulum, the spliceosome, etc. (Figures 6 and 7). Furthermore, Ashapkin et al. [85], in a review study regarding genomic and epigenomic mechanisms of host plants against the parasitic weed *Cuscuta*, mentioned that *Cuscuta japonica* exhibited distinct morphological features after attachment to *Ficus macrocarpa* (host plant) and *Mangifera indica* (non-host plant). A KEGG PEA was carried out, revealing various important biochemical pathways, significantly enriched by DEGs. Several pathways are also demonstrated in our study (Figure 7, Supplementary Figure S2), such as "flavonoid biosynthesis", "plant–pathogen interaction", "arginine and proline metabolism", "MARK signaling", "phenylalanine metabolism", "arginine biosynthesis", and "carbon fixation in photosynthetic organisms". The results presented by Ashapkin et al. [85] indicated that many metabolites and signal pathways are responsible for rendering host resistance to dodders. To determine *Striga* germination rates, one of the most serious parasitic weeds, infesting various *Sorghum* genotypes, Irafasha et al. [86] categorized their samples by hierarchical clustering using self-organizing maps (SOMs). The PEA of SOMs of DEGs revealed various pathways that control several important biological regulations such as seed dormancy break and germination. Specific gene expression patterns in the ABA biosynthesis pathway are also mapped in the carotenoid biosynthesis pathway involving enzymes like PSY, which is also involved in our study. Furthermore, it is also referred that the observations reveal possible interactions between SL, ABA, and other hormone pathways [86]. These similarities indicate that plant resistance mechanisms against biotic and abiotic stressors could, to some extent, exhibit shared mechanistic components and interactions.

Ten out of fourteen genes assessed via qPCR in our study do not appear to be involved in significantly enriched pathways. We should emphasize, however, that 11,000 out of 31,221 total genes in *S. lycopersicum* (35%) are currently annotated to KEGG pathways. This underscores the fact that our knowledge of the genes' roles in molecular mechanisms is still incomplete, and this phenomenon is exacerbated in plant organisms. The validated genes of unknown roles that we included in our study could provide a starting point towards assessing poorly defined mechanisms of resistance against broomrape and discovering novel ones. To date, research on broomrape resistance is mainly focused on the mechanism of SL accumulation in the roots of the host. Our findings suggest that, apart from the unarguable importance of SL metabolism, diverse cellular processes are potentially involved in tomatoes' responses to broomrapes. Further understanding of such biological processes will expand our knowledge of host resistance to broomrapes and will pave the way to alternative management strategies against parasitism in tomatoes and other crops of high economic and nutritional value.

5. Conclusions

Our present research study focused on early-stage infection of tomato with *P. ramosa* and explored the involvement of candidate genes that could serve as molecular markers for future pre-breeding programs. We emphasized exploring further potential mechanisms regarding broomrape resistance by analyzing a large array of genes not previously studied in host–parasite interaction processes and host responses. We investigated herein, for the first time, early-stage broomrape parasitism of two resistant tomato lines, IL6-2 and IL6-3, as compared to the susceptible commercial variety 'Formula' and unveiled a series of genes that show remarkable differential expression between resistant and susceptible genotypes upon broomrape infection. Meticulous analysis of gene expression patterns and PEA datasets combined with the mining of bibliographical data has highlighted the potential involvement of the encoded proteins in crucial biochemical pathways implicated in parasitism and resistant mechanisms. Our studies will be expanded to investigate molecular responses to infection along a post-infection time course including additional timepoints. The knowledge to be acquired together with the valuable findings from this study on the genes described above will be used to enrich our understanding of broomrape parasitism at the molecular level and contribute to the development of sustainable solutions for counteracting the devastating effects of broomrape infection in tomatoes and other important crops.

Supplementary Materials: The following supporting information can be downloaded at https://www.mdpi.com/xxx/s1. Supplementary File S1: Table S1: Tomato plant height (cm) on four different days after inoculation (DAI) with broomrape seeds; Table S2: Tomato plant weight (kg) on four different days after inoculation (DAI) with broomrape seeds; Table S3: Soil plant analysis development (SPAD) for the tomato plants on four different days after inoculation (DAI) with broomrape seeds; Table S4: Relative Water Content (RWC) of tomato leaves on 30 DAI; Table S5: Genomic/transcriptomic alignment rates for *S. lycopersicum* libraries; Table S6: Genomic/transcriptomic alignment rates for introgression line libraries; Table S7: List of the 14 DEGs selected for q PCR validation and information extracted after bioinformatic analysis; Table S8: Primers used for qPCR validation of 14 DEGs; Figure S1: Relative gene expression for the genes A: *ACLA2*, B: *D14*, C: *GABA*, D: *PSY* and E: *ZRP1*, with no significant differences; Figure S2: Genes within each shared pathway for the two species (*S. lycopersicum* X *S. pennellii*) were merged into a single set. The top 20 enriched pathways for each IL (6-2 & 6-3 respectively) are presented; Supplementary File S2: The DEGs, displayed both up- and down-regulation across the different genotypes.

Author Contributions: Conceptualization, E.T. and A.K. (Aliki Kapazoglou); methodology, M.G.; software, A.K. (Anastasios Katsileros), V.K. and S.T.; validation, E.T., M.G. and A.K. (Aliki Kapazoglou); formal analysis, E.T.; investigation, E.T., A.K. (Aliki Kapazoglou) and M.G.; resources, E.T. and D.C.; data curation, M.G., V.K. and S.T.; writing—original draft preparation, E.T., A.K. (Aliki Kapazoglou), M.G. and V.K.; writing—review and editing, E.T., A.K. (Aliki Kapazoglou), M.G., A.G.H. and V.K.; visualization, E.T.; supervision, E.T.; project administration, E.T. All authors have read and agreed to the published version of the manuscript.

Funding: The research work was supported by the Hellenic Foundation for Research and Innovation (HFRI) under the 3rd Call for HFRI PhD Fellowships (Proposal Number: 06001). Additionally, at the initial stages of the experimental procedures, the research work was supported by the research project: "Zeroparasitic: Innovative sustainable solutions for broomrapes: prevention and integrated pest management approaches to overcome parasitism in Mediterranean cropping systems." Project ID: 1485.

Institutional Review Board Statement: Not applicable.

Data Availability Statement: The original contributions presented in the study are included in the article and Supplementary Materials. Additionally, the RNA-Seq data reported in this paper have been deposited in the Gene Expression Omnibus (GEO) database under the accession number GSE245472.

Acknowledgments: The authors want to thank Ioanna Kendrick for English editing. They also would like to thank Francisco Pérez Alfocea and Purificacon Andrea Martinez Melgareio from "Centro de

Edafología y BiologíaAplicada del Segura (CEBAS)", Murcia, Spain, for providing the initial genetic material of the ILs studied in this research work.

Conflicts of Interest: The authors declare no conflicts of interest.

References

1. Varanasi, A.; Prasad, P.V.V.; Jugulam, M. Chapter Three—Impact of Climate Change Factors on Weeds and Herbicide Efficacy. In *Advances in Agronomy*; Sparks, D.L., Ed.; Academic Press: Cambridge, MA, USA, 2016; Volume 135, pp. 107–146.
2. Matzrafi, M.; Seiwert, B.; Reemtsma, T.; Rubin, B.; Peleg, Z. Climate Change Increases the Risk of Herbicide-Resistant Weeds Due to Enhanced Detoxification. *Planta* **2016**, *244*, 1217–1227. [CrossRef]
3. Dor, E.; Goldwasser, Y. "Parasitic Weeds: Biology and Control" Special Issue Editors Summary. *Plants* **2022**, *11*, 1891. [CrossRef] [PubMed]
4. Gerakari, M.; Cheimona, N.; Tani, E.; Travlos, I.; Chachalis, D.; Loddo, D.; Mathiassen, S.K.; Gitsopoulos, T.K.; Scarabel, L.; Panozzo, S.; et al. Biochemical and Rapid Molecular Analyses to Identify Glyphosate Resistance in *Lolium* spp. *Agronomy* **2022**, *12*, 40. [CrossRef]
5. Ramesh, K.; Matloob, A.; Aslam, F.; Florentine, S.K.; Chauhan, B.S. Weeds in a Changing Climate: Vulnerabilities, Consequences, and Implications for Future Weed Management. *Front. Plant Sci.* **2017**, *8*, 95. [CrossRef] [PubMed]
6. Cheng, X.; Floková, K.; Bouwmeester, H.; Ruyter-Spira, C. The Role of Endogenous Strigolactones and Their Interaction with ABA during the Infection Process of the Parasitic Weed *Phelipanche ramosa* in Tomato Plants. *Front. Plant Sci.* **2017**, *8*, 392. [CrossRef] [PubMed]
7. Hershenhorn, J.; Eizenberg, H.; Dor, E.; Kapulnik, Y.; Goldwasser, Y. *Phelipanche aegyptiaca* Management in Tomato. *Weed Res.* **2009**, *49*, 34–47. [CrossRef]
8. Fernández-Aparicio, M.; Reboud, X.; Gibot-Leclerc, S. Broomrape Weeds. Underground Mechanisms of Parasitism and Associated Strategies for Their Control: A Review. *Front. Plant Sci.* **2016**, *7*, 135. [CrossRef]
9. Negewo, T.; Ahmed, S.; Tessema, T.; Tana, T. Biological Characteristics, Impacts, and Management of Crenate Broomrape (*Orobanche crenata*) in Faba Bean (*Vicia faba*): A Review. *Front. Agron.* **2022**, *4*, 708187. [CrossRef]
10. Goldwasser, Y.; Kleifeld, Y. Recent Approaches to Orobanche Management. In *Weed Biology and Management*; Inderjit, Ed.; Springer: Dordrecht, The Netherlands, 2004; pp. 439–466. ISBN 978-94-017-0552-3.
11. Joel, D.M.; Hershenhorn, J.; Eizenberg, H.; Aly, R.; Ejeta, G.; Rich, P.J.; Ransom, J.K.; Sauerborn, J.; Rubiales, D. Biology and Management of Weedy Root Parasites. In *Horticultural Reviews*; John Wiley & Sons, Ltd.: Hoboken, NJ, USA, 2007; pp. 267–349. ISBN 978-0-470-16801-1.
12. Tsialtas, J.T.; Eleftherohorinos, I.G. First Report of Branched Broomrape (*Orobanche ramosa*) on Oilseed Rape (*Brassica napus*), Wild Mustard (*Sinapis Arvensis*), and Wild Vetch (*Vicia* Spp.) in Northern Greece. *Plant Dis.* **2011**, *95*, 1322. [CrossRef]
13. Parker, C. The Parasitic Weeds of the Orobanchaceae. In *Parasitic Orobanchaceae: Parasitic Mechanisms and Control Strategies*; Joel, D.M., Gressel, J., Musselman, L.J., Eds.; Springer: Berlin/Heidelberg, Germany, 2013; pp. 313–344. ISBN 978-3-642-38146-1.
14. Bergougnoux, V. The History of Tomato: From Domestication to Biopharming. *Biotechnol. Adv.* **2014**, *32*, 170–189. [CrossRef]
15. Disciglio, G.; Lops, F.; Carlucci, A.; Gatta, G.; Tarantino, A.; Frabboni, L.; Carriero, F.; Tarantino, E. Effects of Different Methods to Control the Parasitic Weed *Phelipanche ramosa* (L.) Pomel in Processing Tomato Crops. *Ital. J. Agron.* **2016**, *11*, 39–46. [CrossRef]
16. El-Halmouch, Y.; Benharrat, H.; Thalouarn, P. Effect of Root Exudates from Different Tomato Genotypes on Broomrape (*O. aegyptiaca*) Seed Germination and Tubercle Development. *Crop Prot.* **2006**, *25*, 501–507. [CrossRef]
17. Cvejić, S.; Radanović, A.; Dedić, B.; Jocković, M.; Jocić, S.; Miladinović, D. Genetic and Genomic Tools in Sunflower Breeding for Broomrape Resistance. *Genes* **2020**, *11*, 152. [CrossRef] [PubMed]
18. Kostov, K.; Batchvarova, R.; Slavov, S. Application of Chemical Mutagenesis to Increase the Resistance of Tomato to *Orobanche ramosa* L. *Bulg. J. Agric. Sci.* **2007**, *13*, 505–513.
19. Draie, R. Differential Responses of Commercial Tomato Rootstocks to Branched Broomrape. *Res. Plant Sci.* **2017**, *5*, 15–25. [CrossRef]
20. Bari, V.K.; Nassar, J.A.; Kheredin, S.M.; Gal-On, A.; Ron, M.; Britt, A.; Steele, D.; Yoder, J.; Aly, R. CRISPR/Cas9-Mediated Mutagenesis of CAROTENOID CLEAVAGE DIOXYGENASE 8 in Tomato Provides Resistance against the Parasitic Weed *Phelipanche aegyptiaca*. *Sci. Rep.* **2019**, *9*, 11438. [CrossRef] [PubMed]
21. Bari, V.K.; Nassar, J.A.; Meir, A.; Aly, R. Targeted Mutagenesis of Two Homologous ATP-Binding Cassette Subfamily G (*ABCG*) Genes in Tomato Confers Resistance to Parasitic Weed *Phelipanche aegyptiaca*. *J. Plant Res.* **2021**, *134*, 585–597. [CrossRef] [PubMed]
22. Orellana, S.; Yañez, M.; Espinoza, A.; Verdugo, I.; González, E.; Ruiz-Lara, S.; Casaretto, J.A. The Transcription Factor SlAREB1 Confers Drought, Salt Stress Tolerance and Regulates Biotic and Abiotic Stress-Related Genes in Tomato. *Plant Cell Environ.* **2010**, *33*, 2191–2208. [CrossRef] [PubMed]
23. Bolger, A.; Scossa, F.; Bolger, M.E.; Lanz, C.; Maumus, F.; Tohge, T.; Quesneville, H.; Alseekh, S.; Sørensen, I.; Lichtenstein, G.; et al. The Genome of the Stress-Tolerant Wild Tomato Species *Solanum pennellii*. *Nat. Genet.* **2014**, *46*, 1034–1038. [CrossRef]
24. Kapazoglou, A.; Gerakari, M.; Lazaridi, E.; Kleftogianni, K.; Sarri, E.; Tani, E.; Bebeli, P.J. Crop Wild Relatives: A Valuable Source of Tolerance to Various Abiotic Stresses. *Plants* **2023**, *12*, 328. [CrossRef]

25. Chitwood, D.H.; Kumar, R.; Headland, L.R.; Ranjan, A.; Covington, M.F.; Ichihashi, Y.; Fulop, D.; Jiménez-Gómez, J.M.; Peng, J.; Maloof, J.N.; et al. A Quantitative Genetic Basis for Leaf Morphology in a Set of Precisely Defined Tomato Introgression Lines. *Plant Cell* **2013**, *25*, 2465–2481. [CrossRef] [PubMed]
26. Pessoa, H.P.; Dariva, F.D.; Copati, M.G.F.; Paula, R.G.d.; Dias, F.d.O.; Gomes, C.N. Uncovering Tomato Candidate Genes Associated with Drought Tolerance Using *Solanum pennellii* Introgression Lines. *PLoS ONE* **2023**, *18*, e0287178. [CrossRef] [PubMed]
27. Bielecka, M.; Pencakowski, B.; Nicoletti, R. Using Next-Generation Sequencing Technology to Explore Genetic Pathways in Endophytic Fungi in the Syntheses of Plant Bioactive Metabolites. *Agriculture* **2022**, *12*, 187. [CrossRef]
28. García-Campos, M.A.; Espinal-Enríquez, J.; Hernández-Lemus, E. Pathway Analysis: State of the Art. *Front. Physiol.* **2015**, *6*, 383. [CrossRef] [PubMed]
29. Tan, G.; Liu, K.; Kang, J.; Zhang, Y.; Hu, L.; Li, C. Transcriptome Analysis of the Compatible Interaction of Tomato with *Verticillium dahliae* Using RNA-Sequencing. *Front. Plant Sci.* **2015**, *6*, 428. [CrossRef] [PubMed]
30. Jhu, M.-Y.; Farhi, M.; Wang, L.; Zumstein, K.; Sinha, N.R. Investigating Host and Parasitic Plant Interaction by Tissue-Specific Gene Analyses on Tomato and *Cuscuta campestris* Interface at Three Haustorial Developmental Stages. *Front. Plant Sci.* **2022**, *12*, 764843. [CrossRef] [PubMed]
31. Eshed, Y.; Zamir, D. An Introgression Line Population of *Lycopersicon pennellii* in the Cultivated Tomato Enables the Identification and Fine Mapping of Yield-Associated QTL. *Genetics* **1995**, *141*, 1147–1162. [CrossRef] [PubMed]
32. Bai, J.; Wei, Q.; Shu, J.; Gan, Z.; Li, B.; Yan, D.; Huang, Z.; Guo, Y.; Wang, X.; Zhang, L.; et al. Exploration of Resistance to *Phelipanche aegyptiaca* in Tomato. *Pest Manag. Sci.* **2020**, *76*, 3806–3821. [CrossRef]
33. Yan, Y.-S.; Chen, X.-Y.; Yang, K.; Sun, Z.-X.; Fu, Y.-P.; Zhang, Y.-M.; Fang, R.-X. Overexpression of an *F-Box* Protein Gene Reduces Abiotic Stress Tolerance and Promotes Root Growth in Rice. *Mol. Plant* **2011**, *4*, 190–197. [CrossRef]
34. Misra, V.A.; Wafula, E.K.; Wang, Y.; de Pamphilis, C.W.; Timko, M.P. Genome-Wide Identification of MST, SUT and SWEET Family Sugar Transporters in Root Parasitic Angiosperms and Analysis of Their Expression during Host Parasitism. *BMC Plant Biol.* **2019**, *19*, 196. [CrossRef]
35. Asakura, H.; Yamakawa, T.; Tamura, T.; Ueda, R.; Taira, S.; Saito, Y.; Abe, K.; Asakura, T. Transcriptomic and Metabolomic Analyses Provide Insights into the Upregulation of Fatty Acid and Phospholipid Metabolism in Tomato Fruit under Drought Stress. *J. Agric. Food Chem.* **2021**, *69*, 2894–2905. [CrossRef] [PubMed]
36. Babraham Bioinformatics—FastQC A Quality Control Tool for High Throughput Sequence Data. Available online: https://www.bioinformatics.babraham.ac.uk/projects/fastqc/ (accessed on 10 January 2024).
37. Martin, M. Cutadapt Removes Adapter Sequences from High-Throughput Sequencing Reads. *EMBnet J.* **2011**, *17*, 10–12. [CrossRef]
38. Dobin, A.; Davis, C.A.; Schlesinger, F.; Drenkow, J.; Zaleski, C.; Jha, S.; Batut, P.; Chaisson, M.; Gingeras, T.R. STAR: Ultrafast Universal RNA-Seq Aligner. *Bioinformatics* **2013**, *29*, 15–21. [CrossRef] [PubMed]
39. RSEM: Accurate Transcript Quantification from RNA-Seq Data with or without a Reference Genome | BMC Bioinformatics. Available online: https://link.springer.com/article/10.1186/1471-2105-12-323 (accessed on 23 January 2024).
40. Szymański, J.; Bocobza, S.; Panda, S.; Sonawane, P.; Cárdenas, P.D.; Lashbrooke, J.; Kamble, A.; Shahaf, N.; Meir, S.; Bovy, A.; et al. Analysis of Wild Tomato Introgression Lines Elucidates the Genetic Basis of Transcriptome and Metabolome Variation Underlying Fruit Traits and Pathogen Response. *Nat. Genet.* **2020**, *52*, 1111–1121. [CrossRef]
41. Robinson, M.D.; McCarthy, D.J.; Smyth, G.K. edgeR: A Bioconductor Package for Differential Expression Analysis of Digital Gene Expression Data. *Bioinformatics* **2010**, *26*, 139–140. [CrossRef]
42. Benjamini, Y.; Hochberg, Y. Controlling the False Discovery Rate: A Practical and Powerful Approach to Multiple Testing. *J. R. Stat. Soc. Ser. B (Methodol.)* **1995**, *57*, 289–300. [CrossRef]
43. Ritchie, M.E.; Phipson, B.; Wu, D.; Hu, Y.; Law, C.W.; Shi, W.; Smyth, G.K. Limma Powers Differential Expression Analyses for RNA-Sequencing and Microarray Studies. *Nucleic Acids Res.* **2015**, *43*, e47. [CrossRef] [PubMed]
44. Wilkinson, L. Ggplot2: Elegant Graphics for Data Analysis by WICKHAM, H. *Biometrics* **2011**, *67*, 678–679. [CrossRef]
45. Kassambara, A. Ggpubr: "ggplot2" Based Publication Ready Plots 2023. Available online: https://www.mdpi.com/authors/references (accessed on 13 February 2024).
46. Warnes, G.R.; Bolker, B.; Bonebakker, L.; Gentleman, R.; Huber, W.; Liaw, A.; Lumley, T.; Maechler, M.; Magnusson, A.; Moeller, S.; et al. Gplots: Various R Programming Tools for Plotting Data 2024. Available online: https://CRAN.R-project.org/package=gplots (accessed on 14 February 2024).
47. Auguie, B.; Antonov, A. gridExtra: Miscellaneous Functions for "Grid" Graphics 2017. 2017. Available online: https://CRAN.R-project.org/package=gridExtra (accessed on 14 February 2024).
48. VennDiagram: A Package for the Generation of Highly-Customizable Venn and Euler Diagrams in R | BMC Bioinformatics | Full Text. Available online: https://bmcbioinformatics.biomedcentral.com/articles/10.1186/1471-2105-12-35 (accessed on 15 February 2024).
49. Livak, K.J.; Schmittgen, T.D. Analysis of Relative Gene Expression Data Using Real-Time Quantitative PCR and the $2^{-\Delta\Delta CT}$ Method. *Methods* **2001**, *25*, 402–408. [CrossRef]
50. Barker, E.R.; Press, M.C.; Scholes, J.D.; Quick, W.P. Interactions between the Parasitic Angiosperm *Orobanche aegyptiaca* and Its Tomato Host: Growth and Biomass Allocation. *New Phytol.* **1996**, *133*, 637–642. [CrossRef]

51. Kanehisa, M.; Goto, S. KEGG: Kyoto Encyclopedia of Genes and Genomes. *Nucleic Acids Res.* **2000**, *28*, 27–30. [CrossRef] [PubMed]
52. Albert, M.; Axtell, M.J.; Timko, M.P. Mechanisms of Resistance and Virulence in Parasitic Plant–Host Interactions. *Plant Physiol.* **2021**, *185*, 1282–1291. [CrossRef] [PubMed]
53. Julio, E.; Malpica, A.; Cotucheau, J.; Bachet, S.; Volpatti, R.; Decorps, C.; Dorlhac de Borne, F. RNA-Seq Analysis of Orobanche Resistance in *Nicotiana tabacum*: Development of Molecular Markers for Breeding Recessive Tolerance from 'Wika' Tobacco Variety. *Euphytica* **2019**, *216*, 6. [CrossRef]
54. Yang, C.; Fu, F.; Zhang, N.; Wang, J.; Hu, L.; Islam, F.; Bai, Q.; Yun, X.; Zhou, W. Transcriptional Profiling of Underground Interaction of Two Contrasting Sunflower Cultivars with the Root Parasitic Weed *Orobanche cumana*. *Plant Soil* **2020**, *450*, 303–321. [CrossRef]
55. Zhao, Y.; Bi, K.; Gao, Z.; Chen, T.; Liu, H.; Xie, J.; Cheng, J.; Fu, Y.; Jiang, D. Transcriptome Analysis of *Arabidopsis thaliana* in Response to *Plasmodiophora brassicae* during Early Infection. *Front. Microbiol.* **2017**, *8*, 673. [CrossRef]
56. Uluisik, S.; Oney-Birol, S. Uncovering Candidate Genes Involved in Postharvest Ripening of Tomato Using the *Solanum pennellii* Introgression Line Population by Integrating Phenotypic Data, RNA-Seq, and SNP Analyses. *Sci. Hortic.* **2021**, *288*, 110321. [CrossRef]
57. Yao, Z.; Tian, F.; Cao, X.; Xu, Y.; Chen, M.; Xiang, B.; Zhao, S. Global Transcriptomic Analysis Reveals the Mechanism of *Phelipanche aegyptiaca* Seed Germination. *Int. J. Mol. Sci.* **2016**, *17*, 1139. [CrossRef] [PubMed]
58. Zhao, T.; Xu, X. Genome-Wide Analyses of the Genetic Screening of C2H2-Type Zinc Finger Transcription Factors and Abiotic and Biotic Stress Responses in Tomato (*Solanum lycopersicum*) Based on RNA-Seq Data. *Front. Genet.* **2020**, *11*, 540. [CrossRef] [PubMed]
59. Fernández-Aparicio, M.; Kisugi, T.; Xie, X.; Rubiales, D.; Yoneyama, K. Low Strigolactone Root Exudation: A Novel Mechanism of Broomrape (*Orobanche* and *Phelipanche* Spp.) Resistance Available for Faba Bean Breeding. *J. Agric. Food Chem.* **2014**, *62*, 7063–7071. [CrossRef]
60. Bouwmeester, H.; Li, C.; Thiombiano, B.; Rahimi, M.; Dong, L. Adaptation of the Parasitic Plant Lifecycle: Germination Is Controlled by Essential Host Signaling Molecules. *Plant Physiol.* **2021**, *185*, 1292–1308. [CrossRef]
61. Huang, Q.; Lei, Z.; Xiang, L.; Zhang, W.; Zhang, L.; Gao, Y. Transcriptomic Analysis of Sunflower (*Helianthus annuus*) Roots Resistance to *Orobanche cumana* at the Seedling Stage. *Horticulturae* **2022**, *8*, 701. [CrossRef]
62. Wallis, J.G.; Watts, J.L.; Browse, J. Polyunsaturated Fatty Acid Synthesis: What Will They Think of Next? *Trends Biochem. Sci.* **2002**, *27*, 467–473. [CrossRef] [PubMed]
63. Shi, Y.; Yue, X.; An, L. Integrated Regulation Triggered by a Cryophyte ω-3 Desaturase Gene Confers Multiple-Stress Tolerance in Tobacco. *J. Exp. Bot.* **2018**, *69*, 2131–2148. [CrossRef]
64. Hirano, T.; Matsuzawa, T.; Takegawa, K.; Sato, M.H. Loss-of-Function and Gain-of-Function Mutations in FAB1A/B Impair Endomembrane Homeostasis, Conferring Pleiotropic Developmental Abnormalities in *Arabidopsis*. *Plant Physiol.* **2011**, *155*, 797–807. [CrossRef] [PubMed]
65. Shaheen, N.; Khan, U.M.; Farooq, A.; Zafar, U.B.; Khan, S.H.; Ahmad, S.; Azhar, M.T.; Atif, R.M.; Rana, I.A.; Seo, H. Comparative Transcriptomic and Evolutionary Analysis of *FAD-like* Genes of Brassica Species Revealed Their Role in Fatty Acid Biosynthesis and Stress Tolerance. *BMC Plant Biol.* **2023**, *23*, 250. [CrossRef]
66. Han, Y.; Lu, M.; Yue, S.; Li, K.; Dong, M.; Liu, L.; Wang, H.; Shang, F. Comparative Methylomics and Chromatin Accessibility Analysis in *Osmanthus fragrans* Uncovers Regulation of Genic Transcription and Mechanisms of Key Floral Scent Production. *Hortic. Res.* **2022**, *9*, uhac096. [CrossRef]
67. Li, J.; Li, X.; Han, P.; Liu, H.; Gong, J.; Zhou, W.; Shi, B.; Liu, A.; Xu, L. Genome Wide Investigation of *bHLH* Genes and Expression Analysis under Different Biotic and Abiotic Stresses in *Helianthus annuus* L. *Int. J. Biol. Macromol.* **2021**, *189*, 72–83. [CrossRef]
68. Jiang, L.; Tian, X.; Li, S.; Fu, Y.; Xu, J.; Wang, G. The AabHLH35 Transcription Factor Identified from *Anthurium andraeanum* Is Involved in Cold and Drought Tolerance. *Plants* **2019**, *8*, 216. [CrossRef]
69. Mo, F.; Zhang, N.; Qiu, Y.; Meng, L.; Cheng, M.; Liu, J.; Yao, L.; Lv, R.; Liu, Y.; Zhang, Y.; et al. Molecular Characterization, Gene Evolution and Expression Analysis of the *F-Box* Gene Family in Tomato (*Solanum lycopersicum*). *Genes* **2021**, *12*, 417. [CrossRef]
70. Sadanandom, A.; Bailey, M.; Ewan, R.; Lee, J.; Nelis, S. The Ubiquitin–Proteasome System: Central Modifier of Plant Signalling. *New Phytol.* **2012**, *196*, 13–28. [CrossRef]
71. Guo, H.; Ecker, J.R. Plant Responses to Ethylene Gas Are Mediated by SCFEBF1/EBF2-Dependent Proteolysis of EIN3 Transcription Factor. *Cell* **2003**, *115*, 667–677. [CrossRef] [PubMed]
72. Sasaki, A.; Itoh, H.; Gomi, K.; Ueguchi-Tanaka, M.; Ishiyama, K.; Kobayashi, M.; Jeong, D.-H.; An, G.; Kitano, H.; Ashikari, M.; et al. Accumulation of Phosphorylated Repressor for Gibberellin Signaling in an F-Box Mutant. *Science* **2003**, *299*, 1896–1898. [CrossRef] [PubMed]
73. Kepinski, S.; Leyser, O. The *Arabidopsis* F-Box Protein TIR1 Is an Auxin Receptor. *Nature* **2005**, *435*, 446–451. [CrossRef]
74. Piisilä, M.; Keceli, M.A.; Brader, G.; Jakobson, L.; Jõesaar, I.; Sipari, N.; Kollist, H.; Palva, E.T.; Kariola, T. The F-Box Protein MAX2 Contributes to Resistance to Bacterial Phytopathogens in *Arabidopsis thaliana*. *BMC Plant Biol.* **2015**, *15*, 53. [CrossRef] [PubMed]
75. Bai, Y.; Pavan, S.; Zheng, Z.; Zappel, N.F.; Reinstädler, A.; Lotti, C.; De Giovanni, C.; Ricciardi, L.; Lindhout, P.; Visser, R.; et al. Naturally Occurring Broad-Spectrum Powdery Mildew Resistance in a Central American Tomato Accession Is Caused by Loss of Mlo Function. *Mol. Plant-Microbe Interact.* **2008**, *21*, 30–39. [CrossRef] [PubMed]

76. Kusch, S.; Panstruga, R. Mlo-Based Resistance: An Apparently Universal "Weapon" to Defeat Powdery Mildew Disease. *Mol. Plant-Microbe Interact.* **2017**, *30*, 179–189. [CrossRef]
77. Takayama, M.; Koike, S.; Kusano, M.; Matsukura, C.; Saito, K.; Ariizumi, T.; Ezura, H. Tomato Glutamate Decarboxylase Genes *SlGAD2* and *SlGAD3* Play Key Roles in Regulating γ-Aminobutyric Acid Levels in Tomato (*Solanum lycopersicum*). *Plant Cell Physiol.* **2015**, *56*, 1533–1545. [CrossRef]
78. Kim, M.C.; Chung, W.S.; Yun, D.-J.; Cho, M.J. Calcium and Calmodulin-Mediated Regulation of Gene Expression in Plants. *Mol. Plant* **2009**, *2*, 13–21. [CrossRef]
79. Rajani, M.S.; Bedair, M.F.; Li, H.; Duff, S.M.G. Phenotypic Effects from the Expression of a Deregulated AtGAD1 Transgene and GABA Pathway Suppression Mutants in Maize. *PLoS ONE* **2021**, *16*, e0259365. [CrossRef]
80. Pinosa, F.; Buhot, N.; Kwaaitaal, M.; Fahlberg, P.; Thordal-Christensen, H.; Ellerström, M.; Andersson, M.X. *Arabidopsis* Phospholipase Dδ Is Involved in Basal Defense and Nonhost Resistance to Powdery Mildew Fungi. *Plant Physiol.* **2013**, *163*, 896–906. [CrossRef]
81. Seong, S.Y.; Shim, J.S.; Bang, S.W.; Kim, J.-K. Overexpression of OsC3H10, a CCCH-Zinc Finger, Improves Drought Tolerance in Rice by Regulating Stress-Related Genes. *Plants* **2020**, *9*, 1298. [CrossRef] [PubMed]
82. Chicco, D.; Agapito, G. Nine Quick Tips for Pathway Enrichment Analysis. *PLoS Comput. Biol.* **2022**, *18*, e1010348. [CrossRef] [PubMed]
83. Dai, Q.; Geng, L.; Lu, M.; Jin, W.; Nan, X.; He, P.; Yao, Y. Comparative Transcriptome Analysis of the Different Tissues between the Cultivated and Wild Tomato. *PLoS ONE* **2017**, *12*, e0172411. [CrossRef]
84. Wang, H.; Zhou, P.; Zhu, W.; Wang, F. De Novo Comparative Transcriptome Analysis of Genes Differentially Expressed in the Scion of Homografted and Heterografted Tomato Seedlings. *Sci. Rep.* **2019**, *9*, 20240. [CrossRef] [PubMed]
85. Ashapkin, V.V.; Kutueva, L.I.; Aleksandrushkina, N.I.; Vanyushin, B.F.; Teofanova, D.R.; Zagorchev, L.I. Genomic and Epigenomic Mechanisms of the Interaction between Parasitic and Host Plants. *Int. J. Mol. Sci.* **2023**, *24*, 2647. [CrossRef] [PubMed]
86. Irafaha, G.; Mutinda, S.; Mobegi, F.; Hale, B.; Omwenga, G.; Wijeratne, A.J.; Wicke, S.; Bellis, E.S.; Runo, S. A Transcriptome Atlas of *Striga hermonthica* Germination: Implications for Managing An Intractable Parasitic Plant. *Plants People Planet* **2022**. [CrossRef]

current issues in molecular biology

MDPI

Article

Biochemical and Molecular Basis of Chemically Induced Defense Activation in Maize against Banded Leaf and Sheath Blight Disease

Shah Mahmood Hamidi [1], Shweta Meshram [2,*], Aundy Kumar [1], Archana Singh [3], Rajbir Yadav [4] and Robin Gogoi [1,*]

[1] Indian Council of Agricultural Research—Plant Pathology, Indian Agricultural Research Institute, New Delhi 110012, India; shahhamidi@yahoo.com (S.M.H.); kumar@iari.res.in (A.K.)

[2] Department of Plant Pathology, School of Agriculture, Lovely Professional University, Phagwara 144402, Punjab, India

[3] Indian Council of Agricultural Research—Biochemistry, Indian Agricultural Research Institute, New Delhi 110012, India; sarchana19@gmail.com

[4] Indian Council of Agricultural Research—Genetics, Indian Agricultural Research Institute, New Delhi 110012, India; rajbiryadav@yahoo.com

* Correspondence: shweta.26662@lpu.co.in (S.M.); robin.gogoi@icar.gov.in (R.G.)

Abstract: Maize is the third most vital global cereal, playing a key role in the world economy and plant genetics research. Despite its leadership in production, maize faces a severe threat from banded leaf and sheath blight, necessitating the urgent development of eco-friendly management strategies. This study aimed to understand the resistance mechanisms against banded leaf and sheath blight (BLSB) in maize hybrid "Vivek QPM-9". Seven fungicides at recommended doses (1000 and 500 ppm) and two plant defense inducers, salicylic acid (SA) and jasmonic acid (JA) at concentrations of 50 and 100 ppm, were applied. Fungicides, notably Azoxystrobin and Trifloxystrobin + Tebuconazole, demonstrated superior efficacy against BLSB, while Pencycuron showed limited effectiveness. Field-sprayed Azoxystrobin exhibited the lowest BLSB infection, correlating with heightened antioxidant enzyme activity (SOD, CAT, POX, β-1,3-glucanase, PPO, PAL), similar to the Validamycin-treated plants. The expression of defense-related genes after seed priming with SA and JA was assessed via qRT-PCR. Lower SA concentrations down-regulated SOD, PPO, and APX genes but up-regulated CAT and β-1,3-glucanase genes. JA at lower doses up-regulated CAT and APX genes, while higher doses up-regulated PPO and β-1,3-glucanase genes; SOD gene expression was suppressed at both JA doses. This investigation elucidates the effectiveness of certain fungicides and plant defense inducers in mitigating BLSB in maize hybrids and sheds light on the intricate gene expression mechanisms governing defense responses against this pathogen.

Keywords: fungicides; *Rhizoctonia solani* f. sp. *sasakii*; defense inducers; antioxidant enzymes activity; defense gene expression

Citation: Hamidi, S.M.; Meshram, S.; Kumar, A.; Singh, A.; Yadav, R.; Gogoi, R. Biochemical and Molecular Basis of Chemically Induced Defense Activation in Maize against Banded Leaf and Sheath Blight Disease. *Curr. Issues Mol. Biol.* **2024**, *46*, 3063–3080. https://doi.org/10.3390/cimb46040192

Academic Editor: Vijai Bhadauria

Received: 18 January 2024
Revised: 23 March 2024
Accepted: 25 March 2024
Published: 2 April 2024

1. Introduction

The importance of maize (*Zea mays* L.) as a vital food and industrial crop in India is evident, with an annual production of 27.82 million metric tons (MMTs) and an average national yield of 3.02 tons per hectare [1,2]. The production of maize is impacted by various biotic stresses, with susceptibility to diseases being a significant deterrent to achieving high grain yields [3]. Banded leaf and sheath blight (BLSB) stands out as a major disease that is particularly prevalent in tropical regions, notably in south and south-east Asian countries, including India. Under favorable conditions, the disease can result in losses of up to one hundred percent [1,4–7].

To manage BLSB, various strategies can be employed, and chemical approaches involving fungicides have proven to be comparatively effective. However, these methods are not eco-friendly, necessitating the exploration of alternative strategies. The resistance of the host plant plays a crucial role in disease management. Despite continuous efforts, developing resistance against BLSB remains a challenge, as no naturally resistant host plants have been identified. Limited genetic variation for resistance further complicates breeding programs [8]. Unfortunately, true resistance sources against BLSB are scarce, and genetic variability for resistance in maize is limited [8–11]. In the absence of BLSB-resistant maize varieties, chemically induced disease resistance is considered an alternative approach.

Salicylic acid (SA) functions as a vital plant growth regulator, playing a significant role in the signaling pathways triggered by diverse biotic and abiotic stresses [12]. It has been identified as an endogenous regulatory signal molecule, activating the general defense mechanisms of plants [13–15]. On the other hand, jasmonic acid (JA) and its derivatives are pivotal in plant responses to various stresses [16]. Systemic acquired resistance (SAR) is a highly desirable form of resistance that protects against a broad spectrum of related or unrelated pathogens. SAR involves the generation of multiple signals at the site of primary infection, which arms distal portions against subsequent secondary infections [17]. SA is not only essential for pathogen-induced SAR, as none of the identified chemical inducers can induce SAR in SA-deficient backgrounds.

In this context, exploring chemically induced resistance using compounds like salicylic acid (SA) and jasmonic acid (JA) emerges as a promising strategy [18,19]. Understanding the role of these inducers and their application methods is crucial in eliciting systemic acquired resistance (SAR) and activating defense mechanisms in plants [20]. However, the use of fungicides and chemical defense activators in maize, particularly their impact on biotic stress conditions, remains inadequately explored [21].

Reactive oxygen species (ROS) such as H_2O_2 are continuously generated in plant tissues as by-products of several metabolic processes [22]. To cope with ROS, plant cells possess an antioxidative system consisting of both enzymatic and non-enzymatic antioxidants [23]. SOD is considered to be the first line of defense against ROS [24]. CAT and APX are responsible for the scavenging of H_2O_2. CAT converts H_2O_2 to H_2O and O_2 and APX catalyzes the reduction of H_2O_2 using ascorbate as an electron donor. Other peroxidases, including guaiacol peroxidase (GPX), are also involved in H_2O_2 elimination [25]. Peroxidase (POX) oxidizes phenolics to more toxic quinones and generates hydrogen peroxide. The last step in the synthesis of lignin and suberin has been proposed to be catalyzed by peroxidases. PAL catalyzes the first step of the phenylpropanoid pathway, leading to the synthesis of a wide variety of secondary metabolites including flavonoids, coumarins, hydroxycinnamoyl esters, and lignin [26]. Due to the nature and defense-related functions of these metabolites, the activation of PAL against abiotic and biotic stresses has been considered a part of the defensive mechanisms of plants [27–30].

This investigation aimed to uncover effective BLSB management using plant defense inducers and understand the mechanisms behind chemically induced disease resistance in maize. We investigated the potential of SA and JA as inducers of SAR against BLSB, shedding light on their mechanisms of action and optimal application strategies [18,19,31]. This research also focused on elucidating plants' defense mechanisms involving pre-existing physical and chemical barriers, along with inducible defense responses [31]. Enzymes such as superoxide dismutases (SODs) play a critical role in defense against oxidative stress. Different SOD isoforms respond differently to inhibitors, and their localization in various cell compartments underscores their importance in defense mechanisms [32–34]. The objective was to assess the impact of seven fungicides and two plant defense inducers, namely salicylic acid (SA) and jasmonic acid (JA), against *Rhizoctonia solani* f. sp. *sasakii*. This comparative analysis aims to provide insights into the potential level of pathogen control achievable in vitro before transitioning to field applications. This dual approach aimed to provide a comprehensive understanding of the treatment's impact on the pathogen under different growth conditions.

2. Materials and Methods

2.1. Layout of Experiments

The in vivo research work was conducted at the main field and net house of ICAR-IARI, Pusa campus, using a susceptible maize variety: "Vivek QPM-9". The variety was procured from ICAR-Vivekananda Parvatia Krishi Anusandhan Sansthan, Almora, Uttarakhand, India, and raised during the kharif seasons of 2020, and 2021. The plot size was 4.5 m^2 and contained 3 rows (row length: 3 m, R to R 75 cm); 13 to 15 plants/row were maintained. The in vitro experiments were conducted in the laboratories using the poisoned food technique [35]. Three replications were maintained in each treatment of fungicides in both PDA and PDB culture media and were then incubated at 28 °C in a BOD incubator. Each treatment, comprising fungicides in both PDA and PDB culture media, was replicated three times and maintained at 28 °C in a BOD incubator for the culture of *R. solani* f. sp. *sasakii* (Source Indian Type Culture Collection, ITCC Accession No. 6881), Division of Plant Pathology, IARI, New Delhi.

2.2. Isolation of the Pathogen and Mass Multiplication of Inoculum

The original culture for the research was obtained from ITCC (Accession No. 6881). The long-term stored culture was further cultured in potato dextrose agar (PDA), and the actively growing culture (mycelia) was used in the inoculation of maize seedlings, to pass through the host providing better revival with aggressiveness. The fungus was re-isolated on PDA and mass cultured in barley grains, which were used in the inoculation of the maize crops following a standard method [36]. The pure cultures were maintained in PDA slants at 4 ± 1 °C for use in this study. The mass culture of the pathogen was prepared using the method described by Ahuja and Payak [37]. Barley grains were soaked overnight in tap water, and the water was then drained out. Forty grams of water-soaked seeds were placed in 250 mL Erlenmeyer flasks, plugged, and autoclaved two times at 121.6 °C (15 lb) for 20 min. The sterilized barley grains were inoculated with a small quantity of 5–8-days-old fungal culture. The flasks were incubated at 27 ± 1 °C for 10–15 days and shaken at every 2–3 day interval to ensure uniform fungal growth in the grains (Table 1).

Table 1. List of fungicides used in this study along with their trade name.

S. No.	Common Name and (a.i. Formulation)	Trade Name	Chemical Name	Empirical Formula	Source
1	Hexaconazole 5% SC	Contaf Plus	2-(2,4-Dichlorophenyl)-1-(1H-1,2,4-triazol-1-yl) hexan-2-ol	$C_{14}H_{17}C_{l2}N_3O$	TATA Rallis India
2	Carbendazim 50% (*w*/*w*) WP	Bavistin	Methyl 1-2 benzimidazole carbamate	$C_9H_9N_3O_2$	BASF India
3	Validamycin 3% (*w*/*w*) L	Sheathmar-3	2,3-Dihydroxy-6 (hydroxymethyl)-4-[-4,5,6-trihydroxy 3 (hydroxymethyl) cyclohex-2-en-1 yl] amino cyclohexyl β-D glucopyranoside.	$C_{20}H_{35}NO_{13}$	Dhanuka Agritech
4	Tebuconazole 25.9% (m/m) EC	Folicur	1-(4-Chlorophenyl)-4,4-dimethyl-3-(1H, 1,2,4-triazol-1-ylmethyl) pentan-3-ol	$C_{16}H_{22}ClN_3O$	Bayer Crop Science
5	Trifloxystrobin 25% + Tebuconazole 50% (m/m) WG	Nativo	Benzeneacetic acid, (E,E)-alpha-(methoxylmino)-2-((((1-(3-trifluoromethyl) phenyl) ethylidene) amino) oxy) methy)-,methyl ester methyl	$C_{20}H_{19}F_3N_2O_4$ and $C_{16}H_{22}ClN_3O$	Bayer Crop Science
6	Azoxystrobin 23% (*w*/*w*) SC	Amistar	(E)-2-[2-[6-(2-Cyanophenoxy) pyrimidin-4-yl] oxyphenyl]-3-methoxyprop-2-enoate	$C_{22}H_{17}N_3O_5$	Syngenta India
7	Pencycuron 22.9% (*w*/*w*)	Monceren	1-(4-chlorbenzyl)-1-cyclopentyl-3-phenylurea	$C_{19}H_{21}ClN_2O$	Bayer Crop Science

2.3. In Vitro Evaluation of Fungicides against *Rhizoctonia solani f. sp. sasakii*

The fungicides that were mentioned in 2.4 were evaluated in vitro at two different concentrations, 500 and 1000 ppm. The required quantity of the fungicides was put into the molten PDA medium and mixed thoroughly. The medium was poured into each Petri dish (90 mm diameter). After solidification, 5 mm discs of the 4–5-days-old actively grown fungus (mycelium) were cut out using a sterilized cork borer. The mycelium disc was placed at the center of Petri plates, sealed with parafilm and incubated at 28 ± 1 °C. The radial growth of the fungus was recorded treatment wise when complete growth was attained in the control (untreated) plates. In the case of liquid media (PDB), the desired amount of the fungicides was mixed with 40 mL PDB in 100 mL conical flaks. Then, a 5 mm disc of the fungus mycelium was put in the poisoned PDB. The flasks were incubated at 28 ± 1 °C for 14 days in a shaker incubator (ISF-1-W, Kuhner, Birsfelden, Switzerland) maintained at 150 rpm. The mycelial ball was harvested by filtering through Whatman filter paper. Then, the mycelial ball was placed on the glass Petri plates covered with a tissue paper and kept there for 4–6 h to air dry. The ball was further dried overnight in an electric oven at 50 °C. The properly dried mycelia were weighed using an electronic balance (Sartorius, Göttingen, Germany). Percent (%) inhibition was calculated for each treatment [38] using the formula $I = (C - T)/C \times 100$, where I = Percent inhibition (%), C = Colony diameter in control (mm), and T = Colony diameter in treatment (mm).

2.4. Plant Inoculation with the Pathogen and Spray of Fungicides

Inoculation was performed on the 35-days-old maize plants [39] with a barley grain culture of *R. solani.* f. sp. *sasakii*. Three barley grains bearing fungal mycelia were carefully inserted by using fingers between the stalk and sheath at the second or third internode from the soil surface [37]. The fungicides, namely Hexaconazole at 0.1%, Carbendazim at 0.1%, Validamycin at 0.1%, Tebuconazole at 0.05%, Trifloxystrobin 25% + Tebuconazole 50% at 0.05%, Azoxystrobin at 0.05%, and Pencycuron at 0.1%, were sprayed on individual plants 3 days after inoculation (DAI) using hand sprayers (Super Garden, Coimbatore, India). All precautions were taken at the time of spraying to avoid drifting to the adjacent treatments (Figure 1). Leaf and sheath samples showing typical symptoms of BLSB, such as dark lesions with characteristic banding on the leaves and sheath necrosis, were collected for further study (Figure 1).

2.5. Evaluation of Fungicides against BLSB Disease and Grain Yield

The efficacy of the fungicides on BLSB disease of maize was determined. The inoculated and fungicide-sprayed plants were monitored for the appearance of symptoms and disease progression regularly on a weekly basis. The percent (%) disease severity, virulence index, and length of infected area in each plant were recorded on 30 DAI. Records of the disease intensity were calculated using a 1–5 disease rating scale (Table 2) [40]. As per the new rating scale, the percent disease index (PDI) was calculated using the formula given by McKinney [41].

Table 2. Area under the disease progress curve (AUDPC) values for different disease severity categories at 30 days after inoculation.

PDI	AUDPC	Disease Severity (According to 1–5 Scale)
20.0	1050	Resistant (R) (Score: ≤2.0) (1 to 1.5) (PDI ≤ 40.0)
30.0		
50.0	1575	Moderately resistant (MR) (Score: 2.1–3.0) (PDI 40.1–60.0)
60.0		

Table 2. *Cont.*

PDI	AUDPC	Disease Severity (According to 1–5 Scale)
70.0	2175	Moderately susceptible (MS) (Score:3.1–4.0) (PDI 60.1–80.0)
80.0		
90.0	2775	Susceptible (S) (Score: ≥4.0) (PDI ≥ 80.0)
100.0		

Note: The initial assessment was conducted at 0 days post-inoculation, followed by a subsequent assessment at 30 days. The average PDI for each interval was then multiplied by the corresponding time duration. The resulting values were summed to compute the AUDPC, employing the trapezoidal rule. ANOVA at 5% level of significance and post hoc Tukey's HSD test statistics reveal that the treatment groups differ significantly. This comparative analysis offers insights into susceptibility and resistance levels across the treatments (F Statistic = 10.1453, *p*-value 0.0243).

Figure 1. Symptoms of banded leaf and sheath blight (BLSB) on (**A**) leaf, (**B**) sheath, and (**C**) cob. (**D**) Artificially inoculated disease field. (**E**) Measurement of lesion length. (**F**) Treated seeds sown in pots and kept under net house conditions.

According to the PDI disease rating plants could be classified as resistant (R), moderately resistant (MR), moderately susceptible (MS), and susceptible (S). For grain yield (q/ha), the total weight (kg) of the fully matured harvested cobs was recorded and plotted by treatment. The shelling of the grains was initially performed and the moisture content of the grains was recorded by using a moisture meter (AgraTronix MT-PRO, Agratronix Corporate, Streetsboro, OH, USA). Grain yield was calculated by applying three formulas [42].

2.6. Assay of Enzyme Activity in Maize Plants Treated with Fungicides

The enzyme activity in the fungicide-treated maize plants was evaluated. For the healthy group, young leaves of 30–35-days-old plants were collected at specific intervals: 3 days before fungicide spray, i.e., at the time of pathogen inoculation; on the day of fungicide application (day 0); and subsequently 5 days later. In the early morning, upper leaves were cut using sterilized scissors, promptly placed in the pre-labeled ice bags, and stored at −80 °C.

Subsequently, superoxide dismutase (SOD) maize leaf tissue (1 g) was ground in 5 mL grinding media (0.1 M phosphate buffer, pH 7.5, with 0.5 mM EDTA). After centrifugation, the supernatant was used for the SOD assay. The reaction mixture contained phosphate buffer, L-methionine, NBT, riboflavin, and enzyme extract. Absorbance was measured at 560 nm and a standard curve was prepared using known concentrations of a substance relevant to the SOD assay [43]. Catalase (CAT) activity was measured by following Aebi's method (1984) with modifications. Protein estimation was performed using the Bradford method. CAT activity was assayed in a 3 mL reaction mixture (50 mM potassium phosphate buffer, 12.5 mM hydrogen peroxide, 100 µL enzyme extract), and the decomposed H_2O_2 was quantified at 240 nm; activity was calculated within a specific range of 5–12 U/mg and was prepared using known concentrations of bovine serum albumin (BSA) [44–46]. A peroxidase (POX) leaf sample (1 g) was ground in 5 mL grinding media. The supernatant obtained after centrifugation was used for the POX assay; activity was determined at 470 nm using a reaction mixture containing phosphate buffer, guaiacol, enzyme extract, and H_2O_2, and then a standard curve was prepared using known concentrations of guaiacol [47]. Phenylalanine ammonia-lyase (PAL) activity was assayed with a reaction mixture containing L-phenylalanine, borate buffer, and enzyme extract; then absorbance was measured at 290 nm and a standard curve was prepared using known concentrations of trans-cinnamic acid, which is the product of the PAL reaction [48,49]. Polyphenol oxidase (PPO) activity was estimated by recording absorbance at 495 nm in a reaction mixture containing phosphate buffer, proline, and catechol and a standard curve using known concentrations of catechol [50]. β-1,3-glucanase activity was determined by incubating enzyme extract with laminarin solution and measuring the absorbance at 500 nm after adding dinitro salicylic reagent and using a standard curve of D-glucose [51] (Supplementary Figures S1–S3 Annexure I).

2.7. Expression Study of Salicylic Acid (SA) and Jasmonic Acid (JA) Genes of Maize during Infection

Maize seeds (variety Vivek QPM 9) were treated with 50 and 100 ppm concentrations of SA and JA and sown in the 10 cm diameter pots under net house condition Figure 1 shows that the average natural temperature during kharif season ranges from 22 to 25 °C. Two seedlings were maintained in each pot. For gene expression analysis, ten-day-old seedlings were collected. Specific primers for each gene were designed by using BioEdit 7.2.5 and IDT Primer Quest software (eu.idtdna.com/Primerquest/Home/Index, accessed on 20 May 2018), with reference sequences for the maize genes obtained from NCBI (Table 3).

Table 3. PCR primers used in this study.

Gene	Primer Details (5′-3′)	Size (bp)	Accession No
Superoxide dismutase	ZM_SOD (F *) 5′-AGT CAC CCA CCC CAT CCA AG-3′ ZM_SOD (R #) 5′-GTG CGG AGG AAT AGG GAG C-3′	146	NC_050102.1
β-1,3 glucanase	ZM_Glucan (F) 5′-ATG GCG AGG CAG GGT GTC-3′ ZM_Glucan (R) 5′-ACG CCG ATG GAT TGG ACT C-3′	188	NC_050098.1
Polyphenol oxidase	ZM_PPO (F) 5′-CGT CCA AGA AGA CCA CCG T-3′ ZM_PPO (R) 5′-ACT GGA CAG GCC GTT GAG CA-3′	146	NC_050105.1

Table 3. *Cont.*

Gene	Primer Details (5′-3′)	Size (bp)	Accession No
Ascorbate peroxidase	ZM_APX (F) 5′-ACC ATG AAG ACC CCC GTC GA-3′ ZM_APX (R) 5′-GGT AGA AGT CAG CGT AGG ATA G-3′	118	NC_050100.1
Catalase	ZM_CAT (F) 5′-ACG TGC GCC GAC TTC CTG-3′ ZM_CAT (R) 5′-GAA GAA GAC GGG GAA GTT GTT-3′	180	NC_050099.1
Phenylalanine ammonialyase	ZM_PAL (F) 5′-TCG AAC TGC AAC CGA AAG A-3′ ZM_PAL (R) 5′-CAG CCA GGA TTG CCA GAA TA-3′	108	NC_050096.1

* Forward, # Revers.

2.8. Primer Validation, RNA Isolation, cDNA Synthesis, and qRT-PCR Analysis of Maize Genes

The designed primers of SOD (ZM_SOD), β-1,3-glucanase (ZM_Glucan), PPO (ZM_PPO), ascorbate peroxidase (ZM_APX), CAT (ZM_CAT), and PAL were validated following Williams's [52] protocol with minor adjustments in annealing temperature. RNA was isolated from the SA- and JA-treated and control maize seedlings. The Pure LinkTM RNA Mini Kit was used for extraction, purification, and quantification via a NanoDrop spectrophotometer. First-strand cDNA synthesis was performed by using the ImProm-II™ Reverse Transcription System kit. SYBR green-based qRT-PCR analysis was carried out by employing the primers listed in Table 2. Gene expression was confirmed via agarose gel electrophoresis (Supplementary Figure S4). The amplification conditions consisted of an initial denaturation step at 95 °C for 10 min, followed by 40 cycles of denaturation at 95 °C for 15 s, annealing at 60 °C for 30 s, and extension at 72 °C for 30 s. For each gene, qRT-PCR reactions were performed in two technical replicates across two independent biological replicates. Relative gene expression was quantified using the comparative Ct method ($2^{-\Delta\Delta Ct}$) given by Livak [52]. Melt curve has been provided in supplemental file Supplementary Table S1.

2.9. Statistical Analysis

All the field and net house experiments were conducted in a randomized block design (RBD). The laboratory experiments were performed in a completely randomized design (CRD). Statistical analysis was conducted using analysis of variance (ANOVA) and the post hoc Tukey honestly significant difference (HSD) test, methods suitable for identifying pairwise differences when multiple groups are involved. We used online software (https://astatsa.com/OneWay_Anova_with_TukeyHSD/ accessed on 1 December 2023).

3. Results

The present study aimed to provide insights into managing BLSB disease by employing plant defense inducers and comprehending the biochemical mechanisms behind disease resistance in maize. Considering the non-availability of BLSB-resistant maize varieties, chemically induced disease resistance was explored as an alternative. Initially, seven fungicides were assessed in vitro against *R. solani* f. sp. *sasakii*. Furthermore, both fungicides and plant defense inducers were evaluated in vivo to determine their efficacy in restricting BLSB disease in maize.

3.1. In Vitro Evaluation of Fungicides against Rhizoctonia solani f. sp. sasakii

Seven different fungicides at two dosages (500 and 1000 ppm) were evaluated against the pathogen in PDA as a positive control. At 1000 ppm, 100% growth inhibition was observed with treatment using the fungicides Hexaconazole, Carbendazim, Validamycin, Tebuconazole, Trifloxystrobin + Tebuconazole (Nativo), and Azoxystrobin as compared to the control. The lowest growth inhibition was observed in the treatment with Pencycuron (52.04%). At 500 ppm, the highest growth inhibition of the pathogen was observed in the treatments with Hexaconazole, Validamycin, Tebuconazole, and Trifloxystrobin + Tebuconazole (Nativo) (100%), followed by Azoxystrobin (92.19%) and Carbendazim (84.63%).

The lowest growth inhibition was observed in the treatment with Pencycuron (48.52%) (Table 3, Figure 2).

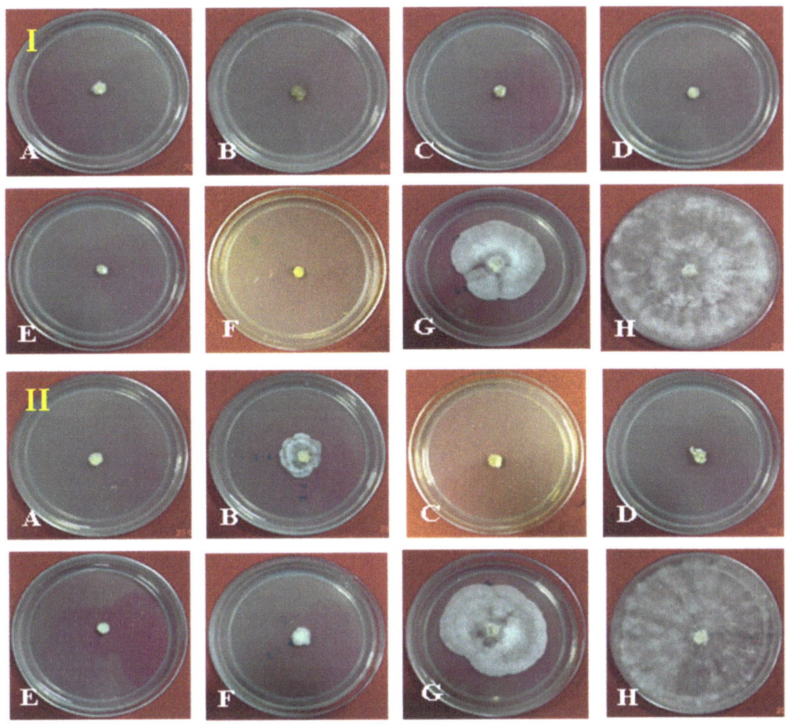

Figure 2. Effect of fungicides (1000 ppm, (**I**) and 500 ppm, (**II**)) on redial growth of *Rhizoctonia solani* f. sp. *sasakii* in vitro: (**A**) Hexaconazole, (**B**) Carbendazim, (**C**) Validamycin, (**D**) Tebuconazole, (**E**) Trifloxystrobin + Tebuconazole (Nativo), (**F**) Azoxystrobin, (**G**) Pencycuron, and (**H**) control (untreated).

The percent reduction in the mycelial mass of the pathogen was also observed to be different in different fungicidal treatments as compared to the untreated control in PDB. At 1000 ppm concentration maximum reduction in mycelial mass was observed in the cases of Hexaconazole, Carbendazim, Validamycin, Tebuconazole, Trifloxystrobin + Tebuconazole (Nativo), and Azoxystrobin (100%), whereas the lowest reduction was observed in the Pencycuron treatment (95.30%). At 500 ppm, the maximum reduction in mycelial mass was observed in the treatments with Hexaconazole, Carbendazim, Validamycin, Tebuconazole, Trifloxystrobin + Tebuconazole (Nativo), and Azoxystrobin (100%), whereas the lowest reduction was observed in the Pencycuron treatment (94.71%) (Table 4, Figure 3).

Table 4. Effect of fungicides on radial growth and mycelial mass of *Rhizoctonia solani* f. sp. *sasakii* inn PDA and PDB.

Treatments	Fungicides	Potato Dextrose Agar (PDA)				Potato Dextrose Broth (PDB)			
		Radial Growth * (mm) at (1000 ppm)	Inhibition (%) at (1000 ppm)	Radial Growth * (mm) at (500 ppm)	Inhibition (%) at (500 ppm)	Mycelial Weight * (mg) at (1000 ppm)	Reduction (%) at (1000 ppm)	Mycelial Weight * (mg) at (500 ppm)	Reduction (%) at (500 ppm)
T1	Hexaconazole	0	100 (90.00) #	0	100.00 (90.00)	0	100.00 (90.00) #	0	100.00 (90.00)
T2	Carbendazim	0	100 −90	13.5	84.63 (66.94)	0	100.00 (90.00)	0	100.00 (90.00)
T3	Validamycin	0	100 −90	0	100.00 (90.00)	0	100.00 (90.00)	0	100.00 (90.00)
T4	Tebuconazole	0	100 −90	0	100.00 (90.00)	0	100.00 (90.00)	0	100.00 (90.00)
T5	Tri. + Teb. (Nativo)	0	100 −90	0	100.00 (90.00)	0	100.00 (90.00)	0	100.00 (90.00)
T6	Azoxystrobin	0	100 −90	13.83	92.19 (76.67)	0	100.00 (90.00)	0	100.00 (90.00)
T7	Pencycuron	43.17	52.04 −46.2	46.33	48.52 (44.12)	15.17	95.3 −77.46	23.7	94.71 −76.78
T8	Control (untreated)	90	0 0	90	0 0	324.27	0 0	461.87	0 0
C. D. (5%)		--	5.05	--	7.61	--	0.32	--	1.3
C. V.		--	3.95	--	6.37	--	0.24	--	0.96

* Data of the table are the means of three replications. # Data within parentheses are angular transformed values.

Figure 3. Superoxide dismutase (SOD), catalase (CAT), peroxidase (POX), polyphenol oxidase (PPO), phenylalanine ammonia-lyase (PAL), and β-1,3-glucnase activity in maize (Vivek QPM-9) inoculated with *R. solani* f. sp. *sasakii* after fungicide application. Activity recorded at different inoculation days in an of interval 0 to 5th day after inoculation. The post hoc Tukey's HSD test ($p < 0.01$) indicates a significant difference (*) in enzyme activity.

3.2. Effect of Fungicides on BLSB Disease of Maize and Grain Yield

The maximum lesion length was observed on the plants sprayed with Hexaconazole (0.1%) (45.76 cm), followed by the plants sprayed with Pencycuron (0.1%) (44.83 cm), Trifloxystrobin + Tebuconazole (Nativo, 0.05%) (37.20 cm), and Carbendazim (0.1%) (36.72 cm). The lowest length of disease lesions was observed in the plants sprayed with Azoxystrobin (0.05%) (31.24 cm), Tebuconazole (0.05%) (32.22 cm), and Validamycin (0.1%) (33.94 cm). The lesion length in untreated (water sprayed) plants was 55.38 cm.

The maximum disease score on the 1–5 scale was observed in the plants sprayed with Pencycuron (0.1%) (3.91), followed by plants sprayed with Carbendazim (0.1%) (3.49), Hexaconazole (0.1%) (3.41), Validamycin (0.1%) (3.23), and Tebuconazole (0.05%) (3.14), whereas the lowest disease scores were observed in plants sprayed with Azoxystrobin (0.05%) (2.81) and plants sprayed with Trifloxystrobin + Tebuconazole (Nativo, 0.05%) (2.97).

The highest percent disease index (PDI) was recorded in the plants sprayed with Pencycuron (0.1%) (77.73%), followed by the plants sprayed with Carbendazim (0.1%) (69.72%), Hexaconazole (0.1%) (68.27%), Validamycin (0.1%) (64.55), and Tebuconazole (0.05%) (62.79%), whereas the lowest percent disease index was recorded in the plants treated with Azoxystrobin (0.05%) (56.2%) and the plants treated with Trifloxystrobin + Tebuconazole (Nativo, 0.05%) (59.36%). The percent disease index for the untreated control plants was recorded as 95.10%.

The highest grain yield was obtained in the plants treated with Azoxystrobin (0.05%) (58.35 q/ha). followed by the plants treated with Tebuconazole (0.05%) (56.61 q/ha), Validamycin (0.1%) (53.93 q/ha), and Carbendazim (0.1%) (50.16 q/ha), whereas the lowest grain yield was recorded in the plants treated with Pencycuron (0.1%) (47.12 q/ha), Trifloxystrobin + Tebuconazole (Nativo, 0.05%) (48.66 q/ha), and the plants treated with Hexaconazole (0.1%) (49.44 q/ha). In the untreated control plants, the grain yield was recorded as 44.93 q/ha (Table 5).

Table 5. Efficacy of fungicides on BLSB disease and grain yield of maize (Vivek QPM-9) under field conditions.

Tr.	Fungicide	Lesion Length * (cm)	Disease Score * (1–5 Scale)	PDI * (%)	Yield $^{\pi}$ (Q/ha)
T1	Hexaconazole (0.1%)	45.76	3.41	68.27 (56.44) #	49.44
T2	Carbendazim (0.1%)	36.72	3.49	69.72 (57.01)	50.16
T3	Validamycin (0.1%)	33.94	3.23	64.55 (53.56)	53.93
T4	Tebuconazole (0.05%)	32.22	3.14	62.79 (52.62)	56.61
T5	Trifloxystrobin + Tebuconazole (0.05%)	37.20	2.97	59.36 (50.38)	48.66
T6	Azoxystrobin (0.05%)	31.24	2.81	56.20 (48.61)	58.35
T7	Pencycuron (0.1%)	44.83	3.91	77.73 (62.01)	47.12
T8	Control (water)	55.38	5.12	95.10 (77.20)	44.93
C. D. (5%)		12.42	1.19	14.60	N/A
C. V.		17.71	19.19	14.43	28.97

* Data of the table are the means of three replications, PDI: percent disease index. # Data within parentheses are angular transformed values. $^{\pi}$ Yield is calculated based on 3 m^2 area of the experimental plots.

3.3. Estimation of Biochemical Defense-Related Enzymes in Maize Treated with Fungicides

This study examined the impact of fungicides on maize's biochemical defense enzymes against *R. solani* f. sp. *sasakii* (Figure 3).

Superoxide dismutase (SOD) activity was the highest with Azoxystrobin (0.05%) at 3 DAI, followed by Validamycin (0.1%), Hexaconazole (0.1%), and Pencycuron (0.1%) at 2 and 3 DAI. Trifloxystrobin + Tebuconazole showed induction at 5 DAI. Tebuconazole (0.05%) and Carbendazim (0.1%) had lower SOD elevations. Catalase (CAT) activity peaked in the control group at 0 and 4 DAI. Azoxystrobin (0.05%), Validamycin (0.1%), Tebuconazole (0.05%), Carbendazim (0.1%), and Pencycuron (0.1%) showed either equal or lower activity, especially at 1 and 4 DAI.

Peroxidase (POX) activity was the highest with Validamycin (0.1%) and Carbendazim (0.1%) at 1 and 5 DAI, while Hexaconazole (0.1%) exhibited the least activity on the 3rd day. Polyphenol oxidase (PPO) activity peaked with Validamycin (0.1%) and Hexaconazole

(0.1%) at 1 and 3 DAI. Azoxystrobin (0.05%), Tebuconazole (0.05%), and Pencycuron (0.1%) showed lower activity levels.

Phenylalanine ammonia-lyase (PAL) activity was notably higher with Hexaconazole (0.1%) and Validamycin (0.1%) on the 1st and 3rd days. The other treatments showed similar patterns to the control. β-1,3-glucanase (β-1,3-G) activity was notably high with Trifloxystrobin + Tebuconazole (Nativo, 0.05%) at 5 DAI, and Tebuconazole (0.05%) at 1 and 5 DAI. Conversely, the Pencycuron (0.1%), Azoxystrobin (0.05%), Validamycin (0.1%), and Tebuconazole (0.05%) treatments all showed lower activity compared to the control.

3.4. Effects of Salicylic Acid and Jasmonic Acid Seed Priming on Expression of Defense-Related Genes in Maize

This study investigated the influence of seed treatments with salicylic acid (SA) and jasmonic acid (JA) on the relative expression of defense-related genes in maize (Figure 4).

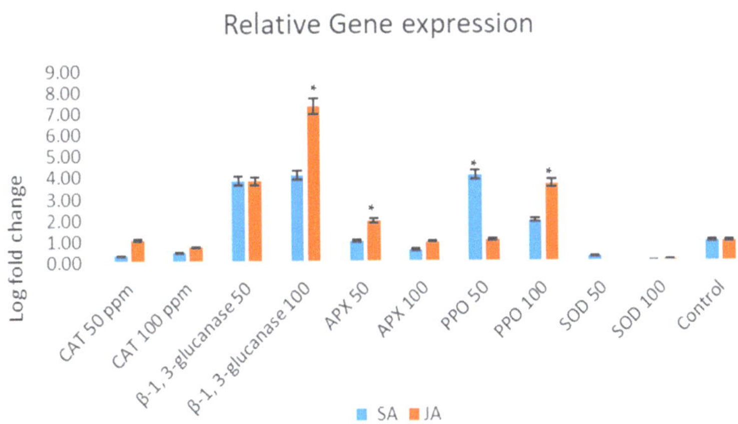

Figure 4. Relative expression of selected defense genes in maize treated with salicylic acid (SA) and jasmonic acid (JA). Enzymes include superoxide dismutase (SOD), polyphenol oxidase (PPO), ascorbate peroxidase (APX), catalase (CAT), and β-1,3-glucanase (β-1,3-gluc). The post hoc Tukey's HSD test ($p < 0.01$) indicates a significant difference (*) in gene expression between the samples treated with salicylic acid and jasmonic acid.

The relative expression of the maize superoxide dismutase (SOD) gene was modulated by the SA and JA treatments. Elevated SA concentrations led to a reduction in SOD expression, indicating a suppressive effect of SA on maize SOD expression. In contrast, escalating doses of JA resulted in a heightened expression of the SOD gene. The expression of the maize polyphenol oxidase (PPO) gene responded differently to the SA treatment, with increased concentrations of SA further diminishing gene expression. Conversely, the JA treatment elicited a robust expression of the PPO gene, suggesting its potential involvement in maize defense mechanisms. Particularly, at a concentration of 100 ppm, JA induced a substantial four-fold increase in PPO gene expression compared to the 50 ppm JA treatment.

In contrast, the relative expression of the maize ascorbate peroxidase (APX) gene exhibited divergent responses to SA and JA. Higher SA concentrations (100 ppm) halved APX expression relative to the control, while the JA treatment at 50 ppm nearly doubled APX expression. Intriguingly, APX expression remained stable at JA concentrations of 100 ppm. The expression of the maize catalase (CAT) gene showed a dose-dependent response to SA and JA treatments. CAT gene expression increased by approximately 3-fold and 3.2-fold with the SA and JA treatments at 50 ppm, respectively. However, as the SA/JA concentrations increased, there was a proportional decrease in CAT gene expression, indicating a dose–response relationship.

Distinct expression patterns were observed for the maize β-1,3-glucanase (β-1,3-gluc) gene in response to the SA and JA treatments. At 50 ppm, β-1,3-glucanase expression was markedly elevated. While escalating SA concentrations suppressed β-1,3-glucanase expression, the JA treatment led to its upregulation. Notably, the JA treatment at 100 ppm resulted in a remarkable 27-fold increase in gene expression, highlighting the involvement of the JA pathway in enhancing defense against maize pathogens (Figure 4).

Overall, the assay results reveal differential enzyme induction patterns across treatments, with Azoxystrobin exhibiting a superior induction of SOD activity, Validamycin and Carbendazim showing the highest POX activities, and Hexaconazole and Validamycin demonstrating elevated PAL activities compared to the other treatments.

4. Discussion

The current study aimed to investigate the management of BLSB disease in maize using plant defense inducers while exploring the biochemical and molecular mechanisms of disease resistance. Previous research has demonstrated the efficacy of Carbendazim against *Rhizoctonia solani*, the causal agent of BLSB, both in vitro and in vivo [53]. Similar outcomes have been reported, with reduced BLSB severity following foliar spray of Carbendazim on maize [54]. Another study reported the minimal effect of Strobilurins (Azoxystrobin) on *R. solani* mycelial growth in vitro [55]. Our findings indicate that, among the fungicides tested, Strobilurins and Carbendazim were more effective, aligning with previous research documenting excellent disease control with Strobilurins, Triazoles, and Benzimidazoles [56].

In addition to gene expression analysis, we evaluated the activity of key defense enzymes. These findings suggest that different fungicides may elicit distinct defense responses in maize, emphasizing the importance of selecting appropriate fungicides for effective disease management. A similar enzymatic induction was also recently reported in maize against another foliar pathogen, *Colletotrichum graminicola*, which resulted in local and systemic resistance, emphasizing the enzyme-mediated defense mechanisms in maize [57].

Superoxide dismutase (SOD) activity has delivered effective mitigations of reactive oxygen species during pathogenic infection in several studies [23,24]. Azoxystrobin, a quinone outside inhibitor (QoI), has been demonstrated to posses a capacity to activate reactive oxygen species, scavenge H_2O_2, and enhance the synthesis of secondary metabolites, as is reflected in the heightened activities of defense-related enzymes like SOD, catalase (CAT), and β-1,3-glucanase, akin to the antifungal antibiotic Validamycin. Conversely, salicylic Acid (SA) application on maize seeds has negatively impacted the expression of SOD and polyphenol oxidase (PPO) genes in a concentration-dependent manner, suggesting a potential damage-induced accumulation of H_2O_2 [22]. Some studies have also highlighted that SOD has minimal effects on JA induction [58]. The constitutive expressions of pathogenesis-related proteins and antioxidant enzyme activities, including superoxide dismutase (SOD), peroxidase (POD), and β-1,3-glucanase, play a pivotal role in triggering maize resistance against various pathogens [59]. These proteins and enzymes are integral components of the plant's defense machinery, contributing to the mitigation of oxidative stress and the degradation of fungal cell walls [22,59,60].

The nuanced responses of APX and CAT genes to different SA concentrations suggest a finely tuned balance in maize antioxidant defenses, with higher SA doses potentially inducing oxidative stress mitigation. Similar findings have been observed by previous studies in various forms of oxidative stress caused by abiotic factors in plants [25]. Furthermore, the dose-dependent decrease in CAT gene expression with rising SA levels implies a dynamic regulation of hydrogen peroxide scavenging mechanisms, which supports the results of previous studies [61] performed during plant stress and development. Meanwhile, the notable increase in β-1,3-glucanase gene expression at 50 ppm SA underscores the complex interplay between SA signaling and defense response activation in maize. These findings aligns with a recent study conducted in maize, where the phyllosphere microbiome modulated the physiology of plants through enzymatic interplay, which ultimately showed defense against the foliar pathogen *Exserohilum turcicum* [62].

Jasmonic acid (JA) application has displayed similar complexities, impacting the expression of genes associated with defense mechanisms [60]. JA negatively affected the expression of the SOD gene in maize, with increasing concentrations leading to an enhanced expression. This dynamic modulation indicates the sophisticated and adaptable nature of plant defense mechanisms under JA influence [58,63]. Additionally, JA application increased the expression of the β-1,3-glucanase gene, emphasizing its role in modulating defense against maize pathogens. The JA 100 ppm dose recorded a 27-fold increase in gene expression, suggesting the involvement of the JA pathway in enhancing defense against maize pathogens. This response is similar to that seen with other maize pathogens such as *maize Fusarium verticillioides* [64] and *Colletotrichum graminicola* [65].

This study reveals the multifaceted interactions between plant defense inducers and the intricate regulation of defense-related genes, providing valuable insights into the complex dynamics of plant responses to varying concentrations of these inducers. The findings contribute to our understanding of the biochemical and molecular mechanisms that govern disease resistance in maize (Figure 5). Our findings offer practical guidance for enhancing maize disease management, specifically against BLSB, by optimizing fungicide selection and application methods. The theoretical implications include advancing our understanding of plant defense mechanisms and plant–fungicide interactions, contributing to the broader agricultural science. However, the limitations include the focus on specific fungicides and enzymes, potentially limiting the generalizability of our findings to broader disease management contexts in maize. Therefore, future research could investigate the synergistic effects of fungicide combinations, integrate biological control methods, and assess the long-term impacts on soil health for sustainable maize disease management.

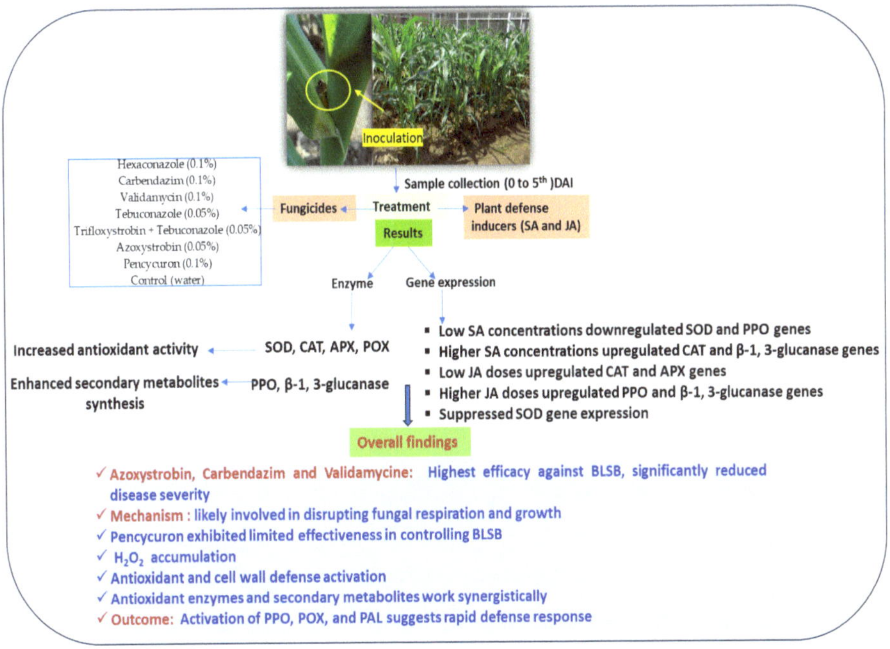

Figure 5. Summary of research on maize defense mechanisms against banded leaf and sheath blight. Investigation of the impact of fungicides and plant defense inducers on biochemical and molecular responses, providing insights into disease resistance in maize in present study.

5. Conclusions

In conclusion, this research provides valuable insights into the management of banded leaf and sheath blight (BLSB) disease in maize. This study explores the biochemical elicitors that induce defense, offering potential avenues for disease control. Both salicylic acid (SA) and jasmonic acid (JA) have been demonstrated to show efficacy in suppressing the necrotrophic soil-borne phytopathogen *R. solani*. While seed priming with these inducers contributed to healthy seed germination, they did not significantly enhance the overall growth, development, and defense induction in maize plants. The artificial application of SA and JA was as effective and able to induce defense responses as chemical application. Hence, we suggest them as a better alternative to chemicals. Considering the resource-intensive nature of maize cultivation, this study sheds light on the prospect of reducing fungicide usage through a plant defense inducer-mediated host resistance approach in disease management. This not only holds promise for effective disease control but also aligns with the goal of promoting a safer and environmentally conscious agricultural environment.

Supplementary Materials: The following supporting information can be downloaded at: https://www.mdpi.com/article/10.3390/cimb46040192/s1, Figure S1: trans-cinnamic acid standard curve; Figure S2: Lowry method standard curve; Figure S3: Standard curve of dextrose; Figure S4: Agrose gel picture of validated primes, Melt curve: Figures S5–S9 and Tables S1–S5. HSD Post Hoc analysis: Table S6.

Author Contributions: Conceptualization, S.M.H., A.K., A.S., R.Y. and R.G.; Formal analysis, S.M.H., S.M., R.Y. and R.G.; Investigation, S.M.H., S.M., A.K., R.Y. and R.G.; Methodology, S.M.H.; Resources, A.K.; Supervision, A.S. and R.G.; Validation, S.M.H. and R.G.; Writing—original draft, S.M.H.; Writing—review and editing, S.M. and R.G. All authors have read and agreed to the published version of the manuscript.

Funding: This research received no external funding.

Institutional Review Board Statement: Not applicable.

Informed Consent Statement: Not applicable.

Data Availability Statement: Data is contained within the article and Supplementary Material.

Conflicts of Interest: The authors declare no conflict of interest.

References

1. Ranum, P.; Peña-Rosas, J.P.; Garcia-Casal, M.N. Global maize production, utilization, and consumption. *Ann. N. Y. Acad. Sci.* **2014**, *1312*, 105–112. [CrossRef]
2. Food and Agricultural Organization of the United Nations. Production, Crops. 2018. Available online: http://www.fao.org/faostat/en/#data/QC (accessed on 5 April 2020).
3. Saxena, S.C. Ear rot and banded leaf and sheath blight of maize. In Proceedings of the 20th Training "Seed Health Management for Better Productivity", New Delhi, India, 28 March–17 April 2008; G. B. Pant University of Agriculture and Technology: Uttarakhand, India, 2008; pp. 119–123. Available online: https://docplayer.net/13170412-Indian-council-of-agricultural-research-new-delhi.html (accessed on 10 March 2018).
4. Hochholdinger, F.; Tuberosa, R. Genetic and genomic dissection of maize root development and architecture. *Curr. Opin. Plant Biol.* **2009**, *12*, 172–177. [CrossRef]
5. Singh, A.; Shahi, J.P. Banded leaf and sheath blight: An emerging disease of maize (*Zea mays* L). *Maydica* **2012**, *57*, 215–219.
6. Devi, B.; Thakur, B.R. Integrated management of banded leaf and sheath blight of maize caused by *Rhizoctonia solani* f. sp. *sasakii*. *Proc. Natl. Acad. Sci. India Sect. B Biol. Sci.* **2018**, *88*, 769–777. [CrossRef]
7. Anonymous. The Biology of *Zea mays* L. ssp *mays* (Maize or Corn). Office of the Gene Technology Regulator. 2017. Available online: https://www.ogtr.gov.au/resources/publications/biology-zea-mays-l-ssp-mays-maize-or-corn (accessed on 9 April 2019).
8. Hooda, K.S.; Khokhar, M.K.; Parmar, H.; Gogoi, R.; Joshi, D.; Sharma, S.S.; Yadav, O.P. Banded leaf and sheath blight of maize: Historical perspectives, current status and future directions. *Proc. Natl. Acad. Sci. India Sect. B Biol. Sci.* **2017**, *87*, 1041–1052. [CrossRef]
9. Sharma, R.C.; Srinivas, P.; Basta, B.K. Banded leaf and sheath blight of maize its epidemiology and management. In *Proceedings of a Maize Symposium*; NARC and CIMMYT: Kathmandu, Nepal, 2002; pp. 108–112.

10. Sharma, R.C. Banded leaf and sheath blight (*Rhizoctonia solani* f. sp *sasakii*) of maize. In *Stresses of Maize in Tropics*; Zaidi, P.H., Singh, N.N., Eds.; Director of Maize Research: Vishnupur, India, 2005; pp. 159–171.
11. Rani, D.V.; Reddy, N.P.; Devi, U.G. Management of maize banded leaf and sheath blight with fungicides and biocontrol agents. *Ann. Biol. Res.* **2013**, *4*, 179–184.
12. Ashraf, M.; Akram, N.A.; Arteca, R.N.; Foolad, M.R. The physiological, biochemical and molecular roles of brassinosteroids and salicylic acid in plant processes and salt tolerance. *Crit. Rev. Plant Sci.* **2010**, *29*, 162–190. [CrossRef]
13. Sakhabutdinova, A.R.; Fatkhutdinova, D.R.; Shakirova, F.M. Effect of salicylic acid on the activity of antioxidant enzymes in wheat under conditions of salination. *Appl. Biochem. Microbiol.* **2004**, *40*, 501–505. [CrossRef]
14. Kumar, R.R.; Sharma, S.K.; Goswami, S.; Singh, K.; Gadpayle, K.A.; Singh, G.P.; Pathak, H.; Rai, R.D. Transcript profiling and biochemical characterization of mitochondrial superoxide dismutase (mtSOD) in wheat (*Triticum aestivum*) under different exogenous stresses. *Aust. J. Crop Sci.* **2013**, *7*, 414–424.
15. Kaya, C.; Ugurlar, F.; Ashraf, M.; Ahmad, P. Salicylic acid interacts with other plant growth regulators and signal molecules in response to stressful environments in plants. *Plant Physiol. Biochem.* **2023**, *196*, 431–443. [CrossRef]
16. Wang, Y.; Mostafa, S.; Zeng, W.; Jin, B. Function and mechanism of jasmonic acid in plant responses to abiotic and biotic stresses. *Int. J. Mol. Sci.* **2021**, *22*, 8568. [CrossRef] [PubMed]
17. Gao, Q.M.; Kachroo, A.; Kachroo, P. Chemical inducers of systemic immunity in plants. *J. Exp. Bot.* **2014**, *65*, 1849–1855. [CrossRef] [PubMed]
18. Jabnoun-Khiareddine, H.; El-Mohamedy, R.S.; Abdel-Kareem, F.; Abdallah, R.A.; Gueddes-Chahed, M.; Daami-Remadi, M. Variation in chitosan and salicylic acid efficacy towards soil-borne and air-borne fungi and their suppressive effect of tomato wilt severity. *J. Plant Pathol. Microbiol.* **2015**, *6*, 2.
19. Liu, H.; Carvalhais, L.C.; Schenk, P.M.; Dennis, P.G. Effects of jasmonic acid signalling on the wheat microbiome differs between body sites. *Sci. Rep.* **2017**, *7*, 41766. [CrossRef] [PubMed]
20. Ahmad, F.; Singh, A.; Kamal, A. Salicylic acid–mediated defense mechanisms to abiotic stress tolerance. In *Plant Signaling Molecules*; Woodhead Publishing: Cambridge, UK, 2019; pp. 355–369.
21. Mohammed, A. Effects of Amistar and Dithane M-45, a systemic fungicide, on growth parameters and antioxidative enzymes of Maize (*Zea mays* L.). *Res. Rev. J. Bot. Sci.* **2014**, *3*, 13–19.
22. Mohammadi, M.A.; Cheng, Y.; Aslam, M.; Jakada, B.H.; Wai, M.H.; Ye, K.; He, X.; Luo, T.; Ye, L.; Dong, C.; et al. ROS and oxidative response systems in plants under biotic and abiotic stresses: Revisiting the crucial role of phosphite triggered plants defense response. *Front. Microbiol.* **2021**, *12*, 631318. [CrossRef] [PubMed]
23. Maschietto, V.; Lanubile, A.; De Leonardis, S.; Marocco, A.; Paciolla, C. Constitutive expression of pathogenesis-related proteins and antioxydant enzyme activities triggers maize resistance towards *Fusarium verticillioides*. *J. Plant Physiol.* **2016**, *200*, 53–61. [CrossRef] [PubMed]
24. Meshram, S.; Gogoi, R.; Bashyal, B.M.; Kumar, A.; Mandal, P.K.; Hossain, F. Expression analysis of maize genes during *Bipolaris maydis* infection and assessing their role in disease resistance and symptom development. *Indian J. Biotechnol.* **2020**, *19*, 82–93.
25. Rajput, V.D.; Harish Singh, R.K.; Verma, K.K.; Sharma, L.; Quiroz-Figueroa, F.R.; Meena, M.; Gour, V.S.; Minkina, T.; Sushkova, S.; Mandzhieva, S. Recent developments in enzymatic antioxidant defence mechanism in plants with special reference to abiotic stress. *Biology* **2021**, *10*, 267. [CrossRef]
26. Meshram, S.; Patel, J.S.; Yadav, S.K.; Kumar, G.; Singh, D.P.; Singh, H.B.; Sarma, B.K. Trichoderma mediate early and enhanced lignifications in chickpea during *Fusarium oxysporum* f. sp. *ciceris* infection. *J. Basic. Microbiol.* **2019**, *59*, 74–86. [CrossRef]
27. Gajewska, E.; Słaba, M.; Andrzejewska, R.; Skłodowska, M. Nickel-induced inhibition of wheat root growth is related to H_2O_2 production, but not to lipid peroxidation. *Plant Growth Regul.* **2006**, *49*, 95–103.
28. MacDonald, M.J.; D'Cunha, G.B. A modern view of phenylalanine ammonia lyase. *Biochem. Cell Biol.* **2007**, *85*, 273–282. [CrossRef] [PubMed]
29. Ghasemi, F.; Heidari, R.; Jameii, R.; Purakbar, L. Responses of growth and antioxidative enzymes to various concentrations of nickel in *Zea mays* leaves and roots. *Rom. J. Biol. Plant Biol.* **2013**, *58*, 37–49.
30. Tripathi, D.; Raikhy, G.; Kumar, D. Chemical elicitors of systemic acquired resistance—Salicylic acid and its functional analogs. *Curr. Plant Biol.* **2019**, *17*, 48–59. [CrossRef]
31. Oostendorp, M.; Kunz, W.; Dietrich, B.; Staub, T. Induced disease resistance in plants by chemicals. *Eur. J. Plant Pathol.* **2001**, *107*, 19–28. [CrossRef]
32. Saleem, M.; Fariduddin, Q.; Castroverde, C.D. Salicylic acid: A key regulator of redox signalling and plant immunity. *Plant Physiol. Biochem.* **2021**, *168*, 381–397. [CrossRef]
33. Zhang, S.W.; Yang, Y.; Wu, Z.M.; Li, K.T. Induced defense responses against *Rhizoctonia solani* in rice seedling by a novel antifungalmycin N2 from *Streptomyces* sp. N2. *Australas. Plant Pathol.* **2020**, *49*, 267–276. [CrossRef]
34. Littlejohn, G.R.; Breen, S.; Smirnoff, N.; Grant, M. Chloroplast immunity illuminated. *New Phytol.* **2021**, *229*, 3088–3107. [CrossRef] [PubMed]
35. Kora, A.J.; Mounika, J.; Jagadeeshwar, R. Rice leaf extract synthesized silver nanoparticles: An in vitro fungicidal evaluation against *Rhizoctonia solani*, the causative agent of sheath blight disease in rice. *Fungal Biol.* **2020**, *124*, 671–681. [CrossRef]

36. Burgess, L.W.; Phan, H.T.; Knight, T.E.; Tesoriero, L. *Diagnostic Manual for Plant Diseases in Vietnam*; Australian Centre for International Agricultural Research: Canberra, Australia, 2008.
37. Ahuja, S.C.; Payak, M.M. A field inoculation technique for evaluating maize germplasm to banded leaf and sheath blight. *Indian Phytopathol.* **1978**, *31*, 517–520.
38. Vincent, J.M. Distortion of fungal hyphae in the presence of certain inhibitors. *Nature* **1947**, *159*, 850. [CrossRef] [PubMed]
39. Payak, M.M.; Sharma, R.C. Disease rating scales in maize in India. In *Techniques of Scoring for Resistance to Important Diseases of Maize*; Indian Agricultural Research Institute: New Delhi, India, 1983; pp. 1–4.
40. Ahuja, S.C.; Payak, M.M. A rating scale for banded leaf and sheath blight of maize. *Indian Phytopathol.* **1983**, *36*, 338–340.
41. McKinney, H.H. A new system of grading plant diseases. *J. Agric. Res.* **1923**, *26*, 195–218.
42. Chikkappa, G.K. Genetic and Molecular Characterization of Early Maize (*Zea mays* L.) Parental Lines and Their Hybrids. Master's Thesis, Division of Genetics, Indian Agricultural Research Institute, New Delhi, India, 2005; pp. 22–23.
43. Dhindsa, R.S.; Plumb-Dhindsa, P.A.; Thorpe, T.A. Leaf senescence: Correlated with increased levels of membrane permeability and lipid peroxidation, and decreased levels of superoxide dismutase and catalase. *J. Exp. Bot.* **1981**, *32*, 93–101. [CrossRef]
44. Aebi, H. [13] Catalase in vitro. In *Methods in Enzymology*; Academic Press: Cambridge, MA, USA, 1984; Volume 105, pp. 121–126.
45. Bradford, M.M. A rapid and sensitive method for the quantitation of microgram quantities of protein utilizing the principle of protein-dye binding. *Anal. Biochem.* **1976**, *72*, 248–254. [CrossRef]
46. Lowry, O.; Rosebrough, N.; Farr, A.L.; Randall, R. Protein measurement with the Folin phenol reagent. *J. Biol. Chem.* **1951**, *193*, 265–275. [CrossRef] [PubMed]
47. Castillo, F.J.; Penel, C.; Greppin, H. Peroxidase release induced by ozone in Sedum album leaves: Involvement of Ca^{2+}. *Plant Physiol.* **1984**, *74*, 846–851. [CrossRef] [PubMed]
48. Koukol, J.; Conn, E.E. The metabolism of aromatic compounds in higher plants: IV. Purification and properties of the phenylalanine deaminase of *Hordeum vulgare*. *J. Biol. Chem.* **1961**, *236*, 2692–2698. [CrossRef] [PubMed]
49. Zucker, M. Induction of phenylalanine deaminase by light and its relation to chlorogenic acid synthesis in potato tuber tissue. *Plant Physiol.* **1965**, *40*, 779. [CrossRef]
50. Macri', F.; Di Lenna, P.; Vianello, A. Preliminary Research on Peroxidase, Polyphenoloxidase Activity and Phenol Content in Healthy and Infected Corn Leaves, Susceptible and Resistant to *Helminthosporium maydis* Race, T. *Riv. Patol. Veg.* **1974**, *10*, 109–121.
51. Abeles, F.B.; Forrence, L.E. Temporal and hormonal control of β-1,3-glucanase in *Phaseolus vulgaris* L. *Plant Physiol.* **1970**, *45*, 395–400. [CrossRef] [PubMed]
52. Williams, J.G.; Kubelik, A.R.; Livak, K.J.; Rafalski, J.A.; Tingey, S.V. DNA polymorphisms amplified by arbitrary primers are useful as genetic markers. *Nucleic Acids Res.* **1990**, *18*, 6531–6535. [CrossRef] [PubMed]
53. Mohanty, S.; Mahapatra, S.; Khandual, A.; Koshale, K.; Mukherjee, A. Impact of fungicides on *Rhizoctonia solani* Kuhn causing sheath blight disease of rice. *Int. J. Chem. Stud.* **2020**, *8*, 2759–2762. [CrossRef]
54. Rajput, L.S. Studies on Banded Leaf and Sheath Blight of Maize Caused by *Rhizoctonia solani* f. sp. *sasakii* EXNER. Ph.D. Thesis, University of Agricultural Sciences, Dharwad, India, 2013.
55. Meyer, M.C.; Bueno, C.J.; De Souza, N.L.; Yorinori, J.T. Effect of doses of fungicides and plant resistance activators on the control of Rhizoctonia foliar blight of soybean, and on *Rhizoctonia solani* AG1–IA in vitro development. *Crop Prot.* **2006**, *25*, 848–854. [CrossRef]
56. Utiamada, C.M. Eficiencia de fungicidas em aplicacao foliar no controle da mela (*Tanatephorus cucumeris*/*Rhizoctonia solani*), na cultura da soja. In Proceedings of the Reuniao de Pesquisa de Soja Da Regiao Central Do Brasil, Cuiaba, Brazil, 28–30 August 2000; Resumos. Embrapa Soja: Cuiaba, Brazil, 2000.
57. Miranda, V.D.; Porto, W.F.; Fernandes, G.D.; Pogue, R.; Nolasco, D.O.; Araujo, A.C.; Cota, L.V.; Freitas, C.G.; Dias, S.C.; Franco, O.L. Comparative transcriptomic analysis indicates genes associated with local and systemic resistance to *Colletotrichum graminicola* in maize. *Sci. Rep.* **2017**, *7*, 2483. [CrossRef] [PubMed]
58. Rudić, J.; Dragićević, M.B.; Momčilović, I.; Simonović, A.D.; Pantelić, D. In Silico study of superoxide dismutase gene family in potato and effects of elevated temperature and salicylic acid on gene expression. *Antioxidants* **2022**, *11*, 488. [CrossRef] [PubMed]
59. Liu, M.; Sui, Y.; Yu, C.; Wang, X.; Zhang, W.; Wang, B.; Yan, J.; Duan, L. Coronatine-Induced Maize Defense against *Gibberella* Stalk Rot by Activating Antioxidants and Phytohormone Signaling. *J. Fungi* **2023**, *9*, 1155. [CrossRef] [PubMed]
60. Ghorbel, M.; Brini, F.; Sharma, A.; Landi, M. Role of jasmonic acid in plants: The molecular point of view. *Plant Cell Rep.* **2021**, *40*, 1471–1494. [CrossRef]
61. Huang, H.; Ullah, F.; Zhao, Y. Mechanisms of ROS regulation of plant development and stress responses. *Front. Plant Sci.* **2019**, *10*, 440478. [CrossRef]
62. Vanacore, M.F.; Sartori, M.; Giordanino, F.; Barros, G.; Nesci, A.; García, D. Physiological Effects of Microbial Biocontrol Agents in the Maize Phyllosphere. *Plants* **2023**, *12*, 4082. [CrossRef]
63. Yan, Y.; Christensen, S.; Isakeit, T.; Engelberth, J.; Meeley, R.; Hayward, A.; Emery, R.N.; Kolomiets, M.V. Disruption of OPR7 and OPR8 reveals the versatile functions of jasmonic acid in maize development and defense. *Plant Cell* **2012**, *24*, 1420–1436. [CrossRef] [PubMed]

64. Battilani, P.; Lanubile, A.; Scala, V.; Reverberi, M.; Gregori, R.; Falavigna, C.; Dall'asta, C.; Park, Y.S.; Bennett, J.; Borrego, E.J.; et al. Oxylipins from both pathogen and host antagonize jasmonic acid-mediated defence via the 9-lipoxygenase pathway in *Fusarium verticillioides* infection of maize. *Mol. Plant Pathol.* **2018**, *19*, 2162–2176. [CrossRef] [PubMed]
65. Gorman, Z.; Christensen, S.A.; Yan, Y.; He, Y.; Borrego, E.; Kolomiets, M.V. Green leaf volatiles and jasmonic acid enhance susceptibility to anthracnose diseases caused by *Colletotrichum graminicola* in maize. *Mol. Plant Pathol.* **2020**, *21*, 702–715. [CrossRef] [PubMed]

current issues in molecular biology

MDPI

Article

Optimization of Plant Growth Regulators for In Vitro Mass Propagation of a Disease-Free 'Shine Muscat' Grapevine Cultivar

Si-Hong Kim [1,2,†], Mewuleddeg Zebro [3,†], Dong-Cheol Jang [1,4], Jeong-Eun Sim [3], Han-Kyeol Park [1], Kyeong-Yeon Kim [1], Hyung-Min Bae [5], Shimeles Tilahun [6,7,*] and Sung-Min Park [1,4,*]

1 Interdisciplinary Program in Smart Agriculture, Kangwon National University, Chuncheon 24341, Republic of Korea; tlghdek@kangwon.ac.kr (S.-H.K.); jdc@kangwon.ac.kr (D.-C.J.); fallphk@kangwon.ac.kr (H.-K.P.); rlaruddus09@naver.com (K.-Y.K.)
2 Smart Farm Research Center, KIST Gangneung, Institute of National Products, 679 Saimdang-ro, Gangneung 25451, Republic of Korea
3 Department of Plant Science, Gangneung-Wonju National University, Gangneung 25457, Republic of Korea; zmewuledeg@gmail.com (M.Z.); jamma2659@naver.com (J.-E.S.)
4 Department of Horticulture, Kangwon National University, Chuncheon 24341, Republic of Korea
5 Novagreen Business Centre, Kangwon National University, Chunchen 24341, Republic of Korea; novagreen2109@gmail.com
6 Agriculture and Life Science Research Institute, Kangwon National University, Chuncheon 24341, Republic of Korea
7 Department of Horticulture and Plant Sciences, Jimma University, Jimma 378, Ethiopia
* Correspondence: shimeles@kangwon.ac.kr (S.T.); parksm@kangwon.ac.kr (S.-M.P.); Tel.: +82-332506424 (S.M.P.)
† These authors contributed equally to this work.

Abstract: This study addresses the propagation challenges faced by 'Shine Muscat', a newly introduced premium grapevine cultivar in South Korea, where multiple viral infections pose considerable economic loss. The primary objective was to establish a robust in vitro propagation method for producing disease-free grapes and to identify effective plant growth regulators to facilitate large-scale mass cultivation. After experimentation, 2.0 μM 6-benzyladenine (BA) exhibited superior shoot formation in the Murashige and Skoog medium compared with kinetin and thidiazuron. Conversely, α-naphthaleneacetic acid (NAA) hindered shoot growth and induced callus formation, while indole-3-butyric acid (IBA) and indole-3-acetic acid (IAA) demonstrated favorable root formation, with IBA showing better results overall. Furthermore, inter simple sequence repeat analysis confirmed the genetic stability of in vitro-cultivated seedlings using 2.0 μM BA and 1.0 μM IBA, validating the suitability of the developed propagation method for generating disease-free 'Shine Muscat' grapes. These findings offer promising prospects for commercial grape cultivation, ensuring a consistent supply of healthy grapes in the market.

Keywords: auxin; cytokinin; ISSR marker; in vitro; 'Shine Muscat'

Citation: Kim, S.-H.; Zebro, M.; Jang, D.-C.; Sim, J.-E.; Park, H.-K.; Kim, K.-Y.; Bae, H.-M.; Tilahun, S.; Park, S.-M. Optimization of Plant Growth Regulators for In Vitro Mass Propagation of a Disease-Free 'Shine Muscat' Grapevine Cultivar. *Curr Issues Mol. Biol.* **2023**, *45*, 7721–7733. https://doi.org/10.3390/cimb45100487

Academic Editor: Jiyu Zhang

Received: 11 September 2023
Revised: 20 September 2023
Accepted: 21 September 2023
Published: 22 September 2023

1. Introduction

Grapevines, belonging to *Vitis* species, are economically important fruits with a wide range of varieties [1]. They are known for their high content of beneficial compounds such as resveratrol and anthocyanins, which possess anticancer, antioxidant, and anti-inflammatory properties [2,3]. However, grapevine viruses, fruit quality deterioration, and seedling necrosis present significant challenges to grape cultivation, particularly in Central and Southeast Asi [4]. To address these challenges, breeders are developing hybrids of European grapes (*V. vinifera*) and American grapes (*V. labrusca*) to enhance disease and pest resistance in vines.

'Shine Muscat', a diploid-bred grape cultivated in Japan, closely resembles European species and the yellow–green Alexandrian-type grape [5]. In Korea, the cultivation of Shine

Muscat, an interspecific diploid hybrid of *V. labruscana* Bailry and *V. vinifera* L., has been rapidly expanding due to increasing consumer preference [6,7]. This cultivar is known for its high carbohydrate content and can be enjoyed by peeling the skin, similar to European grapes. 'Shine Muscat' has gained popularity in Korea and shares similar characteristics with cultivars developed in Japan [7,8].

While the cultivation areas of commonly grown grape varieties such as 'Campbell Early', 'Geobong', and 'MBA' are decreasing, there is a notable increase in the cultivation areas of new high-quality varieties like 'Shine Muscat' [6]. However, a study by Kim et al. [9] reported a high prevalence of grapevine viruses (91.0%) in the primary grape-producing regions of Korea, regardless of the specific region or cultivar.

To ensure the competitiveness of the grapevine industry, establishing a virus-free and disease-free grapevine cultivation system is crucial. In vitro culture plays a vital role in this regard, and it requires the identification of appropriate medium composition [10]. The in vitro culture method consists of two stages: shoot growth and rooting. Previous studies have highlighted the influence of different types and concentrations of plant growth regulators (PGRs) on shoot growth, rooting efficiency, and genotype variations in grapevines [11,12]. Therefore, the efficiency of in vitro propagation can vary depending on the genetic characteristics of the species, necessitating the evaluation and selection of optimal PGR conditions.

In *Vitis* species and other plants, specific cytokinins and auxins play distinct roles in promoting shoot, bud, or root formation during in vitro culture. Cytokinins, such as 6-benzyladenine (BA), kinetin, and zeatin, induce shoot proliferation by promoting the development of multiple shoots from explants, including shoot tips and nodal segments [13,14]. They stimulate cell division and meristem growth, resulting in the formation of new shoots [15]. Additionally, cytokinins help in bud initiation and outgrowth, leading to the development of lateral shoots [16]. On the other hand, auxins like indole-3-butyric acid (IBA), indole-3-acetic acid (IAA), and naphthalene acetic acid (NAA) are crucial for root induction and elongation [17]. They stimulate the formation of adventitious roots from various explants, including shoots and leaf tissues [18]. Moreover, endogenous auxins in *Vitis* species regulate the intricate balance between shoot and root development by influencing various growth processes, such as apical dominance, root formation, and adventitious root initiation [19]. Their precise transport and distribution within the plant, facilitated by PIN proteins, contribute to the spatial arrangement of shoots and roots [20,21]. Understanding the role of endogenous auxins is essential for optimizing growth and development strategies, such as in vitro micropropagation, where exogenous application of auxins can manipulate shoot and root formation for desired outcomes [22]. The balance between cytokinins and auxins is essential in regulating shoot-to-root conversion, as a higher cytokinin-to-auxin ratio promotes shoot development, while a higher auxin-to-cytokinin ratio induces rooting [23,24].

Additionally, in vitro plant propagation methods have been associated with the occurrence of somaclonal mutations [25–27]. Since grape is a perennial crop, genetic variations resulting from in vitro propagation can only be observed at maturity and fruiting stages [28]. Therefore, early detection of genetic variations in in vitro-raised plants is crucial. Molecular markers like ISSR (inter-simple sequence repeats), ISS (interspersed repetitive sequence), RAPD (random amplified polymorphic DNA), and AFLP (amplified fragment length polymorphism) are commonly used to assess genetic variation in plants. ISSR involves amplifying specific regions between microsatellite repeats, providing a cost-effective and easy-to-use method to evaluate genetic diversity [29]. ISS, on the other hand, targets interspersed repetitive elements, enabling researchers to study different genomic regions [30]. RAPD utilizes random primers to amplify DNA fragments, making it a quick and less expensive technique for genetic diversity analysis [31]. AFLP utilizes restriction enzymes and PCR to create DNA fragments, offering high reproducibility and sensitivity [32].

In this study, the genetic stability of 'Shine Muscat' was assessed using inter simple sequence repeat (ISSR) markers. ISSR markers provide a reliable and easily performed method for evaluating genetic stability in grapes as demonstrated in previous studies [33,34]. Hence, the objective of this study was to determine the optimal PGR composition for mass propagation using in vitro culture and to assess the genetic stability of in vitro-raised plants using ISSR markers, aiming to facilitate the large-scale commercial use of 'Shine Muscat'.

2. Materials and Methods

2.1. Culture Establishment and Media Composition

Fifty healthy 'Shine Muscat' plants, cultivated in a greenhouse for one year, were selected as the source materials for this study. Nodal explants, each carrying a 2 cm axillary bud, were carefully collected from actively growing young shoots, located approximately three to six nodes below the tip. Aseptic techniques were used to establish the in vitro culture. Initially, the nodal segments were washed with running tap water for ten minutes, followed by surface sterilization. This involved immersing the segments in 70% ethanol for 1 min and subsequently treating them with 2% (*w/v*) NaOCl for 10 min. To remove NaOCl, the segments were rinsed five times with sterile distilled water in a fume hood.

To initiate the culture, the sterilized nodal segments were placed in Pyrex rimless glass culture tubes (22 mm × 175 mm) containing 1× MS medium. The medium was supplemented with various concentrations of cytokinins (6-benzyladenine, kinetin, and thidiazuron) ranging from 1.0 to 16.0 μM and compared with the control (MS medium). Additionally, sucrose was added to the medium at a concentration of 30 g L^{-1}, and Difco Bacto agar was used to solidify it at a concentration of 8 g L^{-1}. After an initial culture period of four weeks, shoot organogenesis was assessed for each treatment, with thirty explants included in each group. Throughout the culture duration, controlled conditions were maintained, including a temperature of 25 ± 1 °C, a light intensity of 36.59 μmol $m^{-2}s^{-1}$ photosynthetic photon flux density (PPFD), and a photoperiod of 16 h of daylight.

2.2. Shoot Multiplication and Shoot-Related Data Collection

Following the initial four weeks of culture, micro-cuttings obtained from the first-generation in vitro shoots were transferred to another culture tube. The subculture medium composition remained unaltered, except for the omission of kinetin (KIN), and was compared with the control (MS medium). Subsequently, these subcultured micro-cuttings were subjected to the same controlled culture conditions as previously described, aiming to identify the most favorable conditions for multiple shoot proliferation. The evaluation involved recording the number of shoots and nodes per explant, as well as measuring the main shoot length after three weeks of subculture. Each treatment for shoot multiplication was replicated four times, and each replication comprised six explants.

2.3. Root Induction, Root-Related Data Collection, and Acclimatization

To investigate the influence of plant growth regulators (PGRs) on root formation, the nodal segments of the shoots were cut to 1 cm, and the shoots were subsequently subcultured in MS medium supplemented with 2.0 μM BA. Micro-cuttings measuring over 0.5 cm in length were then transferred to culture vessels containing different types of auxins or PGR-free media. The experiment involved testing three types of auxins (IAA, IBA, NAA) at concentrations of 0.25, 0.5, 1.0, 2.0, and 4.0 μM to identify the most effective PGR conditions for root induction compared with the control (MS medium). The same culture conditions utilized for shoot proliferation were maintained throughout the experiment. After four weeks of rooting treatments, data on growth parameters were collected following the methodology established by Kim et al. [9]. Each root induction treatment was replicated four times, with six explants included in each replication.

Subsequently, plants that successfully developed roots were transferred to a 72-hole plug tray, utilizing a peat moss-to-perlite ratio of 1:1 (*v/v*), and covered with a plastic container. To ensure successful acclimatization, the humidity level was maintained at 90–95% and gradually reduced over three weeks. During this period, the plants were kept at a constant temperature of 23 ± 1 °C and received a light intensity of 53.66 μmol $m^{-2}s^{-1}$

PPFD for 12 h daily. Once the plants were sufficiently hardened off, they were transplanted into 1 L plastic pots filled with a suitable potting medium and placed in the glasshouse at Kangwon National University, Republic of Korea.

2.4. Genetic Stability Analysis Using ISSR Markers

In this study, ten explants derived from the optimal in vitro culture concentration were randomly chosen for evaluating genetic stability utilizing ISSR markers. Total genomic DNA was isolated from the young leaf tissue of each plant following the TaKaRa MiniBEST Plant Genomic DNA Extraction Kit instructions. The concentration of the extracted DNA was diluted to 50 ng μL^{-1}, as determined with a nanodrop spectrophotometer (MicroDigital, Seongnam, Republic of Korea).

For the ISSR analysis, a preliminary experiment with 15 ISSR primers (UBC primer Set No. 9, University of British Columbia, Canada) was conducted, and subsequently, 10 primers yielding clear bands were selected for further ISSR analysis (Table 1). The PCR reaction solution consisted of 2 µL of genomic DNA, Accupower premix (Bioneer, Daejeon, Republic of Korea), and 10 pmol primers, making a total volume of 20 µL. PCR amplifications were carried out in a Dice® Touch thermal cycler (TaKaRa, Otsu, Japan) with the following cycling parameters: initial denaturation at 94 °C for 5 min, followed by 35 cycles of denaturation at 94 °C for 30 s, annealing at 48, 50, or 55 °C for 45 s, amplification at 72 °C for 2 min, and a final extension step of 7 min at 72 °C. The resulting PCR amplicons were separated using 1.5% agarose gel electrophoresis and visualized using a GD-1000 gel documentation system (Axygen, San Francisco, CA, USA). Specific fragments within the size range of 100 to 2000 base pairs, consistently generated and clearly resolved, were selected for analysis. These fragments were then utilized to identify the presence or absence of ISSR markers in each sample.

Table 1. List of ISSR primers used in this study and number of band classes generated.

ISSR Primer	Annealing Temperature (°C)	Nucleotide Sequence (5′-3′)	No. of Distinct Band Classes	Total Number of Bands Amplified	% Similarity
UBC 808	50	$(AG)_8C$	8	80	100
UBC 812	50	$(GA)_8C$	8	80	100
UBC 815	55	$(CT)_8G$	4	40	100
UBC 823	50	$(TC)_8C$	4	40	100
UBC 825	55	$(AC)_8T$	9	90	100
UBC 836	50	$(AG)_7ACYA$	8	80	100
UBC 840	50	$(GA)_8Y$	8	80	100
UBC 873	50	$(GACA)_4$	13	130	100
UBC 878	55	$(GGAT)_4$	8	80	100
UBC 881	53	$GGGT(GGGGT)_2G$	8	80	100

To assess the genetic relationships among the samples, Jaccard's similarity coefficient was calculated to generate a similarity matrix. A cluster analysis was performed using the unweighted pair group method with arithmetic averages (UPGMA), and a dendrogram was constructed using NTSYS-PC Ver. 2.1 software.

2.5. Statistical Analysis

Statistical analyses were conducted using SPSS 25.0, a software program (SPSS Inc., Chicago, IL, USA). Analysis of variance was used to assess the effects of the treatments. When the treatment effects were statistically significant ($p < 0.05$), means were compared using Duncan's multiple range test.

3. Results and Discussion

This study addresses the propagation challenges faced by the 'Shine Muscat' grapevine cultivar in South Korea. The primary objectives were to establish a robust in vitro propagation method for producing disease-free grapes and to identify effective plant growth regulators to facilitate large-scale mass cultivation. An overall summary of this study is shown in Figure 1.

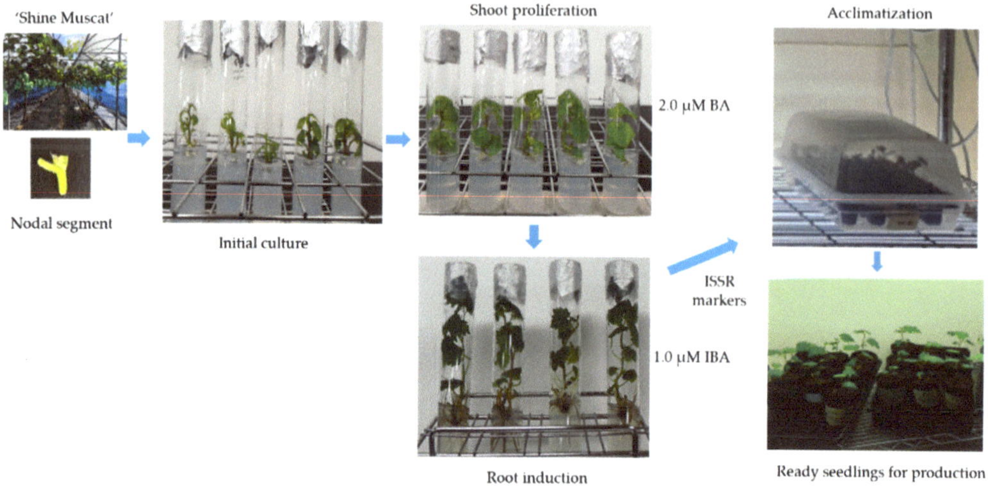

Figure 1. Graphic summary of this study.

3.1. Assessment of the Impact of Hormones on the Initiation of Culture

The outcomes of the initial responses of the nodal segments to different cytokinins after a four-week incubation period are presented in Table 2. Within two weeks, sprouting buds became apparent in the nodal segments. The regeneration rate of the nodal segments varied based on the type and concentration of cytokinins used in the experiment, ranging from 30.4% to 100% (Table 2). Nodal segments cultured in a cytokinin-free medium did not display any shoot or bud formation. However, when nodal segments were cultivated in media containing cytokinins, callus formation was observed after ten days, and the development of new axillary buds was frequently noted after three weeks. This investigation contradicts the study conducted by Murashige and Skoog [35]. This discrepancy could arise from variations in plant responses to nutrients and growth regulators. Unique physiological and biochemical traits inherent to different plant species might contribute to their distinct reactions to combinations of nutrients and cytokinins. Another potential explanation lies in the intricate nature of plant morphogenesis, governed by a multitude of factors such as signaling pathways, gene expression, and hormonal interactions [36]. Nevertheless, additional research is required to validate the results obtained in the present study. Throughout the four-week duration, the medium supplemented with BA exhibited the highest response in terms of shoot and node formation. The regeneration rate with varying concentrations of BA in the culture medium ranged from 90.4% to 97.8%, and there were no significant differences observed. Notably, an increase in BA concentration led to a corresponding increase in the number of shoots and nodes per nodal segment.

The initial culture efficiency of the medium with TDZ addition ranged from 86.4% to 94.5%, which was slightly lower than that of the BA-supplemented medium, but the difference was not significant. In contrast, the medium treated with KIN displayed a lower survival rate of 37.8% to 67.4% compared with the cytokinin-containing medium, and it showed no significant difference from the non-treated section without cytokinins. Prior studies on various types of vines have also demonstrated that KINs do not yield favorable results compared with other cytokinin hormones [37,38]. Therefore, the application of various concentrations of KIN in this study did not have a positive effect on the regeneration or sprout induction of 'Shine Muscat' compared with BA and TDZ.

Table 2. Effect of cytokinin type and concentration of plant growth regulators on culture establishment in the 'Shine Muscat' cultivar.

Plant Growth Regulators/Concentrations	Bud Induction Rate/Nodal Segment (%)	No. of Shoots/ Nodal segment	No. of Nodes/ Nodal Segment	Length of Main Shoot (cm)
Control (MS medium)	54.3 b	0.72 d	2.31 d	1.31 c
BA 1.0 μM	90.4 a	1.05 bc	3.79 abc	2.07 ab
BA 2.0 μM	97.8 a	1.21 abc	4.29 a	2.37 a
BA 4.0 μM	92.3 a	1.31 ab	4.32 a	2.21 ab
BA 8.0 μM	93.7 a	1.38 a	4.43 a	1.93 ab
BA 16.0 μM	94.3 a	1.29 ab	4.57 a	1.62 bc
KIN 1.0 μM	53.7 b	0.67 d	2.27 d	1.69 bc
KIN 2.0 μM	60.0 b	0.65 d	2.84 bcd	1.81 bc
KIN 4.0 μM	43.7 b	0.57 d	2.47 d	2.24 ab
KIN 8.0 μM	37.5 b	0.51 d	2.34 d	1.91 ab
KIN 16.0 μM	30.4 b	0.43 d	1.97 b	1.57 bc
TDZ 1.0 μM	89.4 a	0.97 c	3.64 abc	1.97 ab
TDZ 2.0 μM	92.3 a	1.04 bc	4.21 a	2.35 a
TDZ 4.0 μM	94.5 a	1.20 abc	4.32 a	2.32 a
TDZ 8.0 μM	92.4 a	1.09 bc	4.01 a	2.26 ab
TDZ 16.0 μM	86.4 a	0.94 c	3.81 abc	2.14 ab

Note: Means followed by the same letter within each column are not significantly different at $p < 0.05$.

In the present study, both BA and TDZ showed promising results, with BA exhibiting the highest response in terms of shoot and node formation. This finding aligns with the studies conducted on '*Dracocephalum forrestii*' and 'Rhododendron', where cytokinins have been widely used and proven effective in inducing shoot proliferation [14,15]. Cytokinins play a crucial role in promoting cell division, bud formation, and shoot elongation, which are essential for successful micropropagation [39]. On the other hand, KIN was less effective in promoting regeneration and bud formation in 'Shine Muscat'. KINs have been reported to have limited effects on shoot induction and multiplication compared with other cytokinins [40]. The lower survival rate observed with KIN treatments in this study is in line with previous studies that have suggested its limited utility for in vitro propagation [41].

3.2. *Assessment of the Impact of Hormones on the Growth and Multiplication of Shoots*

Additional investigations were undertaken to identify the most effective concentrations of cytokinins, specifically, BA and TDZ, for shoot multiplication, building on the positive outcomes observed during the initial culture phase. The subculture responses of the explants obtained from the initial cultures displayed variations depending on the type and concentration of cytokinins, as outlined in Table 3. Although there were no significant differences between BA and TDZ, media supplemented with BA demonstrated a good response (0.2% to 3.7%). Although the untreated plot induced 0.41 shoots, the medium containing BA and TDZ led to the induction of 1.04 to 1.89 shoots. Notably, the formation of additional buds and nodes exhibited significant differences based on the type and concentration of cytokinin used. Comparatively, media supplemented with cytokinins resulted in a higher number of shoots and nodes compared with cytokinin-free media at all concentrations. This effect could be due to the stimulation of endogenous auxins, as cytokinins induce the synthesis of auxins.

TDZ showed a positive impact on the main shoot development, and there were no adverse effects on shoot and node formation. The subculture responses in media containing TDZ ranged from 83.3% to 89.7% in terms of shoot induction, with the number of shoots, nodes, and length of the main shoot varying from 1.04 to 1.58, 4.32 to 5.27, and 1.48 to 1.58, respectively. Other criteria exhibited an increasing trend as the TDZ concentration reached 8.0 μM, followed by a decline at higher concentrations. It is important to note that high cytokinin concentration with TDZ could lead to shoot fascination and inhibition of shoot elongation in grapevines, although these aspects were not investigated in this study. Nevertheless, previous studies have reported the positive effects of TDZ on multiple shoot formation in various Vitis species, suggesting its potential usefulness in the in vitro propagation of 'Shine Muscat' [22,23]. Therefore, these findings indicate that TDZ, like BA, can offer valuable benefits for the in vitro propagation of 'Shine Muscat'.

Table 3. Effect of cytokinin type and concentration of plant growth regulators on shoot multiplication in the 'Shine Muscat' cultivar.

Plant Growth Regulators/Concentrations	Bud Induction Rate/Nodal Segment (%)	No. of Shoots/ Nodal Segment	No. of Nodes/ Nodal Segment	Length of Main Shoot (cm)
Control (MS medium)	42.1 b	0.41 d	1.38 c	0.51 b
BA 1.0 μM	93.2 a	1.88 abc	5.24 ab	1.53 ab
BA 2.0 μM	92.1 a	2.14 a	6.51 a	1.69 a
BA 4.0 μM	84.3 a	1.89 abc	5.87 ab	1.45 ab
BA 8.0 μM	82.9 a	1.54 bc	5.54 ab	1.33 ab
BA 16.0 μM	81.5 a	1.31 bcd	5.42 ab	1.21 ab
TDZ 1.0 μM	89.7 a	1.04 c	4.32 b	1.48 ab
TDZ 2.0 μM	85.4 a	1.24 bc	4.47 ab	1.51 ab
TDZ 4.0 μM	83.3 a	1.31 bcd	4.87 ab	1.55 ab
TDZ 8.0 μM	81.9 a	1.58 bc	5.27 ab	1.58 ab
TDZ 16.0 μM	81.3 a	1.39 bcd	5.09 ab	1.49 ab

Note: Means followed by the same letter within each column are not significantly different at $p < 0.05$.

Throughout the initial culture period, subculture responses in media containing BA exhibited consistent performance across all concentrations. Moreover, the number of buds and nodes increased at low concentrations, especially at 2.0 μM BA. Among the various hormonal conditions tested, the highest number of shoots and nodes per explant was achieved with media containing 2.0 μM BA, resulting in 2.14 shoots and 6.51 nodes per explant. These values were 1.35- and 1.24-fold higher, respectively, than TDZ at 8.0 μM. The optimal types and concentrations of cytokinins for in vitro propagation can vary significantly among different plant species [24,25]. However, previous studies have emphasized the positive effects of BA on multiple shoot formation in various plants [26,27]. In grapevines, the ideal concentration of BA for shoot multiplication ranged from 2.22 to 11.1 μM. Concentrations above this range inhibited shoot growth and promoted callus formation, reducing the number of shoots and nodes [28]. Another study reported that a BA concentration of 2.5 μM yielded the best results in terms of growth per explant, while concentrations higher than 2.5 μM decreased these values [42]. Importantly, in this study, the initial culture efficiency at 2.0 μM BA was satisfactory, and when subcultured with 2.0 μM BA, the shoots derived from these cultures exhibited uniform growth and healthy green leaves. These findings strongly support the appropriateness of using 2.0 μM BA for sustainable in vitro propagation of 'Shine Muscat'.

This research emphasizes the significance of cytokinins, specifically BA and TDZ, in promoting shoot multiplication in 'Shine Muscat' grapevines. Both BA and TDZ supplementation led to a higher number of shoots and nodes compared with media without cytokinins. These findings align with previous studies that extensively used cytokinins for in vitro shoot multiplication [13,14,16]. Furthermore, this study's findings on 'Shine Muscat' grapevines validate the beneficial effects of TDZ on main shoot development and multiple shoot formation, consistent with earlier research [43,44]. TDZ is well-known for its ability to stimulate shoot proliferation in various woody plants, as demonstrated in previous studies [45,46].

3.3. Investigation of the Influence of Hormones on the Initiation and Development of Roots

'Shine Muscat' explants were subcultured using 2.0 μM BA, and the response to root growth was evaluated by treating them with different types and concentrations of auxins. The outcomes demonstrated variations in root formation based on the type and concentration of auxins used, as presented in Table 4. When plants produced with 2.0 μM BA were cultured in a medium without any plant growth regulator, a lower rate of root formation was observed. Additionally, the lower part of the explants turned black within two weeks or experienced decreased survival rates toward the late stage of culture in the auxin-free medium.

Table 4. Effect of auxin type and concentration of plant growth regulators on root induction in the 'Shine Muscat' cultivar.

Plant Growth Regulators/Concentrations	% Rooting	% Callusing	Root Number	Root Length (cm)	Shoot Length (cm)
Control (MS medium)	37.4 b	0 d	0.39 d	0.68 c	0.57 c
IAA 0.25 μM	69.7 ab	0 d	2.18 bcd	1.37 a	1.11 abcd
IAA 0.50 μM	70.5 ab	0 d	2.49 abcd	1.24 ab	1.18 abcd
IAA 1.0 μM	70.8 ab	0 d	2.59 abcd	1.18 abc	1.23 abcd
IAA 2.0 μM	67.8 ab	0 d	2.71 abcd	1.04 abc	1.46 abc
IAA 4.0 μM	65.3 ab	0 d	2.54 abcd	0.87 bc	1.39 abc
IBA 0.25 μM	81.1 a	0 d	2.29 bcd	1.12 abc	1.53 ab
IBA 0.50 μM	83.7 a	0 d	2.79 abcd	1.24 ab	1.47 abc
IBA 1.0 μM	86.8 a	0 d	3.42 ab	1.29 ab	1.63 a
IBA 2.0 μM	70.4 ab	11.8 cd	2.57 abcd	0.91 bc	1.22 abcd
IBA 4.0 μM	69.7 ab	24.8 c	1.84 c	0.89 bc	1.09 bcd
NAA 0.25 μM	66.1 ab	47.5 b	1.80 c	0.84 bc	1.08 bcd
NAA 0.50 μM	68.2 ab	63.7 ab	2.61 abcd	1.04 abc	0.89 de
NAA 1.0 μM	70.3 ab	70.5 ab	2.97 abcd	1.12 abc	0.00 f
NAA 2.0 μM	71.8 ab	73.8 ab	3.71 a	0.94 abc	0.00 f
NAA 4.0 μM	75.7 ab	80.2 a	3.34 abc	0.73 c	0.00 f

Note: Means followed by the same letter within each column are not significantly different at $p < 0.05$.

Among the three auxins tested at various concentrations, 1.0 μM IBA displayed the highest rooting percentage (86.8%) and the second-highest number of roots (3.42) without any callus formation (Table 4). At the same concentration level, IAA and NAA produced 2.59 and 2.97 roots with a reduced rooting rate of 16 and 16.5%, respectively, compared with IBA. However, NAA treatments induced callus formation, and higher concentrations of NAA significantly inhibited shoot growth compared with the other auxins. Conversely, IAA did not lead to callus formation or growth inhibition and hyperhydricity at lower concentrations, but the rooting rate and other important criteria were lower than those of IBA.

In this study, the introduction of exogenous auxin led to the initiation of new primordia formation, specifically promoting the development of "xylem pole" vasculature. The establishment of this specialized vasculature played a crucial role in facilitating directional auxin transport for de novo root formation [47]. The observed differences in root formation among the auxins in 'Shine Muscat' are consistent with findings from other studies on woody plants, including Vitis sp. [48,49]. The choice of the appropriate auxin for root induction varies among plant species, and IBA is often preferred due to its more consistent and reliable rooting response [50]. The negative effects of NAA on shoot and root induction align with reports of its toxicity and callus formation issues in certain plant species [51].

Extensive studies have demonstrated that the rooting of explants in woody plants is greatly influenced by the type and concentration of auxins used [29]. Optimal conditions for auxin-induced rooting can vary among plant species and genotypes. While NAA application has shown positive results for root induction in some species, it has also been associated with issues such as toxicity and callus formation, leading to unsuccessful acclimatization of plantlets in other studies [30,31]. Our study observed similar side effects with NAA application in 'Shine Muscat', including callus formation and failure of shoot and root induction, which worsened at higher concentrations. These findings are consistent with previous reports and suggest a potential genetic sensitivity of 'Shine Muscat' to NAA toxicity. Additionally, the IAA application was found to be less effective than IBA for root induction in 'Shine Muscat'. These results could be attributed to the physical properties of

IAA, as it can decompose under the medium and light conditions used, leading to reduced interactions between cytokinins and auxin hormones [32]. While further investigations are required to fully understand these observations, our results suggest that the application of IBA may be more effective for the in vitro propagation of 'Shine Muscat', offering stability, persistence, and protection against toxicity when following our protocol.

This study highlights the significance of using suitable hormones for the successful cultivation of plants during in vitro propagation. Figure 2 illustrates the effects of different plant growth regulators, specifically cytokinins and auxins, on shoot development and root growth. A survival test involving fifty plantlets of 'Shine Muscat' produced with 2.0 μM BA for shoot induction and 1.0 μM IBA during acclimatization (unpublished data) showed that only three plantlets died, indicating no significant survival issues. These findings underscore the effectiveness of utilizing 2.0 μM BA and 1.0 μM IBA for shoot multiplication and root induction in the in vitro propagation of 'Shine Muscat'.

Figure 2. Photographs showing the effects of different plant growth regulators, specifically, cytokinins and auxins, on shoot development and root growth: (**A**) KIN at 16.0 μM resulted in the lowest shoot number; (**B**) BA at 2 μM led to the highest shoot number; (**C**) NAA at 4.0 μM resulted in the highest percentage of callus formation; (**D**) IBA at 1.0 μM led to the highest percentage of root formation.

3.4. Evaluation of Genetic Stability of 'Shine Muscat' Using ISSR Markers

To assess the genetic stability of ten in vitro-grown plants, ISSR markers were used, and their profiles were compared to the parental plants. Out of the 10 ISSR primers tested, all successfully amplified genomic DNA, resulting in 78 distinct and scorable bands ranging in size from 100 to 2000 bp, with an average of 7.4 bands per primer. The ISSR analysis revealed homogeneity between the regenerated plants and the parental plants, as shown in Figure 3. These results indicate that 'Shine Muscat' maintained its genetic stability throughout in vitro propagation, with no noticeable morphological differences observed.

These findings align with previous studies that have shown genetic uniformity among in vitro-raised plantlets of various plant species [33,34].

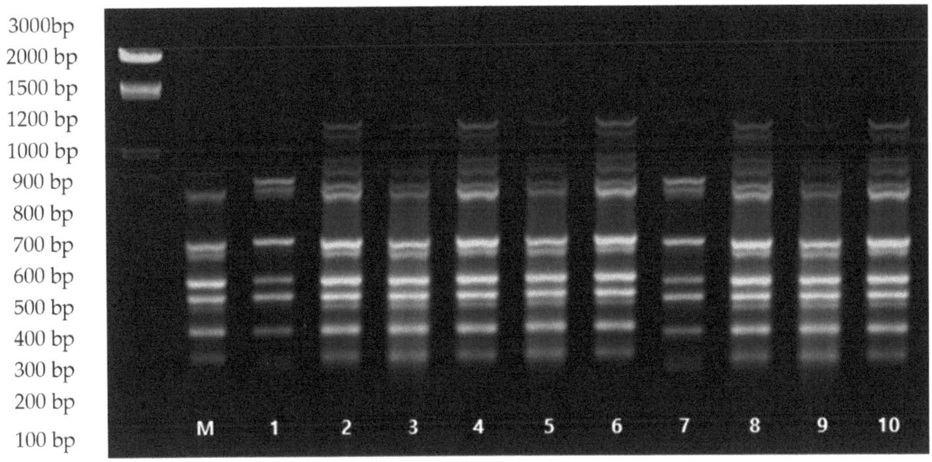

Figure 3. ISSR amplification profiles of the 'Shine Muscat' grape cultivar using primer UBC-873. M: mother plant, 1–10: in vitro-raised plants using 2.0 μM BA and 1.0 μM IBA.

The occurrence of variations during in vitro propagation depends significantly on the choice of explants and the regeneration method used [35]. Nodal segments are commonly preferred due to their high regeneration efficiency, and plants derived from adventitious buds within the nodal segment show a low risk of genetic variation [24,36]. However, the excessive use of synthetic plant growth hormones can potentially induce somatic mutations in the obtained tissues [37]. In this study, ISSR markers proved to be valuable tools for assessing the genetic fidelity of the in vitro-regenerated plants. The genetic profiles determined using DNA markers demonstrated that genetic fidelity was well-maintained during somatic embryogenesis, with only rare instances of deviation [38]. Moreover, the analysis of the hormonal conditions using ISSR markers revealed no differences between the in vitro-obtained plants and the parental plant, indicating that the hormone treatment used did not lead to significant genetic deviations. Thus, the results of the ISSR analysis provide robust evidence supporting the genetic stability of 'Shine Muscat' during in vitro propagation. The homogeneity observed between the regenerated plants and the parental plants confirms the effectiveness of the in vitro culture method using nodal segments as explants. This finding is consistent with previous studies reporting genetic stability among in vitro-raised plantlets of various plant species [33,34], which is crucial for preserving desired traits during commercial propagation of elite varieties.

The use of ISSR markers as a tool for evaluating genetic stability is particularly valuable due to their advantages over other molecular marker techniques like RAPD, ISS, or AFLP. ISSR markers do not require prior knowledge of the target DNA sequence and are highly versatile, making them applicable to a wide range of plant species [29,52].

4. Conclusions

The successful production of genetically identical plants using in vitro plant propagation holds significant importance. In our study, we discovered that using 2.0 μM BA for shoot induction and 1.0 μM IBA for root induction in 'Shine Muscat' resulted in highly effective plantlet production. This hormone treatment led to a higher number of shoots and roots compared with the hormone-free medium and other types of growth hormone applications. Notably, when using 2.0 μM BA for shoot induction and 1.0 μM IBA for root induction, no instances of somaclonal mutation were observed between the in vitro-grown plants. We utilized ISSR markers to identify somatic mutations, and the results confirmed the genetic stability of the propagated plants. Consequently, our established protocol for in vitro plant propagation can be reliably used to ensure the essential genetic stability required for the commercial multiplication of the 'Shine Muscat' cultivar.

Author Contributions: S.-H.K. and M.Z.: Conducted the experiments and wrote the original manuscript draft. J.-E.S., H.-K.P., K.-Y.K. and H.-M.B.: Helped conduct the experiment and write the original manuscript draft. S.T.: Writing—review and editing. D.-C.J. and S.-M.P.: Conceptualization, supervision, funding acquisition, and writing—review and editing. All authors have read and agreed to the published version of the manuscript.

Funding: This study was supported by a 2023 Research Grant from Kangwon National University and financial support from the Ministry of Education (NRF- 2022R1I1A1A01054769).

Institutional Review Board Statement: Not applicable.

Informed Consent Statement: Not applicable.

Data Availability Statement: All data sets are available upon reasonable request from the corresponding author.

Conflicts of Interest: The authors declare no conflict of interest.

References

1. Fortes, A.M.; Pais, M.S. Grape (*Vitis* Species). In *Nutritional Composition of Fruit Cultivars*; Elsevier: Amsterdam, The Netherlands, 2016; pp. 257–286.
2. Doshi, P.; Adsule, P.; Banerjee, K.; Oulkar, D. Phenolic Compounds, Antioxidant Activity and Insulinotropic Effect of Extracts Prepared from Grape (*Vitis vinifera* L.) Byproducts. *J. Food Sci. Technol.* **2015**, *52*, 181–190. [CrossRef] [PubMed]
3. Carmona-Jiménez, Y.; Palma, M.; Guillén-Sánchez, D.A.; García-Moreno, M.V. Study of the Cluster Thinning Grape as a Source of Phenolic Compounds and Evaluation of Its Antioxidant Potential. *Biomolecules* **2021**, *11*, 227. [CrossRef] [PubMed]
4. Mannini, F.; Digiaro, M. The Effects of Viruses and Viral Diseases on Grapes and Wine. In *Grapevine Viruses: Molecular Biology, Diagnostics and Management*; Springer: Cham, Switzerland, 2017; pp. 453–482.
5. Shirasawa, K.; Hirakawa, H.; Azuma, A.; Taniguchi, F.; Yamamoto, T.; Sato, A.; Ghelfi, A.; Isobe, S.N. De Novo Whole-Genome Assembly in an Interspecific Hybrid Table Grape,'Shine Muscat'. *DNA Res.* **2022**, *29*, dsac040. [CrossRef]
6. Lim, Y.-S.; Hassan, O.; Chang, T. First Report of Anthracnose of Shine Muscat Caused by Colletotrichum Fructicola in Korea. *Mycobiology* **2020**, *48*, 75–79. [CrossRef] [PubMed]
7. Yoon, H.-K.; Kang, K.-J.; Park, H.; Lee, U.-S.; Seo, J.-H. Comparison of Fruit Quality According to Harvest Time of Shine Muscat Grape (*Vitis labruscana* Bailey× *V. vinifera* L.). In Proceedings of the XXXI International Horticultural Congress (IHC2022): International Symposium on Integrative Approaches to Product Quality in 1353, Angers, France, 14–20 August 2022; pp. 317–322.
8. Yamada, M.; Yamane, H.; Sato, A.; Hirakawa, N.; Iwanami, H.; Yoshinaga, K.; Ozawa, T.; Mitani, N.; Shiraishi, M.; Yoshioka, M. New Grape Cultivar 'Shine Muscat'. *Bull. Natl. Inst. Fruit Tree Sci.* **2008**, *7*, 21–38.
9. Si-Hong, K.I.M.; Jeong, S.-H.; Jae-Yun, H.E.O. Incidence of 14 Grapevine Viruses in Korean Vineyards. *Not. Bot. Horti Agrobot. Cluj-Napoca* **2021**, *49*, 1–7.
10. Hamdeni, I.; Louhaichi, M.; Slim, S.; Boulila, A.; Bettaieb, T. Incorporation of Organic Growth Additives to Enhance in Vitro Tissue Culture for Producing Genetically Stable Plants. *Plants* **2022**, *11*, 3087. [CrossRef] [PubMed]
11. Poudel, P.R.; Kataoka, I.; Mochioka, R. Effect of Plant Growth Regulators on in Vitro Propagation of *Vitis ficifolia* Var. Ganebu and Its Interspecific Hybrid Grape. *Asian J. Plant Sci.* **2005**, *4*, 466–471. [CrossRef]
12. Yerbolova, L.S.; Ryabushkina, N.A.; Oleichenko, S.N.; Kampitova, G.A.; Galiakparov, N.N. The Effect of Growth Regulators on in Vitro Culture of Some *Vitis vinifera* L. Cultivars. *World Appl. Sci. J.* **2013**, *23*, 76–80.
13. Weremczuk-Jeżyna, I.; Kuźma, Ł.; Kiss, A.K.; Grzegorczyk-Karolak, I. Effect of Cytokinins on Shoots Proliferation and Rosmarinic and Salvianolic Acid B Production in Shoot Culture of *Dracocephalum forrestii* WW Smith. *Acta Physiol. Plant.* **2018**, *40*, 189. [CrossRef]
14. Nowakowska, K.; Pińkowska, A.; Siedlecka, E.; Pacholczak, A. The Effect of Cytokinins on Shoot Proliferation, Biochemical Changes and Genetic Stability of *Rhododendron* 'Kazimierz Odnowiciel' in the in Vitro Cultures. *Plant Cell Tissue Organ Cult.* **2022**, *149*, 675–684. [CrossRef]
15. Chapman, E.J.; Estelle, M. Cytokinin and Auxin Intersection in Root Meristems. *Genome Biol.* **2009**, *10*, 210. [CrossRef]
16. Qiu, Y.; Guan, S.C.; Wen, C.; Li, P.; Gao, Z.; Chen, X. Auxin and Cytokinin Coordinate the Dormancy and Outgrowth of Axillary Bud in Strawberry Runner. *BMC Plant Biol.* **2019**, *19*, 528. [CrossRef] [PubMed]
17. Edelmann, H.G. Plant Root Development: Is the Classical Theory for Auxin-Regulated Root Growth False? *Protoplasma* **2022**, *259*, 823–832. [CrossRef] [PubMed]
18. Cleland, R.E. Auxin and Cell Elongation. In *Plant Hormones and Their Role in Plant Growth and Development*; Springer: Dordrecht, The Netherlands, 1987; pp. 132–148.

19. Ya, R.; Li, J.; Zhang, N.; Yu, Q.; Xu, W. Phenotypically Abnormal Cotyledonary *Vitis vinifera* Embryos Differ in Anatomy, Endogenous Hormone Levels and Transcriptome Profiles. *Tree Physiol.* **2023**, *43*, 467–485. [CrossRef]
20. Ganguly, A.; Lee, S.H.; Cho, M.; Lee, O.R.; Yoo, H.; Cho, H.-T. Differential Auxin-Transporting Activities of PIN-FORMED Proteins in *Arabidopsis* Root Hair Cells. *Plant Physiol.* **2010**, *153*, 1046–1061. [CrossRef]
21. Bennett, T.; Hines, G.; van Rongen, M.; Waldie, T.; Sawchuk, M.G.; Scarpella, E.; Ljung, K.; Leyser, O. Connective Auxin Transport in the Shoot Facilitates Communication between Shoot Apices. *PLoS Biol.* **2016**, *14*, e1002446. [CrossRef]
22. Ribeiro, Y.R. de S.; Aragão, V.P.M.; Sousa, K.R. de; Macedo, A.F.; Floh, E.I.S.; Silveira, V.; Santa-Catarina, C. Involvement of Differentially Accumulated Proteins and Endogenous Auxin in Adventitious Root Formation in Micropropagated Shoot Cuttings of *Cedrela fissilis* Vellozo (Meliaceae). *Plant Cell Tissue Organ. Cult.* **2022**, *148*, 119–135.
23. Jing, H.; Strader, L.C. Interplay of Auxin and Cytokinin in Lateral Root Development. *Int. J. Mol. Sci.* **2019**, *20*, 486. [CrossRef]
24. Kurepa, J.; Smalle, J.A. Auxin/Cytokinin Antagonistic Control of the Shoot/Root Growth Ratio and Its Relevance for Adaptation to Drought and Nutrient Deficiency Stresses. *Int. J. Mol. Sci.* **2022**, *23*, 1933. [CrossRef]
25. Sato, M.; Hosokawa, M.; Doi, M. Somaclonal Variation Is Induced de Novo via the Tissue Culture Process: A Study Quantifying Mutated Cells in Saintpaulia. *PLoS ONE* **2011**, *6*, e23541. [CrossRef] [PubMed]
26. Samarfard, S.; Kadir, M.A.; Kadzimin, S.B.; Saud, H.M.; Ravanfar, S.A.; Danaee, M. *In Vitro* Propagation and Detection of Somaclonal Variation in Phalaenopsis Gigantea as Affected by Chitosan and Thidiazuron Combinations. *HortScience* **2014**, *49*, 82–88. [CrossRef]
27. Krishna, H.; Alizadeh, M.; Singh, D.; Singh, U.; Chauhan, N.; Eftekhari, M.; Sadh, R.K. Somaclonal Variations and Their Applications in Horticultural Crops Improvement. *3 Biotech* **2016**, *6*, 54. [CrossRef]
28. Aljuaid, B.S.; Ismail, I.A.; Attia, A.O.; El Dessoky, S.D. Genetic Stability of in Vitro Propagated Grapevine (*Vitis vinifera* L.) Cv. Al-Bayadi. *J. Agric. Crop.* **2022**, *8*, 12–19. [CrossRef]
29. Gemmill, C.E.C.; Grierson, E.R.P. Inter-Simple Sequence Repeats (ISSR), Microsatellite-Primed Genomic Profiling Using Universal Primers. *Mol. Plant Taxon. Methods Protoc.* **2021**, *2222*, 249–262.
30. Johnson, L.J.; Brookfield, J.F.Y. A Test of the Master Gene Hypothesis for Interspersed Repetitive DNA Sequences. *Mol. Biol. Evol.* **2006**, *23*, 235–239. [CrossRef]
31. Babu, K.N.; Sheeja, T.E.; Minoo, D.; Rajesh, M.K.; Samsudeen, K.; Suraby, E.J.; Kumar, I.P.V. Random Amplified Polymorphic DNA (RAPD) and Derived Techniques. *Mol. Plant Taxon. Methods Protoc.* **2021**, *1115*, 219–247.
32. Sheeja, T.E.; Kumar, I.P.V.; Giridhari, A.; Minoo, D.; Rajesh, M.K.; Babu, K.N. Amplified Fragment Length Polymorphism: Applications and Recent Developments. *Mol. Plant Taxon. Methods Protoc.* **2021**, *2222*, 187–218.
33. Motha, K.; Singh, S.K.; Singh, R.; Ram, C.; Srivastav, M.; Verma, M.K.; Alizadeh, M.; Bhardwaj, C.; Dev, R. Comparative *in Vitro* Propagation of Stress Tolerant Grape (*Vitis* spp.) Rootstocks and Assessment of Clonal Fidelity of Plantlets. *Indian J. Hortic.* **2017**, *74*, 317–325. [CrossRef]
34. Benke, A.P.; Krishna, R.; Samarth, R.R.; Dhumal, S.S.; Ansari, W.A.; Shelke, P.V.; Dukare, S.S.; Singh, M. Development of an Embryo Germination Protocol for Shy-Seeded Grape (*Vitis vinifera* L.). *Plant Genet. Resour.* **2021**, *19*, 252–260. [CrossRef]
35. Classic Murashige, T.; Skoog, F. A Revised Medium for Rapid Growth and Bioassays with Tobacco Tissue Cultures. *Physiol. Plant* **1962**, *15*, 473–497. [CrossRef]
36. Marconi, M.; Wabnik, K. Shaping the Organ: A Biologist Guide to Quantitative Models of Plant Morphogenesis. *Front. Plant Sci.* **2021**, *12*, 2171. [CrossRef]
37. Jamwal, M.; Singh, B.; Sharma, N.; Kumar, R.; Sharma, A.; Sharma, R.M.; Parmar, A.M. In Vitro Regeneration of Grape (*Vitis vinifera* L.) Cv. Perlette. *World J. Agric. Sci.* **2013**, *9*, 161–166.
38. Joon-Ho, K.; Young-Sik, P.; Si-Hong, K.I.M.; Jae-Yun, H.E.O. Evaluation of Genetic Stability and Effects of Plant Growth Regulators for *in Vitro* Propagation of Underutilized *Vitis amurensis* 'Cheongsan'. *Not. Bot. Horti Agrobot. Cluj-Napoca* **2019**, *47*, 987–994.
39. San José, M.C.; Cernadas, M.J.; Janeiro, L.V. Optimization of Micropropagation Protocols in Some Woody Plants Using Meta-Topolin. In *Meta-Topolin: A Growth Regulator for Plant Biotechnology and Agriculture*; Springer: Singapore, 2021; pp. 221–240.
40. Murvanidze, N.; Nisler, J.; Leroux, O.; Werbrouck, S.P.O. Cytokinin Oxidase/Dehydrogenase Inhibitors Stimulate 2iP to Induce Direct Somatic Embryogenesis in Coffea Arabica. *Plant Growth Regul.* **2021**, *94*, 195–200. [CrossRef]
41. Kaviani, B.; Barandan, A.; Tymoszuk, A.; Kulus, D. Optimization of in Vitro Propagation of Pear (*Pyrus communis* L.) 'Pyrodwarf®(S)' Rootstock. *Agronomy* **2023**, *13*, 268. [CrossRef]
42. Ouyang, Y.; Chen, Y.; Lü, J.; Teixeira da Silva, J.A.; Zhang, X.; Ma, G. Somatic embryogenesis and enhanced shoot organogenesis in *Metabriggsia ovalifolia* WT Wang. *Sci. Rep.* **2016**, *6*, 24662. [CrossRef]
43. Nautiyal, A.; Rashid, A.; Agnihotri, A. Induction of Multiple Shoots in Oryza Sativa: Roles of Thidiazuron, 6-Benzylaminopurine, Decapitation, Flooding, and Ethrel® Treatments. *Vitro Cell. Dev. Biol.* **2022**, *58*, 1126–1137. [CrossRef]
44. Yucesan, B.; Turker, A.U.; Gurel, E. TDZ-Induced High Frequency Plant Regeneration through Multiple Shoot Formation in Witloof Chicory (*Cichorium intybus* L.). *Plant Cell. Tissue Organ Cult.* **2007**, *91*, 243–250. [CrossRef]
45. Vinoth, A.; Ravindhran, R. In Vitro Morphogenesis of Woody Plants Using Thidiazuron. In *Thidiazuron: From Urea Derivative to Plant Growth Regulator*; Springer: Singapore, 2018; pp. 211–229.
46. Novikova, T.I.; Zaytseva, Y.G. *TDZ-Induced Morphogenesis Pathways in Woody Plant Culture BT-Thidiazuron: From Urea Derivative to Plant Growth Regulator*; Springer: Singapore, 2018.

47. Ikeuchi, M.; Favero, D.S.; Sakamoto, Y.; Iwase, A.; Coleman, D.; Rymen, B.; Sugimoto, K. Molecular Mechanisms of Plant Regeneration. *Annu. Rev. Plant Biol.* **2019**, *70*, 377–406. [CrossRef]
48. Smart, D.R.; Kocsis, L.; Andrew Walker, M.; Stockert, C. Dormant Buds and Adventitious Root Formation by *Vitis* and Other Woody Plants. *J. Plant Growth Regul.* **2002**, *21*, 296–314. [CrossRef]
49. Zhao, Y.; Chen, Y.; Jiang, C.; Lu, M.-Z.; Zhang, J. Exogenous Hormones Supplementation Improve Adventitious Root Formation in Woody Plants. *Front. Bioeng. Biotechnol.* **2022**, *10*, 1009531. [CrossRef] [PubMed]
50. Frick, E.M.; Strader, L.C. Roles for IBA-Derived Auxin in Plant Development. *J. Exp. Bot.* **2018**, *69*, 169–177. [CrossRef]
51. Matsuoka, H.; Hinata, K. NAA-Induced Organogenesis and Embryogenesis in Hypocotyl Callus of *Solarium melongena* L. *J. Exp. Bot.* **1979**, *30*, 363–370. [CrossRef]
52. Villalobos-Olivera, A.; Ferreira, C.F.; Yanes-Paz, E.; Lorente, G.Y.; Souza, F.V.; Engelmann, F.; Martínez-Montero, M.E.; Lorenzo, J.C. Inter Simple Sequence Repeat (ISSR) Markers Reveal DNA Stability in Pineapple Plantlets after Shoot Tip Cryopreservation. *Vegetos* **2022**, *35*, 360–366. [CrossRef]

current issues in molecular biology

MDPI

Brief Report

Yeast One-Hybrid Screening to Identify Transcription Factors for *IbMYB1-4* in the Purple-Fleshed Sweet Potato (*Ipomoea batatas* [L.] Lam.)

Danwen Fu [1,2], Shaohua Yang [1], Rui Liu [2] and Feng Gao [1,*]

1 Guangdong Provincial Key Lab of Biotechnology for Plant Development, School of Life Sciences, South China Normal University, Guangzhou 510631, China; danwenfu@yeah.net (D.F.); 15914375650@163.com (S.Y.)
2 Institute of Nanfan & Seed Industry, Guangdong Academy of Sciences, Guangzhou 510316, China; liurui701@126.com
* Correspondence: peak0041@vip.sina.com

Abstract: IbMYB1 is a transcription factor involved in the biosynthesis of anthocyanin in the purple-fleshed sweet potato. So far, few studies have investigated transcription factors that are upstream of the promoter *IbMYB1-4*. In this study, a yeast one-hybrid screening aimed at identifying transcription factors upstream of the promoter *IbMYB1-4* was performed in the storage roots of the purple-fleshed sweet potato, and IbPDC, IbERF1, and IbPGP19 were identified as upstream binding proteins for the promoter *IbMYB1-4*. A dual luciferase reporter assay, and yeast one-hybrid assays, were employed to confirm the interaction of these binding proteins with promoters. IbERF1 was found to be an upstream transcription factor for the promoter *IbMYB1*, and is implicated in the biosynthesis of anthocyanin in the purple-fleshed sweet potato. IbERF1 plays a major role in the biosynthesis of anthocyanin in the purple-fleshed sweet potato.

Keywords: anthocyanin biosynthesis; IbERF1; IbPDC; IbPGP19; upstream transcription factors

Citation: Fu, D.; Yang, S.; Liu, R.; Gao, F. Yeast One-Hybrid Screening to Identify Transcription Factors for *IbMYB1-4* in the Purple-Fleshed Sweet Potato (*Ipomoea batatas* [L.] Lam.). *Curr. Issues Mol. Biol.* **2023**, *45*, 5765–5775. https://doi.org/10.3390/cimb45070364

Academic Editor: Angel Llamas

Received: 16 June 2023
Revised: 5 July 2023
Accepted: 10 July 2023
Published: 12 July 2023

1. Introduction

The sweet potato is a major food crop in China, and is mainly grown in tropical and subtropical regions. High levels of anthocyanin with specific physiological activity and functions are present in the purple-fleshed sweet potato (PFSP). Anthocyanin has anticancer and antioxidation functions, and plays a role in lowering blood pressure and preventing arteriosclerosis [1]. The accumulation of anthocyanin is regulated via the expression levels of several genes, including *CHI*, *CHS*, *DFR*, *UF3GT*, *ANS*, and *F3H*. Transcription factors (TFs) regulate the expression of these structural genes. These trans-acting factors comprise a group of DNA-binding proteins that bind specifically to cis-acting elements of eukaryotic genes, and regulate the activation or inhibition of gene transcription [2]. The biosynthesis of anthocyanin in the PFSP is regulated by the TF complex MYB–bHLH–WD40 (MBW) [3,4]. The *IbMYB1* gene expression was reported to correlate positively with that of related structural genes in anthocyanin biosynthesis. The *IbMYB1* promoter (length 2183 bp) was cut into four segments (−2183 bp~−1684 bp, −1683 bp~−1194 bp, −1193 bp~−694 bp, −693 bp~−1 bp). A previous study investigated the *IbMYB1-1* promoter for upstream TFs [5]. It was noted that high concentrations of AbA did not suppress the Leu leakiness with the promoter *IbMYB1-2* and the promoter *IbMYB1-3*, with 400 ng/mL being the minimum inhibitory AbA concentration for the PAbAi-*PIbMYB1-4* strains. However, little is known regarding TFs that are upstream of *PIbMYB1-4* (the promoter of *IbMYB1-4*) in the anthocyanin synthesis pathway of the PFSP storage roots. In this study, a yeast one-hybrid screening aimed at identifying transcription factors upstream of the promoter *IbMYB1-4* was performed in the storage roots of the purple-fleshed sweet potato. IbPDC, IbERF1, and IbPGP19 were identified as upstream binding proteins for the promoter

IbMYB1-4. IbERF1 was found to be an upstream transcription factor for the promoter *IbMYB1*, and is implicated in the biosynthesis of anthocyanin in the purple-fleshed sweet potato.

The AP2/ERF families are ethylene response factors (ERFs) involved in multiple physiological processes during the growth of plants, and in their response to biological, as well as nonbiological, stress [6,7]. The endogenous hormone ethylene is involved in growth and development in the roots, leaves, flowers, and fruit [8]. In the last, the endogenous hormone ethylene controls color changes in fruit [9]. ERFs have been reported to regulate color changes in the fruit peel, and anthocyanin biosynthesis, in orange [10], banana and mango [9], pear [11], apple [12], purple tea [13], carrot [14], and hybrid poplar plants [15]. So far, however, ERFs have not been implicated in the biosynthesis of anthocyanin in the storage roots of the PFSP. Instead, IbERF1 has been found to regulate anthocyanin biosynthesis in the PFSP by binding to the *IbMYB1-4* promoter.

2. Material and Methods

2.1. Plant Material

The PFSP cv. A5 and the white-fleshed sweet potato (WFSP) cv. Yubeibai were cultivated in a garden at South China Normal University, Guangdong, China. Arabidopsis (*Arabidopsis thaliana*) was used for dual luciferase and subcellular localization assays. It was grown at 20 ± 2 °C in a chamber, using a day/night cycle (16 h/8 h), and a constant light intensity (100 μmol/m^2/s).

2.2. Extraction of Genomic DNA and RNA, Gene Isolation, and Sequence Analysis

The root tissue (0.5 g) was ground into a powder with liquid nitrogen using a mortar and pestle. The DNA extraction was then carried out using a plant DNA kit (Tiangen, Beijing, China), and RNA extraction was conducted using a Hipure plant RNA kit (Magen, Guangzhou, China). To eliminate the possibility of DNA contamination, the RNA underwent DNase I digestion using an RNase-free kit (TaKaRa, Shiga, Japan). A BioPhotometer Plus (Eppendorf, Hamburg, Germany) was used to estimate the DNA concentrations and purity by measuring the absorbance at 230, 260, and 280 nm. The DNA samples showed $1.8 \leq OD_{260}/OD_{280} \leq 1.9$ and $OD_{260}/OD_{230} \geq 2.0$, while the RNA samples showed $1.9 \leq OD_{260}/OD_{280} \leq 2.0$ and $OD_{260}/OD_{230} \geq 2.0$.

A GoScript™ Reverse Transcription System (Promega, Madison, WI, USA) was used for cDNA synthesis, with the promoter and genomic fragments subsequently cloned and sequenced. Agarose gels (1.2%) were used to analyze the PCR products. The fragments were ligated into plasmid, transformed into *Escherichia coli* DH5α competent cells, then sent to be sequenced by Sangon Biotech, Shanghai, China.

2.3. Yeast One-Hybrid (Y1H) Screening

Y1H screening was performed using Matchmaker Gold Y1H Library Screening, as described previously [16]. The RNA was extracted from the storage roots of cv. A5 to build a prey cDNA library, and a cDNA pool was inserted separately into the prey vector pGADT7-Rec. The *IbMYB1-4* promoter was inserted into pAbAi, resulting in pAbAi-bait. The plasmids containing pAbAi-bait were subsequently linearized, and transformed into Y1HGold cells. Colonies growing in a synthetic dextrose medium lacking uracil were then selected. After the minimal inhibitory concentration of aureobasidin A (AbA) for bait strains was identified, the linear pGADT7-Rec vector was co-transformed into bait yeast strains, and then selected using synthetic dextrose (SD)/-Leu/AbA plates. The primers used in the Y1H screening assay are shown in Supplementary Table S1.

2.4. Y1H Assay

Y1H was also performed using the Matchmaker Gold Y1H System, as described in another previous study [17]. The Y1H assays were performed to identify the interactions of IbERF1, IbPGP19, and IbPDC with the *IbMYB1-4* promoter. In brief, fragments of the promoter were ligated into the pAbAi vector, while upstream regulators were cloned into pGADT7. The pAbAi was linearized and transformed into Y1HGold cells. The *IbMYB1-4* promoter was inserted into the pAbAi, to create pAbAi-bait. The complete CDSs for IbERF1, IbPGP19, and IbPDC were inserted separately into the pGADT7 vector, to create prey-AD vectors. These were transferred into the bait strain, and cultured on SD/-Leu/AbA plates. The primers used for the Y1H assay are shown in Supplementary Table S1.

2.5. Yeast Two-Hybrid (Y2H) Assay

Y2H assays were performed using the Matchmaker Gold Y2H kit (TaKaRa, Dalian, China), as recommended by the manufacturer, but with minor modifications. In brief, the PEG/lithium acetate method was used to simultaneously transform the pGADT7 (AD) and pGBKT7 (BD) vectors into Y2HGold cells. The transcriptional activity for IbERF1, IbPGP19, and IbPDC was investigated using the Y2H assay. The full-length coding sequences for *IbERF1*, *IbPGP19*, and *IbPDC* were cloned into pGBKT7, to construct bait-BD vectors. The PGBKT7-bait and PGADT7-empty vectors were co-transferred into Y2HGold cells. The positive control was pGBKT7-53 co-transformed with pGADT7-53, while the negative control was pGBKT7 co-transformed with pGADT7-53. The primers used in the Y2H assay are shown in Supplementary Table S1.

2.6. Dual-Luciferase Assays

Dual-luciferase assays were performed, to quantify the transactivation abilities of IbERF1, IbPGP19, and IbPDC with the *IbMYB1-4* promoter. In brief, the full-length cDNAs for *IbERF1*, *IbPGP19*, and *IbPDC* were inserted into the vector pGreen II 0029 62-SK (Shanghai Qi Ming Biotechnology Co., Ltd., Shanghai, China), while the *IbMYB1-4* promoter was inserted into the vector pGreen II 0800-LUC (Shanghai LMAI Biotechnology Co., Ltd., Shanghai, China). Both these constructs were then transformed into *arabidopsis* protoplasts, as described in a previous publication [18]. The LUC and REN enzyme activity ratio was quantified using the E1910 Dual-Luciferase® Reporter Assay (Promega). Three independent experiments were performed for each of the interactions between the TFs and promoters, and three replicates were used for each experiment. The positive control was a Renilla luciferase gene driven by the 35S promoter in a luciferase vector. Mixtures containing each TF with the empty vector 62-SK were also used on promoters as a control. The primers for the dual-luciferase assay are shown in Supplementary Table S1.

2.7. Analysis of Subcellular Localization

For the investigation of subcellular localization, upstream TF CDSs lacking a stop codon were amplified, and cloned into pCambia1300 (Honorgene Co., Ltd. Guangzhou China), with *BamH* I and *Hind* II. This vector contains the promoter for *UBQ*, as well as the *GFP* gene. Both constructs were transformed into *Arabidopsis* protoplasts, in accordance with a previous publication, while protoplasts were prepared, as described previously, but with slight modifications [18]. The protoplasmic cell culturing time at normal temperature was extended from 20 h to 24 h. The GFP fluorescence was visualized, using confocal microscopy with a Zeiss LSM710 instrument. The primers used in the investigation of subcellular localization are shown in Supplementary Table S1.

2.8. Real-Time Quantitative PCR (RT-qPCR)

RT-qPCR was used to evaluate the expression of the upstream TF IbERF1, the TFs IbMYB1, IbbHLH2, and IbWD40, and the structural genes *IbCHI, IbCHS, IbF3H, IbF3′H, IbDFR, IbANS*, and *IbUF3GT*, in the roots (fibrous, thick, storage) of the PFSP cv. A5, and the WFSP *cv.* Yubeibai. The Prime Script™ RT Master Mix (Takara) was employed, to synthesize the first-strand cDNA from the total RNA, and SYBR® Premix Ex Taq™ II (Takara) was used to perform RT-qPCR. The 20 μL reaction volume contained 10 μL of SYBR® Premix Ex Taq™ II, each primer at 0.5 μM, and 100 ng of template cDNA. The amplification program consisted of one cycle for 10 s at 95 °C; then 40 cycles for 5 s at 95 °C, and 30 s at 60 °C, using a Bio-Rad CFX96 Real-Time PCR system (BIO-RAD, Hercules, CA, USA), as recommended by the manufacturer. The internal control was *IbG14*, and the calculations were performed using the Ct analysis method. The primers for the RT-qPCR experiments are shown in Supplementary Table S1.

2.9. Statistical Analyses

Three replicates of each biological sample were evaluated, using one-way analysis of variance (ANOVA). The SPSS 21.0 statistics software package (SPSS Inc., Chicago, IL, USA) was used to determine the significant differences, with Tukey's honest test ($p < 0.05$). Sigmaplot 12.3 was used to draw the figures.

3. Results

3.1. Screening of the IbMYB1 Promoter for Upstream TFs

The *IbMYB1* promoter (length 2183 bp) was cut into four segments (−2183 bp~−1684 bp, −1683 bp~−1194 bp, −1193 bp~−694 bp, and −693 bp~−1 bp). A previous study investigated the *IbMYB1-1* promoter for upstream TFs [18]. It was noted that high concentrations of AbA did not suppress the Leu leakiness with *PIbMYB1-2* and *PIbMYB1-3*, with 400 ng/mL being the minimum inhibitory AbA concentration for the PAbAi-*PIbMYB1-4* strains. A total of 519 positive colonies were screened for *PIbMYB1-4*. The Y1H assay identified 130 binding proteins for the *IbMYB1-4* promoter. Supplementary Table S2 shows the analysis of the gene sequences for *PIbMYB1-4*. The IbERF1, IbPGP19, and IbPDC proteins were found to interact with the promoter of *IbMYB1-4*. This gene is involved in the biosynthesis of anthocyanin.

3.2. Transcriptional Activities of IbERF1, IbPGP19, and IbPDC

Y2H assays were carried out to determine the transcriptional activities of IbERF1, IbPGP19, and IbPDC. The pGADT7-53+pGBKT7-53 and pGBKT7-*IbERF1/IbPGP19/IbPDC*+ pGADT7-empty transformed strains were found to grow on SD/-Trp, SD/-His-AbA plus, and SD/-His-AbA X-a-Gal plus plates, with the color of the emerging yeast colony being blue. The pGADT7-53+PGBKT7 transformed strains were unable to grow on SD/-Trp, SD/-His-AbA plus, and SD/-His-AbA X-a-Gal plus plates (Figure 1). These results suggest that IbERF1 had transcriptional activity.

3.3. Interactions of IbERF1, IbPGP19, and IbPDC with PIbMYB1-4

Y1H assays were performed to determine whether IbERF1, IbPGP19, and IbPDC interact with *PIbMYB1-4*. The positive control was found to grow on the SD/-Leu/AbA plates, but not the negative control. All the transformed strains also grew on the SD/-Leu/AbA plates (Figure 2), thereby confirming the interactions of IbERF1, IbPGP19, and IbPDC with *PIbMYB1-4*.

A dual-luciferase assay was also performed to validate the interactions of IbERF1, IbPGP19, and IbPDC with *PIbMYB1-4*. IbERF1, IbPGP19, and IbPDC each had significant activation effects on *PIbMYB1-4*, as shown in Figure 3. Hence, these results further confirm the interactions of IbERF1, IbPGP19, and IbPDC with *PIbMYB1-4*.

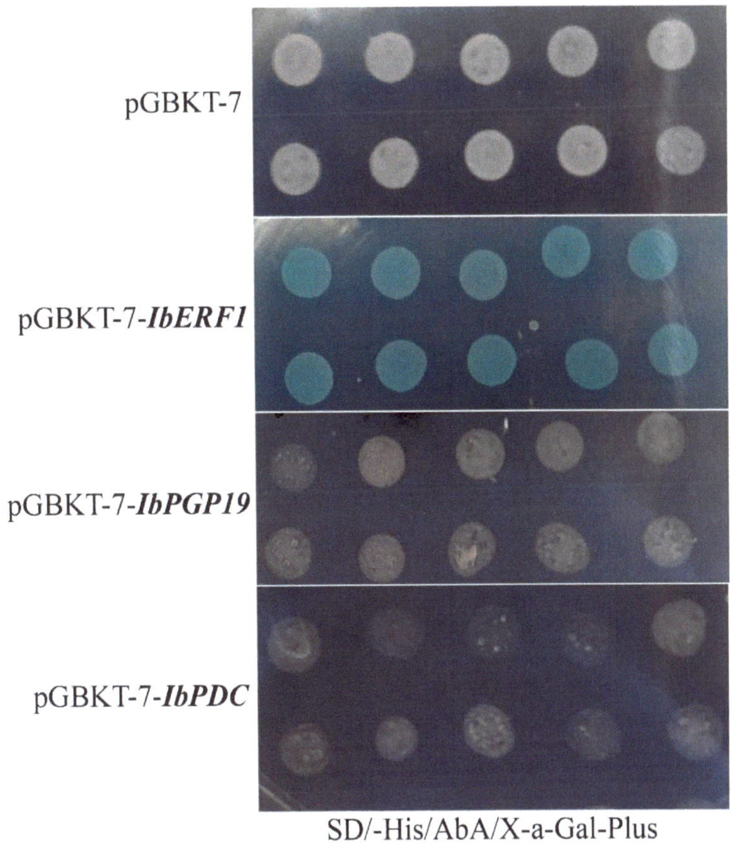

Figure 1. Transcriptional activation of IbERF1, IbPGP19, and IbPDC in yeast cells. The Y2H Gold strains successfully transformed with the corresponding vectors, and grown on SD/-His/AbA/X-α-Gal plates for 3–5 days at 30 °C. The ten points of each treatment were grown on a Petri dish. The growth status of yeast cells evaluated by an X-α-Gal assay was used to monitor transcription activation.

3.4. Subcellular Localization of IbERF1, IbPGP19, and IbPDC

IbERF1, IbPGP19, and IbPDC were predicted to localize in nuclei. In order to test this, green fluorescent protein (GFP) was fused to the C-terminus of IbERF1, IbPGP19, and IbPDC, and transiently expressed in *Arabidopsis* protoplasts under the control of the UBQ promoter. The GFP control showed green fluorescence in the nucleus and cytoplasm. The IbERF1/IbPGP19/IbPDC-GFP fusions exhibited green fluorescence in the nucleus only (Figure 4), indicating that IbERF1, IbPGP19, and IbPDC are nuclear TFs that regulate the expression of downstream genes.

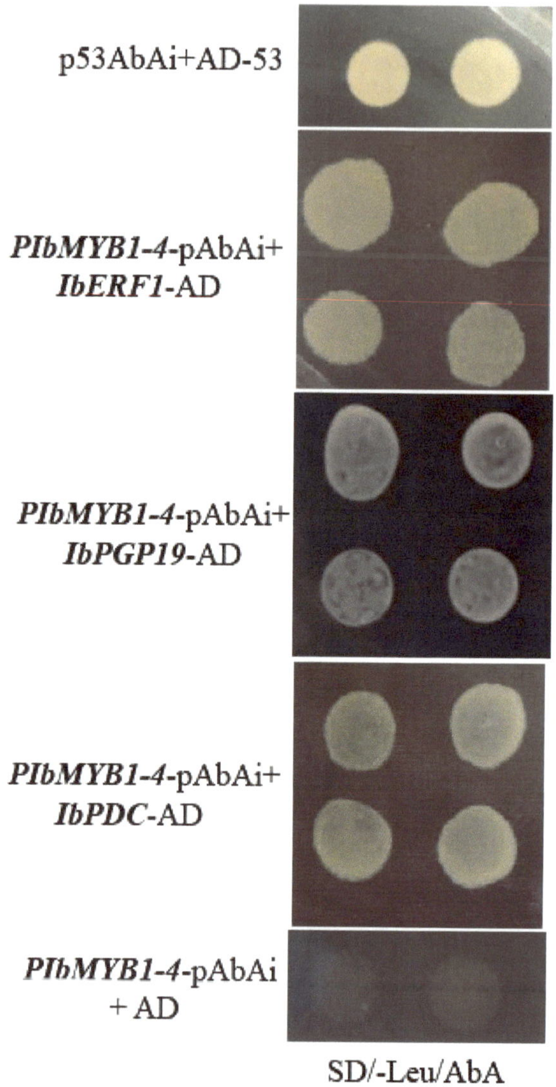

Figure 2. The interaction of IbERF1, IbPGP19, and IbPDC with *PIbMYB1-4*, studied using a yeast one-hybrid assay. The Y1H Gold strains successfully transformed with corresponding vectors, and grown on SD/-Leu/AbA plates for 3–5 days at 30 °C. The four points of each treatment were grown on a Petri dish, and the interactions were confirmed using the yeast cell growth status.

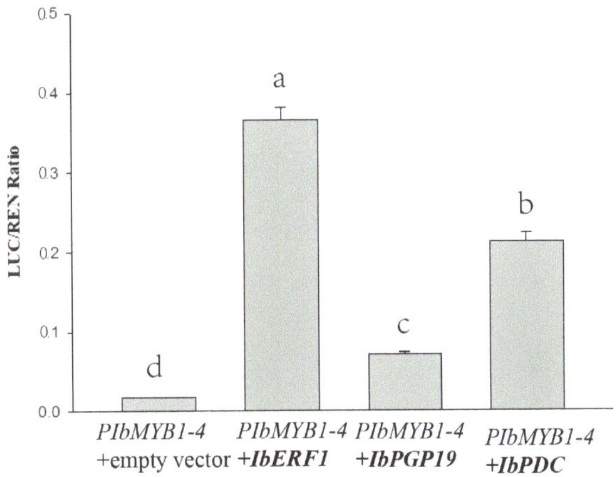

Figure 3. The IbERF1, IbPGP19, and IbPDC proteins activate the *IbMYB1-4* promoters in dual-luciferase assays. The error bars represent the standard deviation (SD). The significance tests are shown as a, b, c, and d. The different lowercase letters in the chart indicate that there is a significant difference ($p < 0.01$).

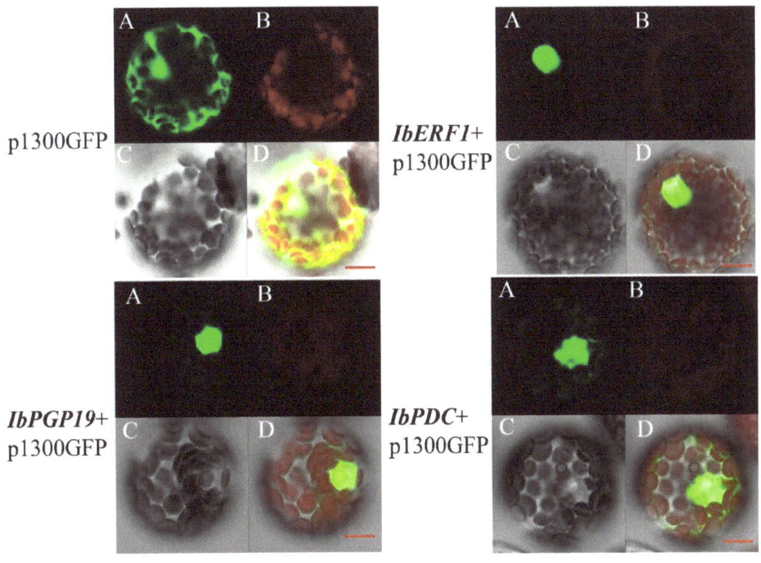

Figure 4. The subcellular localization of the IbERF1, IbPGP19, and IbPDC proteins in *Arabidopsis* protoplasts. The fusion proteins and the GFP control were expressed transiently in *Arabidopsis* protoplasts. The bars represent 20 µm. (**A**) GFP, (**B**) chloroplast, (**C**) light field, (**D**) merged graph.

3.5. Expression Characteristics of IbERF1

Real-time PCR was employed to evaluate the expression of the upstream TFs, TFs, and structural genes involved in the biosynthesis of anthocyanin at the different root stages of the PFSP and WFSP (Figure 5). The expression of the TFs and structural genes at the different root stages of the PFSP was found to be higher than in the WFSP.

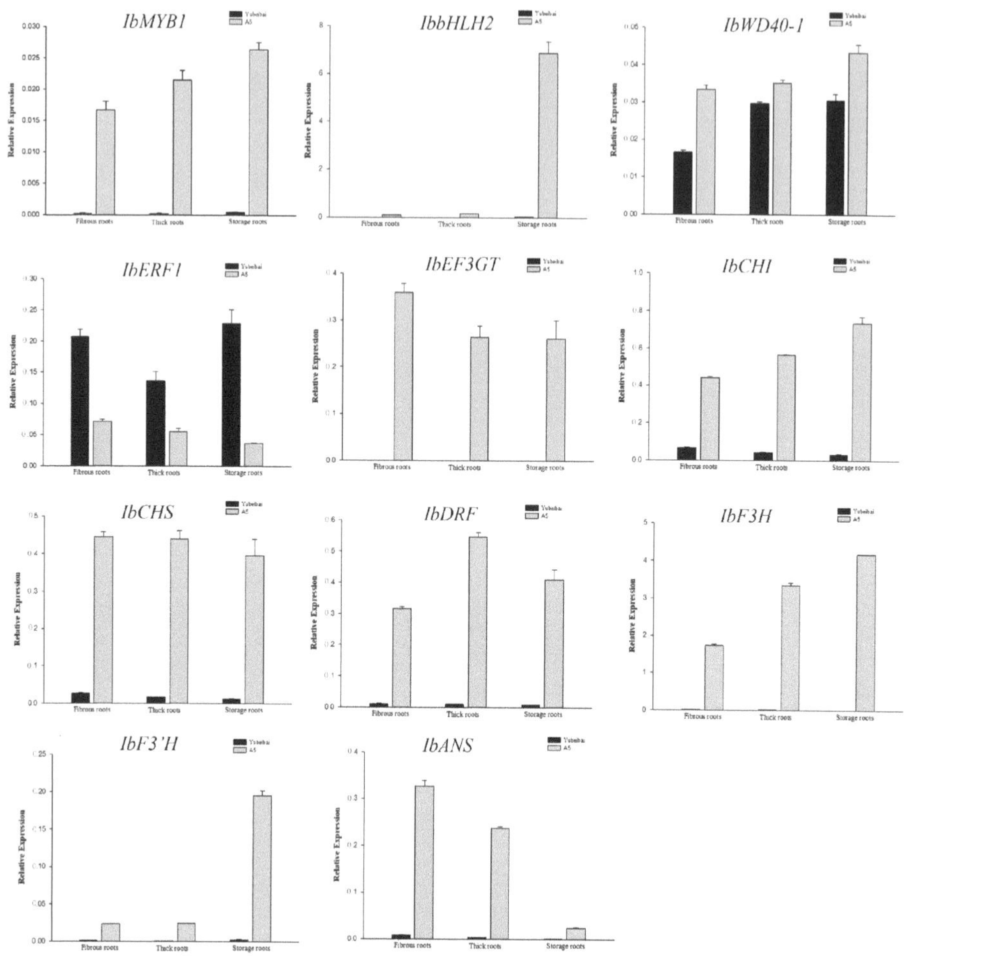

Figure 5. The relative expressions levels of *IbERF1*, *IbMYB1*, and the structural genes involved in the biosynthesis of anthocyanin at the different root stages in the purple- and white-fleshed sweet potato: the fibrous root (diameter < 2 mm), thick root (2 mm < diameter > 5 mm), and storage root (diameter > 5 mm).

4. Discussion

Anthocyanin biosynthesis in plants is regulated via the MBW complex. Other TFs, including ERFs, COP1, WRKYs, SPL9, NACs, and DELLA proteins, are also thought to affect the activity of the MBW complex, and hence the biosynthesis of anthocyanin [19–23]. IbMYB1 has previously been implicated in the biosynthesis of anthocyanin in the PFSP [24]. In the present study, the upstream TFs for *PIbMYB1* involved in anthocyanin biosynthesis in the PFSP's storage roots were screened using a Y1H assay. IbERF1 was found to be an upstream TF for *PIbMYB1-4* in the synthesis of anthocyanin. The interactions between IbERF1 and *PIbMYB1-4* were confirmed using dual-luciferase and Y1H assays. Moreover, IbERF1 was found in the nucleus, using a subcellular localization method. This indicates that IbERF1 is a nuclear protein, and functions as a TF in regulating the expression of downstream genes.

The ethylene response factors AP2/ERFs are a major TF family, involved in multiple physiological processes, including plant growth, and the response to biological and nonbiological stress [5,6]. It has been shown previously that TF ERFs regulate the biosynthesis of anthocyanin in plants. In suspension-cultured carrot cells, DcERF1 can upregulate the *DcPAL3* promoter activity, while DcERF2 functions in different ways in anthocyanin biosynthesis [25]. In Salvia miltiorrhiza flowers, the TF AP2/ERFs can regulate anthocyanin biosynthesis, due to the significantly different expression levels between purple and white flowers, as seen in the transcriptome data [26]. TF ERFs have been found to regulate the light- and ethylene-induced biosynthesis of anthocyanin in different species. In plums, seven PsERFs were positively correlated with PsMYB10, and with many structural genes involved in anthocyanin biosynthesis [27]. In *Arabidopsis*, a double mutant (*aterf4* and *aterf8*) reduced the light-induced anthocyanin accumulation, indicating the positive regulation of anthocyanin biosynthesis by AtERF4 and AtERF8 [25]. In pears, TF ERFs regulate the light-induced biosynthesis of anthocyanin [28].

ERF interacts with MYB proteins, and binds to MYB promoters, to regulate the biosynthesis of anthocyanin in plants [24]. In apples, MdERF1B interacts with the MdMYB1, MdMYB9, and MdMYB11 proteins to regulate the biosynthesis of anthocyanin. Furthermore, MdERF1B also binds to the *MdMYB1*, *MdMYB9*, and *MdMYB11* promoters to regulate anthocyanin biosynthesis [14]. MdERF38 binds to the *MdMYB1* promoter to regulate the drought-stress-induced biosynthesis of anthocyanin at the posttranslational level [13]. In pears, PpERF3 interacts with PpbHLH3 and PpMYB114 in the ERF3-MYB114-bHLH3 complex to regulate the red pear color [12], while Pp4ERF24 and Pp12ERF96 regulate the light-induced biosynthesis of anthocyanin [11]. In strawberries, FaERF9 and FaMYB98 combine to form the ERF-MYB complex, and activate the FaQR promoter, thereby increasing the furaneol content in cultivated strawberries [29]. IbERF1 was found here to interact with the *IbMYB1-4* promotor, and thus regulate the biosynthesis of anthocyanin in the PFSP. These results help to clarify the underlying mechanism for anthocyanin biosynthesis, and improve our understanding of the regulatory networks in the PFSP.

5. Conclusions

In this research, IbERF1 was identified as a TF for *IbMYB1-4* in the regulation of anthocyanin biosynthesis in the PFSP.

Supplementary Materials: The following supporting information can be downloaded at: https://www.mdpi.com/article/10.3390/cimb45070364/s1, Table S1: The primer sets used in this study; Table S2: Information of genes interacted with *pIbMYB1-4* by yeast one hybrid screening.

Author Contributions: F.G. designed the study. S.Y. and R.L. interpreted the results. D.F. performed experiments, wrote the manuscript, analyzed the results, and conceived the study. All authors have read and agreed to the published version of the manuscript.

Funding: This work was funded by the Guangdong Basic and Applied Basic Research Foundation (2022A1515110300), and Shanghai Municipal Afforestation and City Appearance and Environmental Sanitation Administration (G222411).

Institutional Review Board Statement: Not applicable for studies not involving humans or animals.

Informed Consent Statement: I would like to declare on behalf of my co-authors that the work described was original research that has not been published previously, and not under consideration for publication elsewhere, in whole or in part. All authors have seen a copy of the paper and have approved its submission.

Data Availability Statement: The data and materials in this manuscript can be accessed.

Conflicts of Interest: No conflict of interest exist regarding the submission of this manuscript, and the manuscript has been approved by all authors for publication.

References

1. Kano, M.; Takayanagi, T.; Harada, K.; Makino, K.; Ishikawa, F. Antioxidative activity of anthocyanins from purple sweet potato, *Ipomoera batatas* cultivar Ayamurasaki. *Biosci. Biotechnol. Biochem.* **2005**, *69*, 979–988. [CrossRef] [PubMed]
2. Lambert, S.A.; Jolma, A.; Campitelli, L.F.; Das, P.K.; Yin, Y.M.; Albu, M.H.; Chen, X.T.; Taipale, J.; Hughes, T.R. The Human Transcription Factors. *Cell* **2018**, *172*, 650–665. [CrossRef] [PubMed]
3. Ramsay, N.A.; Glover, B.J. MYB-bHLH-WD40 protein complex and the evolution of cellular diversity. *Trends Plant Sci.* **2015**, *10*, 63–70. [CrossRef] [PubMed]
4. Schaart, J.G.; Dubos, C.; Romero, D.L.F.I.; Houwelingen, A.M.M.L.; Vos, R.C.H.D.; Jonker, H.H.; Xu, W.; Routaboul, J.M.; Lepiniec, L.; Bovy, A.G. Identification and characterization of MYB–bHLH–WD40 regulatory complexes controlling proanthocyanidin biosynthesis in strawberry(*Fragaria* × *ananassa*) fruits. *New Phytol.* **2013**, *197*, 454–467. [CrossRef] [PubMed]
5. Fu, D.W.; Hui, Y.K.; Chen, Y.H.; Yang, S.H.; Gao, F. Screening of Upstream Transcript Factors of *IbMYB1-1* by Yeast One-hybrid in Purple-fleshed Sweet Potato. *Trop. Plant Biol.* **2022**, *15*, 148–156. [CrossRef]
6. Johni, D.; Yogita, N.S.; Banashree, S.; Hari, P.; Deka, B.; Dhanawantari, L.; Channake, S. Ethylene Response Factor (ERF) Family Proteins in Abiotic Stresses and CRISPR–Cas9 Genome Editing of ERFs for Multiple Abiotic Stress Tolerance in Crop Plants: A Review. *Mol. Biotech.* **2019**, *61*, 153–172.
7. Vimolmangkang, S.; Han, Y.P.; Wei, G.C.; Korban, S.S. An apple MYB transcription factor, MdMYB3, is involved in regulation of anthocyanin biosynthesis and flower development. *BMC Plant Biol.* **2013**, *13*, 176. [CrossRef]
8. Ueda, H.; Kusaba, M. Strigolactone regulates leaf senescence in concert with ethylene in *Arabidopsis*. *Plant Physio.* **2015**, *169*, 138–147. [CrossRef]
9. Xie, X.L.; Yin, X.R.; Chen, K.S. Roles of APETALA2/Ethylene responsive factors in regulation of fruit quality. *Crit. Rev. Plant Sci.* **2016**, *35*, 120–130. [CrossRef]
10. Lia, S.J.; Xie, X.L.; Liua, S.C.; Chena, K.S.; Yina, X.R. Auto- and mutual-regulation between two CitERFs contribute to ethylene induced citrus fruit degreening. *Food Chem.* **2019**, *299*, 125–163. [CrossRef]
11. Ni, J.B.; Apekshika, T.P.; Gao, Y.H.; Yu, W.J.; Tao, R.Y.; Teng, Y.W.; Bai, S.L. Ethylene-activated PpERF105 induces the expression of the repressor-type R2R3-MYB gene PpMYB140 to inhibit anthocyanin biosynthesis in red pear fruit. *Plant J.* **2021**, *105*, 167–181. [CrossRef]
12. An, J.P.; Zhang, X.W.; Bi, S.Q.; You, C.X.; Wang, X.F.; Hao, Y.J. The ERF transcription factor MdERF38 promotes drought stress-induced anthocyanin biosynthesis in apple. *Plant J.* **2019**, *101*, 573–589. [CrossRef]
13. Tony, K.M.; Mamta, M.; Romit, S.; Ram, K.S. Transcriptional analysis reveals key insights into seasonal induced anthocyanin degradation and leaf color transition in purple tea (*Camellia sinensis* (L.) O. Kuntze). *Sci. Rep.* **2021**, *11*, 1244–1258.
14. Soichi, K.; Yukie, C.; Masayuki, K.; Kazuhiro, M.; Yoshihiro, O. Upregulation of the promoter activity of the carrot (*Daucus carota*) phenylalanine ammonia-lyase gene (*DcPAL3*) is caused by new members of the transcriptional regulatory proteins, DcERF1 and DcERF2, which bind to the GCC-box homolog and act as an activator to the *DcPAL3* promoter. *J. Plant Res.* **2008**, *121*, 499–508.
15. Kim, M.H.; Cho, J.S.; Bae, E.K.; Choi, Y.I.; Seok, H.E.; You, J.L.; Hyoshin, L.; Park, E.J.; Ko, J.H. PtrMYB120 functions as a positive regulator of both anthocyanin and lignin biosynthetic pathway in a hybrid poplar. *Tree Physiol.* **2021**, *41*, 2409–2423. [CrossRef]
16. Shim, J.S.; Jung, C.; Lee, S.; Min, K.; Lee, Y.W.; Choi, Y.; Lee, J.S.; Song, J.T.; Kim, J.K.; Choi, Y.D. AtMYB44 regulates WRKY70 expression and modulates antagonistic interaction between salicylic acid and jasmonic acid signaling. *Plant J.* **2013**, *73*, 483–495. [CrossRef]
17. Vidal, M.; Legrain, P. Yeast forward and reverse n-hybrid systems. *Nucleic Acids Res.* **1999**, *27*, 919–929. [CrossRef]
18. Fu, D.W.; Zhang, J.C.; Wang, D.C.; Gao, F. Molecular cloning, functional characterization and transcriptional regulation of the promoter of dihydroflavonal-4-reductase gene from purple-fleshed sweet potato. *South Afr. J Bot.* **2020**, *131*, 206–213. [CrossRef]
19. Morishita, T.; Kojima, Y.; Maruta, T.; Nishizawa-Yokoi, A.; Yabuta, Y.; Shigeoka, S. *Arabidopsis* NAC transcription factor, ANAC078, regulates flavonoid biosynthesis under high-light. *Plant Cell Physiol.* **2009**, *50*, 2210–2222. [CrossRef]
20. Jakaria, C.O.; Hu, X.J.; Lin, Q.; Wang, Z.D. Comparative Transcriptomic Profiling to Understand Pre- and Post-Ripening Hormonal Regulations and Anthocyanin Biosynthesis in Early Ripening Apple Fruit. *Molecules* **2018**, *23*, 1908.
21. Maier, A.; Schrader, A.; Kokkelink, L.; Falke, C.; Welter, B.; Iniesto, E.; Rubio, V.; Uhrig, J.F.; Hulskamp, M. Light and the E3 ubiquitin ligase COP1/SPA control the protein stability of the MYB transcription factors PAP1 and PAP2 involved in anthocyanin accumulation in *Arabidopsis*. *Plant J.* **2013**, *74*, 638–651. [CrossRef] [PubMed]
22. Rameneni, J.J.; Choi, S.R.; Chhapekar, S.S.; Kim, M.S.; Singh, S.; Yi, S.Y.; Oh, S.H. Red Chinese Cabbage Transcriptome Analysis Reveals Structural Genes and Multiple Transcription Factors Regulating Reddish Purple Color. *Int. J. Mol. Sci.* **2020**, *21*, 2901. [CrossRef] [PubMed]
23. Alabd, A.; Ahmad, M.; Zhang, X.; Gao, Y.H.; Peng Zhang, L. Light-responsive transcription factor PpWRKY44 induces anthocyanin accumulation by regulating PpMYB10 expression in pear. *Hortic. Res.* **2022**, *9*, 122. [CrossRef]
24. Park, S.C.; Kim, Y.H.; Kim, S.H.; Jeong, Y.J.; Kim, C.Y.; Lee, J.S.; Baef, J.Y.; Ahn, M.J.; Jeong, J.C.; Lee, H.S.; et al. Overexpression of the IbMYB1 gene in an orange-fleshed sweet potato cultivar produces a dual-pigmented transgenic sweet potato with improved antioxidant activity. *Physiol. Plant.* **2015**, *153*, 525–537. [CrossRef] [PubMed]
25. Koyama, T.; Sato, F. The function of Ethylene Response Factor genes in the light-induced anthocyanin production of *Arabidopsis thaliana* leaves. *Plant Biotechnol.* **2018**, *35*, 87–91. [CrossRef]

26. Jiang, T.; Zhang, M.D.; Wen, C.X.; Xie, X.L.; Tian, W.; Wen, S.Q.; Lu, R.K.; Liu, L.D. Integrated metabolomic and transcriptomic analysis of the anthocyanin regulatory networks in *Salvia miltiorrhiza Bge*. Flowers. *BMC Plant Biol.* **2020**, *20*, 349. [CrossRef]
27. Cheng, Y.D.; Liu, L.Q.; Yuan, C.; Guan, J.F. Molecular Characterization of Ethylene Regulated Anthocyanin Biosynthesis in Plums During Fruit Ripening. *Plant Mol. Biol. Rep.* **2016**, *34*, 777–785. [CrossRef]
28. Bai, S.; Sun, Y.; Qian, M.; Yang, F.; Ni, J.; Tao, R.; Li, L.; Shu, Q.; Zhang, D. Transcriptome analysis of bagging-treated red Chinese sand pear peels reveals light-responsive pathway functions in anthocyanin accumulation. *Sci. Rep.* **2017**, *7*, 63. [CrossRef]
29. Zhang, Y.Y.; Yin, X.R.; Xiao, Y.W.; Zhang, Z.Y.; Li, S.J.; Liu, X.F.; Zhang, B.; Yang, X.F.; Donald, G. An Ethylene Response Factor-Myb transcription complex regulates furaneol biosynthesis by activating Quinone Oxidoreductase expression in strawberry. *Plant Physiol.* **2018**, *178*, 189–201. [CrossRef]

current issues in molecular biology

MDPI

Review

Research Progress on Anthocyanin-Mediated Regulation of 'Black' Phenotypes of Plant Organs

Fei Wang, Jinliao Chen, Ruonan Tang, Ruixin Wang, Sagheer Ahmad, Zhongjian Liu * and Donghui Peng *

Key Laboratory of National Forestry and Grassland Administration for Orchid Conservation and Utilization, College of Landscape Architecture and Art, Fujian Agriculture and Forestry University, Fuzhou 350002, China; wangfei_icon@163.com (F.W.); fjchenjl@126.com (J.C.); trn7020@163.com (R.T.); 13559107013@163.com (R.W.); sagheerhortii@gmail.com (S.A.)

* Correspondence: zjliu@fafu.edu.cn (Z.L.); fjpdh@fafu.edu.cn (D.P.)

Abstract: The color pattern is one of the most important characteristics of plants. Black stands out among the vibrant colors due to its rare and distinctive nature. While some plant organs appear black, they are, in fact, dark purple. Anthocyanins are the key compounds responsible for the diverse hues in plant organs. Cyanidin plays an important role in the deposition of black pigments in various plant organs, such as flower, leaf, and fruit. A number of structural genes and transcription factors are involved in the metabolism of anthocyanins in black organs. It has been shown that the high expression of R2R3-MYB transcription factors, such as *PeMYB7*, *PeMYB11*, and *CsMYB90*, regulates black pigmentation in plants. This review provides a comprehensive overview of the anthocyanin pathways that are involved in the regulation of black pigments in plant organs, including flower, leaf, and fruit. It is a great starting point for further investigation into the molecular regulation mechanism of plant color and the development of novel cultivars with black plant organs.

Keywords: color patterns; plant organs; anthocyanins; structural genes; transcription factors

Citation: Wang, F.; Chen, J.; Tang, R.; Wang, R.; Ahmad, S.; Liu, Z.; Peng, D. Research Progress on Anthocyanin-Mediated Regulation of 'Black' Phenotypes of Plant Organs. *Curr. Issues Mol. Biol.* **2023**, *45*, 7242–7256. https://doi.org/10.3390/cimb45090458

Academic Editors: Shimeles Tilahun and Angel Llamas

Received: 2 August 2023
Revised: 21 August 2023
Accepted: 30 August 2023
Published: 1 September 2023

1. Introduction

The significance of color diversity is recognized in the coevolution between plants and pollinators, such as insects and birds [1,2]. The variety of colors of the flowers of *Delphinium*, ranging from white to pink, scarlet, blue and purple, gives this plant great ornamental potential [3]. In *Chrysanthemum indicum*, a yellow flower is a good source of usual quercitrin and myricetin, which is important for the development of possible pharmaceuticals [4]. Except for the vibrant and vivid color patterns produced by most of the plants, a dark color can also be seen in some plant organs. For example, 'Queen of Night' (horticultural hybrid tulip) and *Lisianthius nigrescens* produce flowers with a dark purple color [5]. *Prunus cistena* 'Pissardii' possess black leaves [6] and *Aronia melanocarpa* produce black berries [7]. Color is an essential trait of plants, and the ornamental plant cultivars with multiple colors will be more diversified in the future. In particular, novel plant varieties with unique colors will increase in popularity. Therefore, understanding the mechanism of plant color patterns will be useful for breeding plants with a wide range of colors and for studying plant evolution [8–11].

Studies have found that various colors of plant organs are generally caused by the types and amount of accumulation of specific flavonoids, carotenoids, and alkaloids [12–14]. Anthocyanins are among the most important flavonoid compounds that are commonly found in numerous plants and fruits and play a vital role in the pigmentation of plant organs [12,15]. For instance, anthocyanins significantly affect the color of the fuchsia flower of chrysanthemum, the dark purple fruit of eggplant and the pink flower of lily [16–18]. The color of reddish leaf in poinsettia, and red and black berries in grape exhibit a significant correlation with the accumulation of anthocyanins [19,20].

Previous studies have shown that the variation in gene expression in the flavonoid biosynthesis pathway leads to a distinct accumulation of anthocyanin in plant organs, resulting in color polymorphism [21]. The reduction of cyanidin accumulation during fruit maturation in *Ananas comosusis* due to downregulation of *AcHOX21* and *AcMYB12*, and the fluctuations in the endogenous levels of JA (Jasmonic acid), GA3 (Gibberellic acid) and auxins drive the discoloration of *A. comosus* peel due to anthocyanin-mediated discoloration [22]. The *NsMYB1* gene promotes the accumulation of anthocyanin in the black fruit of *Nitraria sibirica* Pall. [23]. The purple leaves of *Dendrobium bigibbum* are associated with *MYB2*, and the transient overexpression of *DbMYB2* significantly enhances anthocyanin accumulation in tobacco [24]. In evergreen azaleas, a diverse range of anthocyanins can be observed in purple flowers in contrast to red flowers, while no anthocyanins are detected in the white petals [25]. Moreover, the dark color is attributed to the accumulation of anthocyanin in both the embryos and the seed coats of *Glycine max* [26]. These research findings suggest that the intensification of color is strongly associated with an increase in levels of anthocyanin. Moreover, gibberellins, sugars and light are crucial elements that are necessary for the activation of anthocyanin gene transcription and the accumulation of pigments [27]. The presence of sunlight can enhance the absorption of anthocyanins, particularly in the skin of apples and grapes, while the absence of light can cause the opposite effect [28,29].

The presence of black color in plants is a rare and attractive characteristic, and there exist some studies that have examined the molecular basis of this color in plants. This review examines the studies on the accumulation of anthocyanins and the regulation metabolism, which are responsible for the dark colors in plants. It also broadens our comprehension of the black color patterns found in various plant parts.

2. Synthesis Pathways and Regulation of Anthocyanin Metabolism

2.1. Biosynthesis of Anthocyanin

Anthocyanins are in the forms of anthocyanidin glycosides, which endow a variety of colors to plant organs, mainly ranging from red to purple and blue [30–33]. The colors provide plants with distinct visual effects through diverse biosynthetic pathways (examples in Table 1). Previous studies have shown that anthocyanins are derived from a branch of the flavonoid metabolism pathway in plants, and their biosynthesis takes place in three distinct phases [34–36].

Stage 1: Phenylalanine → 4-coumaryl-CoA. The primary enzymes, namely phenylalanine ammonia lyase (*PAL*), cinnamate 4-hydroxylase (*C4H*), and 4-coumaroyl-CoA ligase (*4CL*), catalyze the synthesis of phenylalanine, thereby generating 4-coumaroyl-CoA, which serves as the primary substrate for plant anthocyanin biosynthesis. This process is common to many secondary metabolisms in plants [37].

Stage 2: 4-coumaryl-coA and malonyl-CoA → dihydrokaempferol. In this stage, the synthesis of dihydrokaempferol is catalyzed by three different enzymes, namely *CHS*, *CHI*, and *F3H*. This process is a pivotal reaction in flavonoid metabolism, and the genes responsible for the synthesis of these three enzymes are referred to as early biosynthetic genes (EBGs) [12,38].

Stage 3: dihydrokaempferol, dihydroquercetin and dihydromyricetin → various anthocyanins. The enzyme dihydroflavonol 4-reductase (*DFR*) catalyzes the production of dihydrokaempferol, dihydroquercetin, and dihydromyricetin, thereby generating the corresponding leucoanthocyanidins. Then, the leucoanthocyanidins are transformed into anthocyanins with the catalytic action of anthocyanidin synthase (*ANS*) and UDP-glucose flavonoid glucosyltransferase (*UFGT*). This synthesis stage of anthocyanins is represented by the genes that regulate the synthesis of *DFR*, *ANS* and *UFGT*, which are referred to as late biosynthetic genes (LBGs) [39].

Table 1. Components of common anthocyanins and their coloration in different organs of plants.

Plants	Main Anthocyanins	Color	Plant Organ	Reference
Lisanthius nigrescense	delphinidin-3-*O*-rhamnol(1–6)galactoside, delphinidin-5-*O*-glucoside	black	corolla	[5]
Cosmos atrosanguineus	cyanidin-3-*O*-glucoside, cyanidin-3-*O*-rutinoside	black	flower	[40]
Cercis canadensis	cyanidin-3-glucoside and malvidin-3-glucoside	purple	flower	[41]
Dahlia variabilis	cyanidin-3-(6″-malonylglucoside)-5-glucoside	black	flower	[42]
Cyclamen purpurascens	cyanidin-3-*O*-rutinoside, cyanidin-3-*O*-glucoside, delphinidin-3-*O*-glucoside, malvidin-3-*O*-glucoside, peonidin-3-*O*-rutinoside	red	flower	[43]
Phacelia campanularia	phacelianin(dicaffeoyl anthocyanin): 3-*O*-(6-*O*-(4′-*O*-(6-*O*-(4′-*O*-β-D-glucopyranosyl-(*E*)-caffeoyl)-β-D-glucopyranosyl)-(*E*)-caffeoyl)-β-D-glucopyranosyl)-5-*O*-(6-*O*-malonyl-β-D-glucopyranosyl)delphinidin	blue	flower	[44]
Loropetalum chinense var. *rubrum*	petunidin-3,5-diglucoside	dark purple	leaf	[45]
eggplant	delphinidin-3-*p*-coumaroyl-rutinoside-5-glucoside	dark purple	fruit	[17]
Crataegus maximowiczii	cyanidin-3-*O*-glucoside, cyanidin-3-*O*-galactoside	black	fruit	[46]
soybean	cyanidin-3-glucoside and delphinidin-3-glucoside	black	seed	[47]
Zea mays L. *sinensis kulesh*	pelargonidin-3-*O*-glucoside	black	seed	[48]

2.2. Regulation of Anthocyanin Metabolism

The distribution of anthocyanins varies based on plant species, plant tissues, developmental stages, and environmental factors [49]. Anthocyanins are water-soluble compounds that are produced in the cytoplasm and subsequently transported to the vacuole and other parts of plants [50–52]. Until now, the membrane transporters involved in the anthocyanin transport have been confirmed, including ATP-binding cassette, multidrug and toxic compound extrusion (MATE), bilitranslocase homolog (BTL), and vesicle-mediated transport [53]. Although the major transporters have been adequately identified, further investigation is required to determine the molecular mechanism of anthocyanin transport from the synthesis site to the storage site.

The accumulation of anthocyanin in plants is regulated by a series of structural genes [54]. During the development of *Malus hupehensis*, the color of flowers undergoes a transition from red to white due to a decrease in the expression of anthocyanin biosynthesis genes [55]. In the rose variety 'Rhapsody in Blue', transient overexpression of *RhF3′H* and *RhGT74F2* has a significant impact on the accumulation of anthocyanins in the blue-purple petals [56]. The expression of two *CsUFGTs* genes exhibits a positive correlation with the substantial accumulation of anthocyanin compounds in the purple-leaf tea plant [57]. Moreover, it has been demonstrated that the expression profiles of *CHS*, *F3H*, *DFR*, *ANS*, and *UFGT* exhibit a positive correlation with the accumulation of anthocyanin in apples [28,58]. However, the expression profiles of these genes vary in plants based on tissue types, growth stages, and varieties.

Three transcription factor families, including MYB, bHLH, and WD40, play a crucial role in the regulation of anthocyanin accumulation [59]. The majority of MYB transcription factors exert a positive influence on the biosynthesis of anthocyanin in plants [60,61]. However, *CmMYB7* is a negative regulator of anthocyanin biosynthesis in 'Jinba', a white flowering chrysanthemum cultivar [62]. The *CPC* (Cross-Pathway Control Protein), which is closely associated with epidermis development, has the ability to decrease anthocyanin content in plants through inhibiting the expression of LBGs in the anthocyanin synthesis pathway [63]. The decrease in anthocyanin content observed in *Petunia hybrida* is attributed to the overexpression of *PhMYB27*, which has the ability to prevent the formation of MBW complexes or convert activation complexes into repressive complexes [64].

Numerous studies have revealed that the production of anthocyanin is significantly affected by pH, sugars, temperature, sunlight, and other factors [65–67]. The color of anthocyanins is dependent on the pH of the solution; this is because of the molecular structure of anthocyanins having an ionic nature [68]. Under acidic conditions, some of the anthocyanins appear red. Anthocyanins have a purple hue in neutral pH while the color changes to blue in an increasing pH condition [33]. Decreases in orchard temperatures result in a change in the color of the apple pericarp, indicating that the temperature has an impact on the biosynthesis of anthocyanin [69,70]. For example, the presence of high temperatures has the potential to significantly enhance the expression of numerous genes associated with anthocyanin biosynthesis, including but not limited to *PAL1*, *ANS*, *3GT*, *CHS2*, *UA5*, *DF4R*, *CHI*, *UA3GT2* and *UA3GHT5* in strawberry [71]. Moreover, elevated temperatures can enhance the absorption of anthocyanins from the endoplasmic reticulum to the vacuole by triggering the reactivation of *Mate TT12* genes, further deepening the color of fruit in strawberries. However, high temperatures can also reduce the amount of pigment in fruits by inhibiting the expression of genes and enzymatic activity involved in the production of anthocyanins [72–74]. Solfanelli et al. [75] studied the role of sugar in the synthesis of anthocyanins in plants. They found a significantly high expression of *CHS*, *CHI*, *F3H*, *F3′H*, and *FLS* at low concentrations of sucrose, whereas a concentrated sucrose solution only induced the expression of *DFR*, *LDOX*, and *UF3GT*. The photoperiod directly affects the expression of structural genes, which in turn regulate anthocyanin accumulation [76]. Exogenous gibberellin promotes the accumulation of anthocyanins in *P. hybrida* corolla by inducing the expression of *CHS* [77].

In addition to this, the molecular modification of anthocyanins can affect the formation of color. The process of glycosylation and methylation of anthocyanins results in a redder hue, whereas the accumulation of acylated anthocyanins results in a highly stable blue hue [78]. The balance between biosynthesis and degradation is what determines the accumulation of anthocyanins in plants [79]. The changes in pH, temperature, co-pigmentation, oxygen, and enzymes may affect the stability of anthocyanins, which is influenced by a variety of factors [80,81]. The occurrence of sporadic accumulation and disappearance of anthocyanin during plant development or changes in environmental conditions suggest that anthocyanin degradation is regulated in accordance with its requirements in plants [82]. High temperature increases the expression of some anthocyanin-degrading genes, such as laccase-9 and laccase-14, and also stimulates anthocyanin degradation by enhancing the activity of POD enzymes [83]. Despite the extensive research conducted on anthocyanin biosynthesis, the knowledge regarding its degradation remains limited [84,85]. The color of fruits, flowers, and leaves in plants holds significant ornamental value as ornamental plant, and economic value as in a variety of agricultural products. So, a comprehensive assessment of anthocyanin degradation may provide new insights into ways to inhibit the process and consequently enhance pigmentation in conditions of low synthesis.

2.3. Color Modification

Color is one of the most important characteristics of many plant types. But some plants have limited color ranges because of the genetics of the species, and genetic modification technology is the sole efficacious approach to overcome this limitation [86,87]. For example, through genetic modification, the flower color of *Phalaenopsis* spp. and *Cyclamen persicum* can be changed from pink to light pink, from purple to red or pink, respectively [88,89]. There are violet carnations, roses, and chrysanthemums that have been developed by expressing a petunia, pansy, or campanula flavonoid 3-,5-hydroxylase gene, and genetically modified carnation and rose varieties have been commercialized [90]. In addition, transcription factors regulating the anthocyanin pathway have been identified, and as further knowledge is gained regarding the spatial regulation of flavonoid biosynthesis, there will be potential for the genetic modification of pigmentation patterns in more plants [91–93].

3. Black Organs in Plants

3.1. Black Flower

The flowers of most angiosperms are bright-colored, which makes them more attractive to pollinators. Despite this, the species that produce black flowers hold a great significance (as illustrated in Figure 1). In fact, there is no plant in nature that is purely black. Although certain plant organs may appear black to the naked eye, they actually possess a dark shade of purple owing to the substantial accumulation of anthocyanins [40,41,46]. In 1996, a variety of *Phalaenopsis aphrodite* with black spots on petals was discovered, which is an important breeding resource for generating color variation in flowering plants [94,95]. *Tulipa Julia* has black patches on the underside of its petals, and the intense violet flowers of 'Queen of Night' (a hybrid tulip) appear black under certain lighting conditions [5]. The *L. species*, belonging to the Gentianaceae family, is a distinctive black-flowered species in the plant kingdom, renowned for its striking black tubular blossoms that can reach up to 5 cm in length [5]. According to a study conducted by Shibata et al. [96], only five varieties of *Tulipa gesneriana* were found to possess black flowers out of a total of 107 varieties.

Figure 1. Examples of dark purple flowers. (**a**) Flower of *Tacca chantrieri*. (**b**) Flower of *Dahlia pinnata*. (**c**) Spathe of *Zantedeschia aethiopica*. (**d**) Flower of *Clowesia jumbo*.

3.2. Black Leaf and Fruit

The majority of plants lack black foliage, but a few species still possess this characteristic, such as *L. chinense* var. *rubrum* and *Prunus cerasifera*, which have dark purple leaves (Figure 2a,b). Black fruits such as *P. cerasifera* and *Morus alba* var. *alba*, which possess a high concentration of anthocyanins and appear dark purple (Figure 2b–d), have the potential to serve as effective antioxidants and health supplements [97]. They possess remarkable antioxidant properties in removing free radicals from the body, enhancing blood vessel flexibility, preventing cardiovascular diseases and cancer [98]. The leaves of *P. cistena* 'Pissardii' and *P. cerasifera* exhibt deep purple and aubergine hues, respectively, and possess significant ornamental value [6,99].

In recent years, black fruits have experienced a significant increase in demand owing to their potential utilization as a food colorant and as a source of valuable natural anthocyanins [100]. Black berries (*A. melanocarpa*) possess a high level of anthocyanin content, which significantly enhances their nutritional value [7,100,101]. The fruits of *Lycium ruthenicum* and *Morus nigra* are also purple-black or purple-red because they contain abundant anthocyanins [102,103]. Furthermore, there exist vegetables and crops in nature that exhibit a black hue, such as eggplant [17], black carrot [104], black seed soybean [47,105], black rapeseed [37] and black rice [106].

(a) (b) (c) (d)

Figure 2. Examples of dark purple leaf and fruit. (**a**) Leaf of *L. chinense* var. *rubrum*. (**b**) Leaf and fruit of *P. cerasifera*. (**c**) Fruit of *M. alba* var. *alba*. (**d**) Fruit of *T. chantrieri*.

4. Regulation of Anthocyanin Metabolism in Black Organs in Plants

4.1. Components of Anthocyanins

As depicted in Figure 3, some studies have demonstrated that cyanidin, pelargonidin, and delphinidin are the common anthocyanins found in the dark plant organs [107–109]. The black flower color of *D. variabilis* is caused by the substantial accumulation of cyanidin-3-(6′′-malonylglucoside)-5-glucoside [108]. The purple-violet flowers of transgenic chrysanthemum are caused by the accumulation of delphinidin in ray florets, which is caused by the B-ring hydroxylation of anthocyanin, which transforms cyanidin to delphinidin, resulting in the flower color changing from magenta to purple or pink to violet [110]. In the black flowers of *Tulipa* 'Queen of Night', three primary anthocyanins are identified, namely delphinidin (50%), cyanidin (29%) and pelargonidin (21%), and delphinidin 3-glucoside is the most common type of delphinidin pigment [96]. The *p*-coumaroyltriglycoside of delphinidin is a predominant constituent of the dark purple flowers of *Viola tricolor* 'Jet Black' [111]. The flowers of black *C. atrosanguineus* contain two primary anthocyanins, namely cyanidin-3-*O*-glucoside and cyanidin-3-*O*-rutinoside, and the total anthocyanin content in the black variety is approximately 3~4 times higher than that in the red variety [40]. The black corolla of *L. nigrescense* contains one major pigment and one minor pigment, and the contents of delphinidin-3-*O*-rhamnol(1–6)galactoside and delphinidin 5-*O*-glucoside account for 24% of the petals' dry weight [5].

The leaves of *P. cistena* 'pissardii' appear deep purple when they are exposed to strong sunlight [6]. Coexistence and interaction among cyanidin galactoside, cyanidin and chlorophyll are the main causes of the purplish red leaf of *P. cerasifera* [99]. Petunidin-3-*O*-glucoside, anthocyanin-3-*O*-galactoside, and anthocyanin-3-*O*-glucoside are the main anthocyanins that cause the purple leaf phenotype of the tea plant [112]. Zhao et al. [113] have demonstrated that a millet variety (B100) exhibits purple leaves during the seedling and maturity stages. Purple pigments are mainly distributed in the leaf epidermis. The purple leaf color of *L. chinense* var. *rubrum* is influenced by the petunidin-3,5-diglucoside [45].

Five cyanidin derivatives have been identified from blackberry, namely cyanidin-3-rutinoside, cyanidin-3-(malonyl)-glucoside, cyanidin-3-xyloside, cyanidin-3-glucoside and cyanidin-3-dioxaloylglucoside [114,115]. A previous study revealed the presence of four significant anthocyanins in the purplish black berries of *A. melanocarpa*, including cyanidin-3-*O*-galactoside (68.68%), cyanidin-3-*O*-arabinoside (25.62%), cyanidin-3-*O*-glucoside (5.28%) and cyanidin-3-*O*-xyloside (0.42%) [7]. The purple black fruits of *L. ruthenicum* are loaded with petunidin derivatives, which have high ornamental and economic significance [102]. The maturation stage of the mulberry fruit drives the gradual change in color from light red to blackish purple due to the accumulation of anthocyanins. A study conducted on 11 genotypes of *Morus alba*, comprising five black, four white, and two pink multiple fruit varieties, revealed that the abundance of anthocyanins in black fruits ranges from 45.42 to 208.74 mg per 100 g [116], while the amount of cyanidin-3-glucoside in the fresh fruit of *M. nigra* was very high, at 704.1 mg per 100 g [103]. The black-colored fruits of *C. maximowiczii* are closely associated with the accumulation of cyanidin, pelargonidin, peonidin and delphinidin derivatives, particularly cyanidin-3-*O*-glucoside and cyanidin-3-*O*-galactoside. And the contents of delphinidin-3-*O*-galactoside, pelargonidin-3-*O*-arabinoside, pelargonidin-3-*O*-glucoside, peonidin-3-*O*-arabinoside and peonidin-3-*O*-glucoside in black peel are twice as high as in red peel [46].

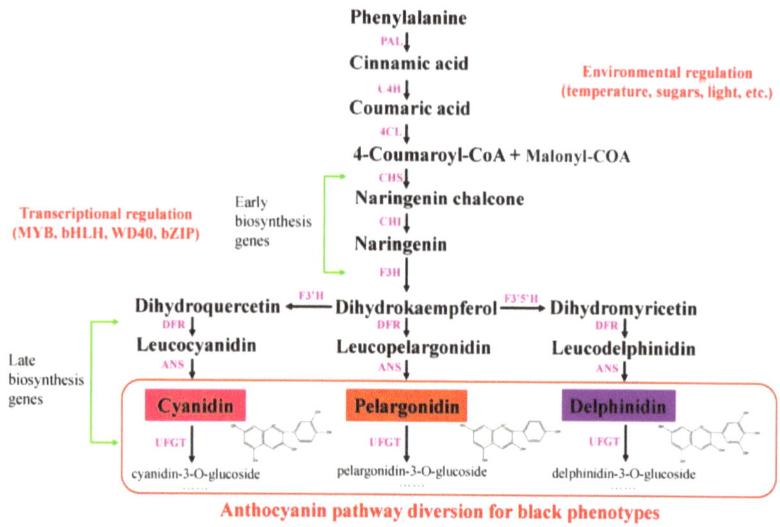

Figure 3. A simplified schematic of the anthocyanin pathway leading to black phenotypes in plant. PAL: Phenylalanine ammonia lyase. C4H: cinnamate 4-hydroxylase. 4CL: 4-coumaroyl-CoA ligase; CHS: Chalcone synthase; CHI: Chalcone isomerase; F3H: Flavanone-3-hydroxylase; F3′H: Flavanone-3′-hydroxylase; F3′5′H: Flavanone-3′,5′-hydroxylase; DFR: Dihydroflavonol-4-reductase; ANS: Anthocyanidin synthase; UFGT: UDP-glucose flavonoid 3-O-glucosyltransferase.

Previous studies have revealed that the dark purple of eggplant is attributed to delphinidin-3-p-coumaroylrutinoside-5-glucoside [17], whereas the accumulation of black pigments in soybean seeds is caused by cyanidin-3-glucoside and delphinidin-3-glucoside [47,105]. In *Capsicum annuum*, only a single anthocyanin (delphinidin-3-*p*-coumaroyl-rutinoside-5-glucoside) is found in the violet fruit, black fruit, and black leaves. The distinctive black pigmentation is caused by the high concentrations of delphinidin, combined with chlorophyll and other carotenoid pigments [117]. The dark purple color of *Daucus carota* subsp. *sativus* (a black variety of carrot) is due to an acylated form of anthocyanin, namely [cyanidin-3-(*p*-coumaroyl)-diglucoside-5-glucoside] [104,118]. The content of black pigment in the black rapeseed seed coat is significantly higher than that of other tissues [37]. Four anthocyanins were identified from black rice, namely cyanidin-3-rutinoside, peonidin-3-glucoside, cyanidin-3,5-diglucoside and cyanidin-3-glucoside [106,119].

In general, the genetic background of the species or variety determines the constituents of anthocyanins in plants [120]. Furthermore, certain studies have suggested that the development of plant color is correlated with the structure of organ tissue, pigment distribution and its types. So, it is possible that it will be regulated through genetic engineering, which has made rapid progress in color breeding of plants for its advantages over traditional breeding technologies [12]. Nonetheless, the mechanism of anthocyanin synthesis and metabolism is highly intricate, encompassing numerous metabolic steps and enzymes. Therefore, there exist numerous structural genes and regulatory genes associated with anthocyanin pigments (37–40). At present, scientists have conducted a comprehensive examination of the synthetic pathway of anthocyanins, which are commonly present in plants, and their associated genes. However, it remains challenging to alter a specific trait of plants to generate distinctive color phenotypes of the species and to breed novel varieties with stable inheritance within a brief timeframe.

4.2. Structural Genes

The accumulation of pigments in different plant organs is regulated by many structural genes, which are involved in the synthesis of anthocyanins. For example, the knockdown of *F3H* by RNAi in torenia with blue flower produces white flowers [121]. The accumulation of anthocyanins and flavonols in the white and red flower species are caused by the expression of *DFR* and *FLS* genes, and heterologous *FLS* expression in transgenic tobacco promotes flavonol biosynthesis and blocks anthocyanin accumulation, leading to white flowers [122]. In addition, *FNS* and *IVS* are key genes involved in anthocyanin biosynthesis and regulation in black flower plants. In the black flower variety of *D. pinnata*, DvIVS-1 promoter has high activity but the expression of *DvFNS* is significantly decreased. Moreover, artificial silencing of *FLS* or *FNS* results in increased accumulation of anthocyanin in *P. hybrida* [123]. This suggests that the silencing of *DvFNS* can lead to the loss of flavonoids and eliminates competition for substrates, so that substrates originally used for flavonoid synthesis can be used for anthocyanin synthesis [42]. Then, the DvIVS-1 promoter helps to synthesize large amounts of anthocyanins in black flowers of *D. pinnata*. A high expression of a number of genes (such as *RsCHI1*, *RsFLS1*, *RsANS2*, and *RsAT2*) contributes to the deep blackish crimson flowers of a variety of *Rhododendron sanguineum* [21]. Some research shows that flower color deepens with the increase in anthocyanin content [124,125]. Hence, high expression levels of structural genes are primarily useful for deep hues by promoting the production of anthocyanin [126]. The study on the reddish-purple color in the petals of *Rhododendron simsii* flowers have shown that co-pigmentation, normally with flavonols contributed by *RsFLS*, may result in high accumulation of anthocyanins that shift color toward deep blackish crimson [127].

The expressions of nine genes (*PAL, 4CL, DFR, LDOX-1, LDOX-2, AT, UFGT, GT, 5GT*) related to anthocyanin synthesis are significantly higher in purple leaf of foxtail millet (B100) at maturity stage than the green leaf variety (YG1). The expression of three genes (*DFR, LDOX-2* and *AT*) in purple leaf of B100 are significantly higher at seedling stage and maturity stage [113]. A previous study confirmed the key structural genes of anthocyanin biosynthesis in purple leaf of ZK, including two *F3′H* genes, two *ANS* genes with positive correlations and three PPO genes with negative correlations [112]. Zong et al. [128] detected the transcriptional product of *AN2* in the black fruit of *Lycium barbarum*. The genetic diversity analysis of *AN2* gene also shows that yellow, white, purple, and red cultivars of *Lycium chinense* originate from *L. barbarum*. The overexpression of *IbMYB1-2* can significantly increase anthocyanin content in the root tuber of transgenic sweet potato [129]. In the anthocyanin biosynthesis pathway, the high expression of *F3′H* and *UFGT* genes promotes a high accumulation of cyanidin derivatives, producing *Crataegus pinnatifida* with black fruit [46].

4.3. Transcription Factors

The regulation of anthocyanin accumulation is jointly regulated by transcription activators and transcription repressors, primarily comprising MYB, bHLH, WD40, and bZIP [59]. *PeMYB11* is one of them, and it is a major R2R3-MYB TF that is highly expressed in the black flowers of *Phalaenopsis equestris* [95]. The HORT1 (Harlequin Orchid RetroTransposon 1) can lead to a strong expression of *PeMYB11*. Therefore, the flowers of *P.* Yushan Little Pearl variety, which contains HORT1, have an excellent anthocyanin accumulation capacity [95,130]. The purple spots on the sepals of *Phalaenopsis aphrodite* 'Panda' are regulated by *PeMYB7*, *PeMYB11*, miR156, and miR858 [130]. Moreover, it should be noted that miR156 and miR858 are the primary siRNAs of *PeMYB7* and *PeMYB11*, respectively, and both cause a significant increase in the expression of genes associated with the anthocyanin biosynthesis pathway (*PeCHl*, *PeANS*, *PeC4H*, *PeF3H*, *PeF3′H*, *Pe3Hl*, and *Pe4CL2*) in spot tissues [130]. In addition, the abnormal expression of bHLH or MYB results in the appearance of dark purple leaf and flower in transgenic petunia plants [131,132]. Anthocyanin accumulation in purple leaves of ZK is strongly correlated with *CsMYB90*, and *CsMYB90* overexpression in transgenic tobacco plants with dark purple callus is also observed [112].

Research has demonstrated that the bZIP family principally functions as a positive regulator of anthocyanin biosynthesis. However, Tu et al. [133] have discovered that *VvbZIP36* is a negative regulator of anthocyanin biosynthesis in *Vitis vinifera* and plays an important role in balancing the synthesis of stilbenes (α-viniferin), lignans, flavonols, and anthocyanins. The insertion of a precursor DNA transposon into the regulatory region of Purple (Pr), which belongs to the R2R3-MYB TF encoding genes, results in the up-regulation of Pr expression, thereby causing the accumulation of dark color in *Brassica oleracea* [134]. It appears that the purple color of *Ipomoea batatas* is caused by the dominant expression of *IbMYB1* [135]. In conclusion, the enhanced expression of regulatory transcription factors in the anthocyanin biosynthesis pathway may be responsible for the appearance of black flowers, leaves, and fruits in plants. However, the specific molecular mechanism is still to be elucidated.

4.4. Other Factors

The stability of anthocyanins can be increased by modification to form stable structures. In most plants, only *O*-glycosylation occurs for anthocyanins. In grapes, the structures of the individual anthocyanins include both 3-*O*-monoglucosides and 3,5-*O*-diglucosides. Diglucosidic anthocyanins are more stable than their monoglucosidic counterparts, whereas monoglucosidic anthocyanins tend to have deeper colors than their diglucosidic counterparts [136]. Furthermore, prolonged exposure to high temperatures and prolonged exposure to sunlight significantly affect the stability of anthocyanins in plants [137]. For example, anthocyanin stability against heat stress is increased by the methoxylation and acylation of cyanidin-3-*O*-glucoside from blackberries [138]. Diacylated anthocyanins provide significantly higher blue color stability to red cabbage at 50 °C compared to non-acylated anthocyanins [139]. In black carrot (*Daucus carota* L.), acylated anthocyanins remain more stable during temperature increases of 20–50 °C than non-acylated anthocyanins from blackberry (*Rubus glaucus* Benth.) [140]. And the level of anthocyanins from black carrots remains relatively stable until 90 °C, probably due to di-acylation of the anthocyanin structure [141,142]. Anthocyanins are protected from hydration by acylation, thereby making them more stable, because acylated anthocyanins are generated after the acylation of glycosyl groups of anthocyanins [143]. However, the acylated anthocyanins in black carrot are decomposed under extreme heat stress (95 °C), indicating that the stability of acylated anthocyanins is rapidly decreased [144]. Acylated anthocyanins are found in flowers and vegetables, whereas non-acylated anthocyanins are mostly distributed in fruits [145].

Some encoding enzymes used in biosynthesis and genes responsible for regulating black pigmentation have been identified using advanced molecular biology techniques. But the effect of accumulation and stability on various factors (e.g., pH, anthocyanin transport) deserves further investigation.

5. Conclusions and Future Directions

In this review, the molecular mechanism of anthocyanin-mediated black pigmentation in plants is analyzed. Cyanidin is the key factor in black pigmentation and induces black color in ornamental and fruit crops, but the current investigations into black pigmentation in plants are inadequate. Further research on the temporal variation of gene expression in diverse species, organs, and tissues, the interactions between transcription factors and genes, and the effects of anthocyanin transport and of environmental factors on black pigmentation are still lacking. Therefore, it is imperative to select materials of wild-type or self-crossing origin, possessing original color and relatively stable homozygous genotypes, in order to conduct further investigations on the molecular regulation mechanism of black color in plants. It also holds a great significance to achieve color modifications to obtain black color in ornamental plants, fruits, and vegetable crops through the utilization of genetic engineering technology in the future.

Author Contributions: Conceptualization, F.W., Z.L. and D.P.; writing—original draft preparation, F.W.; investigation and visualization, R.T. and R.W.; writing—review and editing, F.W., J.C. and S.A.; Supervision, Z.L. and D.P.; funding acquisition, D.P. All authors have read and agreed to the published version of the manuscript.

Funding: This research was supported by the Forestry Peak Discipline Construction Project of Fujian Agriculture and Forestry University (72202200205).

Institutional Review Board Statement: Not applicable.

Informed Consent Statement: Not applicable.

Data Availability Statement: The datasets analyzed in this study are available in the manuscript text.

Acknowledgments: The authors thank Yunxiao Guan of laboratory for valuable comments of the manuscript.

Conflicts of Interest: The authors declare no conflict of interest.

References

1. Bradshaw, H.D.; Schemske, D.W. Allele substitution at a flower colour locus produces a pollinator shift in monkeyflowers. *Nature* **2003**, *426*, 176–178. [CrossRef]
2. Streisfeld, M.; Kohn, J. Environment and pollinator-mediated selection on parapatric floral races of *Mimulus aurantiacus*. *J. Evol. Biol.* **2007**, *20*, 122–132. [CrossRef] [PubMed]
3. Kolar, F.R.; Ghatge, S.R.; Nimbalkar, M.S.; Dixit, G.B. Mutational changes in *Delphinium malabaricum* (Huth.) Munz.: A potential ornamental plant. *J. Hortic. Res.* **2015**, *23*, 5–15. [CrossRef]
4. Wu, L.Y.; Gao, H.Z.; Wang, X.L.; Ye, J.H.; Lu, J.L.; Liang, Y.R. Analysis of chemical composition of *Chrysanthemum indicum* flowers by GC/MS and HPLC. *J. Med. Plants Res.* **2010**, *4*, 421–426.
5. Markham, K.R.; Bloor, S.J.; Nicholson, R.; Rivera, R.; Shemluck, M.; Kevan, P.G.; Michener, C. Black flower coloration in wild *Lisianthius nigrescens*: Its chemistry and ecological consequences. *Z. Für Naturforschung C* **2004**, *59*, 625–630. [CrossRef] [PubMed]
6. Wang, Q.J. The Rule of Anthocyanin Synthesis and Infulence Factors in the Leaves of Red-leaf Tree Species in Prunus. Master's Thesis, Shandong Agricultural University, Taian, China, 2007.
7. Gao, N.; Sun, X.; Li, D.; Gong, E.; Tian, J.; Si, X.; Jiao, X.; Xing, J.; Wang, Y.; Meng, X.; et al. Optimization of anthocyanidins conversion using chokeberry pomace rich in polymeric proanthocyanidins and cellular antioxidant activity analysis. *Food Sci. Technol.* **2020**, *133*, 109889. [CrossRef]
8. Noman, A.; Aqeel, M.; Deng, J.; Khalid, N.; Sanaullah, T.; Shuilin, H. Biotechnological advancements for improving floral attributes in ornamental plants. *Front. Plant Sci.* **2017**, *8*, 530. [CrossRef]
9. Whibley, A.C.; Langlade, N.B.; Andalo, C.; Hanna, A.I.; Bangham, A.; Thébaud, C.; Coen, E. Evolutionary paths underlying flower color variation in *Antirrhinum*. *Science* **2006**, *313*, 963–966. [CrossRef]
10. Hopkins, R.; Rausher, M.D. Pollinator-mediated selection on flower color allele drives reinforcement. *Science* **2012**, *335*, 1090–1092. [CrossRef]
11. Mu, J.; Yang, Y.; Luo, Y.; Su, R.; Niklas, K. Pollinator preference and pollen viability mediated by flower color synergistically determine seed set in an Alpine annual herb. *Ecol. Evol.* **2017**, *7*, 2947–2955. [CrossRef]
12. Zhao, D.Q.; Tao, J. Recent advances on the development and regulation of flower color in ornamental plants. *Front. Plant Sci.* **2015**, *6*, 261. [CrossRef] [PubMed]
13. Yuan, H.; Zhang, J.; Nageswaran, D.; Li, L. Carotenoid metabolism and regulation in horticultural crops. *Hortic. Res.* **2015**, *2*, 15036. [CrossRef] [PubMed]
14. Christinet, L.; Burdet, F.X.; Zaiko, M.; Hinz, U.; Zrÿd, J.-P. Characterization and functional identification of a novel plant 4, 5-extradiol dioxygenase involved in betalain pigment biosynthesis in *Portulaca grandiflora*. *Plant Physiol.* **2004**, *134*, 265–274. [CrossRef] [PubMed]
15. Koo, Y.; Poethig, R.S. Expression pattern analysis of three R2R3-MYB transcription factors for the production of anthocyanin in different vegetative stages of *Arabidopsis* leaves. *Appl. Biol. Chem.* **2021**, *64*, 5. [CrossRef]
16. Kawase, K.; Tsukamoto, Y. Studies on flower color in *Chrysanthemum morifolium* Ramat III. Quantitative effects of major pigments on flower color variation, and measurement of color qualities of petals with a color difference meter. *J. Jpn. Soc. Hortic. Sci.* **1976**, *45*, 65–75. [CrossRef]
17. Noda, Y.; Kneyuki, T.; Igarashi, K.; Mori, A.; Packer, L. Antioxidant activity of nasunin, an anthocyanin in eggplant peels. *Toxicology* **2000**, *148*, 119–123. [CrossRef]
18. Yamagishi, M. How genes paint lily flowers: Regulation of colouration and pigmentation patterning. *Sci. Hortic.* **2013**, *163*, 27–36. [CrossRef]

19. Moustaka, J.; Tanou, G.; Giannakoula, A.; Adamakis, I.-D.S.; Panteris, E.; Eleftheriou, E.P.; Moustakas, M. Anthocyanin accumulation in poinsettia leaves and its functional role in photo-oxidative stress. *Environ. Exp. Bot.* **2020**, *175*, 104065. [CrossRef]
20. Boss, P.; Davies, C. Molecular biology of anthocyanin accumulation in grape berries. In *Grapevine Molecular Physiology & Biotechnology*; Springer: Berlin/Heidelberg, Germany, 2009; pp. 263–292.
21. Ye, L.-J.; Möller, M.; Luo, Y.-H.; Zou, J.-Y.; Zheng, W.; Wang, Y.-H.; Liu, J.; Zhu, A.-D.; Hu, J.-Y.; Li, D.-Z. Differential expressions of anthocyanin synthesis genes underlie flower color divergence in a sympatric *Rhododendron sanguineum* complex. *BMC Plant Biol.* **2021**, *21*, 204. [CrossRef]
22. Luan, A.; Zhang, W.; Yang, M.; Zhong, Z.; Wu, J.; He, Y.; He, J. Unveiling the molecular mechanism involving anthocyanins in pineapple peel discoloration during fruit maturation. *Food Chem.* **2023**, *412*, 135482. [CrossRef]
23. Bao, X.; Zong, Y.; Hu, N.; Li, S.; Liu, B.; Wang, H. Functional R2R3-MYB transcription factor NsMYB1, regulating anthocyanin biosynthesis, was relative to the fruit color differentiation in *Nitraria sibirica* Pall. *BMC Plant Biol.* **2022**, *22*, 186. [CrossRef]
24. Lim, G.-H.; Kim, S.W.; Ryu, J.; Kang, S.-Y.; Kim, J.-B.; Kim, S.H. Upregulation of the MYB2 Transcription Factor is Associated with Increased Accumulation of Anthocyanin in the Leaves of *Dendrobium bigibbum*. *Int. J. Mol. Sci.* **2020**, *21*, 5653. [CrossRef]
25. Mizuta, D.; Ban, T.; Miyajima, I.; Nakatsuka, A.; Kobayashi, N. Comparison of flower color with anthocyanin composition patterns in evergreen azalea. *Sci. Hortic.* **2009**, *122*, 594–602. [CrossRef]
26. Lu, N.; Rao, X.; Li, Y.; Jun, J.H.; Dixon, R.A. Dissecting the transcriptional regulation of proanthocyanidin and anthocyanin biosynthesis in soybean (*Glycine max*). *Plant Biotechnol. J.* **2021**, *19*, 1429–1442. [CrossRef]
27. Weiss, D. Regulation of flower pigmentation and growth: Multiple signaling pathways control anthocyanin synthesis in expanding petals. *Physiol. Plant.* **2000**, *110*, 152–157. [CrossRef]
28. Takos, A.M.; Jaffé, F.W.; Jacob, S.R.; Bogs, J.; Robinson, S.P.; Walker, A.R. Light-induced expression of a MYB gene regulates anthocyanin biosynthesis in red apples. *Plant Physiol.* **2006**, *142*, 1216–1232. [CrossRef]
29. Azuma, A.; Yakushiji, H.; Koshita, Y.; Kobayashi, S. Flavonoid biosynthesis-related genes in grape skin are differentially regulated by temperature and light conditions. *Planta* **2012**, *236*, 1067–1080. [CrossRef] [PubMed]
30. De Pascual-Teresa, S.; Santos-Buelga, C.; Rivas-Gonzalo, J.C. LC–MS analysis of anthocyanins from purple corn cob. *J. Sci. Food Agric.* **2002**, *82*, 1003–1006. [CrossRef]
31. Di Paola-Naranjo, R.D.; Sánchez-Sánchez, J.; González-Paramás, A.M.; Rivas-Gonzalo, J.C. Liquid chromatographic–mass spectrometric analysis of anthocyanin composition of dark blue bee pollen from *Echium plantagineum*. *J. Chromatogr. A* **2004**, *1054*, 205–210. [CrossRef] [PubMed]
32. Lohachoompol, V.; Mulholland, M.; Srzednicki, G.; Craske, J. Determination of anthocyanins in various cultivars of highbush and rabbiteye blueberries. *Food Chem.* **2008**, *111*, 249–254. [CrossRef]
33. Khoo, H.E.; Azlan, A.; Tang, S.T.; Lim, S.M. Anthocyanidins and anthocyanins: Colored pigments as food, pharmaceutical ingredients, and the potential health benefits. *Food Nutr. Res.* **2017**, *61*, 1361779. [CrossRef]
34. Grotewold, E. The genetics and biochemistry of floral pigments. *Annu. Rev. Plant Biol.* **2006**, *57*, 761–780. [CrossRef]
35. Tanaka, Y.; Sasaki, N.; Ohmiya, A. Biosynthesis of plant pigments: Anthocyanins, betalains and carotenoids. *Plant J.* **2008**, *54*, 733–749. [CrossRef]
36. Zhang, J.; Li, S.; An, H.; Zhang, X.; Zhou, B. Integrated transcriptome and metabolome analysis reveals the anthocyanin biosynthesis mechanisms in blueberry (*Vaccinium corymbosum* L.) leaves under different light qualities. *Front. Plant Sci.* **2022**, *13*, 1073332. [CrossRef] [PubMed]
37. Zhang, X.; Chen, J.; Chen, L.; Wang, H.; Li, J. Imbibition behavior and flooding tolerance of rapeseed seed (*Brassica napus* L.) with different testa color. *Genet. Resour. Crop Evol.* **2008**, *55*, 1175–1184. [CrossRef]
38. Huang, J.-X.; Qu, L.-J.; Yang, J.; Yin, H.; Gu, H.-Y. A preliminary study on the origin and evolution of chalcone synthase (CHS) gene in angiosperms. *Acta Bot. Sin.* **2004**, *46*, 10–19.
39. Pelletier, M.K.; Murrell, J.R.; Shirley, B.W. Characterization of flavonol synthase and leucoanthocyanidin dioxygenase genes in *Arabidopsis* (Further evidence for differential regulation of "early" and "late" genes). *Plant Physiol.* **1997**, *113*, 1437–1445. [CrossRef] [PubMed]
40. Amamiya, K.; Iwashina, T. Qualitative and quantitative analysis of flower pigments in chocolate Cosmos, *Cosmos atrosanguineus*, and its hybrids. *Nat. Prod. Commun.* **2016**, *11*, 1934578X1601100122. [CrossRef]
41. Perkins-Veazie, P.; Ma, G.; Schwickerath, J.; Meyer, E.; Chen, H. The Absence of Malvidin-3-Glucoside in Petiole Tissue Predicts Rare Red-Type Flower of Eastern Redbud (*Cercis canadensis* L.). *Agriculture* **2023**, *13*, 598. [CrossRef]
42. Deguchi, A.; Tatsuzawa, F.; Hosokawa, M.; Doi, M.; Ohno, S. Quantitative evaluation of the contribution of four major anthocyanins to black flower coloring of dahlia petals. *Hortic. J.* **2016**, *85*, 340–350. [CrossRef]
43. Xia, D.; He, G.; Wang, K.; Wang, T.; Zhu, Z.; Niu, Z.; Shi, G.; Liu, G. Anthocyanins Profiling Analysis and RNA-Seq Revealed the Dominating Pigments and Coloring Mechanism in *Cyclamen* Flowers. *Biology* **2022**, *11*, 1721. [CrossRef]
44. Mori, M.; Kondo, T.; Toki, K.; Yoshida, K. Structure of anthocyanin from the blue petals of *Phacelia campanularia* and its blue flower color development. *Phytochemistry* **2006**, *67*, 622–629. [CrossRef]
45. Zhang, X.; Zhang, L.; Zhang, D.; Su, D.; Li, W.; Wang, X.; Chen, Q.; Cai, W.; Xu, L.; Cao, F. Comprehensive analysis of metabolome and transcriptome reveals the mechanism of color formation in different leave of *Loropetalum Chinense* var. Rubrum. *BMC Plant Biol.* **2023**, *23*, 133. [CrossRef]

46. Zhang, X.; Wang, J.; Li, P.; Sun, C.; Dong, W. Integrative metabolome and transcriptome analyses reveals the black fruit coloring mechanism of *Crataegus maximowiczii* C.K. Schneid. *Plant Physiol. Biochem.* **2023**, *194*, 111–121. [CrossRef]
47. Nagai, I. A genetico-physiological study on the formation of anthocyanin and brown pigments in plants. *J. Coll. Agric. Imp. Univ. Tokyo* **1921**, *8*, 1–92.
48. Hu, X.; Liu, J.; Shan, Q.; Bai, S.; Li, W.; Wen, T.; Guo, X.; Hu, J. The Accumulation and Biosynthesis of Anthocyanin in Black, White, and Yellow Waxy Corns (*Zea mays* L. sinensis kulesh) during Kernel Maturation. *Foods* **2023**, *12*, 1486. [CrossRef]
49. Zhang, Q.; Zhai, J.; Chen, G.; Lin, W.; Peng, C. The changing distribution of anthocyanin in *Mikania Micrantha* leaves as an adaption to low-temperature environments. *Plants* **2019**, *8*, 456. [CrossRef]
50. Dooner, H.K.; Robbins, T.P.; Jorgensen, R.A. Genetic and developmental control of anthocyanin biosynthesis. *Annu. Rev. Genet.* **1991**, *25*, 173–199. [CrossRef]
51. Jeong, S.T.; Goto-Yamamoto, N.; Kobayashi, S.; Esaka, M. Effects of plant hormones and shading on the accumulation of anthocyanins and the expression of anthocyanin biosynthetic genes in grape berry skins. *Plant Sci.* **2004**, *167*, 247–252. [CrossRef]
52. Koes, R.; Verweij, W.; Quattrocchio, F. Flavonoids: A colorful model for the regulation and evolution of biochemical pathways. *Trends Plant Sci.* **2005**, *10*, 236–242. [CrossRef]
53. Kaur, S.; Sharma, N.; Kapoor, P.; Chunduri, V.; Pandey, A.K.; Garg, M. Spotlight on the overlapping routes and partners for anthocyanin transport in plants. *Physiol. Plant.* **2021**, *171*, 868–881. [CrossRef]
54. Dubos, C.; Le Gourrierec, J.; Baudry, A.; Huep, G.; Lanet, E.; Debeaujon, I.; Routaboul, J.M.; Alboresi, A.; Weisshaar, B.; Lepiniec, L. MYBL2 is a new regulator of flavonoid biosynthesis in *Arabidopsis thaliana*. *Plant J.* **2008**, *55*, 940–953. [CrossRef] [PubMed]
55. Han, M.; Yang, C.; Zhou, J.; Zhu, J.; Meng, J.; Shen, T.; Xin, Z.; Li, H. Analysis of flavonoids and anthocyanin biosynthesis-related genes expression reveals the mechanism of petal color fading of *Malus hupehensis* (Rosaceae). *Braz. J. Bot.* **2020**, *43*, 81–89. [CrossRef]
56. Jiang, S.-H.; Wang, H.-H.; Zhang, R.; Yang, Z.-Y.; He, G.-R.; Ming, F. Transcriptomic-based analysis to identify candidate genes for blue color rose breeding. *Plant Mol. Biol.* **2023**, *111*, 439–454. [CrossRef] [PubMed]
57. Chen, X.; Wang, P.; Zheng, Y.; Gu, M.; Lin, X.; Wang, S.; Jin, S.; Ye, N. Comparison of metabolome and transcriptome of flavonoid biosynthesis pathway in a purple-leaf tea germplasm Jinmingzao and a green-leaf tea germplasm Huangdan reveals their relationship with genetic mechanisms of color formation. *Int. J. Mol. Sci.* **2020**, *21*, 4167. [CrossRef]
58. Kondo, S.; Hiraoka, K.; Kobayashi, S.; Honda, C.; Terahara, N. Changes in the expression of anthocyanin biosynthetic genes during apple development. *J. Am. Soc. Hortic. Sci.* **2002**, *127*, 971–976. [CrossRef]
59. Zhao, L.; Gao, L.; Wang, H.; Chen, X.; Wang, Y.; Yang, H.; Wei, C.; Wan, X.; Xia, T. The R2R3-MYB, bHLH, WD40, and related transcription factors in flavonoid biosynthesis. *Funct. Integr. Genom.* **2013**, *13*, 75–98. [CrossRef]
60. Borevitz, J.O.; Xia, Y.; Blount, J.; Dixon, R.A.; Lamb, C. Activation tagging identifies a conserved MYB regulator of phenylpropanoid biosynthesis. *Plant Cell* **2000**, *12*, 2383–2393. [CrossRef]
61. Gonzalez, A.; Zhao, M.; Leavitt, J.M.; Lloyd, A.M. Regulation of the anthocyanin biosynthetic pathway by the TTG1/bHLH/Myb transcriptional complex in *Arabidopsis* seedlings. *Plant J.* **2008**, *53*, 814–827. [CrossRef] [PubMed]
62. Xiang, L.; Liu, X.; Li, H.; Yin, X.; Grierson, D.; Li, F.; Chen, K. CmMYB# 7, an R3 MYB transcription factor, acts as a negative regulator of anthocyanin biosynthesis in chrysanthemum. *J. Exp. Bot.* **2019**, *70*, 3111–3123.
63. Zhu, H.-F.; Fitzsimmons, K.; Khandelwal, A.; Kranz, R.G. CPC, a single-repeat R3 MYB, is a negative regulator of anthocyanin biosynthesis in *Arabidopsis*. *Mol. Plant* **2009**, *2*, 790–802. [CrossRef]
64. Albert, N.W.; Davies, K.M.; Lewis, D.H.; Zhang, H.; Montefiori, M.; Brendolise, C.; Boase, M.R.; Ngo, H.; Jameson, P.E.; Schwinn, K.E. A conserved network of transcriptional activators and repressors regulates anthocyanin pigmentation in eudicots. *Plant Cell* **2014**, *26*, 962–980. [CrossRef]
65. Wu, X.; Prior, R.L. Identification and characterization of anthocyanins by high-performance liquid chromatography—Electrospray ionization—Tandem mass spectrometry in common foods in the United States: Vegetables, nuts, and grains. *J. Agric. Food Chem.* **2005**, *53*, 3101–3113. [CrossRef]
66. Chen, S.-Y.; Tang, Y.-M.; Hu, Y.-Y.; Wang, Y.; Sun, B.; Wang, X.-R.; Tang, H.-R.; Chen, Q. FaTT12-1, a multidrug and toxin extrusion (MATE) member involved in proanthocyanidin transport in strawberry fruits. *Sci. Hortic.* **2018**, *231*, 158–165. [CrossRef]
67. Yoshida, K.; Mori, M.; Kondo, T. Blue flower color development by anthocyanins: From chemical structure to cell physiology. *Nat. Prod. Rep.* **2009**, *26*, 884–915. [CrossRef] [PubMed]
68. Turturică, M.; Oancea, A.M.; Râpeanu, G.; Bahrim, G. Anthocyanins: Naturally occuring fruit pigments with functional properties. *Ann. Univ. Dunarea Jos Galati. Fascicle VI-Food Technol.* **2015**, *39*, 9–24.
69. Iglesias, I.; Salvia, J.; Torguet, L.; Cabús, C. Orchard cooling with overtree microsprinkler irrigation to improve fruit colour and quality of 'Topred Delicious' apples. *Sci. Hortic.* **2002**, *93*, 39–51. [CrossRef]
70. Iglesias, I.; Salvia, J.; Torguet, L.; Montserrat, R. The evaporative cooling effects of overtree microsprinkler irrigation on 'Mondial Gala'apples. *Sci. Hortic.* **2005**, *103*, 267–287. [CrossRef]
71. Zhang, L.; Wang, L.; Zeng, X.; Chen, R.; Yang, S.; Pan, S. Comparative transcriptome analysis reveals fruit discoloration mechanisms in postharvest strawberries in response to high ambient temperature. *Food Chem. X* **2019**, *2*, 100025. [CrossRef] [PubMed]

72. Iglesias, I.; Graell, J.; Faro, D.; Larrigaudiere, C.; Recasens, I.; Echeverria, G.; Vendrell, M. Efecto del sistema de riego en la coloracion de los frutos, contenido de antocianos y actividad de la fenilalanina amonioliasa (pal), en manzanas cv. 'starking delicious'. *Investig. Agr. Prod. Prot. Veg* **1999**, *14*, 1–2.
73. Palmer, J.; Lozano, L.; Chagné, D.; Volz, R.; Lin-Wang, K.; Bonany, J.; Micheletti, D.; Troggio, M.; White, A.; Kumar, S. Physiological, molecular and genetic control of apple skin colouration under hot temperature environments. In Proceedings of the XXVIII International Horticultural Congress on Science and Horticulture for People (IHC2010): International Symposium on 929, Lisbon, Portugal, 22–27 August 2010; pp. 81–87.
74. Lin-Wang, K.; Micheletti, D.; Palmer, J.; Volz, R.; Lozano, L.; Espley, R.; Hellens, R.P.; Chagne, D.; Rowan, D.D.; Troggio, M. High temperature reduces apple fruit colour via modulation of the anthocyanin regulatory complex. *Plant Cell Environ.* **2011**, *34*, 1176–1190. [CrossRef]
75. Solfanelli, C.; Poggi, A.; Loreti, E.; Alpi, A.; Perata, P. Sucrose-specific induction of the anthocyanin biosynthetic pathway in *Arabidopsis*. *Plant Physiol.* **2006**, *140*, 637–646. [CrossRef] [PubMed]
76. Jaakola, L.; Hohtola, A. Effect of latitude on flavonoid biosynthesis in plants. *Plant Cell Environ.* **2010**, *33*, 1239–1247. [CrossRef]
77. Weiss, D.; van der Luit, A.; Knegt, E.; Vermeer, E.; Mol, J.N.; Kooter, J.M. Identification of endogenous gibberellins in *Petunia* flowers (induction of anthocyanin biosynthetic gene expression and the antagonistic effect of abscisic acid). *Plant Physiol.* **1995**, *107*, 695–702. [CrossRef]
78. Khoo, H.; Chew, L.; Ismail, A.; Azlan, A. Anthocyanins in purple colored fruits. In *Polyphenols: Chemistry, Dietary Sources and Health Benefits*; Nova Science Publishers: New York City, NY, USA, 2012; pp. 133–152.
79. Mei, Y.; Xie, H.; Liu, S.; Zhu, J.; Zhao, S.; Wei, C. Metabolites and transcriptional profiling analysis reveal the molecular mechanisms of the anthocyanin metabolism in the "Zijuan" tea plant (*Camellia sinensis* var.assamica). *J. Agric. Food Chem.* **2020**, *69*, 414–427. [CrossRef]
80. Enaru, B.; Drețcanu, G.; Pop, T.D.; Stănilă, A.; Diaconeasa, Z. Anthocyanins: Factors affecting their stability and degradation. *Antioxidants* **2021**, *10*, 1967. [CrossRef] [PubMed]
81. Shruthi, V.; Ramachandra, C. Roselle (*Hibiscus sabdariffa* L.) calyces: A potential source of natural color and its health benefits. In *Food Bioactives*; Apple Academic Press: Palm Bay, FL, USA, 2019; pp. 169–190.
82. Oren-Shamir, M. Does anthocyanin degradation play a significant role in determining pigment concentration in plants? *Plant Sci.* **2009**, *177*, 310–316. [CrossRef]
83. Peng, L.; Yang, S.; Chen, R.; Johnb, S.; Ye, J.; Fan, G.; Zhou, H.; Peng, L.; Pana, S. Physiological and quality changes of postharvest strawberries at different storage temperature and their relationships to fruit discoloration. *Int. J. Food Nutr. Sci.* **2017**, *4*, 81–88.
84. Holton, T.A.; Cornish, E.C. Genetics and biochemistry of anthocyanin biosynthesis. *Plant Cell* **1995**, *7*, 1071. [CrossRef]
85. Passeri, V.; Koes, R.; Quattrocchio, F.M. New challenges for the design of high value plant products: Stabilization of anthocyanins in plant vacuoles. *Front. Plant Sci.* **2016**, *7*, 153. [CrossRef]
86. Kempken, F.; Jung, C. *Genetic Modification of Plants: Agriculture, Horticulture and Forestry*; Springer Science & Business Media: Berlin/Heidelberg, Germany, 2009; Volume 64.
87. Tanaka, Y.; Brugliera, F.; Kalc, G.; Senior, M.; Dyson, B.; Nakamura, N.; Katsumoto, Y.; Chandler, S. Flower color modification by engineering of the flavonoid biosynthetic pathway: Practical perspectives. *Biosci. Biotechnol. Biochem.* **2010**, *74*, 1760–1769. [CrossRef] [PubMed]
88. Chen, W.-H.; Hsu, C.-Y.; Cheng, H.-Y.; Chang, H.; Chen, H.-H.; Ger, M.-J. Downregulation of putative UDP-glucose: Flavonoid 3-O-glucosyltransferase gene alters flower coloring in *Phalaenopsis*. *Plant Cell Rep.* **2011**, *30*, 1007–1017. [CrossRef] [PubMed]
89. Boase, M.R.; Lewis, D.H.; Davies, K.M.; Marshall, G.B.; Patel, D.; Schwinn, K.E.; Deroles, S.C. Isolation and antisense suppression of flavonoid 3′, 5′-hydroxylase modifies flower pigments and colour in *cyclamen*. *BMC Plant Biol.* **2010**, *10*, 107. [CrossRef]
90. Okitsu, N.; Noda, N.; Chandler, S.; Tanaka, Y.J. Flower color and its engineering by genetic modification. In *Ornamental Crops*; Springer: Cham, Switzerland, 2018; pp. 29–62.
91. Century, K.; Reuber, T.L.; Ratcliffe, O.J. Regulating the regulators: The future prospects for transcription-factor-based agricultural biotechnology products. *Plant Physiol.* **2008**, *147*, 20–29. [CrossRef]
92. Hichri, I.; Barrieu, F.; Bogs, J.; Kappel, C.; Delrot, S.; Lauvergeat, V. Recent advances in the transcriptional regulation of the flavonoid biosynthetic pathway. *J. Exp. Bot.* **2011**, *62*, 2465–2483. [CrossRef] [PubMed]
93. Han, Y.-J.; Kim, Y.-M.; Lee, J.-Y.; Kim, S.J.; Cho, K.-C.; Chandrasekhar, T.; Song, P.-S.; Woo, Y.-M.; Kim, J.-I. Production of purple-colored creeping bentgrass using maize transcription factor genes Pl and Lc through Agrobacterium-mediated transformation. *Plant Cell Rep.* **2009**, *28*, 397–406. [CrossRef]
94. Chen, T. Development of harlequin flower derived from somaclone mutants of *Phalaenopsis* Golden Peoker 'Brother'. In Proceedings of the 8th Asia Pacific Orchid Conference (APOC8), Tainan, Taiwan, 6–14 March 2004; pp. 128–137.
95. Hsu, C.-C.; Su, C.-J.; Jeng, M.-F.; Chen, W.-H.; Chen, H.-H. A HORT1 retrotransposon insertion in the PeMYB11 promoter causes harlequin/black flowers in *Phalaenopsis* orchids. *Plant Physiol.* **2019**, *180*, 1535–1548. [CrossRef]
96. Shibata, M.; Ishikura, N. Paper chromatographic survey of anthocyanin in tulip-flowers. I. *Jpn. J. Bot.* **1960**, *17*, 230–238.
97. Dhalaria, R.; Verma, R.; Kumar, D.; Puri, S.; Tapwal, A.; Kumar, V.; Nepovimova, E.; Kuca, K. Bioactive compounds of edible fruits with their anti-aging properties: A comprehensive review to prolong human life. *Antioxidants* **2020**, *9*, 1123. [CrossRef] [PubMed]
98. Garcia, C.; Blesso, C.N. Antioxidant properties of anthocyanins and their mechanism of action in atherosclerosis. *Free Radic. Biol. Med.* **2021**, *172*, 152–166. [CrossRef]

99. Yan, G.; HouHua, L.; Ling, L.; YaJie, W.; Man, X. Chromogenic pigments in *Prunus cerasifera* leaves. *J. Zhejiang AF Univ.* **2014**, *31*, 481–487.
100. Slimestad, R.; Torskangerpoll, K.; Nateland, H.S.; Johannessen, T.; Giske, N.H. Flavonoids from black chokeberries, *Aronia melanocarpa*. *J. Food Compos. Anal.* **2005**, *18*, 61–68. [CrossRef]
101. Wawer, I.; Wolniak, M.; Paradowska, K. Solid state NMR study of dietary fiber powders from aronia, bilberry, black currant and apple. *Solid State Nucl. Magn. Reson.* **2006**, *30*, 106–113. [CrossRef] [PubMed]
102. Deng, K.; Ouyang, J.; Hu, N.; Meng, J.; Su, C.; Wang, J.; Wang, H. Improved colorimetric analysis for subtle changes of powdered anthocyanins extracted from *Lycium ruthenicum* Murr. *Food Chem.* **2022**, *371*, 131080. [CrossRef]
103. Kamiloglu, S.; Serali, O.; Unal, N.; Capanoglu, E. Antioxidant activity and polyphenol composition of black mulberry (*Morus nigra* L.) products. *J. Berry Res.* **2013**, *3*, 41–51. [CrossRef]
104. Nath, P.; Dukare, A.; Kumar, S.; Kale, S.; Kannaujia, P. Black carrot (*Daucus carota* subsp. *sativus*) anthocyanin-infused potato chips: Effect on bioactive composition, color attributes, cooking quality, and microbial stability. *J. Food Process. Preserv.* **2022**, *46*, e16180. [CrossRef]
105. Yoshikura, K.; Hamaguchi, Y. Anthocyanins of black soybean. *J. Jpn. Soc. Food Nutr.* **1969**, *22*, 367–370.
106. Hou, Z.; Qin, P.; Zhang, Y.; Cui, S.; Ren, G. Identification of anthocyanins isolated from black rice (*Oryza sativa* L.) and their degradation kinetics. *Food Res. Int.* **2013**, *50*, 691–697. [CrossRef]
107. Ahmad, S.; Chen, J.; Chen, G.; Huang, J.; Zhou, Y.; Zhao, K.; Lan, S.; Liu, Z.; Peng, D. Why Black Flowers? An Extreme Environment and Molecular Perspective of Black Color Accumulation in the Ornamental and Food Crops. *Front. Plant Sci.* **2022**, *13*, 885176. [CrossRef]
108. Deguchi, A.; Ohno, S.; Hosokawa, M.; Tatsuzawa, F.; Doi, M. Endogenous post-transcriptional gene silencing of flavone synthase resulting in high accumulation of anthocyanins in black dahlia cultivars. *Planta* **2013**, *237*, 1325–1335. [CrossRef]
109. Deguchi, A.; Tatsuzawa, F.; Hosokawa, M.; Doi, M.; Ohno, S. Tobacco streak virus (strain dahlia) suppresses post-transcriptional gene silencing of flavone synthase II in black dahlia cultivars and causes a drastic flower color change. *Planta* **2015**, *242*, 663–675. [CrossRef]
110. Noda, N.; Aida, R.; Kishimoto, S.; Ishiguro, K.; Fukuchi-Mizutani, M.; Tanaka, Y.; Ohmiya, A. Genetic engineering of novel bluer-colored chrysanthemums produced by accumulation of delphinidin-based anthocyanins. *Plant Cell Physiol.* **2013**, *54*, 1684–1695. [CrossRef] [PubMed]
111. Takeda, K.; Hayashi, K. Crystallization and Some Properties of the Genuine Anthocyanin Inherent to the Deep Violet Color of Pansy Studies on Anthocyanins. XLVIII. *Proc. Jpn. Acad.* **1965**, *41*, 449–454. [CrossRef]
112. Cai, J.; Lv, L.; Zeng, X.; Zhang, F.; Chen, Y.; Tian, W.; Li, J.; Li, X.; Li, Y. Integrative analysis of metabolomics and transcriptomics reveals molecular mechanisms of anthocyanin metabolism in the Zikui tea plant (*Camellia sinensis* cv. Zikui). *Int. J. Mol. Sci.* **2022**, *23*, 4780. [CrossRef]
113. Zhao, Y.; Li, Y.; Zhen, X.; Zhang, J.; Zhang, Q.; Liu, Z.; Hou, S.; Han, Y.; Zhang, B. Uncovering the mechanism of anthocyanin accumulation in a purple-leaved variety of foxtail millet (*Setaria italica*) by transcriptome analysis. *PeerJ* **2022**, *10*, e14099. [CrossRef]
114. Cho, M.J.; Howard, L.R.; Prior, R.L.; Clark, J.R. Flavonoid glycosides and antioxidant capacity of various blackberry, blueberry and red grape genotypes determined by high-performance liquid chromatography/mass spectrometry. *J. Sci. Food Agric.* **2004**, *84*, 1771–1782. [CrossRef]
115. Jordheim, M.; Enerstvedt, K.H.; Andersen, Ø.M. Identification of cyanidin 3-*O*-β-(6″-(3-Hydroxy-3-methylglutaroyl)glucoside) and other anthocyanins from wild and cultivated blackberries. *J. Agric. Food Chem.* **2011**, *59*, 7436–7440. [CrossRef]
116. Natić, M.M.; Dabić, D.Č.; Papetti, A.; Akšić, M.M.F.; Ognjanov, V.; Ljubojević, M.; Tešić, Ž.L. Analysis and characterisation of phytochemicals in mulberry (*Morus alba* L.) fruits grown in Vojvodina, North Serbia. *Food Chem.* **2015**, *171*, 128–136. [CrossRef]
117. Lightbourn, G.J.; Griesbach, R.J.; Novotny, J.A.; Clevidence, B.A.; Rao, D.D.; Stommel, J.R. Effects of anthocyanin and carotenoid combinations on foliage and immature fruit color of *Capsicum annuum* L. *J. Hered.* **2008**, *99*, 105–111. [CrossRef]
118. Polat, S.; Guclu, G.; Kelebek, H.; Keskin, M.; Selli, S. Comparative elucidation of colour, volatile and phenolic profiles of black carrot (*Daucus carota* L.) pomace and powders prepared by five different drying methods. *Food Chem.* **2022**, *369*, 130941. [CrossRef] [PubMed]
119. Glagoleva, A.; Kukoeva, T.; Mursalimov, S.; Khlestkina, E.; Shoeva, O. Effects of combining the genes controlling anthocyanin and melanin synthesis in the barley grain on pigment accumulation and plant development. *Agronomy* **2022**, *12*, 112. [CrossRef]
120. Carbone, F.; Preuss, A.; De Vos, R.C.; D'AMICO, E.; Perrotta, G.; Bovy, A.G.; Martens, S.; Rosati, C. Developmental, genetic and environmental factors affect the expression of flavonoid genes, enzymes and metabolites in strawberry fruits. *Plant Cell Environ.* **2009**, *32*, 1117–1131. [CrossRef]
121. Ono, E.; Fukuchi-Mizutani, M.; Nakamura, N.; Fukui, Y.; Yonekura-Sakakibara, K.; Yamaguchi, M.; Nakayama, T.; Tanaka, T.; Kusumi, T.; Tanaka, Y. Yellow flowers generated by expression of the aurone biosynthetic pathway. *Proc. Natl. Acad. Sci. USA* **2006**, *103*, 11075–11080. [CrossRef] [PubMed]
122. Luo, P.; Ning, G.; Wang, Z.; Shen, Y.; Jin, H.; Li, P.; Huang, S.; Zhao, J.; Bao, M. Disequilibrium of flavonol synthase and dihydroflavonol-4-reductase expression associated tightly to white vs. red color flower formation in plants. *Front. Plant Sci.* **2016**, *6*, 1257. [CrossRef]

123. Davies, K.M.; Schwinn, K.E.; Deroles, S.C.; Manson, D.G.; Lewis, D.H.; Bloor, S.J.; Bradley, J.M. Enhancing anthocyanin production by altering competition for substrate between flavonol synthase and dihydroflavonol 4-reductase. *Euphytica* **2003**, *131*, 259–268. [CrossRef]
124. Sasaki, N.; Nishizaki, Y.; Uchida, Y.; Wakamatsu, E.; Umemoto, N.; Momose, M.; Okamura, M.; Yoshida, H.; Yamaguchi, M.; Nakayama, M. Identification of the glutathione S-transferase gene responsible for flower color intensity in carnations. *Plant Biotechnol. J.* **2012**, *29*, 223–227. [CrossRef]
125. Miyagawa, N.; Miyahara, T.; Okamoto, M.; Hirose, Y.; Sakaguchi, K.; Hatano, S.; Ozeki, Y. Dihydroflavonol 4-reductase activity is associated with the intensity of flower colors in delphinium. *Plant Biotechnol. J.* **2015**, *32*, 249–255. [CrossRef]
126. Jaakola, L.; Maäättä, K.; Pirttilä, A.M.; Törrönen, R.; Kärenlampi, S.; Hohtola, A. Expression of genes involved in anthocyanin biosynthesis in relation to anthocyanin, proanthocyanidin, and flavonol levels during bilberry fruit development. *Plant Physiol.* **2002**, *130*, 729–739. [CrossRef]
127. Huyen, D.T.T.; Ureshino, K.; Van, D.T.; Miyajima, I. Co-pigmentation of anthocyanin-flavonol in the blotch area of *Rhododendron simsii* Planch. flowers. *Hortic. J.* **2016**, *85*, 232–237. [CrossRef]
128. Zong, Y.; Zhu, X.; Liu, Z.; Xi, X.; Li, G.; Cao, D.; Wei, L.; Li, J.; Liu, B. Functional MYB transcription factor encoding gene AN2 is associated with anthocyanin biosynthesis in *Lycium ruthenicum* Murray. *BMC Plant Biol.* **2019**, *19*, 169. [CrossRef]
129. Hou, W.; Yan, P.; Shi, T.; Lu, P.; Zhao, W.; Yang, H.; Zeng, L.; Yang, J.; Li, Z.; Fan, W. Modulation of anthocyanin accumulation in storage roots of sweetpotato by transcription factor IbMYB1-2 through direct binding to anthocyanin biosynthetic gene promoters. *Plant Physiol. Biochem.* **2023**, *196*, 868–879. [CrossRef] [PubMed]
130. Zhao, A.; Cui, Z.; Li, T.; Pei, H.; Sheng, Y.; Li, X.; Zhao, Y.; Zhou, Y.; Huang, W.; Song, X. mRNA and miRNA expression analysis reveal the regulation for flower spot patterning in *Phalaenopsis* 'Panda'. *Int. J. Mol. Sci.* **2019**, *20*, 4250. [CrossRef]
131. Albert, N.W.; Lewis, D.H.; Zhang, H.; Irving, L.J.; Jameson, P.E.; Davies, K.M. Light-induced vegetative anthocyanin pigmentation in *Petunia*. *J. Exp. Bot.* **2009**, *60*, 2191–2202. [CrossRef]
132. Albert, N.W.; Lewis, D.H.; Zhang, H.; Schwinn, K.E.; Jameson, P.E.; Davies, K.M. Members of an R2R3-MYB transcription factor family in Petunia are developmentally and environmentally regulated to control complex floral and vegetative pigmentation patterning. *Plant J.* **2011**, *65*, 771–784. [CrossRef] [PubMed]
133. Tu, M.; Fang, J.; Zhao, R.; Liu, X.; Yin, W.; Wang, Y.; Wang, X.; Wang, X.; Fang, Y. CRISPR/Cas9-mediated mutagenesis of VvbZIP36 promotes anthocyanin accumulation in grapevine (*Vitis vinifera*). *Hortic. Res.* **2022**, *9*, uhac022. [CrossRef] [PubMed]
134. Chiu, L.-W.; Zhou, X.; Burke, S.; Wu, X.; Prior, R.L.; Li, L. The purple cauliflower arises from activation of a MYB transcription factor. *Plant Physiol.* **2010**, *154*, 1470–1480. [CrossRef]
135. Mano, H.; Ogasawara, F.; Sato, K.; Higo, H.; Minobe, Y. Isolation of a regulatory gene of anthocyanin biosynthesis in tuberous roots of purple-fleshed sweet potato. *Plant Physiol.* **2007**, *143*, 1252–1268. [CrossRef]
136. Jackson, R.S. *Wine Science: Principles and Applications*; Academic Press: Cambridge, MA, USA, 2008.
137. Patras, A.; Brunton, N.P.; O'Donnell, C.; Tiwari, B.K. Effect of thermal processing on anthocyanin stability in foods; mechanisms and kinetics of degradation. *Trends Food Sci. Technol.* **2010**, *21*, 3–11. [CrossRef]
138. Shahidi, F. *Dried Fruits: Phytochemicals and Health Effects*; John Wiley & Sons: Hoboken, NJ, USA, 2012.
139. Fenger, J.-A.; Robbins, R.J.; Collins, T.M.; Dangles, O. The fate of acylated anthocyanins in mildly heated neutral solution. *Dye. Pigment.* **2020**, *178*, 108326. [CrossRef]
140. Zozio, S.; Pallet, D.; Dornier, M. Evaluation of anthocyanin stability during storage of a coloured drink made from extracts of the Andean blackberry (*Rubus glaucus* Benth.), açai (*Euterpe oleracea* Mart.) and black carrot (*Daucus carota* L.). *Fruits* **2011**, *66*, 203–215. [CrossRef]
141. Rhim, J.-W. Kinetics of thermal degradation of anthocyanin pigment solutions driven from red flower cabbage. *Food Sci. Biotechnol.* **2002**, *11*, 361–364.
142. Kırca, A.; Özkan, M.; Cemeroglu, B. Stability of black carrot anthocyanins in various fruit juices and nectars. *Food Chem.* **2006**, *97*, 598–605. [CrossRef]
143. Goto, T.; Hoshino, T.; Takase, S. A proposed structure of commelinin, a sky-blue anthocyanin complex obtained from the flower petals of Commelina. *Tetrahedron Lett.* **1979**, *20*, 2905–2908. [CrossRef]
144. Sadilova, E.; Carle, R.; Stintzing, F.C. Thermal degradation of anthocyanins and its impact on color and in vitro antioxidant capacity. *Mol. Nutr. Food Res.* **2007**, *51*, 1461–1471. [CrossRef] [PubMed]
145. Vidana Gamage, G.C.; Lim, Y.Y.; Choo, W.S. Sources and relative stabilities of acylated and nonacylated anthocyanins in beverage systems. *J. Food Sci. Technol.* **2022**, *59*, 831–845. [CrossRef] [PubMed]

current issues in molecular biology

MDPI

Article

Transcriptome Analysis Reveals That *FvPAP1* Genes Are Related to the Prolongation of Red-Leaf Period in *Ficus virens*

Qingchao Ma [1,2], Shuhua Zhong [2], Tianci Ma [2], Yajie Yue [2], Shihui Zou [1], Shunzhao Sui [2], Lijiao Ai [1,*] and Yulong Guo [2,*]

1 Chongqing Key Laboratory of Germplasm Innovation and Utilization of Native Plants, Chongqing Landscape and Gardening Research Institute, Chongqing 400715, China; mqcmadman@email.swu.edu.cn (Q.M.)

2 Chongqing Engineering Research Center for Floriculture, College of Horticulture and Landscape, Southwest University, Chongqing 401329, China; zsh15220@email.swu.edu.cn (S.Z.); yyj990928@email.swu.edu.cn (Y.Y.); sszcq@swu.edu.cn (S.S.)

* Correspondence: alj01461@foxmail.com (L.A.); yulong@swu.edu.cn (Y.G.)

Abstract: *Ficus virens* is a deciduous tree that is highly valuable both economically and medicinally. Like other plants with 'red young leaves', the red-leaf period of most *F. virens* trees lasts only a few days, and the red leaves have little ornamental value. However, in recent years, some lines of *F. virens* with bright red young leaves and a prolonged red-leaf period have been utilized for urban greening. To explore the mechanism of the different lengths of the duration of *F. virens* leaves, we analyzed the physiology and changes in gene expression during the development of two varieties of leaves. The detection of anthocyanin in different developmental stages of the *F. virens* leaves showed that the changes in color of the red leaves of *F. virens* were primarily caused by the change in anthocyanin content. A transcriptome analysis showed that the expression of genes related to the biosynthesis of anthocyanin changed significantly during the development of leaves. A *MYB* gene *FvPAP1*, which was consistent with the change in anthocyanin content, was identified. A real-time quantitative reverse transcription PCR analysis and heterologous expression transgenic studies showed that *FvPAP1* promoted the biosynthesis of anthocyanins. The difference in the expression of *FvPAP1* in time and intensity in the young leaves may be the reason for the difference in the duration of the red-leaf period in different lines of *F. virens*. A sequence analysis showed that the cDNA sequence of *FvPAP1* was polymorphic, and possible reasons were discussed. These results can provide insight for similar studies on the mechanism of the formation of red coloring in other woody plant leaves and provide molecular targets to breed new materials with more prolonged red-leaf periods in *F. virens*.

Keywords: *FvPAP1*; anthocyanin; *Ficus virens*; red-leaf period; transcriptome analysis

Citation: Ma, Q.; Zhong, S.; Ma, T.; Yue, Y.; Zou, S.; Sui, S.; Ai, L.; Guo, Y. Transcriptome Analysis Reveals That *FvPAP1* Genes Are Related to the Prolongation of Red-Leaf Period in *Ficus virens*. *Curr. Issues Mol. Biol.* **2024**, *46*, 5724–5743. https://doi.org/10.3390/cimb46060343

Academic Editor: Vijai Bhadauria

Received: 29 April 2024
Revised: 30 May 2024
Accepted: 5 June 2024
Published: 8 June 2024

1. Introduction

Ficus virens is a deciduous tree that is a member of the Moraceae [1]. It is native to south and southwestern China, particularly in Chongqing, Sichuan, Hubei, and other places, and it is the city tree of Chongqing. *F. virens* is highly valuable both medicinally and ecologically [2–8]. It is commonly used in roadside afforestation, vertical greening, riverbank protection, and bonsai gardening [9–12]. Many plants have young red leaves [13], which is a phenomenon known as the 'red young leaves'. Since the young leaves of most plants rapidly turn green as they develop, and the red color simultaneously disappears, the short red-leaf period has no obvious ornamental value [14]. However, some varieties of these plants that remain red for longer periods have also been isolated and propagated, including *Photinia* × *fraseri* [15], *Prunus cerasifera* [16], *Populus deltoids* [17], 'Red Dragon' *Corylus avellana* [18], and *Cornus florida* [19].

The concentrated defoliation of *F. virens* primarily occurs in the spring from March to May in Chongqing. When a *F. virens* tree defoliates, the leaves of the whole tree will quickly fall within a few days. Simultaneously, the new leaves wrapped in the giant stipules will quickly expand, and the new and old leaves will be replaced within a few days [5]. The whole canopy quickly changes from the yellow of the aging leaves to the color of the new leaves, and the entire process is considered to be spectacular. Like most plants with red young leaves, the red-leaf period of *F. virens* only lasts for a short time, which renders the plant relatively invaluable ornamentally. However, with the wide application of *F. virens* in landscaping in recent years, nursery workers have intentionally or unintentionally propagated some lines with obviously prolonged red-leaf periods and enhanced aesthetic value (Figure S1).

The primary material basis for the formation of red leaves is the anthocyanin members of the flavonoids [13]. The genes that encode the anthocyanin biosynthetic enzymes can be divided into early biosynthetic genes (EBGs) and late biosynthetic genes (LBGs) [20]. The EBGs include chalcone synthase (*CHS*), chalcone isomerase (*CHI*), and flavanone 3-hydroxylase (*F3H*), which are located upstream of the anthocyanin biosynthetic pathway. The LBGs include flavonoid 3′-hydroxylase (*F3′H*), flavonoid 3′5′-hydroxylase (*F3′5′H*), dihydroflavonol 4-reductase (*DFR*), and anthocyanin synthase (*ANS*), which are located downstream of the anthocyanin biosynthetic pathway. The EBGs and LBGs are structural genes in the anthocyanin biosynthetic pathway.

The regulatory mechanism of the biosynthesis of anthocyanins in plants is relatively conservative and primarily acts through the MBW complex formed by the combination of the transcriptional regulators MYB, bHLH, and WD40 to bind to the promoter region of each anthocyanin biosynthetic gene to regulate their transcription. In most cases, bHLH and WD40, which constitute the MBW complex, are constitutively expressed. Therefore, the spatiotemporal specificity of the formation of the MBW complex is determined by the specificity of the expression of MYB transcription factors (TFs), which, in turn, determines the spatiotemporal characteristics of anthocyanin biosynthesis [21]. An MYB TF protein is characterized by a highly conserved R domain that binds to DNA. The number of R domains can be divided into four categories, namely, R2R3-MYB, 1R-MYB, 3R-MYB, and 4R-MYB. The MYB activators that regulate anthocyanin biosynthesis are primarily from R2R3-MYB [21,22]. The R2R3-MYB activators, such as *Arabidopsis thaliana* AtMYB75/PAP1, AtMYB90/PAP2, AtMYB113, AtMYB114, PhAN2, and *Petunia hybrida* PhAN4, play a positive regulatory role in the biosynthesis of anthocyanin [23–25], while the MYB repressors, including R2R3-MYB (AtMYB3, AtMYB4, AtMYB7, AtMYB32, and PhMYB27) and R3-MYB (AtMYBL2 and PhMYBx) [26–28], inhibit anthocyanin biosynthesis by competing with the R2R3-MYB activators.

The prolonging of the red-leaf period would be achieved by enhancing the biosynthesis and/or stability of the anthocyanins in which the enhancement of biosynthesis is the basis. Prolonging the red-leaf period of the young leaves in *F. virens* can enhance its ornamental value. However, to our knowledge, research on the mechanism of red leaf formation in *F. virens* has not yet been reported. Based on the fact that the MYB TFs control the expression of anthocyanin biosynthesis in plants in most cases, we hypothesized that prolonging the red young leaves of *F. virens* may be related to the changes in expression of one or several *MYB* genes. To verify this hypothesis, we conducted a transcriptome sequencing analysis on the wild-type (W-type) and red-type (R-type) leaves of *F. virens* with different lengths of red-leaf periods to explore the changes in gene expression during the development of leaves. The structural and regulatory genes of the anthocyanin biosynthetic pathway were identified by their phylogenetic relationship, and a real-time quantitative reverse transcription PCR (qRT-PCR) determined their trends of expression. The R2R3-MYB gene *FvPAP1*, which may promote anthocyanin biosynthesis, was eventually cloned and functionally verified. The results showed that during the development of *F. virens* leaves, the change in color of the red leaves was caused by a change in the anthocyanin content. The structural genes and regulatory genes related to anthocyanin biosynthesis are involved in regulating the extension of the red-leaf period of *F. virens*. *FvPAP1* can activate the biosynthesis of anthocyanin, which may be related to extending the red-leaf period.

These results can provide molecular targets for the selection of *F. virens* plants with a further extension of the red-leaf period.

2. Materials and Methods

2.1. Plant Materials

To observe and conveniently collect materials, this study was conducted with 2-year-old cuttings approximately 2 m high. They were planted in the Beibei District of Chongqing, China (106°18′02″~106°40′57″ E. 29°37′~30°05′08″ N; average altitude: 358 m) under natural conditions. The contents of anthocyanin and chlorophyll were determined, and the transcriptome sequencing was conducted in April 2022. qRT-PCR was performed in April 2023. After the first young leaf was exposed from the stipules, the young leaves of each branch were numbered from the bottom to top. The third leaves were used as material to determine the contents of anthocyanin and chlorophyll and perform a qRT-PCR analysis.

The *P. hybrida* inbred line 'Mitchell Diploid' and *N. tabacum* variety 'K326' were used as the transgenic receptors. *P. hybrida* plants for the transient expression assay were grown in a greenhouse (greenhouse model: LB96Ss4, China) with no exogenous heat and light supplementation. Aseptic *N. tabacum* seedlings were cultured under alternating light and dark conditions at 16:8 h (24°:20 °C).

2.2. Quantitative Analysis of Anthocyanins

The anthocyanin content was determined by the HCl methanol method [25,29]. A total of 0.25 g fresh sample was ground into powder in liquid nitrogen and incubated in 10 mL 1% HCl–methanol (*v*/*v*) at 4 °C in the dark for 1 h, shaking 2–3 times, 6000/7000 rpm, and 7–10 min centrifugation. A volume of 200 µL of the supernatant was placed in 96-well microtiter plates, and the absorbance of the supernatant was measured at 530 nm and 657 nm on a Varioskan Flash Spectral Scanning Multimode Reader (Thermo Fisher Scientific, Waltham, MA, USA). The relative anthocyanin content was then determined to be $(A530 - 0.25 \times A657)/W\ g^{-1}$ fresh weight.

2.3. Quantitative Analysis of Chlorophyll

The chlorophyll content was determined by the acetone ethanol mixing method [30]. A fresh sample of 0.2 g leaves was cut into pieces and placed in a 1:1 ratio of 20 mL acetone–anhydrous ethanol. The mixture was placed at room temperature, and the leaves were soaked for 24 h until they had completely whitened. A volume of 200 µL of the supernatant was placed in a 96-well microtiter plate. The absorbance of the supernatant at 633 nm and 645 nm was measured on the Varioskan Flash Spectral Scanning Multimode Reader. The relative total chlorophyll content was $(8.02 \times A633 + 20.21 \times A645) \cdot 1000\ V/W$ mg fresh weight.

2.4. Transcriptomic Analysis

Based on the Illumina NovaSeq 6000 sequencing platform (Illumina, San Diego, CA, USA), the library was constructed using the Illumina TruSeq^{TM} RNA sample prep Kit method. The total RNA was extracted from the leaves of different developmental stages of *F. virens*. The RNA concentration, purity, and integrity were detected by a NanoDrop 2000 (Thermo Fisher Scientific) and agarose gel electrophoresis, and the RIN value was determined by an Agilent 2100 (Agilent Technologies, Santa Clara, CA, USA). The mRNA was isolated from the total RNA using magnetic beads with Oligo (dT). The mRNA was randomly broken by adding fragmentation buffer, and a small fragment of approximately 300 bp was isolated by magnetic bead screening. Under the action of reverse transcriptase, random hexamers were added to reverse the biosynthesis of one-stranded cDNA, followed by two-stranded biosynthesis. End Repair Mix was added to fill it into a flat end, and an A″ base was then added to the 3′ end to connect the Y-shaped connector. An Illumina platform was used for sequencing.

The transcriptome analysis of 15 samples was completed, and a total of 124.22 Gb Clean Data were obtained. The Clean Data of each sample was >7.04 Gb, and the percentage of Q 30 bases was more than 93.68%. The Clean Reads of each sample were compared with the banyan (*F. benghalensis*) genome sequence, and the alignment rate ranged from 61.8% to 63.51%, which was lower than the generally acceptable 70%. Therefore, we used Trinity (k-mer = 30) to perform the de novo assembly of all the sample clean data and optimize the assembly results.

The level of expression and DEG analysis were based on the default parameters and software of the Meiji Bioplatform [31]. The expression software was RSEM (Version 1.3.1), and the expression index was TPM. The differential expression analysis (DEG) software was DESeq2 (Version 1.38.0), p-adjust < 0.05, and the multiple test correction method was BH. The DEG groups were Rs4 vs. Rs5, Rs5 vs. Rs6, Rs4 vs. Rs6, Ws4 vs. Ws5, Rs4 vs. Ws4, and Rs5 vs. Ws5.

The cluster analysis was analyzed in the form of the mean value of the group; the log base value was 10; the gene clustering algorithm was expanded by hierarchical clustering; the gene clustering method was average; and the gene distance algorithm was Euclidean. A cluster analysis divided all the genes into five sub-clusters, which corresponded to heatmap1, heatmap2, heatmap3, heatmap4, and heatmap5.

A WGCNA visual analysis utilized the default module of the Meiji platform. In the module identification part, the networkType was signed; the soft power was 9; minModuleSize was 30; minKMEtoStay was 0.3; and mergeCutHeight was 0.25. In the module analysis, the correlation coefficient was calculated using Spearman; the anthocyanin content in the corresponding period was selected as the phenotypic data; and the phenotypic data type was continuous. The first 30 connectivity nodes in the screening module were analyzed in the visual analysis, and the connections between the nodes with weight values > 0.02 were screened for analysis.

2.5. Identification of the Structural Genes of the Anthocyanin Pathway

Since thousands of genes were differentially expressed for the adjacent developmental stages, the structural genes related to anthocyanin biosynthesis were primarily conducted on the basis of KEGG pathway annotation. The genes related to anthocyanin biosynthesis were primarily related to MAP00941 and MAP00942 in the KEGG pathway [32].

2.6. Phylogenetic Analysis for the MYB Genes

A phylogenetic analysis using the protein sequences was applied to search for the MYB genes. As previously described, the sequences of 13 MYB proteins related to anthocyanin biosynthesis in *A. thaliana* and petunia were retrieved from GenBank, and their accession numbers were as follows: PhAN2 (BAP28593.1), PhAN4 (WGV46820.1), PhMYB27 (AHX24372.1), PhMYBx (AHX24371.1), AtMYB3 (BAA21618.1), AtMYB4 (BAA21619.1), AtMYB7 (Q42379.1), AtMYB32 (EFH43356.1), AtPAP1/AtMYB75 (Q9FE25.1), AtPAP2/AtMYB90 (Q9ZTC3.1), AtMYB113 (Q9FNV9.1), AtMYB114 (Q9FNV8.1), and AtMYBL2 (NP_001321410.1). Owing to large differences in the C-terminal sequences of MYB proteins even among functionally equivalent members, only the R2R3 domains were used in the MYB phylogenetic analysis. The phylogenetic trees were constructed by MEGA 7 [33]. Parameters for the NJ tree were set as the p-distance model and pairwise deletion with the bootstrap value as 1000.

2.7. Primer Design

Considering the potential existence of errors in the contiguous sequences obtained by de novo transcriptome assembly, we decided to confirm the target gene sequences by PCR cloning and sequencing before designing quantitative PCR primers to analyze the changes in gene expression. If a target unigene assembled by Trinity (k-mer = 30) was hypothesized to have a complete CDS sequence, the primers were first designed to amplify its CDS. If the amplification did not produce any bonds, we assumed that the assembly by Trinity (k-mer = 30) should be incorrect, and new primers were designed by re-referring to the other Trinity (k-mer = 25) assembly result and/or the sequence of homologous genes in the genome sequencing results of *F. microcarpa*. If the target unigene assembled by Trinity (k-mer = 30) did not have a complete CDS sequence, the unigene in the Trinity (k-mer = 25) assembly results and the homologous gene of *F. microcarpa* genome were analyzed to search a complete CDS. If the complete CDS was identified, primers were designed based on the corresponding sequences. The primers were designed and evaluated using Primer Premier 6 (http://www.premierbiosoft.com/primerdesign/index.html, accessed on 20 November 2022). The primers used in this study are shown in Table 1.

Table 1. Primer sequence and application.

Primer	Primer Sequence	Application
FvPAP1-F	ATGGATGGCCGTTCCTCTG	Cloning
FvPAP1-R	TTAATTTCCTTGGTCGAGATCC	
FvANS-F	ATGGTGTTCTCAGAGGCTTC	
FvANS-R	TTATTTGGAAATTCCGTTGCTATG	
FvCHI-F	ATGTCTCCGATGACACTTACA	
FvCHI-R	TCAAACTGCCAACAGTTTCTT	
FvCHS2-F	ATGGCCTCTGTATACGAAATC	
FvCHS2-R	TTAATCAAGGGCAAGACTGTG	
FvF3′H-F	ATGCCTTCTGTCTCGATCATCCT	
FvF3′H-R	TTATGCTTGATATACGTGTTGGGC	
FvUGT1-F	ATGGCAAAATTTCACATCGCC	
FvUGT1-R	TTAACTATGATTTGCCAGCTCTTCC	
FvF3H-F	ATGTCTCCGCCTTCAACTCTC	
FvF3H-R	TTAAGCCAGAATCTGGTCGAGAG	
FvDFR1-F	ATGGTATCGGAGGGCGAGAT	
FvDFR1-R	CTAAGCCTCCACGCCATTG	
FvBZ1-F	ATGGCATCACCACCACCAGC	
FvBZ1-R	CTAAAACGTCGGACGATTCAACGG	
FvCHS-F	ATGGTGACTGTCGAGGAGGTC	
FvCHS-R	CTAGATTGCCACGCTATGGAGC	
FvF3H-QF	GGCAGTGGTGAACTCAAACTACAG	qRT-PCR
FvF3H-QR	CCTGGCAAGCTCAAGGTCCTT	
FvANS-QF	GCCCTCACCTTCATCCTACACAAC	
FvANS-QR	GGAGAATACTCTTGAACTTGCCATTGC	
FvCHI-QF	CACACAGTCACCATCCGGTTCCT	
FvCHI-QR	TTCCGCCAGAATTGAAGCCATGC	
FvCHS1-QF	TGGGAATCTCAGACTGGAACTCTCTT	
FvCHS1-QR	GGCACACTTCCTCCTCATCTCATC	
FvCHS2-QF	TGACGGGCATCTGAGGGAAGTAG	
FvCHS2-QR	TGGTCTCCACCTGATCCAGAATCG	
FvDFR1-QF	CCGTTACATCGCCAGTTCACAC	
FvDFR1-QR	CTGCTTCAACGAACATATCCTCCAAG	
FvF3′H-QF	GGAATGACTTTGAAGTGATACCGTTTGG	
FvF3′H-QR	GCCCGTTGTAGAGTGAGCCCAT	
FvUGT1-QF	GTGGATCTTTAATATCCTCCTCGGTTACC	
FvUGT1-QR	TTCCTCTTCGCCGCCAAACC	
FvBZ1-QF	GATCTCAAGTCCAAATTCAAGAAGTTCCTC	
FvBZ1-QR	ACCATGACCGAACCAAAGCTAATGT	
FvPAP1-QF	GCATGATTGGTACAAATACATCCGAGG	
FvPAP1-QR	CAGGTTGATCTTCAGCTAATTCTTGAGAC	
FvACT-QF	CCTCTACGGCAACATTGTCCTCAG	
FvACT-QR	CTCCGATCCAGACACTGTACTTCCT	
FvUBi-QF	CTCTCCACCTTGTCCTCCGTCTTC	
FvUBi-QR	CCCAAAGCAGCAACGACAACCAT	
FvPAP1-QF	GCATGATTGGTACAAATACATCCGAGG	
FvPAP1-QR	CAGGTTGATCTTCAGCTAATTCTTGAGAC	
FvACT-QF	CCTCTACGGCAACATTGTCCTCAG	
FvACT-QR	CTCCGATCCAGACACTGTACTTCCT	
FvUBi-QF	CTCTCCACCTTGTCCTCCGTCTTC	
FvUBi-QR	CCCAAAGCAGCAACGACAACCAT	

2.8. Cloning of the Transcript Sequences Related to Anthocyanin Biosynthesis

The stage 4 leaves of the R-type and W-type *F. virens* were collected, and the total RNA was extracted using a Boer plant rapid extraction kit (B2114, Chongqing Boer Biotech

Co., Ltd., Chongqing, China). The quality of total RNA was measured by agarose gel electrophoresis, and the concentration of RNA was measured by a NanoDrop 2000 spectrophotometer (Gene Company Limited, Beijing, China). The cDNA was synthesized to clone the gene using the All-in-One First-Strand Synthesis MasterMix (with dsDNase) (Cat: EG15133S, Jiangsu Yugong Biotech Co., Ltd., Lianyungang, China). The PCR was amplified using Top Taq enzyme (AP151-11, TransGen Biotech Co., Ltd., Beijing, China). The amplification parameters were adjusted based on the primer Tm value. The amplification product was ligated to T-Vector pMDTM19 (Simple) (3271, Takara Bio Co., Ltd., Shiga, Japan) for Sanger sequencing.

2.9. qRT-PCR

A total of 2000 ng was used to synthesize 40 μL of cDNA as a template. Real-time PCR was conducted using 2 × TSINGKE® Master qPCR Mix kit (SYBR Green I with UDG) (TSE203, Beijing Tsingke Biotechnology Co., Ltd., Beijing, China). A volume of 10 μL of reaction system with 1 μL of upstream/downstream primers, and 1 μL of cDNA template was used for each PCR reaction. *Actin* (OR682436) and *Ubiquitin* (OR682446) were used as the internal reference genes in *F. virens*, and the *PhSAND* and *RPS13* genes were used as internal reference genes in the *P. hybrida* material [25]. The primers for quantitative PCR were designed in the 3′ non-conserved region of the CDS. The quantitative PCR primers used in the experiment are shown in Table 1. Quantitative PCR was performed on a CFX96 Optical Reaction Module for Real-Time PCR (2023 Bio-Rad, Hercules, CA, USA). Bio-Rad CFX Manager 3.1 software, and the $2^{-\Delta CT}$ method was used to analyze the gene expression [34].

2.10. Construction of the FvPAP1 Expression Vector

The T-Vector plasmid that harbored the complete CDS of *FvPAP1* as a template was used with NocIPAP1-F: catttacgaacgatagccATGGATGGCCGTTCCT and PstIPAP1-R: gctcaccatctgcagactacctccATTTCCTTGGTCGAGATCC to add restriction sites *NocI* and *PstI* to the upstream and downstream of the *FvPAP1* coding region sequence, respectively. The pGMF500 plasmid (Figure S2) and the target fragments were digested with *Pst1* and *Noc1* and ligated. The *FvPAP1* sequence was cloned into the pGMF500 plasmid and placed under the control of the 35S promoter and the NOS terminator. The correct plasmid that was verified by sequencing was designated pGmf-*FvPAP1*. The plasmid was introduced into *Agrobacterium* GV3101 (pSoup) by the heat shock method for plant genetic transformation.

2.11. Transient Expression

The *P. hybrida* petals and leaves were transiently expressed using the *Agrobacterium* injection method as described by Yuan Meng et al. [35]. The *Agrobacterium* that harbored pGmf-*FvPAP1* was selected and cloned into 700 μL of YEB liquid medium that contained the corresponding antibiotics and cultured at 28 °C for 24 h at 200 rpm. A volume of 200 μL of the cultured *Agrobacterium* solution was transferred to 20 mL of YEB medium that contained the corresponding antibiotics. The LB medium contained 15 μM of acetosyringone (AS). It was cultured at 28 °C, 200 rpm until the logarithmic phase of *Agrobacterium* growth (OD_{600} = 0.5–0.6). The cells were centrifuged at 7000 rpm for 10 min at room temperature to collect the cells, and the cells were suspended in the infected fluid that contained 10 mM $MgCl_2$, 10 mM MES, and 150 μM acetosyringone, pH = 5.6 to OD_{600} = 0.8. It was incubated stationary at room temperature for 2~3 h. A 1 mL syringe was used to gently open a small opening on the back of the petals and leaves of *P. hybrida*, and the bacterial liquid was sucked from the syringe tube that removed the syringe and injected from the wound. The injected plants were placed in the dark for 24 h and cultured at 24:20 °C for 16:8 h.

2.12. Genetic Transformation of Nicotiana Tabacum

N. tabacum was transformed genetically using the leaf disc method [36]. The seeds were disinfected with 75% ethanol for 30 s and then mixed with 5% sodium hypochlorite

(NaCIO) and sterile water at a ratio of 3:1 for 10 min. After disinfection, the *N. tabacum* seeds were washed 4–5 times with sterile water and transferred to MS media. When the *N. tabacum* leaves grew to four or five true leaves, the leaves were cut to 0.5 × 0.5 cm and placed in MS + 2.0 mg/L 6-benzylaminopurine (6-BA) + 0.2 mg/L 1-napthaleneacetic acid (NAA) media for pre-culture for 2 d. The *Agrobacterium* GV3101 (pSoup) transformed with pGmf-*FvPAP1* was cultured in YEB liquid media to OD_{600} = 0.8 and then resuspended in the infection solution (MS + 50 mg L^{-1} AS + 50 mg L^{-1} MES) to OD_{600} = 0.4. The pre-cultured *N. tabacum* leaves were shaken at low speed in the dark for 10 min and transferred to MS + 2.0 mg L^{-1} 6-BA + 0.2 mg L^{-1} NAA + 500 mg L^{-1} Cb + 5 mg L^{-1} Basta media for screening culture.

3. Results

3.1. Morphological Characteristics of Ficus virens during Leaf Development

During the development of young leaves, the whole bud is usually wrapped by a large stipule until the length of the bud reaches approximately 10 cm, and the true leaf expands and breaks through the outer stipules that usually show different degrees of red (Figure S3). When the true leaves are wrapped and on the day of expansion, they are green. They become obviously red on the third day after expansion, and the red is most obvious on the seventh day after expansion. The redness of the leaves will then gradually fade until the inherent green of the leaves is presented. The new leaves of most *F. virens* plants are light red, and the red-leaf period last for approximately 10 days. Because most *F. virens* trees manifest this red-leaf period, we regard them as the wild-type (W-type) (Figure 1a). In recent years, the new leaves of some *F. virens* trees have been found to be bright red, and the red state lasts for 20–30 days. This type of *F. virens* is called the red-type (R-type) *F. virens* (Figure 1a) in this study. The anthocyanin and chlorophyll of the leaves of these two varieties of *F. virens* were determined at different stages. The results showed that the chlorophyll content increased with the development of the leaves, and the change in the anthocyanin content was consistent with the red change in the leaves that was observed (Figure 1b). When the expanded leaves appeared red (Stage 2), there were significant differences in the anthocyanin content between the two varieties of *F. virens* at adjacent development stages. This demonstrates that the red formation of the leaves of *F. virens* is related to the biosynthesis of anthocyanins.

3.2. Comprehensive Transcriptomic Analysis

To explore the key genes in the formation of red leaves and explore the possible mechanism of the elongation of the red-leaf period in *F. virens*, the leaves of s4 and s5 from the W-type plants (Ws4 and Ws5) and s4, s5, and s6 from the R-type plants (Rs4, Rs5, and Rs6) (Figure 1a) were collected for transcriptome sequencing. A transcriptome analysis of 15 samples obtained 124.22 Gb of clean data. Trinity (k-mer = 30) was used to assemble clean data from all the samples, and the assembly results were optimized and evaluated. There were 52,200 unigenes, and they were 1018.03 bp long on average.

After comparison with the six major databases, there were 23,125, 11,473, 21,758, 27,063, 20,784, and 19,841 unigenes annotated in the NR, Swiss-Prot, Pfam, Clusters of Orthologous Genes (COG), Gene Ontology (GO), and Kyoto Encyclopedia of Genes and Genomes (KEGG) databases, respectively. There were 7907 genes annotated in all six datasets (Figure 2a). A total of 62.54% of the genes annotated in the NR library were from *Morus notabilis* (Figure 2b), which is also a member of the Moraceae family like *F. virens*.

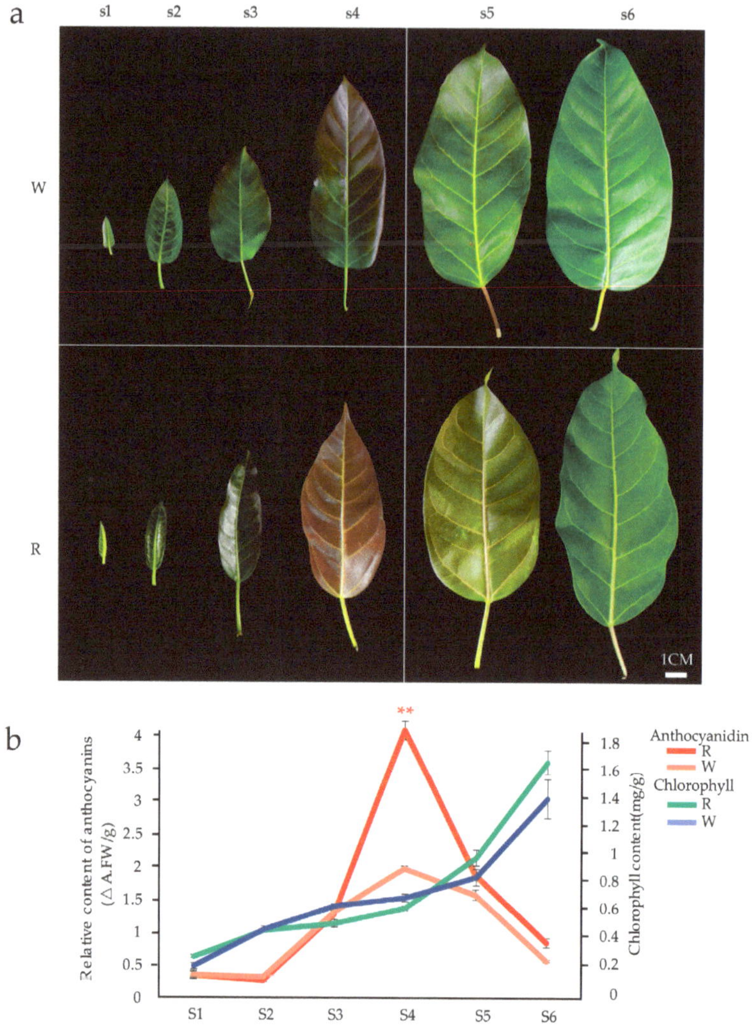

Figure 1. Changes in leaf size and pigment content of two varieties of *Ficus virens* at different developmental stages: (**a**) The morphological changes in the leaves of two types of *F. virens* at different developmental stages. W: wild-type; R: red-type. s1 (stage 1): the stage of bud wrapped by stipules; s2, s3, s4, s5 and s6: the first, third, seventh, thirtieth, and fortieth day after the day of expansion, respectively. Bar = 1 cm. (**b**) The changes in pigment content in the leaves of two kinds of *F. virens* in the sixth stage, including anthocyanin and total chlorophyll. The significance of the difference between adjacent stages was tested by a single-factor analysis of variance and *t*-test. ** $p < 0.01$. In the figure, the anthocyanin content of the W-type and R-type only differed significantly in the s4 period.

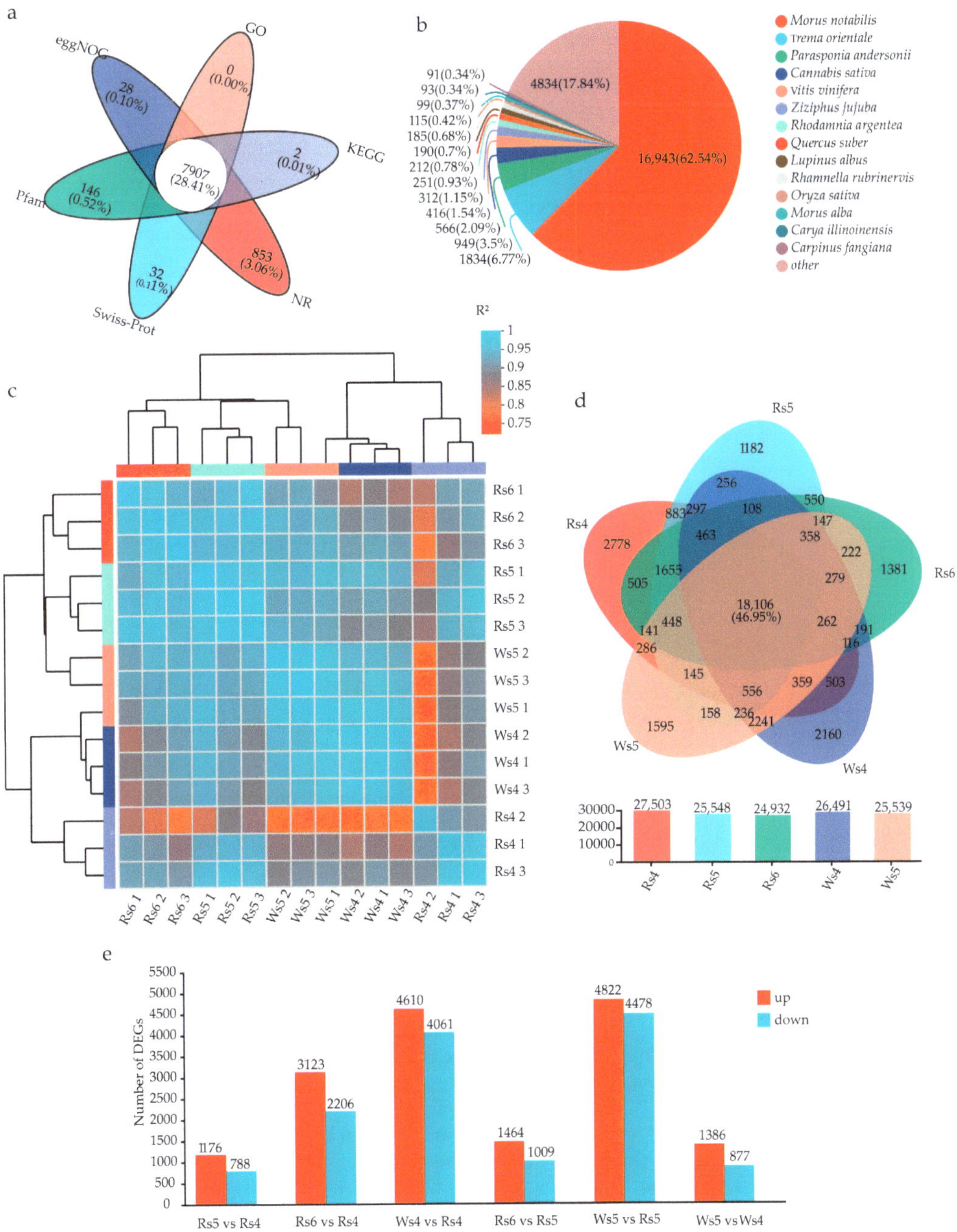

Figure 2. Transcriptome analysis of different leaf development stages of *Ficus virens*: (**a**) Venn diagram of the number of unigenes annotated in the NR, Swiss-Prot, Pfam, COG, GO, and KEGG databases. (**b**) NR annotated species pie chart. (**c**) Sample correlation heat map. The right and lower sides of the map are sample names, and the left and upper sides are sample clustering situations. Color represents

the correlation coefficient among samples. (**d**) Venn diagram showing co-expression of genes in different sample. (**e**) Statistical map of the DEGs between samples. The abscissa represents different comparison groups, and the ordinate represents the corresponding number of up- and downregulated genes. COG: Clusters of Orthologous Genes; DEGs: differentially expressed genes; GO: Gene Ontology; KEGG: Kyoto Encyclopedia of Genes and Genomes.

The repeatability correlation among different samples of the leaves of the two lines was reasonable (Figure 2c). For unigenes with expression > 0.1 transcript per million (TPM), 22,397 unigenes were annotated for the W-type, 20,672 for the R-type, and 18,106 unigenes were co-annotated (Figure 2d). A large number of DEGs between any two different stages of leaf samples were detected in this study (Figure 2e).

3.3. *Identification of the Structural Genes Related to Anthocyanin Biosynthesis and the MYB Genes*

Anthocyanins are flavonoids, and the biosynthesis of anthocyanin is primarily regulated owing to the activities of MAP00941 and MAP00942 in the KEGG pathway. Using the results of assembly at k-mer = 30, a total of 85 structural genes related to anthocyanin biosynthesis expressed in the leaves of the wild-type and R-type were obtained by KEGG. Anthocyanin reductase (*ANR*), UDP-glucuronosyltransferase (*UGT*), and *F3H* are single-copy genes, and other structural genes have multiple copies. *ANS* and *F3′5′H* were not found in the transcriptome database. An analysis of the differentially expressed genes (DEGs) obtained 57 DEGs from the 85 structural genes. We analyzed 141 MYB genes obtained from the transcriptome data and obtained 71 *MYB* DEGs.

Since the activation of LBGs requires the MBW complex [28], we performed a weighted gene co-expression network analysis (WGCNA) visualization and cluster analysis using the 57 DEGs that were shown to encode anthocyanin biosynthetic enzymes and 71 *MYB* DEGs. A WGCNA visualization analysis enabled the identification of the structural genes *CHS*, *ANR*, *F3H*, leucoanthocyanidin reductase (*LAR*), flavonol synthase (*FLS*), and shikimate/quinate hydroxycinnamoyltransferase (*HCT*) that were related to the changes in anthocyanin content. Among them, there were three copies of *CHS*s, and the rest were expressed as one copy (Figure 3a). With the exception of *HCT*, the other genes were the members with the highest level of expression in their respective gene family. A cluster analysis was used to obtain the *MYB* genes that were consistent with the trend of expression of the structural genes related to anthocyanin biosynthesis. The clustering results showed that these genes primarily clustered in the heatmap1 subclass, with a total of 52 genes (Figure 3b). The trend of expression of this subclass was basically consistent with the trend of anthocyanin content in the leaves (Figure 3c). A total of 24 highly expressed anthocyanin structural genes (Table 2) were found in the heatmap1 sub-cluster. These 24 genes contained eight anthocyanin structural genes obtained by WGCNA visualization analysis, which served as the basis for the following *MYB* gene analysis using Heatmap1.

The 28 *MYB* genes that were identified from the Heatmap1 subclass (Figure 3d) were used to construct phylogenetic relationships with the 13 *MYB* genes from *A. thaliana* and *Petunia hybrida*. The results showed that TRINITY_DN7483_c0_g2 clustered with PhAN2 and AtMYB113 activated anthocyanin biosynthesis (Figure 3e). This finding indicated that it may play a role in promoting anthocyanin biosynthesis during the development of *F. virens* leaves. Simultaneously, this gene is also one of the ones related to the change in anthocyanin content identified by the WGCNA visualization analysis (Figure 3a). Therefore, we hypothesized that it may play a role in regulating the red duration of the W-type and R-type *F. virens* leaves. It was further cloned and analyzed together with the eight structural genes (Table 2) obtained from the WGCNA visualization analysis.

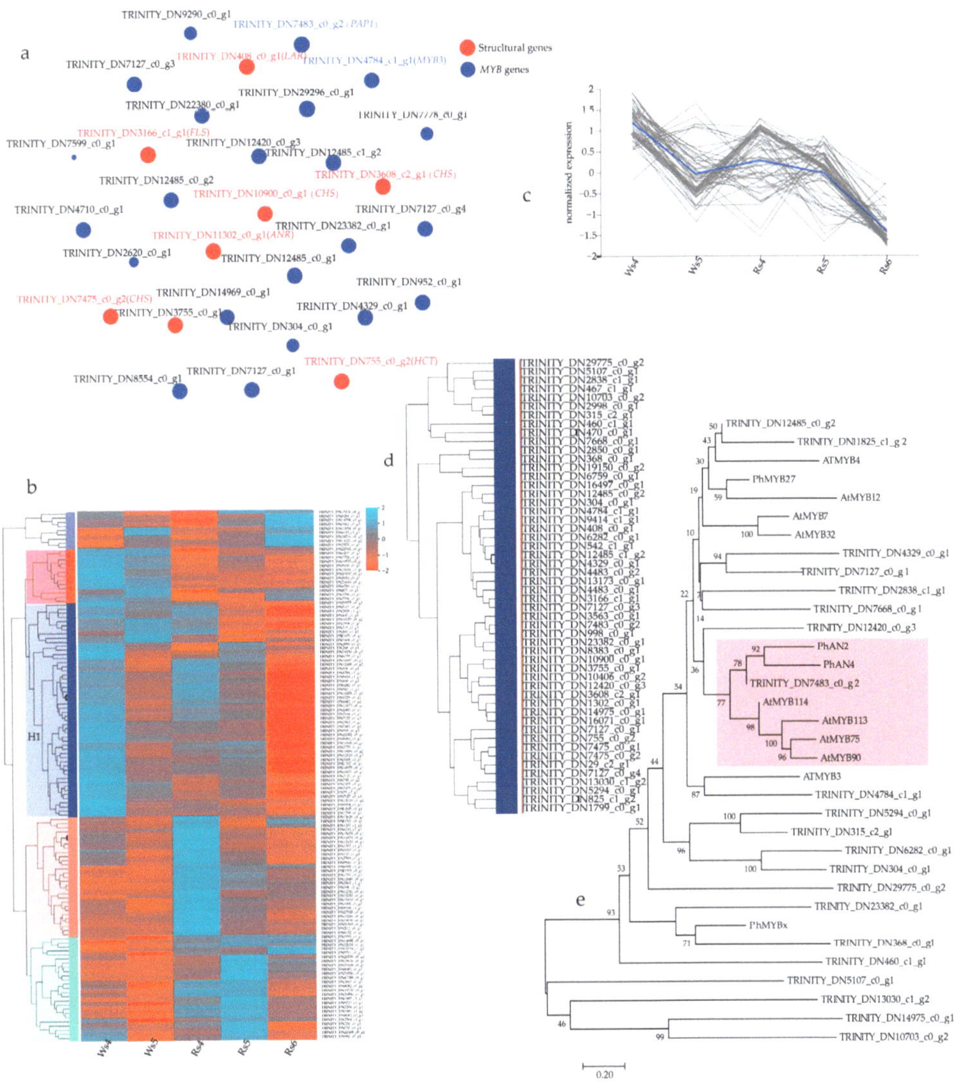

Figure 3. Identification of the structural and regulatory genes related to anthocyanin biosynthesis in *Ficus virens*: (**a**) WGCNA visual network diagram. The correlation among the genes was visualized. Each node in the visualization graph represents a gene. Greater the connectivity of a node indicates more importance. Red is marked as the structural gene related to anthocyanin biosynthesis. (**b**) The clustering heat map of the structural DEGs and regulatory DEGs related to anthocyanin biosynthesis in the transcriptome leaves of *F. virens*. Each column in the figure represents a sample; each row represents a unigene, and the color represents the level of expression of the unigene in the sample. The left side is the tree diagram of unigene clustering. The closeness of two unigene branches indicates closer levels of expression levels. (**c**) Chart of the trend of expression of the sub-cluster heatmap1 genes. Each line (gray) in the figure represents a unigene change trend, and the fitting line (blue) represents the change trend of the average expression of all unigenes in the sub-cluster. (**d**) Sub-cluster heatmap1 detail. (**e**) Phylogenetic analysis of the putative MYB members. WGCNA: weighted gene co-expression network analysis.

Table 2. Candidate genes for anthocyanin biosynthesis in *Ficus virens*.

Candidate Gene	Total No.	Unigene	Name	CDS Length	GenBank Accession Number
ANS	1	TRINITY_DN2320_c1_g1 (k-mer = 25)	*FvANS*	1077 bp	OR682438
ANR	1	TRINITY_DN11302_c0_g1	-	-	
BZ1	1	TRINITY_DN3563_c0_g1	*FvBZ1*	1104 bp	OR682439
CHS	5	TRINITY_DN7475_c0_g2	-	-	
		TRINITY_DN3608_c2_g1	*FvCHS1*	-	OR682437
		TRINITY_DN10900_c0_g1	-	-	
		TRINITY_DN7475_c0_g1	-	-	
		TRINITY_DN579_c1_g1_i1 (k-mer = 25)	*FvCHS2*	1173 bp	OR682441
C4H	2	TRINITY_DN4483_c0_g2	-	-	
		TRINITY_DN4483_c0_g1	-	-	
F3′H	1	TRINITY_DN13173_c0_g1	*FvF3H*	1530 bp	OR682443
C3′H	2	TRINITY_DN9414_c1_g1	-	-	
		TRINITY_DN8383_c0_g1	-	-	
DFR	2	TRINITY_DN16497_c0_g1	*FvDFR1*	1086 bp	OR682442
		TRINITY_DN2850_c0_g1	-	-	
CAMT	1	TRINITY_DN542_c1_g1	-	-	
HCT	4	TRINITY_DN2998_c0_g1	-	-	
		TRINITY_DN1799_c0_g1	-	-	
		TRINITY_DN755_c0_g2	-	-	
		TRINITY_DN29_c2_g1	-	-	
CHI	2	TRINITY_DN16071_c0_g1	-	-	
		TRINITY_DN998_c0_g1	*FvCHI*	702 bp	OR682440
F3H	1	TRINITY_DN3755_c0_g1	*FvF3H*	-	OR682435
FLS	1	TRINITY_DN3166_c1_g1	-	-	
LAR	2	TRINITY_DN10406_c0_g2	-	-	
		TRINITY_DN408_c0_g1	-	-	
UGT	1	TRINITY_DN13686_c0_g1	*FvUGT1*	1374 bp	OR682445
MYB	1	TRINITY_DN7483_c0_g2	*FvPAP1*	939 bp	OR682444

-: no gene cloning, the full length of the target sequence was not cloned.

3.4. Cloning of the Transcript Sequences Related to Anthocyanin Biosynthesis

Considering that the sequences assembled by transcriptome sequencing using next-generation sequencing technology (NGS) may be incorrect and different from the actual sequences, we decided to first clone the complete coding frame of each target gene to more reliably examine the changes in expression and analyze the gene functions. The members with the highest level of expression in each gene related to the anthocyanin biosynthesis family identified by WGCNA and a cluster analysis were selected for gene cloning. The sequences cloned in this study were all transcripts. First, primers were designed based on the sequence assembled by Trinity (k-mer = 30), and their complete coding frame sequences were amplified. The primers used for amplification are shown in Table 1.

The amplification results showed that the amplified fragments of *FvDFR1* (OR682442) and *FvF3′H* (OR682443) were consistent with the length of their respective unigene. The assembled *FvBZ1* unigene was 1104 bp long, while the actual amplified fragment was 1410 bp (OR682439). In comparison with other *CHI* genes from plants, the assembled *CHI* unigene only represent a 5′ partial CDS. To clone the complete CDS, we searched the recently published genomic sequence of the closely related plant *Ficus microcarpa* for homologous genes. The 3′ primer was designed based on the homologous reference sequence (GWHPABKV009265.1), and the complete CDS (OR682440) was obtained. The *UGT* gene was not identified by the clustering analysis and WGCNA described above. We chose to clone *UGT* because it was closely related to the stability of anthocyanin. The assembled

UGT unigene only represents a 5′ partial CDS. The 3′ primers were designed based on the homologous sequence of *F. microcarpa* (GWHPABKV019926.1), and the complete CDS (OR682445) of *UGT* was cloned (Figure S4).

Since the reference sequence obtained by Trinity using the default parameter (k-mer = 30) was not annotated to *ANS* and the other genes, we adjusted the parameter of Trinity (k-mer from 30 to 25) to re-assemble, and K-mer25_TRINITY_DN2320_c1_g1 was annotated as an *ANS* gene. The product that was obtained by PCR amplification using primers based on this sequence was consistent with the predictions (OR682438, Figure S4).

The PCR product obtained using primers designed based on the assembled *CHS* unigene sequence was substantially different from the unigene sequence (homology < 27%). However, the 195 bp PCR products that were amplified with quantitative primers FvCHS1-QF and FvCHS1-QR were consistent with the TRINITY_DN3608_c2_g1 unigene. The primers designed based on TRINITY_DN10900_c0_g1 did not amplify any product. Thus, it is likely that the assembled sequence was incorrect. Since the complete CDS of the candidate CHS genes were not cloned according to the sequence assembled by Trinity (k-mer = 30), we re-designed the PCR primers based on the highest expressed *CHS* sequence K-mer25_TRINITY_DN579_c1_g1 assembled by the parameter-adjusted Trinity (k-mer = 25) method. The cloning and sequencing of the amplified product showed that it was consistent with the unigene (Figure S4), and we designed it as *FvCHS2* (OR682441).

Only one unigene represented *F3H* in the Trinity (k-mer = 30)-assembled results, but no product was amplified using the primers designed based on the unigene sequence. However, a 179 bp fragment was produced with the real-time PCR primers *FvF3H*-QF and *FvF3H*-QR designed according to the unigene sequence, and its sequence was consistent with the origin unigene (Figure S4).

An analysis of the putative protein sequence of the TRINITY_DN7483_c0_g2 *MYB* unigene showed that the R3 domain was incomplete; therefore, we searched the unfiltered original assembly sequence and found that this unigene had eight transcripts. Primers *FvPAP1*-F1 and *FvPAP1*-R1 (Table 1) were designed based on the common sequences at the 5′ and 3′ ends of these eight transcripts to amplify the longest complete CDS sequence of the gene. A total of 1% agarose gel electrophoresis of the amplified product showed a single band (Figure S4). After cloning to the T-vector, the sequencing showed that there were differences among the sequences of different clones. A total of 26 clones were sequenced. The sequencing results showed that the predicted protein sequences of fifteen clones had a complete R2R3 domain, two clones possessed an incomplete R2 domain, and the other nine clones had completely lost the R2R3 domain. A sequence alignment of the nucleotides showed that the deletion of the R2R3 domain in the protein sequence of the seven clones was owing to the insertion of a GCAA 4 base and a large number of deletions of 175 bp in the coding region, which resulted in coding shift changes and premature termination (Figure S5). In addition to these two obvious changes, there were also other indels and base substitutions at other positions of the gene. These results suggest that there may be selective splicing and/or sequence polymorphism at this locus. We registered the sequence T3—FvPAP1—44 (Figure S5), which encodes a gene with the complete R2R3 domain and is the closest to AtPAP2 in GenBank as *FvPAP1* (OR682444).

3.5. Analysis of the Expression of the Anthocyanin Biosynthetic and Regulatory Genes

To further verify the relationship between the changes in expression of the genes related to anthocyanin biosynthesis and content in the transcriptome sequencing results, a qRT-PCR analysis was performed on the six developmental stages of the W-type and R-type *F. virens* leaves. The results showed that the results of qRT-PCR were consistent with the trend of changes for each gene in the transcriptome data (Figure 4).

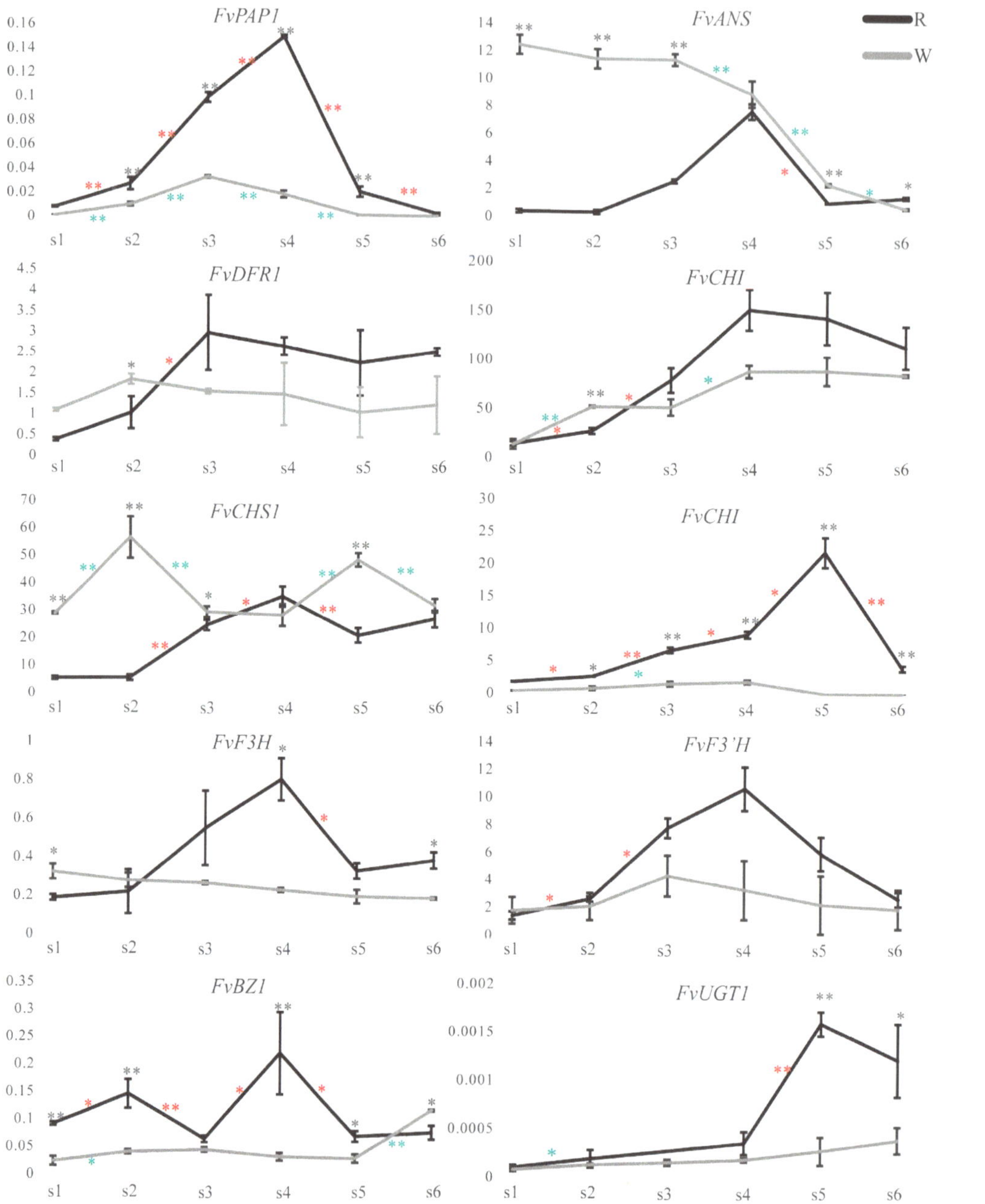

Figure 4. The levels of expression of the genes related to anthocyanin biosynthesis during the development of leaves in the W-type and R-type *Ficus virens* plants. A Student's *t*-test was used to evaluate the significance of the difference between the adjacent stages. The annotation between periods indicates that there are differences between adjacent periods (W-Type * is red *, W-Type * is green *, * $p < 0.05$. ** $p < 0.01$), and the annotation directly above the period indicates the difference between materials (* is black *, * $p < 0.05$, ** $p < 0.01$).

The expression of *FvPAP1* and the change in the content of anthocyanins in the leaves of W-type first increased and then decreased, but the anthocyanins reached their peak at s4, while the highest level of expression of *FvPAP1* was at s3. The levels of expression of *FvPAP1* were consistent with the trend in the changes of anthocyanin during the development of the R-type *F. virens* leaves, and both reached their highest values at the s4 stage (Figure 4).

The levels of expression of *FvPAP1*, *FvDFR1*, *FvCHS2*, and *FvF3′H* showed a trend of increasing first and then decreasing during the development of the W-type *F. virens* leaves. The change in the level of expression of *FvCHS2* was completely consistent with the trend in the change of anthocyanin content. Both reached their highest values at the s4 stage. The change in the level of expression of *FvF3′H* was consistent with the change in the level of expression of *FvPAP1*, which reached its highest value at the s3 stage. The levels of expression of *FvANS*, *FvDFR1*, *FvCHI*, *FvCHS1*, *FvF3H*, *FvBZ1*, and *FvUGT1* did not show the same tendency to vary in parallel with the changes in anthocyanin content or level of expression of *FvPAP1* (Figure 4).

During the development of the R-type *F. virens* leaves, the levels of expression of most genes showed a trend of increasing first and then decreasing. The trend of the change in *FvPAP1* expression was consistent with that in anthocyanin content, and the highest value appeared at the s4 stage. The trend of change in the levels of expression of the *FvCHI*, *FvCHS*, *FvANS*, *FvF3H*, and *FvF3′H* genes was consistent with that of the anthocyanin content and *FvPAP1* expression, and the highest value appeared at the s4 stage. The level of expression of *FvDFR1* reached its peak at the s3 stage, and the levels of expression of *FvCHS2* and *FvUGT1* reached their peak at the s5 stage. The levels of expression of these genes were consistent with the changes in anthocyanin content and *FvPAP1* on the whole. However, the levels of expression of *FvBZ1* formed two peaks at the s2 and s4 stages (Figure 4).

3.6. Functional Verification of FvPAP1

The level of expression of *FvPAP1* was basically consistent with the content of anthocyanin and the levels of expression of *FvF3′H*, *FvANS*, and *FvDFR1*. A phylogenetic tree analysis showed that *FvPAP1* clustered with the *MYB* protein (AtMYB75/PAP1, AtMYB90/PAP2, AtMYB113, AtMYB114, PhAN2, and PhAN4) that activates the biosynthesis of anthocyanin. These suggested that *FvPAP1* may be involved in the activation of anthocyanin biosynthesis in the leaves of *F. virens*. However, whether it can promote the biosynthesis of anthocyanin merits further experimental verification. Because the technology of analyzing gene function by transient expression and stable expression has not been established in *F. virens*, we chose to further verify the function of *FvPAP1* through heterologous expression.

First, we introduced the 35S:*FvPAP1* vector into the petals and leaves of petunia using the *Agrobacterium*-mediated method (injection) for transient expression. A total of 12 petunia petals were injected with 35S:*FvPAP1*, and seven of them began to show obvious purple spots at the injection site 3 days after injection (Figure 5a). Ten petals of *P. hybrida* were injected with the empty vector, and no purple spot appeared until the petals had wilted (approximately 7 days after injection). These results indicate that *FvPAP1* can activate the biosynthesis of anthocyanin in the floral organs.

A total of 27 *P. hybrida* leaves were injected with 35S:*FvPAP1*, and lavender spots appeared on the leaves (Figure 5b) on the fourth day after injection. Another 27 leaves of *P. hybrida* were injected with the empty vector, and no purple spot appeared within 14 days. On the ninth day after injection, the *P. hybrida* leaves were collected to extract the total RNA for qRT-PCR. The quantitative results showed that the levels of expression of the structural genes *PhCHSA*, *PhF3′5′H*, *PhF3H*, *PhDFR*, *PhANS*, *PhPAL1*, and *PhF3′H* in the anthocyanin biosynthetic pathway in the leaves injected with *FvPAP1* were significantly higher than those injected with the empty vector. These results indicate that the *FvPAP1* gene activates the biosynthesis of anthocyanin in the leaves (Figure 5e).

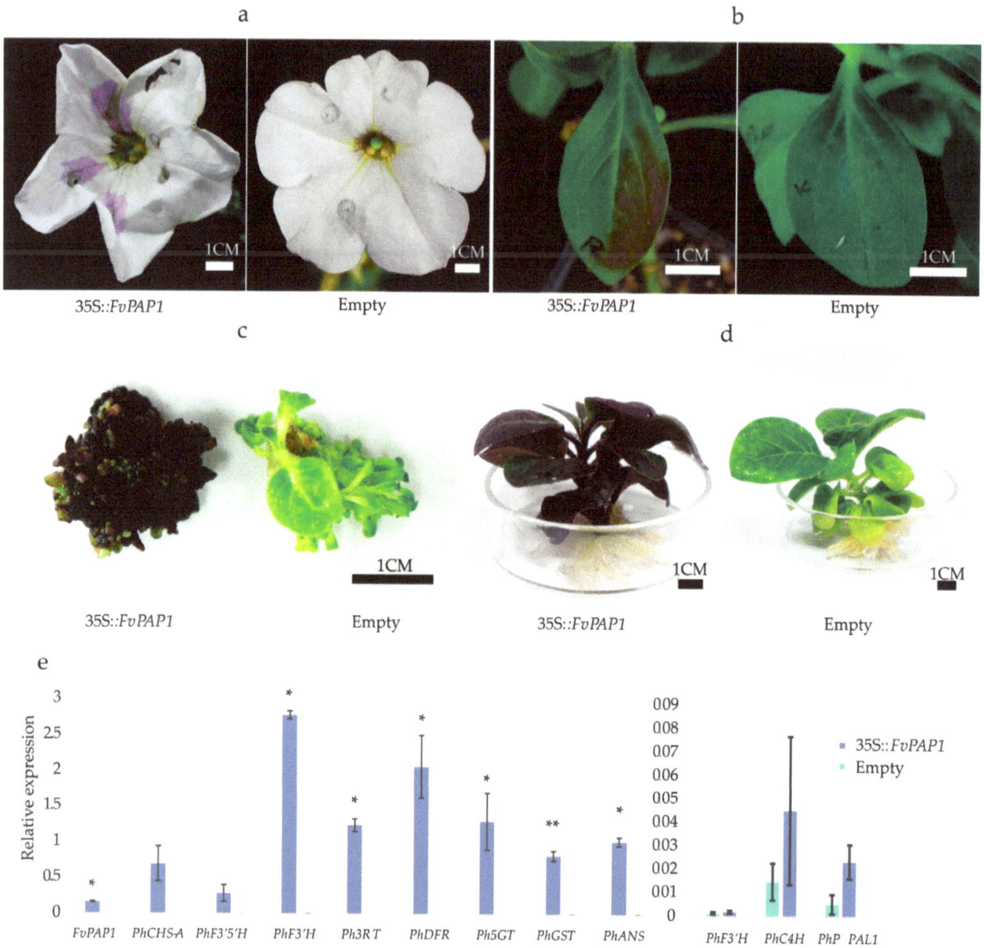

Figure 5. Functional analysis of *FvPAP1*: (**a**) Transient expression of *FvPAP1* in the *Petunia hybrida* flowers. (**b**) Transient expression of *FvPAP1* in the *P hybrida* leaves. (**c**) The purple phenomenon of *FvPAP1* in tobacco callus. (**d**) Purple phenomenon observed in tobacco seedlings after transformation with *FvPAP1*. (**e**) The levels of expression of the anthocyanin structural genes in the *P. hybrida* leaves that transiently expressed *FvPAP1*. The left side of each gene histogram is 35s::*FvPAP1* expression; the right side is empty expression, and both were calculated as the same internal gene reference. The significance was tested by a single-factor analysis of variance and t-test. * $p < 0.05$, ** $p < 0.01$.

We subsequently used the leaf disc method to introduce *FvPAP1* into *Nicotiana tabacum*. A total of 97 resistant calli were obtained after transformation and screening with Basta, and 86 of them were dark purple-red, with obvious characteristics of anthocyanin accumulation. In addition, the adventitious buds that regenerated from the callus were purple-red. More than 100 resistant calli were obtained by transformation of the empty vector, and no purple calli were found (Figure 5c,d). These indicated that *FvPAP1* can stably activate the biosynthesis of anthocyanin in tobacco.

4. Discussion

With the increasing demand for landscape ecological construction, leaf color has attracted increasing amounts of attention because of its substantial ornamental value. The leaves of some woody plants appear red at the young leaf stage, and the red gradually fades with the development of the leaves. If the red phenomenon that occurs during the development of the leaves can be utilized, it will substantially enhance the landscape value. In *Ficus virens*, young buds germinate, and new leaves rapidly expand as soon at the old ones fall from the tree. The replacement of old leaves by new ones is completed in a very short time. The red color on the rapidly unfolding new leaves can provide a colorful scene at this stage and increase the visual effect to result in high ornamental values. We found that the R-type *F. virens* leaves maintained their red state for a longer time than the W-type in the new leaf developmental stage, and there was a corresponding improvement in the landscape effect.

The leaf color of plants is primarily determined by pigments. The detection of pigments in different developmental stages of the *F. virens* leaves showed that as the red color appeared and then disappeared with the development of leaves, the anthocyanin content increased first and then gradually decreased (Figure 1b). This result is consistent with research on the transiently flush young leaves that are red in *Castanopsis fiss* [37], *Syzygium luehmannii* [38], *Hevea brasiliensis* [39], *Juglans regia* [40], and other plants. The levels of expression of *FvCHI*, *FvANS*, *FvDFR1*, *FvF3′H*, and other structural genes related to anthocyanin biosynthesis were significantly higher in the red-leaf period than in the green-leaf period (s3–s5), which indicated that the red appearance of new leaves in *F. virens* was related to the change in the content of anthocyanin. The accumulation of anthocyanin during young leaf development is primarily considered as a defense measure against excess light in many plant species [41,42]. However, the specific role of anthocyanins in the development of young leaves remains to be elucidated.

The levels of expression of the structural genes during anthocyanin biosynthesis directly affects the anthocyanin content, and the MBW complex regulates the expression of the LBGs during the biosynthesis of anthocyanin. The change in the level of expression of *MYB* is key to the change in anthocyanin content. In tea (*Camellia sinensis*), the change in the level of expression of *MYB* is crucial for young red leaves [43]. The MBW complex of IbMYB1/IbMYB2/IbMYB3-IbbHLH2-IbWDR1 in *Ipomoea batatas* activated the expression of the structural genes *IbCHS-D* and *IbDFR-B* related to anthocyanin biosynthesis [44]. In this study, the level of expression of *FvPAP1* increased first and then decreased during the development of leaves in both W-type and R-type *F. virens*, which was consistent with the anthocyanin content. The level of expression of *FvPAP1* reached its highest level at the s3 stage in the WT, while it was highest at the s4 stage in the R-type. The level of expression in the trend of changes of *FvPAP1* was completely consistent with the trend in the change of anthocyanin content (Figure 4), which suggested an important role of *FvPAP1* in the formation of red leaves in *F. virens*.

To verify whether the level of expression of *FvPAP1* can promote anthocyanin biosynthesis, we constructed an overexpression vector, and investigated the function by transient expression in *P. hybrida* and stable expression in *N. tabacum*. The results showed that the transgenic tobacco callus was clearly purple, and the injection site in the petunia petals became purple. This phenomenon did not appear in the control material. These results verified that *FvPAP1* had the function of activating anthocyanin biosynthesis (Figure 5).

During the process of cloning *FvPAP1*, we found that the *FvPAP1* locus has sequence polymorphism. Only 57.6% of the cDNA sequence numbers had a complete R2R3 conserved domain, and the remaining cDNA sequences had some differences, which resulted in partial or complete deletion of the R2R3 domain. Because fragment deletion and indel did not occur in other genes cloned using the same experimental protocol and chemicals, it seems impossible that the sequence variation in the *FvPAP1* gene was caused by PCR amplification. These results suggest that there may be selective splicing and/or sequence polymorphism at this gene locus. The relationship between the duration of red-leaf period and *FvPAP1* polymorphism merits further study.

The prolonging of the red-leaf period is not only related to biosynthesis but also to the degradation of anthocyanin. There are various enzymatic and non-enzymatic factors that affect the stability and concentration of anthocyanin. Anthocyanin may become discolored as a result of active enzyme-driven breakdown processes. In addition to enzymatic factors, non-enzymatic factors (B-ring hydroxylation, glycosylation, metal ions, and pH, among others) and environmental factors (temperature, light, and irradiance among others) also affect the color and stability of anthocyanins and may enhance their vulnerability to the enzymes that degrade anthocyanins [21,45–48]. No significant differences in the levels of expression of the genes related to key genes of the light signaling pathway and anthocyanin degradation between the two varieties of *F. virens* were found during the process of transcriptome analysis in this study. The degradation of anthocyanin may not be the primary factor that causes the difference in the length of red-leaf period in *F. virens*.

Transcriptome sequencing technology has become an important tool to study the phenomena of plant life. However, owing to the lack of high-quality genome sequencing data in most plants, the de novo assembly of expressed sequences is still an important method for the analysis of DEGs and cloning of cDNA in plants. Owing to the complexity of many plant genomes, including a high degree of polyploidization and gene duplication, the de novo assembly of short NGS reads is very challenging, and errors easily occur in the assembled sequences. In this study, the *ANS* gene was not annotated in the results of Trinity assembly with default parameter. Thus, the putative expression boxes of *BZ1* (TRINITY_DN3563_c0_g1), *UGT* (TRINITY_DN13686_c0_g1), and *MYB* (TRINITY_DN7483_c0_g2) were not completely assembled. Primers were designed to amplify *CHS* (TRINITY_DN3608_c2_g1 and TRINITY_DN10900_c0_g1) and *F3H* (TRINITY_DN3755_c0_g1) based on the unigene sequences, but the assembled sequences were not obtained. This indicated that there were differences between the assembled and actual sequences. The *CHS* gene family may have highly expanded, which resulted in difficulty in sequence assembly because 21 *CHS* genes were annotated in the *F. microcarpa* genome, and only one complete CDS of *CHS* gene with a high level of expression was amplified after our substantial research in this study. Therefore, we believe that if a DEG analysis is conducted using NGS technology in a plant with a complex genome, it is necessary to clone the complete CDS sequence of the target gene to confirm whether the assembled sequence is correct before quantitative PCR amplification is performed.

5. Conclusions

The change in color in the red leaves of *Ficus virens* was primarily caused by the change in the content of anthocyanin. *FvPAP1* could stably activate the biosynthesis of anthocyanin. The difference in the level of expression of *FvPAP1* in time and intensity in the young leaves may be the reason for the difference in the duration of the red-leaf period in different lines of *Ficus virens*.

Supplementary Materials: The following supporting information can be downloaded at: https://www.mdpi.com/article/10.3390/cimb46060343/s1, Figure S1: Red-type (R-type) leaves of *Ficus virens*; Figure S2: pGMF500 vector; Figure S3: Bud of *Ficus virens* is wrapped by a large stipule; Figure S4: Electrophoretogram of cloned gene. Marker is DL2000. Gene annotation * is expressed as the result of cloning using qRT-PCR primers. Electrophoretogram of qRT-PCR primers: The upper band was W-type *Ficus virens* as the template, and the lower band was R-type *Ficus virens* as the template. From left to right, the quantitative primers of *FvDFR, FvF3′H, FvCHS, FvPAP1, FvUGT1, FvANS, FvCHI, FvBZI,* and *FvF3H* were cloned, * indicates that the gene is a product of qRT-PCR primer cloning; Figure S5: *FvPAP1* polymorphism: (a) Amino acid sequence of R2R3 domain of *FvPAP1*. (b) Partial sequence of *FvPAP1*. Red arrow represents the insertion of a GCAA 4 base.

Author Contributions: Conceptualization, Y.G.; Data curation, Q.M. and S.Z. (Shihui Zou); Funding acquisition, S.S.; Investigation, Q.M. and S.Z. (Shuhua Zhong); Project administration, L.A.; Supervision, Y.G.; Validation, Q.M., S.Z. (Shuhua Zhong), T.M., and Y.Y.; Writing—original draft, Q.M.; Writing—review and editing, Y.G. All authors have read and agreed to the published version of the manuscript.

Funding: This research was funded by XTZW2021-KF02 Exploration on Molecular Basis of Characteristic Traits of Native Plants.

Data Availability Statement: Data is contained within the article and supplementary material.

Conflicts of Interest: The authors declare no conflicts of interest.

References

1. Corner, E.J.H. *Ficus* (Moraceae) and Hymenoptera (Chalcidoidea): Figs and their pollinators. *Biol. J. Linn. Soc.* **1985**, *25*, 187–195. [CrossRef]
2. Iqbal, D.; Khan, M.S.; Khan, M.S.; Ahmad, S.; Hussain, M.S.; Ali, M. Bioactivity guided fractionation and hypolipidemic property of a novel HMG-CoA reductase inhibitor from *Ficus virens* Ait. *Lipids Health Dis.* **2015**, *14*, 15. [CrossRef] [PubMed]
3. Iqbal, D.; Khan, M.S.; Khan, M.S.; Ahmad, S.; Srivastava, A.K. An in vitro and molecular informatics study to evaluate the antioxidative and β-hydroxy-β-methylglutaryl-CoA reductase inhibitory property of *Ficus virens* Ait. *Phytother. Res.* **2014**, *28*, 899–908. [CrossRef] [PubMed]
4. Hafeez, A.; Jain, U.; Sajwan, P.; Srivastava, S.; Thakur, A. Evaluation of Carrageenan induced anti-inflammatory activity of ethanolic extract of bark of *Ficus virens* Linn. in swiss albino mice. *J. Phytopharm.* **2013**, *2*, 39–43. [CrossRef]
5. Lee, S.H.; Ng, A.B.; Lim, R.C.; Francisco, R.; Ng, W.Q.; Ng, X.Y.; Neo, L.; Yee, A.T.; Chong, K.Y.; Tan, H.T. Status and distribution in Singapore of *Ficus virens* Aiton (Moraceae). *Nat. Singap.* **2013**, *6*, 223–227.
6. Patel, M.H.; Desai, B.; Jha, S.; Patel, D.; Mehta, A.; Garde, Y. Phytochemical, pharmacognosy and ethnobotanical importance of the *Ficus virens* Aiton. *Pharma Innov. J.* **2023**, *12*, 4017–4021.
7. Rai, S.; Sabesan, G.S.; Babu, K. Comparative pharmacognostic studies on the barks of four Ficus species. *Turk. J. Bot.* **2010**, *34*, 215–224. [CrossRef]
8. Chaudhary, I.J.; Rathore, D. Phytomonitoring of dust load and its effect on foliar micromorphological characteristics of urban trees. *Plantica J. Plant Sci.* **2018**, *2*, 170–179.
9. Liu, L.; Guan, D.; Peart, M.R.; Wang, G.; Zhang, H.; Li, Z. The dust retention capacities of urban vegetation-a case study of Guangzhou, South China. *Environ. Sci. Pollut. Res. Int.* **2013**, *20*, 6601–6610. [CrossRef]
10. Soejono, S.; Budiharta, S.; Arisoesilaningsih, E. Proposing local trees diversity for rehabilitation of degraded lowland areas surrounding springs. *Biodiversitas J. Biol. Divers.* **2013**, *14*, 37–42. [CrossRef]
11. Zhang, X.; Zeng, B.; Zhong, Z. Differential responsiveness in stem height and diameter growth of two Ficus tree species in the Three Gorges reservoir region of China as affected by branch removal. *Can. J. For. Res.* **2007**, *37*, 1748–1754. [CrossRef]
12. Wang, Z.; Jiang, X.; Jiang, H.; Yang, L.; Zhou, F.; Lu, C. Application of major tree species for the CAST stereoscopic landscape demonstration project. *J. Landsc. Res.* **2012**, *4*, 65–71.
13. Dominy, N.J.; Lucas, P.W.; Ramsden, L.W.; Riba-Hernandez, P.; Stoner, K.E.; Turner, I.M. Why are young leaves red? *Oikos* **2002**, *98*, 163–176. [CrossRef]
14. Manetas, Y. Why some leaves are anthocyanic and why most anthocyanic leaves are red? *Flora-Morphol. Distrib. Funct. Ecol. Plants* **2006**, *201*, 163–177. [CrossRef]
15. Zheng, X.; Niklas, K.J.; Ratkowsky, D.A.; Jiao, Y.; Ding, H.; Shi, P. Comparison of Leaf Shape between a Photinia Hybrid and One of Its Parents. *Plants* **2022**, *11*, 2370. [CrossRef]
16. Liu, W.; Nisar, M.F.; Wan, C. Characterization of Phenolic Constituents from *Prunus cerasifera* Ldb Leaves. *J. Chem.* **2020**, *2020*, 1–5. [CrossRef]
17. Chen, X.; Liu, H.; Wang, S.; Zhang, C.; Liu, L.; Yang, M.; Zhang, J. Combined transcriptome and proteome analysis provides insights into anthocyanin accumulation in the leaves of red-leaved poplars. *Plant Mol. Biol.* **2021**, *106*, 491–503. [CrossRef] [PubMed]
18. Mehlenbacher, S.A.; Smith, D.C. 'Red Dragon'ornamental hazelnut. *HortScience* **2009**, *44*, 843–844. [CrossRef]
19. Wadl, P.A.; Wang, X.; Pantalone, V.R.; Trigiano, R.N. Inheritance of red foliage in flowering dogwood (*Cornus florida* L.). *Euphytica* **2010**, *176*, 99–104. [CrossRef]
20. Holton, T.A.; Cornish, E.C. Genetics and biochemistry of anthocyanin biosynthesis. *Plant Cell* **1995**, *7*, 1071. [CrossRef]
21. Liu, Y.; Tikunov, Y.; Schouten, R.E.; Marcelis, L.F.; Visser, R.G.; Bovy, A. Anthocyanin biosynthesis and degradation mechanisms in Solanaceous vegetables: A review. *Front. Chem.* **2018**, *6*, 52. [CrossRef] [PubMed]
22. Stracke, R.; Werber, M.; Weisshaar, B. The R2R3-MYB gene family in Arabidopsis thaliana. *Curr. Opin. Plant Biol.* **2001**, *4*, 447–456. [CrossRef] [PubMed]
23. Gonzalez, A.; Zhao, M.; Leavitt, J.M.; Lloyd, A.M. Regulation of the anthocyanin biosynthetic pathway by the TTG1/bHLH/Myb transcriptional complex in Arabidopsis seedlings. *Plant J.* **2008**, *53*, 814–827. [CrossRef] [PubMed]
24. Li, G.; Serek, M.; Gehl, C. Physiological changes besides the enhancement of pigmentation in *Petunia hybrida* caused by overexpression of PhAN2, an R2R3-MYB transcription factor. *Plant Cell Rep.* **2023**, *42*, 609–627. [CrossRef]
25. Zhang, B.; Xu, X.; Huang, R.; Yang, S.; Li, M.; Guo, Y. CRISPR/Cas9-mediated targeted mutation reveals a role for AN4 rather than DPL in regulating venation formation in the corolla tube of *Petunia hybrida*. *Hortic. Res.* **2021**, *8*, 116. [CrossRef] [PubMed]
26. Koes, R.; Verweij, W.; Quattrocchio, F. Flavonoids: A colorful model for the regulation and evolution of biochemical pathways. *Trends Plant Sci.* **2005**, *10*, 236–242. [CrossRef]

27. Albert, N.W.; Davies, K.M.; Lewis, D.H.; Zhang, H.; Montefiori, M.; Brendolise, C.; Boase, M.R.; Ngo, H.; Jameson, P.E.; Schwinn, K.E. A conserved network of transcriptional activators and repressors regulates anthocyanin pigmentation in eudicots. *Plant Cell* **2014**, *26*, 962–980. [CrossRef] [PubMed]
28. Dubos, C.; Stracke, R.; Grotewold, E.; Weisshaar, B.; Martin, C.; Lepiniec, L. MYB transcription factors in Arabidopsis. *Trends Plant Sci.* **2010**, *15*, 573–581. [CrossRef] [PubMed]
29. Viola, I.L.; Camoirano, A.; Gonzalez, D.H. Redox-Dependent Modulation of Anthocyanin Biosynthesis by the TCP Transcription Factor TCP15 during Exposure to High Light Intensity Conditions in Arabidopsis. *Plant Physiol.* **2016**, *170*, 74–85. [CrossRef]
30. Zai, H.; Jing, C.; Zhong, X. *Plant Physiological Experiment*; Harbin University of Technology Press: Harbin, China, 2004; pp. 68–70.
31. Ren, Y.; Yu, G.; Shi, C.; Liu, L.; Guo, Q.; Han, C.; Zhang, D.; Zhang, L.; Liu, B.; Gao, H.; et al. Majorbio Cloud: A one-stop, comprehensive bioinformatic platform for multiomics analyses. *iMeta* **2022**, *1*, e12. [CrossRef]
32. Shen, N.; Wang, T.; Gan, Q.; Liu, S.; Wang, L.; Jin, B. Plant flavonoids: Classification, distribution, biosynthesis, and antioxidant activity. *Food Chem.* **2022**, *383*, 132531. [CrossRef]
33. Kumar, S.; Stecher, G.; Li, M.; Knyaz, C.; Tamura, K. MEGA X: Molecular evolutionary genetics analysis across computing platforms. *Mol. Biol. Evol.* **2018**, *35*, 1547. [CrossRef] [PubMed]
34. Giulietti, A.; Overbergh, L.; Valckx, D.; Decallonne, B.; Bouillon, R.; Mathieu, C. An overview of real-time quantitative PCR: Applications to quantify cytokine gene expression. *Methods* **2001**, *25*, 386–401. [CrossRef] [PubMed]
35. Yuan, M.; Xu, C.Y. BiFC Assay for Detecting Protein-protein Interaction in Tobacco Leaves. *Bio-Protocol* **2018**, *Bio-101*, e1010133. [CrossRef]
36. Horsch, R.; Fry, J.; Hoffmann, N.; Wallroth, M.; Eichholtz, D.; Rogers, S.; Fraley, R. A simple and general method for transferring genes into plants. *Science* **1985**, *227*, 1229–1231. [CrossRef] [PubMed]
37. Zhang, T.J.; Chow, W.S.; Liu, X.T.; Zhang, P.; Liu, N.; Peng, C.L. A magic red coat on the surface of young leaves: Anthocyanins distributed in trichome layer protect *Castanopsis fissa* leaves from photoinhibition. *Tree Physiol.* **2016**, *36*, 1296–1306. [CrossRef]
38. Woodall, G.; Stewart, G. Do anthocyanins play a role in UV protection of the red juvenile leaves of Syzygium? *J. Exp. Bot.* **1998**, *49*, 1447–1450. [CrossRef]
39. Fang, Y.; Mei, H.; Zhou, B.; Xiao, X.; Yang, M.; Huang, Y.; Long, X.; Hu, S.; Tang, C. De novo Transcriptome Analysis Reveals Distinct Defense Mechanisms by Young and Mature Leaves of *Hevea brasiliensis* (Para Rubber Tree). *Sci. Rep.* **2016**, *6*, 33151. [CrossRef] [PubMed]
40. Wang, L.; Li, L.; Zhao, W.; Fan, L.; Meng, H.; Zhang, G.; Wu, W.; Shi, J.; Wu, G. Integrated metabolomic and transcriptomic analysis of the anthocyanin and proanthocyanidin regulatory networks in red walnut natural hybrid progeny leaves. *PeerJ* **2022**, *10*, e14262. [CrossRef]
41. Krause, G.H.; Virgo, A.; Winter, K. High susceptibility to photoinhibition of young leaves of tropical forest trees. *Planta* **1995**, *197*, 583–591. [CrossRef]
42. Ranjan, S.; Singh, R.; Singh, M.; Pathre, U.V.; Shirke, P.A. Characterizing photoinhibition and photosynthesis in juvenile-red versus mature-green leaves of *Jatropha curcas* L. *Plant Physiol. Biochem.* **2014**, *79*, 48–59. [CrossRef] [PubMed]
43. Sun, B.; Zhu, Z.; Cao, P.; Chen, H.; Chen, C.; Zhou, X.; Mao, Y.; Lei, J.; Jiang, Y.; Meng, W. Purple foliage coloration in tea (*Camellia sinensis* L.) arises from activation of the R2R3-MYB transcription factor CsAN1. *Sci. Rep.* **2016**, *6*, 32534. [CrossRef] [PubMed]
44. Deng, J.; Wu, D.; Shi, J.; Balfour, K.; Wang, H.; Zhu, G.; Liu, Y.; Wang, J.; Zhu, Z. Multiple MYB Activators and Repressors Collaboratively Regulate the Juvenile Red Fading in Leaves of Sweetpotato. *Front. Plant Sci.* **2020**, *11*, 941. [CrossRef] [PubMed]
45. Patras, A.; Brunton, N.P.; O'Donnell, C.; Tiwari, B.K. Effect of thermal processing on anthocyanin stability in foods; mechanisms and kinetics of degradation. *Trends Food Sci. Technol.* **2010**, *21*, 3–11. [CrossRef]
46. Zhang, Z.; Pang, X.; Xuewu, D.; Ji, Z.; Jiang, Y. Role of peroxidase in anthocyanin degradation in litchi fruit pericarp. *Food Chem.* **2005**, *90*, 47–52. [CrossRef]
47. Luo, H.; Deng, S.; Fu, W.; Zhang, X.; Zhang, X.; Zhang, Z.; Pang, X. Characterization of Active Anthocyanin Degradation in the Petals of *Rosa chinensis* and *Brunfelsia calycina* Reveals the Effect of Gallated Catechins on Pigment Maintenance. *Int. J. Mol. Sci.* **2017**, *18*, 699. [CrossRef]
48. Oren-Shamir, M.; Nissim-Levi, A. Temperature and gibberellin effects on growth and anthocyanin pigmentation in Photinia leaves. *J. Hortic. Sci. Biotechnol.* **1999**, *74*, 355–360. [CrossRef]

current issues in molecular biology

MDPI

Review

Exploring Plant Meiosis: Insights from the Kinetochore Perspective

Kang-Di Zhou [1,2], Cai-Xia Zhang [1], Fu-Rong Niu [3], Hao-Chen Bai [2], Dan-Dan Wu [4], Jia-Cheng Deng [2], Hong-Yuan Qian [2], Yun-Lei Jiang [2] and Wei Ma [1,*]

1 Science Center for Future Foods, Jiangnan University, 1800 Lihu Road, Wuxi 214122, China; 17637083887@163.com (K.-D.Z.); zhangcaixia8445@126.com (C.-X.Z.)
2 School of Biotechnology, Jiangnan University, 1800 Lihu Road, Wuxi 214122, China; 1022200109@stu.jiangnan.edu.cn (H.-C.B.); 1023200332@stujiangnan.edu.cn (J.-C.D.); 1023210122@stujiangnan.edu.cn (H.-Y.Q.); 1023210125@stu.jiangnan.edu.cn (Y.-L.J.)
3 College of Forestry, Gansu Agricultural University, Lanzhou 730070, China; niufurong@gmail.com
4 State Key Laboratory of Crop Gene Exploration and Utilization in Southwest China, Sichuan Agricultural University, Chengdu 611130, China; wudandan@sicau.edu.cn
* Correspondence: wei.ma03@jiangnan.edu.cn; Tel.: +86-15202663433

Abstract: The central player for chromosome segregation in both mitosis and meiosis is the macromolecular kinetochore structure, which is assembled by >100 structural and regulatory proteins on centromere DNA. Kinetochores play a crucial role in cell division by connecting chromosomal DNA and microtubule polymers. This connection helps in the proper segregation and alignment of chromosomes. Additionally, kinetochores can act as a signaling hub, regulating the start of anaphase through the spindle assembly checkpoint, and controlling the movement of chromosomes during anaphase. However, the role of various kinetochore proteins in plant meiosis has only been recently elucidated, and these proteins differ in their functionality from those found in animals. In this review, our current knowledge of the functioning of plant kinetochore proteins in meiosis will be summarized. In addition, the functional similarities and differences of core kinetochore proteins in meiosis between plants and other species are discussed, and the potential applications of manipulating certain kinetochore genes in meiosis for breeding purposes are explored.

Keywords: kinetochore; meiosis; spindle assembly checkpoint; chromosomal passenger complex; cohesin

Citation: Zhou, K.-D.; Zhang, C.-X.; Niu, F.-R.; Bai, H.-C.; Wu, D.-D.; Deng, J.-C.; Qian, H.-Y.; Jiang, Y.-L.; Ma, W. Exploring Plant Meiosis: Insights from the Kinetochore Perspective. *Curr. Issues Mol. Biol.* **2023**, *45*, 7974–7995. https://doi.org/10.3390/cimb45100504

Academic Editor: Shimeles Tilahun

Received: 11 August 2023
Revised: 12 September 2023
Accepted: 20 September 2023
Published: 28 September 2023

1. Introduction

The macromolecular kinetochore structure, which is assembled by more than 100 structural and regulatory proteins on centromere DNA [reviewed in [1]], is the key player for chromosome segregation in both mitosis and meiosis. During cell division, kinetochores bridge the chromosomal DNA and microtubule polymers to mediate chromosome segregation and alignment [2]. In addition, they can function as a signaling hub to regulate anaphase onset via the spindle assembly checkpoint and control anaphase chromosome movement [2].

Meiosis is a special cell division type in which germ cells can produce gametes in the sexual reproduction stage among eukaryotes. A single round of DNA replication is followed by two sequential rounds of chromosome segregation, termed meiosis I and II (MI and MII). Accurate chromosome segregation during meiosis is fundamental for genetic material stable distribution. In plants, two types of meiotic divisions have been found. In plants with one single size-restricted centromere per chromosome (monocentric chromosomes), homologous chromosomes segregate during meiosis I, and sister chromatids separate in meiosis II (Figure 1). However, unlike monocentric chromosomes, plants such as *Luzula elegans* [3] and *Rhynchospora* [4] harbor holocentric chromosomes which lack a primary constriction site and form holokinetic kinetochores. The kinetochores are distributed along almost the entire poleward surface of the chromatids, to which spindle fibers attach. Due to this special chromosome structure, sister chromatids separate in meiosis I, whereas homologous chromosomes segregate in the second meiotic division in *Luzula elegans* [3] and *Rhynchospora* [4].

Understanding the mechanism of meiosis is critical since the mis-segregation of chromosomes leads to aneuploidy and chromosome instability and is the main cause for miscarriage, birth defects, and infertility in animals. In plants, abnormal chromosome segregation in meiosis can generate aneuploids or cause sterility [5,6]. However, the functions of many kinetochore proteins in plant meiosis have only recently been understood, and these proteins differ functionally from those found in animals.

In this review, we will summarize our current knowledge of how plant kinetochore proteins function in meiosis. In addition, the functional similarities and differences of core kinetochore proteins in meiosis between plants and other species are discussed. Additionally, potential applications for crop breeding practice by modifying some kinetochore genes in meiosis have emerged [5–9].

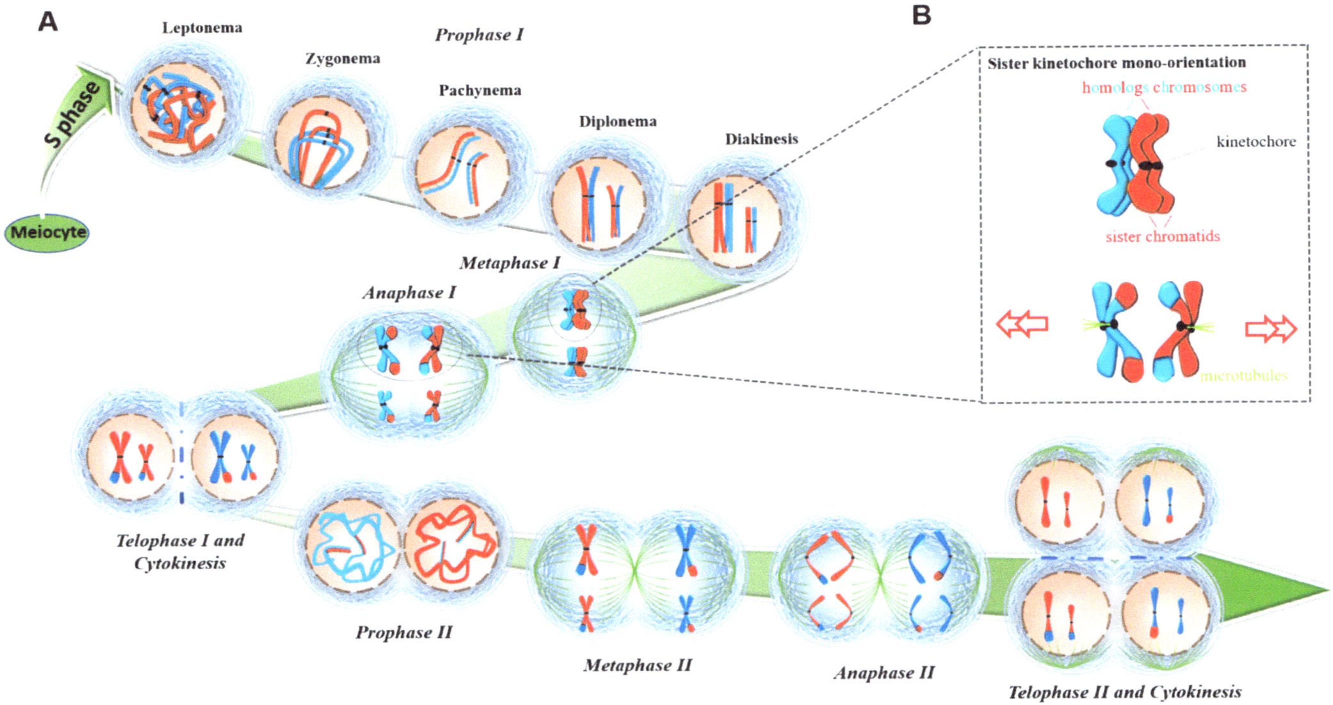

Figure 1. Overview of meiosis. (**A**). Chromosome segregation at different meiotic stages. At leptonema, chromosome axes are formed, and recombination is initiated. At zygonema, the synaptonemal complex is polymerized, where synapsis occurs and recombination proceeds. At pachynema, synapsis is completed, and recombination further progresses. At diplonema, the synaptonemal complexes are disassembled. Homologous chromosomes are connected by chiasmata. At diakinesis, chromosome condensation occurs, and bivalents can be distinguished. After prophase I, nuclear envelopes break down. At metaphase I, bivalents are aligned on the metaphase plate. At anaphase I, the release of arm sister chromatid cohesion allows the migration of chromosomes to two poles. Pericentromeric cohesion is specifically protected. At telophase I and cytokinesis, two nuclei form and chromosomes briefly decondense. In monocotyledons, cytokinesis occurs before meiosis

II starts; in dicotyledons, cytokinesis happens only at telophase II. At metaphase II, sister chromatids are aligned on two metaphase plates. At anaphase II, sister chromatids separate following centromeric cohesion release. At telophase II, four nuclei are formed. At cytokinesis, haploid spores are released. (**B**). The panel shows the chromosome state at metaphase I and anaphase I (rectangle) magnified. Cohesion protection and sister kinetochore fusion occur during metaphase I to ensure the sister chromatids exhibit mono-orientation during anaphase I. For simplicity, only two pairs of homologous chromosomes with different lengths are shown. Each homologous chromosome has two chromosomes (blue and red) which distinguish the different parental origins. The kinetochore is depicted as a black sphere; the transiently accumulating synaptonemal complex is shown in green in prophase I stage; and the spindle is shown in green in other stages. Progression through different meiotic stages is denoted by a green arrow.

2. Kinetochore Structural Proteins

Assembled by more than 100 structural and regulatory proteins on centromere DNA in eukaryotes [reviewed in [9]], the kinetochore is typically divided into two parts: the inner (DNA-proximal) and outer (microtubule-proximal) kinetochore [6] (Figure 2). Here, we will summarize how kinetochore structural proteins function in plant meiosis.

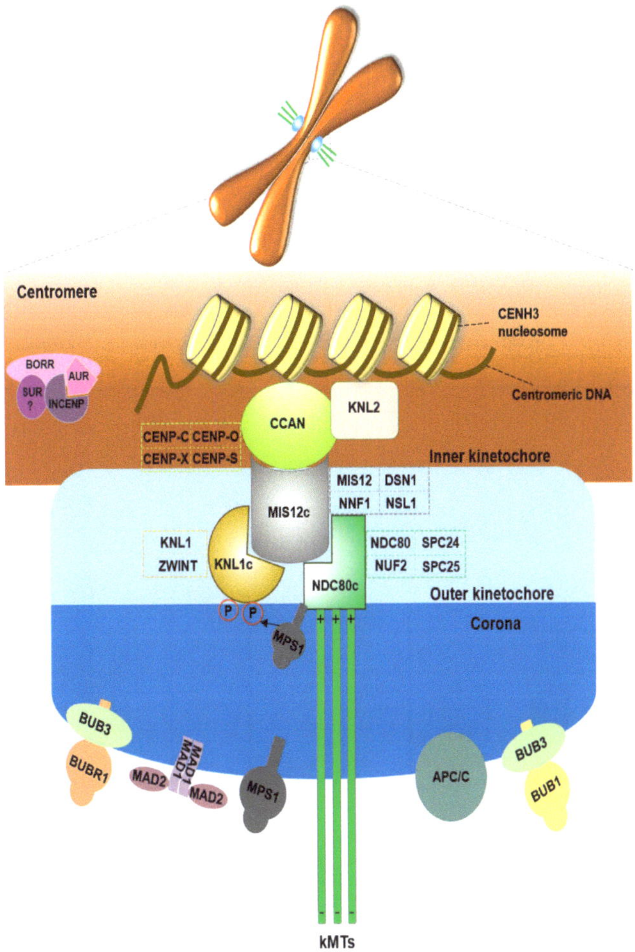

Figure 2. The centromere-kinetochore region.

2.1. CENH3 Protein

Plant centromeres are determined epigenetically by a specific histone H3 variant, called centromere histone 3 CENH3 (first described as human CENP-A [10]). Like other histone subunits, CENH3 carries an N-terminal tail (which protrudes from the nucleosome and is a target for posttranslational modification) and C-terminal Histone Fold Domain (which interacts with DNA and other histones from the nucleosome) (Figure 3). Contrary to conventional histones, CENH3 is evolving rapidly. The N-terminal tail of CENH3 is barely alignable even among closely related species, but the C-terminal Histone Fold Domain is largely conserved [reviewed in [11]]. The C-terminal part of CENH3 seems to be sufficient for mitotic centromere function in plants, but meiotic centromeres neither load nor tolerate impaired CENH3 molecules [12,13]. Ravi and Chan made the construct in which the hypervariable N-terminal tail of CENH3 was swapped out for the tail of the Schematic of the kinetochore showing simplified protein–protein interactions between major kinetochore components in plants. This illustrates the trilaminar structure of the kinetochore, as observed through electron microscopy. From inside to outside, the kinetochore's structure can be broken down into three layers: the centromere (brown), the inner kinetochore (gray), the outer kinetochore (blue) and the kinetochore's outer region, which is known as the fibrous corona (dark blue) [14]. A particular histone H3 variant (CENH3) specifies the position of the kinetochore in the centromere region. Various inner kinetochore parts throughout the cell cycle and connect with kinetochores: CCAN complex (which includes CENP-C, CENP-O, CENP-S, and CENP-X) and KNL2 protein. Particularly during cell division, numerous additional proteins are attracted to the outer kinetochore, including those in multiprotein complexes containing NDC80 (NDC80 complex, including Ndc80-Nuf2-Spc24-Spc25), MIS12 (MIS12 complex, including Mis12-Dsn1-Nnf1-Nsl1), and KNL1 (KNL1 complex, including Knl1 and ZWINT). They offer the spindle assembly checkpoint (SAC) proteins a landing platform, such as MPS1. The NDC80 complex can recruit the MPS1 and seems to be directly involved in microtubule binding. AUR, BORR, SUR ("?" represents the homologs not identified in the plant yet), and INCENP are chromosomal passenger complex (CPC) components that preferentially populate the centromere area and control the integrity of microtubule–kinetochore attachments. The SAC components (MPS1, BUB1, BUB3, BUBR1, MAD1, MAD2) and APC/C can be recruited to the fibrous corona.

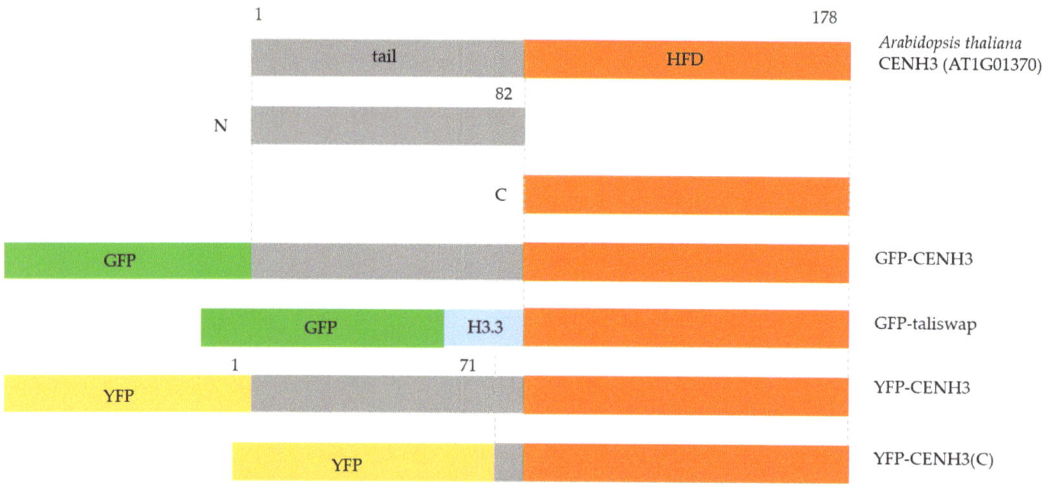

Figure 3. The schematic diagram illustrates the constructs utilized for expressing different CENH3 variants in *Arabidopsis* [13,15].

The conventional H3.3 variant (encoded by At1g13370) carried GFP-tagged variants to the N-terminal tail domain (GFP-tailswap) (Figure 3). The GFP-tailswap expressing *Arabidopsis* plants showed sterility due to defects during sporogenesis, and the CENH3 signal was reduced to an undetectable level in meiocytes [15]. Moreover, *Arabidopsis* plants expressing the C-terminal part of CENH3-fused tag (EYFP-), named EYFP-CENH3(C) (C-terminal of CENH3 containing loop1 region) construct (Figure 3), showed chromosome segregation defects, decreased fertility, and the impaired loading of the tag (YFP-) signal in meiosis [13]. Strikingly, EYFP-CENH3(C) expression can even reduce the amount of endogenous CENH3; the insufficient CENH3 loading leads to the formation of lagging chromosomes and micronuclei [13]. Therefore, in contrast to mitosis, the N-terminal tail of CENH3 plays a key role in different loading mechanisms of CENH3 during plant meiosis [13].

GFP-CENH3, GFP-tailswap, YFP-CENH3, and YFP-CENH3(C) transgenes are described in this review paper. Tail, N-terminal tail domain; HFD, C-terminal histone fold domain; H3.3, the tail of conventional H3.3 variant; GFP, green fluorescent protein; YFP, yellow fluorescent protein.

2.2. Inner Kinetochore

The basic step of kinetochore formation is the incorporation of the CENH3 into nucleosome centromere positions. CENH3 was thought to physically interact with the inner kinetochore Constitutive Centromere-Associated Network (CCAN, which provides a platform for the assembly of centromeres and has been identified as playing a crucial role in recruiting the outer kinetochore protein complex) in most eukaryotes [16]. In vertebrates, CCAN consists of 16 CENP-named proteins: CENP-C, CENP-H, CENP-I, CENP-K, CENP-L, CENP-M, CENP-N, CENP-O, CENP-P, CENP-Q, CENP-U, CENP-R, CENP-T, CENP-W, CENP-S, and CENP-X [reviewed in [17]]. However, 12 out of the 16 CCAN components in vertebrates cannot be identified by homology search techniques in plants [18]. Only CENP-C, CENP-O, CENP-S, and CENP-X have been identified in moss *Physcomitrella patens*, an emerging model system for plant cellular biology [18]. It is unknown whether the CCAN complex occurs in the plants because CENP-S, CENP-X, and CENP-O were proven not to localize to the kinetochores in *Physcomitrella patens* [18]. In fission yeast, CENP-S and CENP-X, which are the CCAN network's components, add the function of kinetochore assembly. By working with the Fanconi anemia pathway's Fanconi Anemia Core Complex (FANCM) DNA translocase, CENP-S and CENP-X might also be involved in DNA repair [19]. The CENP-S and CENP-X present in *Physcomitrella patens* may be related to DNA repair. What function a CENP-O homolog might play on its own is not at all clear, especially when it does not localize to the kinetochore. Moss *Physcomitrella patens*, maize and *Arabidopsis* have been shown to contain CENP-C, the only remaining kinetochore component having a vertebrate CCAN homolog. However, regarding its function in meiosis, a thorough functional investigation is currently lacking [6,20,21]. In fission yeast [22], *Drosophila* [23], mice [24], and fish [25], CENP-C functions not merely as a structural link between the centromere and the kinetochore but also as joining the processes of early prophase homolog synapsis (means the alignment of chromosomes along its length and results in synaptonemal complex formation) to late metaphase kinetochore assembly and signaling in meiosis [reviewed in [26]].

It has been found that in both animals and plants, CENH3 nucleosomes bind not only directly to CENP-C but also to a well-studied inner kinetochore protein KINETOCHORE NULL2 (KNL2, also called M18BP1 in vertebrates) [20,27–30]. KNL2 is responsible for the initiation of CENH3 deposition at the centromeres in humans [31], *C. elegans* [32], and fission yeast [31]. In plants, research has split KNL2 into eudicots αKNL2 (previous characterization) & βKNL2 (new representation) and grasses γKNL2 & δKNL2 (except that the maize genome contains only one copy of the δKNL2 gene) [33] (Figure 4). The conserved and characteristic Swi3-Ada2-NCoR-TFIIIB-associated (SANTA) domain has been found in all known KNL2 proteins. In some organisms, it is also accompanied by a putative SANT/ domain such as human [34] (Figure 4). Another conserved motif, named CENPC-like motif (CENPC-k), was identified on the C-terminal part of the KNL2 homologs in a wide spectrum of eukaryotes. However, human KNL2 does not contain a CENPC motif. A CENP-C binding domain (CBD), located in the middle part of the human protein, is required for the centromere localization instead [35,36] (Figure 4). In plants, research has found that *Arabidopsis* KNL2 can still target centromeres and interact with DNA even if the SANTA domain is deleted [29,30], likely because *Arabidopsis* KNL2 recognizes centromeric nucleosomes by the CENPC-k motif at its C terminus, which is required for KNL2 centromeric localization, similar to CENP-C [29]. All KNL2 act in both mitosis and meiosis [33]. Depletion of KNL2 in different organisms causes defects in CENH3 assembly. Knockout of M18BP1 in human HeLa cells with RNAi abolished centromeric recruitment of newly synthesized CENP-A, leading to chromosome mis-segregation and interphase micronuclei [37]. In *Arabidopsis*, knockout of αKNL2 via a T-DNA insertion showed reduced CENH3 loading at the centromeres and chromosome segregation defects in both mitosis and meiosis [30]. Conversely, αKNL2 expression is stable in CENH3-RNAi transformants, indicating that αKNL2 acts upstream of CENH3 and has a function in the assembly of CENH3 at the centromeres [30]. Moreover, βKNL2 knockout leads to cell cycle disorder, such as abnormal seed development, and a semi-lethal mutation phenotype indicating the defects in meiotic and mitotic chromosome segregation in *Arabidopsis* [33].

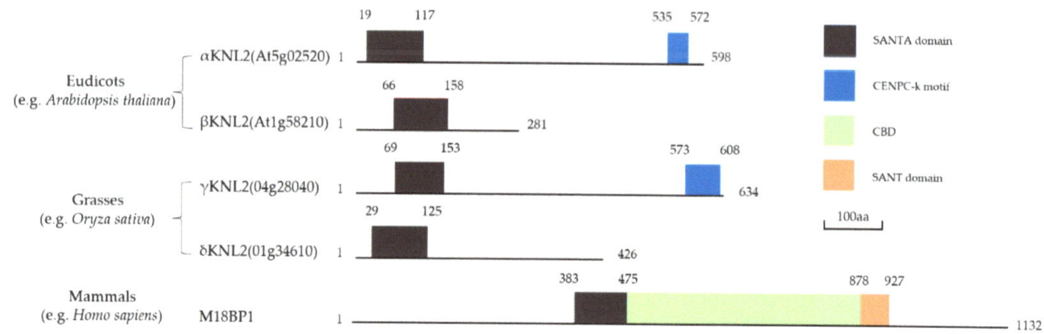

Figure 4. Schematic diagram of KNL2 conserved domain constructs in different species.

The figure shows the KNL2 conserved domain of eudicots (*Arabidopsis thaliana)*, grasses (*Oryza sativa*), and mammals (*Homo sapiens*). There are two KNL2 proteins in *Arabidopsis thaliana*, αKNL2 & βKNL2, and two KNL2 proteins in *Oryza sativa*, γKNL2 & δKNL2. However, the *Homo sapiens* category only has one KNL2—M18BP1. SANTA domain, CENPC-K motif, CBD domain, and SANT domain are all highlighted in different colors: black for the SANTA domain, blue for the CENPC-K motif, cyan for the CBD, and brown for the SANT domain. The graphic demonstrates that KNL2 and KNL2 have comparable domains. The figure shows that the domains of αKNL2 and γKNL2 and those of βKNL2 and δKNL2 are comparable.

2.3. Outer Kinetochore and KMN Network

An outer kinetochore subcomplex called the KMN network (KNL1-MIS12-NDC80) that binds to spindle microtubules is recruited by the CCAN network in the inner kinetochores, which also connects with the centromere [38–41]. The KMN complex includes: KNL1c (consisting of Knl1 and ZWINT), NDC80c (consisting of Ndc80-Nuf2-Spc24-Spc25 proteins), and MIS12c (consisting of Mis12-Dsn1-Nnf1-Nsl1 proteins). MIS12c anchors NDC80c and KNL1c from yeast to humans (Figure 2) [40–47]. Like centromeric proteins, most kinetochore proteins are fast evolving and, hence, display less sequence homology between their respective orthologs despite their functional conservation. Only a few kinetochore proteins from the KMN group that offers high sequence identity with the yeast or mammalian counterparts have been characterized in plants. In *Arabidopsis*, the interaction of necessary for nuclear function 1 (NNF1) and MIS12 has been confirmed in Y2H and co-immunoprecipitation tests [45]. Shin et al. (2018) confirmed that AtNUF2 directly interacted with AtNDC80 and AtSPC25 subunits in *Arabidopsis* by the yeast two-hybrid analysis. Recently, studies of the NDC80 complex in *Arabidopsis* showed that MERISTEM UNSTRUCTURED (SPC24/MUN) interacts conservatively with subunits NDC80, NUF2, and SPC25 using co-immunoprecipitation analysis, and MUN protein has a coiled-coil region and a globular domain at the end, which are typical structural features found in all four components of the NDC80 kinetochore complex [47].

In the KMN group, only NDC80 [40], MIS12 [41], NNF1 [45], NUF2 [46], and SPC24/MUN [47] homologs are functionally characterized in plants. They display conserved functions in chromosome segregation and microtubule organization during cell division. In maize, the localization patterns of Knl1 signals overlap with those of Mis12 and Ndc80, indicating that the deficiency of Knl1 can likely impair kinetochore function. This impairment leads to abnormal chromosome behavior during cell division in early endosperm development, ultimately resulting in defective kernels [38]. During meiosis, maize MIS12 interacts with NDC80, forming a visible MIS12-NDC80 bridge. This bridge fuses sister kinetochores, directing sister kinetochores' mono-orientation behavior during metaphase I and initiating the homolog chromosomes segregation [41]. After systemically reducing the levels of MIS12, the MIS12-NDC80 bridge is broken. This leads to kinetochores orienting and separating randomly in metaphase I, causing chromosome non-disjunction in anaphase I due to the presence of sister chromatid cohesion [41]. Another component of the MIS12 complex, NNF1, *atnnf1-1* mutation in *Arabidopsis* causes embryo mortality, which shows its crucial significance in kinetochore function during cell division [45]. For the NDC80 complex, it has been shown that the mutation of AtNUF2 led to severe mitotic defects, not only in the embryo and endosperm but also in seedlings, resulting in seed abortion and stagnating seedling growth [46]. Moreover, the partially complemented *nuf2-3/-*$^{DD45;ABI3pro::AtNUF2}$ (*nuf2-3/-*^{DA}) root apical meristem cells, along with the aberration of spindle MTs, resulted in blocked root growth [46]. Furthermore, AtSPC25 could co-localize with CENH3 and MT arrays in various phases of mitosis suggesting that the AtSPC25 gene may perform a similar function as AtNUF2 [46]. Recently, studies of another component of the NDC80 complex in *Arabidopsis* showed that the *mun-1* mutant, which is a weak allele because of the insertion of T-DNA in the promoter region of the SPC24 homolog, displays stunted growth, embryo arrest, DNA aneuploidy, and problems in chromosome segregation with a low cell division rate [47]. Additionally, Null mutants of MUN from TALEN and CRISPR/Cas9-mediated mutagenesis demonstrated zygotic embryonic lethality comparable to *nuf2-1*, indicating it is necessary for proper cell division [47].

3. Spindle Assembly Checkpoint (SAC) Proteins

During cell division, the spindle assembly checkpoint (SAC) acts to maintain genome stability by prolonging metaphase "stagnation" until all kinetochores correctly attach to the microtubule spindle apparatus. Once all kinetochores become stably attached to spindle, the SAC is inactive, which allows chromosome segregation and cell division to proceed. The SAC core proteins, including monopolar spindle 1 (MPS1), budding uninhibited by benomyl (BUB: BUB1, BUB3), BUB1-related protein 1 (BUBR1, also called MAD3), and mitotic arrest deficient (MAD: MAD1, MAD2) have been identified in divergent branches of the eukaryotic kingdom, including yeast, animals, and plants (Table 1); for reviews see [48,49]. Therefore, it seems that the SAC appears to be an ancient control mechanism [50]. In mitosis and meiosis, the regulatory mechanisms of the SAC proteins and the relationship between them have all been well studied [51–53]. Here, we will focus on the plant SAC proteins for their mechanisms during meiosis.

3.1. MPS1 Protein

MPS1 protein was originally found in yeast and identified as a kinase [65,66]. MPS1 orthologues can be identified in all supergroups of eukaryotes and in metazoan, with the exception of nematodes [67]. MPS1 protein as an upstream SAC regulator initiates SAC signaling, which has been proven in yeast [68] and human cells [69–72]. MPS1 phosphorylates Spc105/KNL1 at several phosphorylation sites to start a biochemical cascade that recruits almost all other SAC components directly or indirectly to kinetochores and monitors accurate chromosome alignment [69,73–76]. In recent years, the function of MPS1 protein in plant meiosis has been released. Like yeast [77], *Drosophila* [78,79], and mice [80], a mutation of MPS1 leads to chromosome mis-segregation and aneuploidy because of the precocious entry into anaphase I with erroneous kinetochore attachments in *Arabidopsis*, which finally leads to male and female gametophyte abortion [54]. Moreover, MPS1(AtPRD2) is involved in the formation of double strand break (DSB) and spindle structure during meiosis in *Arabidopsis* [55]. However, the rice OsPRD2 does not participate in spindle assembly during meiosis I but shows a conservative role in the meiotic DSB generation [56,57]. Consistent with other DSB defective mutants, later synapsis and recombination were disrupted in *osprd2* [57]. This indicates that the function of MPS1(PRD2) is different among plant genomes of different lineages; further studies need to be explored in different plants.

Table 1. List of SAC and CPC proteins in plant meiosis.

Name	Homologs	Protein Function or Feature	Mutant Phenotype	Reference
SAC proteins				
MPS1	AtMPS1 (*Arabidopsis*)	required for faithful chromosome segregation	chromosome mis-segregation; aneuploidy; precocious into anaphase I	[54]
	AtPRD2 (*Arabidopsis*)	involves the formation of DSB and spindle structure	gametophytes aborted; abnormal meiosis products	[55]
	OsPRD2 (Rice)	meiotic DSB formation	male and female completely sterile; twenty-four univalent	[56,57]
BUB1	BRK1 (Rice)	proper tension between homologous kinetochores	precocious separations of sister chromatids; sterile tetrad	[58]
	ZmBUB1 (Maize)	Bub1-mediated phosphorylation of H2AThr133	decline of anther fertility	[59]
BUB3	ZmBUB3 (Maize)	located at the kinetochore		[59]
MAD2	MAD2 (Maize)	at centromere; relates to the distance between kinetochores		[60]
CPC proteins				
Aurora	α-Aurora β-Aurora (*Arabidopsis*)	catalytic subunit of the CPC	microsporogenesis and defects in polyploid and aneuploid offspring	[61,62]
INCENP	WYR (*Arabidopsis*)	involved in cell cycle control	defects in gametophyte cell division and development	[63]
Borealis	BORR (*Arabidopsis*)	required for proper chromosome segregation and cell division	undeveloped ovules, aborted seeds and embryonic defects	[64]

3.2. BUB Proteins

BUB proteins include BUB1 kinase (also called BMF1 in plants), BUB3, and an unusual pseudo-kinase BUB1-related protein1 (BUBR1, also called MAD3). In yeast and mammals, BUB1 localizes to kinetochores through the BUB3-binding domain. Although BUB1 homologs in plants have also been found, a BUB3-binding region named Gle2-binding-sequence (GLEBS) domain is missing in all plant BUB1 proteins [58,81,82]. In fission yeast, BUB1 is required for localization and the centromeric protection of cohesin subunit Rec8 during meiosis I, which may be because of the interaction between BUB1 and Shugoshin (Sgo) [59,83,84]. It has been shown that the N-terminus of BUB1 is necessary for the targeting of Sgo1 to centromeres and the protection of cohesion, whereas the C-terminal acts together with Sgo2 to promote sister kinetochore co-orientation at the MI stage during fission yeast meiosis [59].

In plants, rice *brk1* (homologs of BUB1 in rice and located at the kinetochore) display a sterile phenotype because of the precocious loss of sister chromatid cohesion at the onset of anaphase I. Then, the tension between homologous kinetochore reduces, which finally leads to a sterile tetrad phenotype [58]. In maize, the ZmBub1 RNAi transgenic line also showed a decline of anther fertility to certain degrees [82]. Moreover, maize ZmBub1 can localize to metaphase I centromeric regions in *sgo1* mutants [82], which suggests that ZmBub1 localization to kinetochores is independent of ZmSgo1 during maize meiosis. Furthermore, it has been shown that in maize *afd1* (homolog of Rec8), mutant ZmBub1 is detectable during the entire meiotic division. Taken together, ZmBub1 recruitment is Rec8- and Sgo1-independent [82]. However, *Arabidopsis bub1* mutants are morphologically similar to wild-type even under microtubule-destabilizing conditions [81].

The roles of Bub3 have been well studied in mouse oocytes, and similarity between mammalian mitosis and meiosis has been illustrated [85]. Overexpression of Bub3 causes meiotic arrest while depletion of Bub3 from kinetochores causes chromosome misalignment and abnormal polar body extrusion, leading to aneuploidy [85]. In plants, three homologs of BUB3: BUB3.1, BUB3.2, and BUB3.3 have been identified in *Arabidopsis*, and they all contain WD40 repeats which are required for mitotic checkpoint complex formation [86,87]. BUB3.1 and BUB3.2 show 88% sequence similarity, and BUB3.3 only shares 37% amino acid identity with BUB3.1 and BUB3.2 [87]. In *Arabidopsis*, the localization pattern of BUB3.1:GFP and BUB3.2:GFP are present at the kinetochores in mitotic cells, BUB3.3:GFP is present in the cytoplasm, indicating a plant specific accumulation pattern that does not correspond to established SAC function in animal and yeast cells [88]. In *Arabidopsis*, loss of BUB3.1 function leads to embryonic lethality, *bub3.2* mutants do not exhibit any defects, and there is missing information about *bub3.3* mutants [87–89]. In maize meiosis, BUB3 signals were determined by immunostaining, which appeared first as weak signals in interphase and then became stronger through prophase I. The signals persisted at the kinetochores throughout meiosis I and II [82].

MAD3/BUBR1 is a paralogous protein of BUB1 and contains a MAD3/BUB1 domain, a GLEBS domain, and an inactive pseudo-kinase domain. MAD3/BUBR1 has been identified in yeast [90], *Drosophila* [90,91], a mouse [92], and plants [93]. In a mouse [92,94] and mice [95], BubR1 depletion by RNAi accelerated meiotic progression, and overexpression of BubR1 caused meiotic arrest. Moreover, cold treatment disrupted spindle microtubules in BubR1-depleted oocytes, suggesting that BubR1 monitors kinetochore–microtubule attachments [92]. Furthermore, BubR1 is essential for maintaining sister chromatid cohesion during meiotic progression in both sexes in *Drosophila* [96]. Two MAD3/BUBR1 homologs (called MAD3.1 and MAD3.2) have been found in *Arabidopsis* and maize; they all lost their GLEBS domain and pseudo-kinase domain (Review in [93]). MAD3.1 (also called BMF2 in plants) localizes to kinetochore under microtubule-destabilizing conditions. In addition, in MAD3.1 and MAD3.2 (also called BMF3 in plants), direct interaction was observed at kinetochores, indicating that the interaction only occurs when the SAC is active [97]. However, their function in meiosis is still unclear in plants.

3.3. MAD Proteins

MAD proteins, including MAD1 and MAD2, have been identified in almost all eukaryotes. The kinetochore localization of MAD1 has not been reported, but MAD2 homologs have been shown to strongly accumulate at kinetochores under microtubule-destabilizing conditions in *Arabidopsis*, maize, and wheat [97,98]. Under SAC-activated conditions, MAD1 localizes predominantly to unattached kinetochores and recruits MAD2 to form the MAD1–MAD2 complex. MAD1 is a coiled-coil protein with a MAD2 interaction motif; the MAD1–MAD2 interaction has been confirmed by gel filtration and co-immunoprecipitation experiments in budding yeast [99]. The loss of MAD1 causes chromosome dislocation and premature entry into anaphase I in yeast [100], *Caenorhabditis elegans* [101], and a mouse [102]. In *Arabidopsis*, localization of MAD1 to the nuclear periphery was also observed, but the relationship between SAC components and DNA repair has not been studied yet [103]. Interestingly, it has been shown that *Arabidopsis* MAD1 regulates flowering time and endo-polyploidization, suggesting that MAD1 is involved in cell cycle control in the timing of reproductive transition [104].

MAD2 was reported to localize to kinetochores during meiosis in a mouse, a rat, maize, and a grasshopper [60,105–107]. However, the patterns of MAD2 dynamics are different among species. Specifically, in meiosis I of maize, MAD2 localizes to unattached kinetochores. After proper attachment, it is lost from the kinetochores, indicating that MAD2 may sense microtubule attachment [107]. However, in mouse spermatogenesis, MAD2 is shown to remain at most kinetochores throughout meiosis I and is lost only during metaphase of meiosis II. The persistence of MAD2 at kinetochores in oocytes was seen during meiosis II [105]. It has been shown that MAD2 is a key SAC protein involved in the regulation of meiotic chromosome segregation. In the absence of MAD2, the meiotic cells showed similar defects as MAD1 mutants during meiosis I in yeast and a mouse [108–110]. In maize, the loss of MAD2 staining in meiosis was not correlated with initial microtubule attachment but was correlated with a measure of tension: the distance between homologous or sister kinetochores (in meiosis I and II, respectively). After the staining of tension-sensitive 3F3/2 (the 3F/2 antibody recognizes a phosphor-epitope that is localized to prometaphase kinetochores until the chromosomes have aligned properly at the metaphase plate), phosphor-epitope was present and absent concomitantly with MAD2 at the meiotic kinetochores [107].

4. Chromosomal Passenger Complex (CPC) Proteins

Accurate chromosome segregation to avoid chromosomal instability and aneuploidy is guaranteed by the SAC. The loading and function of SAC at the kinetochores depend on a few complexes, especially the chromosome passenger complex (CPC). Thus far, core enzyme Aurora kinase and three non-enzymatic kinases including inner centromere protein (INCENP), Borealin, and Survivin in CPC proteins have been identified in most organisms (Table 1). CPC components are preferentially populated in the centromere area (reviewed in [111]), and CPC is known in correcting kinetochore–microtubule attachments [112,113], activating the SAC complex [112], stabilizing the spindle midzone [113,114], and promoting cytokinesis [115] during cell division. Aurora B is the core component in CPC. The N-terminus of auroraB kinase binds to the C-terminus of INCENP, while Survivin and Borealin bind the N-terminus of INCENP (Figure 2).

4.1. Aurora Kinase

The Aurora kinases are a family of highly conserved serine/threonine kinases. Yeasts, including *Saccharomyces cerevisiae* and *Schizosaccharomyces pombe*, contain a single Aurora homolog, Ipl1p (increase in ploidy 1) [116] and Ark1 (Aurora-related kinase 1, an Aurora-B-like kinase) [117], respectively. At least two functionally divergent Aurora members of the multicellular eukaryotes are present: Aurora A and Aurora B. Mammals can have up to three Aurora genes: Aurora A, B, and C [118]. In plants, three homologs of Aurora categorized into two groups were identified in *Arabidopsis* and rice: α-Aurora (AUR1 and AUR2, similar to Aurora A in mammals and locates to the spindle microtubule), β-Aurora (AUR3, similar to Aurora C in mammals and locates to the kinetochore) [61,62,119]. It has been shown that Aurora kinases play an important role in centrosome maturation, spindle assembly, meiotic maturation, and metaphase I spindle orientation [120], as well as regulate SAC activity [121], control the kinetochore orientation during meiosis [122], and function specifically in meiotic spindle attachment in oocytes [123] and during spermatogenesis [124].

It has been shown that all three AtAurora kinases localize to male and female gametophytes, suggesting a possible prominent function during plant meiosis. In contrast to yeast and animals, the role of plant Aurora kinases in meiosis is only starting to emerge. In *Arabidopsis*, the reduced activity of Aurora kinases (AtAurora1, −2, −3) leads to meiotic defects and the formation of unreduced pollen including plants with an increased ploidy level [125,126]. This not only reduces expression of Aurora kinases but also overexpression results in genomic instability [125]. In *Arabidopsis*, overexpression of any member of the AtAurora family is not possible (overexpression of untagged AtAurora kinases is detrimental for plant development), so only plants with tagged [GFP, tandem affinity purification (TAP) tag] or truncated forms of Aurora kinases can be expressed. This indicates that the activity of Aurora kinases is affected by the affinity tag, making their overexpression possible for plants [125]. *Arabidopsis* 35S::AtAurora1-TAP transformants showed increased endopolyploidization, disturbed meiosis, and the formation of small amounts of tetraploid seeds [125]. Overexpression of truncated AtAurora1 resulted in an imbalanced segregation of chromosomes during meiosis and the formation of 39% aneuploid seeds [125]. Thus, in plants, balanced expression of Aurora kinases is essential for the proper execution of mitotic and meiotic divisions, and therefore for overall growth and development [125–127]. Analysis of *aurora1 aurora2* double mutants and Ataurora3 RNAi transformants in the *aurora1* mutant background showed that they exhibited disturbed meiosis and form triploid or a combination of triploid and tetraploid seeds, respectively [125].

4.2. Other CPC Components

INCENP is the largest non-catalytic subunit of the CPC, which binds to all other CPC components directly in yeast and animals. The N-terminal region of INCENP interacts with Borealin and Survivin, while the C-terminal domain, called the IN-box, with four amino acid residues binds to Aurora B [128,129]. Like yeast and animals, a putative *Arabidopsis* ortholog of INCENP, named WYR, contains a characteristic C-terminal domain and a coiled-coil domain; an IN-box aurora B-binding domain has been identified as well. In addition, *Arabidopsis* WYR has a long N-terminal region of unknown function, which is only conserved in plants [130]. In general, INCENP functions in the CPC complex to regulate cell cycles, including chromosome segregation, the spindle assembly checkpoint, and cytokinesis [64]. However, in plants, it is still unclear if INCENP acts as a putative plant CPC because of missing information about its CPC binding partners. WYR has been found to play a role in meiosis; *Arabidopsis wyr-1/+* produces dyads and triads in male sporogenesis, indicating the failure of chromosome segregation (meiotic non-reduction) during meiosis [130].

N-terminal Borealin acts as the INCENP-binding region, while its C-terminal part consists of a homodimerization domain that is involved in stable CPC localization to centromeres [63]. The location of the CPC complex on the chromosomes depends on the phosphorylation status of the histones during mitosis and meiosis [131]. BOREALIN RELATED (BORR, Borealin homolog in *Arabidopsis*) has been identified in *Arabidopsis* and in most branches of the plant kingdom, which co-localizes with INCENP homologs to the central region of the kinetochore and the film-forming body during mitosis and meiosis [130]. BORR is required for proper chromosome segregation during cell division in *Arabidopsis* and is strongly expressed not only in proliferating cells but also in flowering tissues, including male and female reproductive organs [129]. *Borr* mutants showed lagging chromosomes and abnormal cell division, resulting in undeveloped ovules, aborted seeds, and embryonic defects [129], indicating its role in meiosis.

In yeast and animals, Aurora, INCENP, and Borealin, together with Survivin, form the CPC complex [111]. However, Survivin is still missing from the picture in plants, raising the question whether complete and functional canonical CPC exists in plants. In a recent report, Komaki, S. et al. identified *Arabidopsis* BOREALIN RELATED INTERACTOR 1 and 2 (BORI1 and BORI2) as redundant Survivin-like proteins in the context of the CPC in plants [132]. In rat and mouse oocytes, Survivin is a critical regulator of spindle assembly checkpoint activity and chromosome alignment during meiosis [133,134]. However, it is not yet clear whether BORI1 and BORI2 function in plant meiosis.

In total, little is yet known about the role of plant CPC complex, especially in meiosis. Further genetic and biochemical analysis is needed to confirm plant CPC components and to understand their roles in meiosis.

5. Centromeric Cohesion in Meiosis

To ensure that each gamete receives only half the number of parental chromosomes, meiocytes undergo a reductional division separating parental homologs during meiosis I. The kinetochores of sister chromatids fuse and then point toward the same direction (mono-orientation) in meiosis I (Figure 1), allowing homologous chromosomes to correctly segregate to different sides [135,136]. This is partly mediated by sister chromatid cohesion. The cohesin complex forms a ring-like structure that comprises four main subunits: structural maintenance of chromosomes 1 (SMC1) and SMC3 protein, and the α-kleisin SCC1 (also called RAD21 or REC8) and the SCC3 proteins [137–141]. Studies demonstrated that SCC1 is acting on the mitotic cohesin complex and is replaced during meiosis by REC8 in almost all eukaryotes [142–144].

5.1. Sister Chromatid Cohesion

In meiosis I, to make sure homologous chromosomes segregate, sister kinetochores of each homolog have to face toward the same spindle pole. Therefore, sister chromatid cohesion must be released in two steps: in meiosis I, loss of cohesion at the chromosome arm releases chiasmata and enables the reductional segregation [142,145]. Moreover, the cohesion in the centromeric region is retained until metaphase II, which ensures the segregation of sister chromatids during anaphase II [146]. Mutants of Rec8 lead to cohesion protection defect at the centromeric region, which has been proven in many organisms [135,146–149]. Moreover, SMC1 and SMC3 are the vital cohesin proteins during meiosis, and AtSMC3 exists in the cytoplasm and nucleus and the chromosome and nuclear matrix in *Arabidopsis* in meiosis and mitosis cells. During meiosis, at the prophase of meiosis, AtSMC3 was located along the sister chromatids to the axial and lateral elements, and at metaphase I, AtSMC3 was only located in the centromere [150]. Interestingly, tomato SMC1 and SMC3 show similar positioning modes to AtSMC3, but no spindle position was recorded in tomatoes [151]. This indicates that the function of plant cohesin subunits is consistent with other eukaryotic organisms in meiosis (a detailed review of plant cohesin in meiosis can be found in [127,140,152–154]).

5.2. Mono-Orientation

In the last two decades, several meiosis-specific proteins, including budding yeast Spo13 [155], fission yeast Moa1 [156], mouse MEIKIN [157], and *Drosophila* Mtrm [158], have been identified with crucial functions in mono-orientation. These proteins, so-called meiosis I kinase regulators (MOKIRs), mediate the segregation of homologous chromosomes (mono-orientation). The MOKIRs bear no obvious sequence similarity, although their functions are conserved. However, MOKIRs are missing from the picture in the plants, raising the intriguing possibility that MOKIRs have been lost in the green lineage and that the plant meiosis mono-orientation is regulated by different mechanisms. It has been suggested that kinetochore mono-orientation also depends on Rec8 in many species, including yeasts [159], *C. elegans* [160], *Drosophila* [161], and plants [135,162,163]. In plants, mutants of meiosis-specific α-kleisin Rec8 homologs, including *Arabidopsis* SYNAPSIS1(*syn1*--also called DETERMINATE INFERTILE1, *dif1*) [141,164,165], rice *osrad21-4* [163,166], and maize *afd1* (absence of first division) [167] exhibit an equational segregation of chromosomes at metaphase I. In addition to REC8, SCC3 is a desired protein for the mono-orientation during meiosis I [168]. Through a blast search of SCC proteins of many species, its homologous protein AtSCC3 was found in *Arabidopsis*, which involves the male and female meiosis, and its disruption causes the early release of cohesion at anaphase I [135,169].

5.3. Cohesin Protectors

Shugoshins (SGOs, which means 'guardian spirit' in Japanese) were firstly identified in *Drosophila* (MEI-S332) [170], and successively described in yeast, mammals, and plants [171]. A conserved feature of shugoshin proteins is their localization to pericentromeres during meiosis I, consistent with their role in protecting cohesion in this region [170,172–178]. Research in yeast and vertebrates has shown that SGO is phosphorylated by kinase Aurora B and binds as a homodimer with protein phosphatase 2A (PP2A). Then, the complex dephosphorylates the cohesion subunit REC8 to protect it from separase (cysteine protease) during meiosis [179–182]. There are differences in both amino acid sequences and certain accessory functions among SGOs from different species [183,184]. The organisms including fission yeast, plants, and mammals contain two shugoshin-like proteins (Sgo1 and Sgo2, acting differently in mitosis and meiosis I), in contrast to budding yeast and *Drosophila*, which only contain a single shugoshin protein Sgo1 [170,173]. Sgo1 is predominantly responsible for the protection of centromeric cohesion during meiosis in fission yeast and plants [173,176,183,185], whereas Sgo2 carries out this function in mammals [178]. The role of SGO has been identified in *Arabidopsis*, maize, and rice, and the *sgo1* mutants show the precocious separation of sister chromatids due to loss of centromeric cohesion at anaphase I, producing unbalanced chromosomes at MII [176,183–186]. Furthermore, in *Arabidopsis*, *atsgo1 atsgo2* double mutants showed that the frequency of unbalanced meiotic products had significantly increased compared with the single *atsgo1* mutant [176,184–186].

Moreover, a plant-specific protein PATRONUS1 (PANS1,) has been identified in *Arabidopsis* [184,185]. Similar to SGOs, PANS1 is not required for monopolar attachment of sister kinetochores in meiosis I. However, in contrast to the SGOs, *pans1* meiocytes show a premature release of sister chromatid cohesion at metaphase II but not at meiosis I, indicating that the protein is required for the protection of cohesion during interkinesis, at a later stage than SGOs [184,185]. PANS1 may be a regulator of the APC/C complex because of the interaction with some of the APC/C subunits revealed by TAP-TAG and Y2H experiments [184]. In rice and wheat, there are also PANS1 similar proteins identified, named RSS1 and TdRL1, respectively [187–189]. However, the mechanism of PANS1 is still unclear.

6. Concluding Remarks and Perspective

In plants, altered expression of key meiotic regulators at kinetochores results in impaired meiotic divisions followed by the formation of aneuploid or polyploid progenies. For example, manipulating Aurora kinase activity by Aurora inhibitors efficiently resulted in the formation of aneuploids and polyploids [125]. This observation is interesting from an evolutionary point of view because it can be assumed that, during evolution, the influence of biotic and abiotic factors on the activity of meiotic genes resulted in the formation of aneuploids or polyploids that served as a basis for the formation of new species. In addition, this knowledge can be used for applied research because manipulating key meiotic genes can efficiently result in apomixis and the induction of haploids. For example, inactivating a few key meiotic genes including Rec8 activity can induce apomixis into rice [3] and breed seedless watermelon varieties [7]. In addition, it has been demonstrated that modulation of CENH3 can introduce haploids in *Arabidopsis* [190,191] and in crop species such as maize [192], wheat [193], and barley [194]. Detailed studies in this direction might help in understanding the problem of plant genome stability. Compared with animal and yeast kingdoms, many of the meiotic studies regarding the kinetochore perspective in plants have just started. The presence of numerous plant-specific phenotypes in kinetochore meiotic regulator mutants, in contrast to the observed functional divergence between animal homologs, makes it intriguing to further unravel these key kinetochore meiotic regulators in meiosis.

Author Contributions: K.-D.Z., C.-X.Z., F.-R.N., H.-C.B., D.-D.W. and W.M. wrote the manuscript; K.-D.Z., H.-C.B., J.-C.D., H.-Y.Q., Y.-L.J. and W.M. drew the figures. All authors have read and agreed to the published version of the manuscript.

Funding: This work was supported by the National Natural Science Foundation of China (Grant No. 32270097), the Youth Science Foundation of Jiangnan University (No. 2050205), and the Postgraduate Research & Practice Innovation Program of Jiangsu Province for K.D.Z. (No. KYCX23_2567). We thank Zhiyi Lv for the linguistic assistance during the preparation of this manuscript.

Institutional Review Board Statement: The study did not require ethical approval.

Informed Consent Statement: Not applicable.

Data Availability Statement: The authors confirm that the data supporting the findings of this study are available within the article.

Conflicts of Interest: The authors declare no conflict of interest.

References

1. Earnshaw, W.C. Discovering centromere proteins: From cold white hands to the A, B, C of CENPs. *Nat. Rev. Mol. Cell Biol.* **2015**, *16*, 443–449. [CrossRef] [PubMed]
2. Zafarullah, M. An Investigation of Functional Dependency of CENP-C and CENH3 in *Arabidopsis thaliana*. Master's Thesis, University of California, Davis, CA, USA, 2017.
3. Heckmann, S.; Jankowska, M.; Schubert, V.; Kumke, K.; Ma, W.; Houben, A. Alternative meiotic chromatid segregation in the holocentric plant *Luzula elegans*. *Nat. Commun.* **2014**, *5*, 4979. [CrossRef] [PubMed]

4. Cabral, G.; Marques, A.; Schubert, V.; Pedrosa-Harand, A.; Schlögelhofer, P. Chiasmatic and achiasmatic inverted meiosis of plants with holocentric chromosomes. *Nat. Commun.* **2014**, *5*, 5070. [CrossRef] [PubMed]
5. Cao, L.; Li, C.; Li, H.; Wang, Z.; Jiang, Y.; Guo, Y.; Sun, P.; Chen, X.; Li, Q.; Tian, H. Disruption of REC8 in Meiosis I led to watermelon seedless. *Plant Sci.* **2022**, *323*, 111394. [CrossRef]
6. Watts, A.; Kumar, V.; Raipuria, R.K.; Bhattacharya, R. In vivo haploid production in crop plants: Methods and challenges. *Plant Mol. Biol. Rep.* **2018**, *36*, 685–694. [CrossRef]
7. Wang, K. Fixation of hybrid vigor in rice: Synthetic apomixis generated by genome editing. *aBIOTECH* **2020**, *1*, 15–20. [CrossRef]
8. Xie, E.; Li, Y.; Tang, D.; Lv, Y.; Shen, Y.; Cheng, Z. A strategy for generating rice apomixis by gene editing. *J. Integr. Plant Biol.* **2019**, *61*, 911–916. [CrossRef]
9. Meng, D.; Liu, C.; Chen, S.; Jin, W. Haploid induction and its application in maize breeding. *Mol. Breed.* **2021**, *41*, 20. [CrossRef]
10. Earnshaw, W.C.; Rothfield, N. Identification of a family of human centromere proteins using autoimmune sera from patients with scleroderma. *Chromosoma* **1985**, *91*, 313–321. [CrossRef]
11. Malik, H.S.; Henikoff, S. Phylogenomics of the nucleosome. *Nat. Struct. Mol. Biol.* **2003**, *10*, 882–891. [CrossRef]
12. Schubert, V.; Lermontova, I.; Schubert, I. Loading of the centromeric histone H3 variant during meiosis–how does it differ from mitosis? *Chromosoma* **2014**, *123*, 491–497. [CrossRef] [PubMed]
13. Lermontova, I.; Koroleva, O.; Rutten, T.; Fuchs, J.; Schubert, V.; Moraes, I.; Koszegi, D.; Schubert, I. Knockdown of CENH3 in *Arabidopsis* reduces mitotic divisions and causes sterility by disturbed meiotic chromosome segregation. *Plant J.* **2011**, *68*, 40–50. [CrossRef] [PubMed]
14. Jokelainen, P.T. The ultrastructure and spatial organization of the metaphase kinetochore in mitotic rat cells. *J. Ultrastruct. Res.* **1967**, *19*, 19–44. [CrossRef] [PubMed]
15. Ravi, M.; Chan, S.W. Haploid plants produced by centromere-mediated genome elimination. *Nature* **2010**, *464*, 615–618. [CrossRef] [PubMed]
16. Cortes-Silva, N.; Ulmer, J.; Kiuchi, T.; Hsieh, E.; Cornilleau, G.; Ladid, I.; Dingli, F.; Loew, D.; Katsuma, S.; Drinnenberg, I.A. CenH3-independent kinetochore assembly in Lepidoptera requires CCAN, including CENP-T. *Curr. Biol.* **2020**, *30*, 561–572.e510. [CrossRef]
17. Perpelescu, M.; Fukagawa, T. The abcs of cenps. *Chromosoma* **2011**, *120*, 425–446. [CrossRef]
18. Kozgunova, E.; Nishina, M.; Goshima, G. Kinetochore protein depletion underlies cytokinesis failure and somatic polyploidization in the moss *Physcomitrella patens*. *eLife* **2019**, *8*, e43652. [CrossRef]
19. Bhattacharjee, S.; Osman, F.; Feeney, L.; Lorenz, A.; Bryer, C.; Whitby, M.C. MHF1–2/CENP-SX performs distinct roles in centromere metabolism and genetic recombination. *Open Biol.* **2013**, *3*, 130102. [CrossRef]
20. Dawe, R.K.; Reed, L.M.; Yu, H.-G.; Muszynski, M.G.; Hiatt, E.N. A maize homolog of mammalian CENPC is a constitutive component of the inner kinetochore. *Plant Cell* **1999**, *11*, 1227–1238. [CrossRef]
21. Ogura, Y.; Shibata, F.; Sato, H.; Murata, M. Characterization of a CENP-C homolog in *Arabidopsis thaliana*. *Genes Genet. Syst.* **2004**, *79*, 139–144. [CrossRef]
22. Tanaka, K.; Chang, H.L.; Kagami, A.; Watanabe, Y. CENP-C functions as a scaffold for effectors with essential kinetochore functions in mitosis and meiosis. *Dev. Cell* **2009**, *17*, 334–343. [CrossRef] [PubMed]
23. Unhavaithaya, Y.; Orr-Weaver, T.L. Centromere proteins CENP-C and CAL1 functionally interact in meiosis for centromere clustering, pairing, and chromosome segregation. *Proc. Natl. Acad. Sci. USA* **2013**, *110*, 19878–19883. [CrossRef] [PubMed]
24. Fan, J.; Liu, Y.; Zhong, Y. Immunization with CENP-C causes aberrant chromosome segregation during oocyte meiosis in mice. *J. Immunol. Res.* **2021**, *2021*, 4610494. [CrossRef] [PubMed]
25. Kral, L.G. Possible identification of CENP-C in fish and the presence of the CENP-C motif in M18BP1 of vertebrates. *F1000Research* **2016**, *4*, 474. [CrossRef]
26. Fellmeth, J.E.; McKim, K.S. Meiotic CENP-C is a shepherd: Bridging the space between the centromere and the kinetochore in time and space. *Essays Biochem.* **2020**, *64*, 251–261. [CrossRef] [PubMed]
27. Gopalakrishnan, S.; Sullivan, B.A.; Trazzi, S.; Della Valle, G.; Robertson, K.D. DNMT3B interacts with constitutive centromere protein CENP-C to modulate DNA methylation and the histone code at centromeric regions. *Hum. Mol. Genet.* **2009**, *18*, 3178–3193. [CrossRef] [PubMed]
28. Hori, T.; Shang, W.-H.; Hara, M.; Ariyoshi, M.; Arimura, Y.; Fujita, R.; Kurumizaka, H.; Fukagawa, T. Association of M18BP1/KNL2 with CENP-A nucleosome is essential for centromere formation in non-mammalian vertebrates. *Dev. Cell* **2017**, *42*, 181–189. [CrossRef]
29. Sandmann, M.; Talbert, P.; Demidov, D.; Kuhlmann, M.; Rutten, T.; Conrad, U.; Lermontova, I. Targeting of *Arabidopsis* KNL2 to centromeres depends on the conserved CENPC-k motif in its C terminus. *Plant Cell* **2017**, *29*, 144–155. [CrossRef]
30. Lermontova, I.; Kuhlmann, M.; Friedel, S.; Rutten, T.; Heckmann, S.; Sandmann, M.; Demidov, D.; Schubert, V.; Schubert, I. *Arabidopsis* kinetochore null2 is an upstream component for centromeric histone H3 variant cenH3 deposition at centromeres. *Plant Cell* **2013**, *25*, 3389–3404. [CrossRef]
31. Hayashi, T.; Fujita, Y.; Iwasaki, O.; Adachi, Y.; Takahashi, K.; Yanagida, M. Mis16 and Mis18 are required for CENP-A loading and histone deacetylation at centromeres. *Cell* **2004**, *118*, 715–729. [CrossRef]
32. De Rop, V.; Padeganeh, A.; Maddox, P.S. CENP-A: The key player behind centromere identity, propagation, and kinetochore assembly. *Chromosoma* **2012**, *121*, 527–538. [CrossRef] [PubMed]

33. Zuo, S.; Yadala, R.; Yang, F.; Talbert, P.; Fuchs, J.; Schubert, V.; Ahmadli, U.; Rutten, T.; Pecinka, A.; Lysak, M.A. Recurrent plant-specific duplications of KNL2 and its conserved function as a kinetochore assembly factor. *Mol. Biol. Evol.* **2022**, *39*, msac123. [CrossRef] [PubMed]
34. Zhang, D.; Martyniuk, C.J.; Trudeau, V.L. SANTA domain: A novel conserved protein module in Eukaryota with potential involvement in chromatin regulation. *Bioinformatics* **2006**, *22*, 2459–2462. [CrossRef] [PubMed]
35. Dambacher, S.; Deng, W.; Hahn, M.; Sadic, D.; Fröhlich, J.; Nuber, A.; Hoischen, C.; Diekmann, S.; Leonhardt, H.; Schotta, G. CENP-C facilitates the recruitment of M18BP1 to centromeric chromatin. *Nucleus* **2012**, *3*, 101–110. [CrossRef]
36. Stellfox, M.E.; Nardi, I.K.; Knippler, C.M.; Foltz, D.R. Differential binding partners of the Mis18α/β YIPPEE domains regulate Mis18 complex recruitment to centromeres. *Cell Rep.* **2016**, *15*, 2127–2135. [CrossRef]
37. Fujita, Y.; Hayashi, T.; Kiyomitsu, T.; Toyoda, Y.; Kokubu, A.; Obuse, C.; Yanagida, M. Priming of centromere for CENP-A recruitment by human hMis18α, hMis18β, and M18BP1. *Dev. Cell* **2007**, *12*, 17–30. [CrossRef]
38. Su, H.; Liu, Y.; Wang, C.; Liu, Y.; Feng, C.; Sun, Y.; Yuan, J.; Birchler, J.A.; Han, F. Knl1 participates in spindle assembly checkpoint signaling in maize. *Proc. Natl. Acad. Sci. USA* **2021**, *118*, e2022357118. [CrossRef]
39. Neumann, P.; Oliveira, L.C.; Jang, T.S.; Novak, P.; Koblizkova, A.; Schubert, V.; Houben, A.; Macas, J. Disruption of the standard kinetochore in holocentric *Cuscuta* species. *Proc. Natl. Acad. Sci. USA* **2023**, *120*, e2300877120. [CrossRef]
40. Du, Y.; Dawe, R.K. Maize NDC80 is a constitutive feature of the central kinetochore. *Chromosome Res.* **2007**, *15*, 767–775. [CrossRef]
41. Li, X.; Dawe, R.K. Fused sister kinetochores initiate the reductional division in meiosis I. *Nat. Cell Biol.* **2009**, *11*, 1103–1108. [CrossRef]
42. Cheeseman, I.M.; Chappie, J.S.; Wilson-Kubalek, E.M.; Desai, A. The conserved KMN network constitutes the core microtubule-binding site of the kinetochore. *Cell* **2006**, *127*, 983–997. [CrossRef] [PubMed]
43. Petrovic, A.; Mosalaganti, S.; Keller, J.; Mattiuzzo, M.; Overlack, K.; Krenn, V.; De Antoni, A.; Wohlgemuth, S.; Cecatiello, V.; Pasqualato, S. Modular assembly of RWD domains on the Mis12 complex underlies outer kinetochore organization. *Mol. Cell* **2014**, *53*, 591–605. [CrossRef] [PubMed]
44. Petrovic, A.; Pasqualato, S.; Dube, P.; Krenn, V.; Santaguida, S.; Cittaro, D.; Monzani, S.; Massimiliano, L.; Keller, J.; Tarricone, A. The MIS12 complex is a protein interaction hub for outer kinetochore assembly. *J. Cell Biol.* **2010**, *190*, 835–852. [CrossRef] [PubMed]
45. Allipra, S.; Anirudhan, K.; Shivanandan, S.; Raghunathan, A.; Maruthachalam, R. The kinetochore protein NNF1 has a moonlighting role in the vegetative development of *Arabidopsis thaliana*. *Plant J.* **2022**, *109*, 1064–1085. [CrossRef]
46. Li, J.; Wang, Y.; Zou, W.; Jian, L.; Fu, Y.; Zhao, J. AtNUF2 modulates spindle microtubule organization and chromosome segregation during mitosis. *Plant J.* **2021**, *107*, 801–816. [CrossRef]
47. Shin, J.; Jeong, G.; Park, J.Y.; Kim, H.; Lee, I. MUN (MERISTEM UNSTRUCTURED), encoding a SPC24 homolog of NDC80 kinetochore complex, affects development through cell division in *Arabidopsis thaliana*. *Plant J.* **2018**, *93*, 977–991. [CrossRef] [PubMed]
48. Zhou, J.; Liu, Y.; Guo, X.; Birchler, J.A.; Han, F.; Su, H. Centromeres: From chromosome biology to biotechnology applications and synthetic genomes in plants. *Plant Biotechnol. J.* **2022**, *20*, 2051–2063. [CrossRef]
49. London, N.; Biggins, S. Signalling dynamics in the spindle checkpoint response. *Nat. Rev. Mol. Cell Biol.* **2014**, *15*, 736–748. [CrossRef]
50. van Hooff, J.J.; Tromer, E.; van Wijk, L.M.; Snel, B.; Kops, G.J. Evolutionary dynamics of the kinetochore network in eukaryotes as revealed by comparative genomics. *EMBO Rep.* **2017**, *18*, 1559–1571. [CrossRef]
51. Sun, S.-C.; Kim, N.-H. Spindle assembly checkpoint and its regulators in meiosis. *Hum. Reprod. Update* **2012**, *18*, 60–72. [CrossRef]
52. Liu, B.; Lee, Y.-R.J. Spindle assembly and mitosis in plants. *Annu. Rev. Plant Biol.* **2022**, *73*, 227–254. [CrossRef] [PubMed]
53. Marston, A.L.; Wassmann, K. Multiple duties for spindle assembly checkpoint kinases in meiosis. *Front. Cell Dev. Biol.* **2017**, *5*, 109. [CrossRef] [PubMed]
54. Jiang, H.; Wang, F.F.; Wu, Y.T.; Zhou, X.; Huang, X.Y.; Zhu, J.; Gao, J.F.; Dong, R.B.; Cao, K.M.; Yang, Z.N. MULTIPOLAR SPINDLE 1 (MPS1), a novel coiled-coil protein of *Arabidopsis thaliana*, is required for meiotic spindle organization. *Plant J.* **2009**, *59*, 1001–1010. [CrossRef] [PubMed]
55. De Muyt, A.; Pereira, L.; Vezon, D.; Chelysheva, L.; Gendrot, G.; Chambon, A.; Lainé-Choinard, S.; Pelletier, G.; Mercier, R.; Nogue, F. A high throughput genetic screen identifies new early meiotic recombination functions in *Arabidopsis thaliana*. *PLoS Genet.* **2009**, *5*, e1000654. [CrossRef]
56. Xue, Z.; Liu, C.; Shi, W.; Miao, Y.; Shen, Y.; Tang, D.; Li, Y.; You, A.; Xu, Y.; Chong, K. OsMTOPVIB is required for meiotic bipolar spindle assembly. *Proc. Natl. Acad. Sci. USA* **2019**, *116*, 15967–15972. [CrossRef]
57. Wang, C.; Qu, S.; Zhang, J.; Fu, M.; Chen, X.; Liang, W. OsPRD2 is essential for double-strand break formation, but not spindle assembly during rice meiosis. *Front. Plant Sci.* **2022**, *13*, 1122202. [CrossRef]
58. Wang, M.; Tang, D.; Luo, Q.; Jin, Y.; Shen, Y.; Wang, K.; Cheng, Z. BRK1, a Bub1-related kinase, is essential for generating proper tension between homologous kinetochores at metaphase I of rice meiosis. *Plant Cell* **2012**, *24*, 4961–4973. [CrossRef]
59. Vaur, S.; Cubizolles, F.; Plane, G.; Genier, S.; Rabitsch, P.K.; Gregan, J.; Nasmyth, K.; Vanoosthuyse, V.; Hardwick, K.G.; Javerzat, J.-P. Control of Shugoshin function during fission-yeast meiosis. *Curr. Biol.* **2005**, *15*, 2263–2270. [CrossRef]
60. Nicklas, R.B.; Waters, J.C.; Salmon, E.; Ward, S.C. Checkpoint signals in grasshopper meiosis are sensitive to microtubule attachment, but tension is still essential. *J. Cell Sci.* **2001**, *114*, 4173–4183. [CrossRef]

61. Weimer, A.K.; Demidov, D.; Lermontova, I.; Beeckman, T.; Van Damme, D. Aurora kinases throughout plant development. *Trends Plant Sci.* **2016**, *21*, 69–79. [CrossRef]
62. Demidov, D.; Van Damme, D.; Geelen, D.; Blattner, F.R.; Houben, A. Identification and dynamics of two classes of aurora-like kinases in *Arabidopsis* and other plants. *Plant Cell* **2005**, *17*, 836–848. [CrossRef] [PubMed]
63. Jeyaprakash, A.A.; Klein, U.R.; Lindner, D.; Ebert, J.; Nigg, E.A.; Conti, E. Structure of a Survivin–Borealin–INCENP core complex reveals how chromosomal passengers travel together. *Cell* **2007**, *131*, 271–285. [CrossRef] [PubMed]
64. Kaitna, S.; Mendoza, M.; Jantsch-Plunger, V.; Glotzer, M. Incenp and an aurora-like kinase form a complex essential for chromosome segregation and efficient completion of cytokinesis. *Curr. Biol.* **2000**, *10*, 1172–1181. [CrossRef]
65. Winey, M.; Goetsch, L.; Baum, P.; Byers, B. MPS1 and MPS2: Novel yeast genes defining distinct steps of spindle pole body duplication. *J. Cell Biol.* **1991**, *114*, 745–754. [CrossRef]
66. Lauze, E.Y.; Stoelcker, B.; Luca, F.; Weiss, E.; Schutz, A.; Winey, M. Yeast spindle pole body duplication gene MPS1 encodes an essential dual specificity protein kinase. *EMBO J.* **1995**, *14*, 1655–1663. [CrossRef]
67. Espeut, J.; Lara-Gonzalez, P.; Sassine, M.; Shiau, A.K.; Desai, A.; Abrieu, A. Natural loss of Mps1 kinase in nematodes uncovers a role for polo-like kinase 1 in spindle checkpoint initiation. *Cell Rep.* **2015**, *12*, 58–65. [CrossRef] [PubMed]
68. Maure, J.-F.; Kitamura, E.; Tanaka, T.U. Mps1 kinase promotes sister-kinetochore bi-orientation by a tension-dependent mechanism. *Curr. Biol.* **2017**, *17*, 2175–2182. [CrossRef]
69. Hewitt, L.; Tighe, A.; Santaguida, S.; White, A.M.; Jones, C.D.; Musacchio, A.; Green, S.; Taylor, S.S. Sustained Mps1 activity is required in mitosis to recruit O-Mad2 to the Mad1–C-Mad2 core complex. *J. Cell Biol.* **2010**, *190*, 25–34. [CrossRef]
70. Maciejowski, J.; George, K.A.; Terret, M.-E.; Zhang, C.; Shokat, K.M.; Jallepalli, P.V. Mps1 directs the assembly of Cdc20 inhibitory complexes during interphase and mitosis to control M phase timing and spindle checkpoint signaling. *J. Cell Biol.* **2010**, *190*, 89–100. [CrossRef]
71. Santaguida, S.; Tighe, A.; D'Alise, A.M.; Taylor, S.S.; Musacchio, A. Dissecting the role of MPS1 in chromosome biorientation and the spindle checkpoint through the small molecule inhibitor reversine. *J. Cell Biol.* **2010**, *190*, 73–87. [CrossRef]
72. Tighe, A.; Staples, O.; Taylor, S. Mps1 kinase activity restrains anaphase during an unperturbed mitosis and targets Mad2 to kinetochores. *J. Cell Biol.* **2008**, *181*, 893–901. [CrossRef] [PubMed]
73. Benzi, G.; Camasses, A.; Atsunori, Y.; Katou, Y.; Shirahige, K.; Piatti, S. A common molecular mechanism underlies the role of Mps1 in chromosome biorientation and the spindle assembly checkpoint. *EMBO Rep.* **2020**, *21*, e50257. [CrossRef] [PubMed]
74. London, N.; Ceto, S.; Ranish, J.A.; Biggins, S. Phosphoregulation of Spc105 by Mps1 and PP1 regulates Bub1 localization to kinetochores. *Curr. Biol.* **2012**, *22*, 900–906. [CrossRef] [PubMed]
75. Ji, Z.; Gao, H.; Jia, L.; Li, B.; Yu, H. A sequential multi-target Mps1 phosphorylation cascade promotes spindle checkpoint signaling. *eLife* **2017**, *6*, e22513. [CrossRef] [PubMed]
76. Tipton, A.R.; Ji, W.; Sturt-Gillespie, B.; Bekier, M.E.; Wang, K.; Taylor, W.R.; Liu, S.-T. Monopolar spindle 1 (MPS1) kinase promotes production of closed MAD2 (C-MAD2) conformer and assembly of the mitotic checkpoint complex. *J. Biol. Chem.* **2013**, *288*, 35149–35158. [CrossRef] [PubMed]
77. Straight, P.D.; Giddings, T.H., Jr.; Winey, M. Mps1p regulates meiotic spindle pole body duplication in addition to having novel roles during sporulation. *Mol. Biol. Cell.* **2000**, *11*, 3525–3537. [CrossRef]
78. Gilliland, W.D.; Wayson, S.M.; Hawley, R.S. The meiotic defects of mutants in the *Drosophila* mps1 gene reveal a critical role of Mps1 in the segregation of achiasmate homologs. *Curr. Biol.* **2005**, *15*, 672–677. [CrossRef]
79. Gilliland, W.D.; Hughes, S.E.; Cotitta, J.L.; Takeo, S.; Xiang, Y.; Hawley, R.S. The multiple roles of Mps1 in *Drosophila* female meiosis. *PLoS Genet.* **2007**, *3*, e113. [CrossRef]
80. Hached, K.; Xie, S.Z.; Buffin, E.; Cladière, D.; Rachez, C.; Sacras, M.; Sorger, P.K.; Wassmann, K. Mps1 at kinetochores is essential for female mouse meiosis I. *Development* **2011**, *138*, 2261–2271. [CrossRef]
81. Paganelli, L.; Caillaud, M.C.; Quentin, M.; Damiani, I.; Govetto, B.; Lecomte, P.; Karpov, P.A.; Abad, P.; Chaboute, M.E.; Favery, B. Retracted: Three BUB 1 and BUBR 1/MAD 3-related spindle assembly checkpoint proteins are required for accurate mitosis in *Arabidopsis*. *New Phytol.* **2015**, *205*, 202–215. [CrossRef]
82. Su, H.; Liu, Y.; Dong, Q.; Feng, C.; Zhang, J.; Liu, Y.; Birchler, J.A.; Han, F. Dynamic location changes of Bub1-phosphorylated-H2AThr133 with CENH3 nucleosome in maize centromeric regions. *New Phytol.* **2017**, *214*, 682–694. [CrossRef]
83. Miyazaki, S.; Kim, J.; Yamagishi, Y.; Ishiguro, T.; Okada, Y.; Tanno, Y.; Sakuno, T.; Watanabe, Y. Meikin-associated polo-like kinase specifies Bub1 distribution in meiosis I. *Genes Cells* **2017**, *22*, 552–567. [CrossRef]
84. Yamaguchi, S.; Decottignies, A.; Nurse, P. Function of Cdc2p-dependent Bub1p phosphorylation and Bub1p kinase activity in the mitotic and meiotic spindle checkpoint. *EMBO J.* **2003**, *22*, 1075–1087. [CrossRef] [PubMed]
85. Li, M.; Li, S.; Yuan, J.; Wang, Z.-B.; Sun, S.-C.; Schatten, H.; Sun, Q.-Y. Bub3 is a spindle assembly checkpoint protein regulating chromosome segregation during mouse oocyte meiosis. *PLoS ONE* **2009**, *4*, e7701. [CrossRef] [PubMed]
86. Fraschini, R.; Beretta, A.; Sironi, L.; Musacchio, A.; Lucchini, G.; Piatti, S. Bub3 interaction with Mad2, Mad3 and Cdc20 is mediated by WD40 repeats and does not require intact kinetochores. *EMBO J.* **2001**, *20*, 6648–6659. [CrossRef] [PubMed]
87. Lermontova, I.; Fuchs, J.; Schubert, I. The *Arabidopsis* checkpoint protein Bub3. 1 is essential for gametophyte development. *Front. Biosci.* **2008**, *13*, 5202–5211. [CrossRef] [PubMed]
88. Komaki, S.; Schnittger, A. The spindle assembly checkpoint in *Arabidopsis* is rapidly shut off during severe stress. *Dev. Cell* **2017**, *43*, 172–185.e175. [CrossRef]

89. Zhang, H.; Deng, X.; Sun, B.; Lee Van, S.; Kang, Z.; Lin, H.; Lee, Y.-R.J.; Liu, B. Role of the BUB3 protein in phragmoplast microtubule reorganization during cytokinesis. *Nat. Plants* **2018**, *4*, 485–494. [CrossRef]
90. Rancati, G.; Crispo, V.; Lucchini, G.; Piatti, S. Mad3/BubR1 phosphorylation during spindle checkpoint activation depends on both Polo and Aurora kinases in budding yeast. *Cell Cycle* **2005**, *4*, 972–980. [CrossRef]
91. Pérez-Mongiovi, D.; Malmanche, N.; Bousbaa, H.; Sunkel, C. Maternal expression of the checkpoint protein BubR1 is required for synchrony of syncytial nuclear divisions and polar body arrest in *Drosophila melanogaster*. *Development* **2005**, *132*, 4509–4520. [CrossRef]
92. Wei, L.; Liang, X.-W.; Zhang, Q.-H.; Li, M.; Yuan, J.; Li, S.; Sun, S.-C.; Ouyang, Y.-C.; Schatten, H.; Sun, Q.-Y. BubR1 is a spindle assembly checkpoint protein regulating meiotic cell cycle progression of mouse oocyte. *Cell Cycle* **2010**, *9*, 1112–1121. [CrossRef]
93. Komaki, S.; Schnittger, A. The spindle checkpoint in plants—A green variation over a conserved theme? *Curr. Opin. Plant Biol.* **2016**, *34*, 84–91. [CrossRef] [PubMed]
94. Parra, M.T.; Gomez, R.; Viera, A.; Llano, E.; Pendas, A.M.; Rufas, J.S.; Suja, J.A. Sequential assembly of centromeric proteins in male mouse meiosis. *PLoS Genet.* **2009**, *5*, e1000417. [CrossRef] [PubMed]
95. Jeganathan, K.B.; Van Deursen, J. Differential mitotic checkpoint protein requirements in somatic and germ cells. *Biochem. Soc. Trans.* **2006**, *34*, 583–586. [CrossRef] [PubMed]
96. Heng-Yu, F. BubR1, a spindle assembly checkpoint protein in mammalian oocyte meiosis. *Cell Cycle* **2010**, *9*, 1456–1465. [CrossRef]
97. Caillaud, M.-C.; Paganelli, L.; Lecomte, P.; Deslandes, L.; Quentin, M.; Pecrix, Y.; Le Bris, M.; Marfaing, N.; Abad, P.; Favery, B. Spindle assembly checkpoint protein dynamics reveal conserved and unsuspected roles in plant cell division. *PLoS ONE* **2009**, *4*, e6757. [CrossRef] [PubMed]
98. Kimbara, J.; Endo, T.R.; Nasuda, S. Characterization of the genes encoding for MAD2 homologues in wheat. *Chromosome Res.* **2004**, *12*, 703–714. [CrossRef]
99. Chen, R.-H.; Brady, D.M.; Smith, D.; Murray, A.W.; Hardwick, K.G. The spindle checkpoint of budding yeast depends on a tight complex between the Mad1 and Mad2 proteins. *Mol. Biol. Cell.* **1999**, *10*, 2607–2618. [CrossRef]
100. Cheslock, P.S.; Kemp, B.J.; Boumil, R.M.; Dawson, D.S. The roles of MAD1, MAD2 and MAD3 in meiotic progression and the segregation of nonexchange chromosomes. *Nat. Genet.* **2005**, *37*, 756–760. [CrossRef]
101. Kitagawa, R.; Rose, A.M. Components of the spindle-assembly checkpoint are essential in *Caenorhabditis elegans*. *Nat. Cell Biol.* **1999**, *1*, 514–521. [CrossRef]
102. Zhang, D.; Li, M.; Ma, W.; Hou, Y.; Li, Y.-H.; Li, S.-W.; Sun, Q.-Y.; Wang, W.-H. Localization of mitotic arrest deficient 1 (MAD1) in mouse oocytes during the first meiosis and its functions as a spindle checkpoint protein. *Biol. Reprod.* **2005**, *72*, 58–68. [CrossRef] [PubMed]
103. Ding, D.; Muthuswamy, S.; Meier, I. Functional interaction between the *Arabidopsis* orthologs of spindle assembly checkpoint proteins MAD1 and MAD2 and the nucleoporin NUA. *Plant Mol. Biol.* **2012**, *79*, 203–216. [CrossRef] [PubMed]
104. Bao, Z.; Zhang, N.; Hua, J. Endopolyploidization and flowering time are antagonistically regulated by checkpoint component MAD1 and immunity modulator MOS1. *Nat. Commun.* **2014**, *5*, 5628. [CrossRef]
105. Kallio, M.; Eriksson, J.E.; Gorbsky, G.J. Differences in spindle association of the mitotic checkpoint protein Mad2 in mammalian spermatogenesis and oogenesis. *Dev. Biol.* **2000**, *225*, 112–123. [CrossRef]
106. Zhang, D.; Ma, W.; Li, Y.-H.; Hou, Y.; Li, S.-W.; Meng, X.-Q.; Sun, X.-F.; Sun, Q.-Y.; Wang, W.-H. Intra-oocyte localization of MAD2 and its relationship with kinetochores, microtubules, and chromosomes in rat oocytes during meiosis. *Biol. Reprod.* **2004**, *71*, 740–748. [CrossRef] [PubMed]
107. Yu, H.-G.; Muszynski, M.G.; Kelly Dawe, R. The maize homologue of the cell cycle checkpoint protein MAD2 reveals kinetochore substructure and contrasting mitotic and meiotic localization patterns. *J. Cell Biol.* **1999**, *145*, 425–435. [CrossRef]
108. Niault, T.; Hached, K.; Sotillo, R.; Sorger, P.K.; Maro, B.; Benezra, R.; Wassmann, K. Changing Mad2 levels affects chromosome segregation and spindle assembly checkpoint control in female mouse meiosis I. *PLoS ONE* **2007**, *2*, e1165. [CrossRef]
109. Wang, J.Y.; Lei, Z.L.; Nan, C.L.; Yin, S.; Liu, J.; Hou, Y.; Li, Y.L.; Chen, D.Y.; Sun, Q.Y. RNA interference as a tool to study the function of MAD2 in mouse oocyte meiotic maturation. *Mol. Reprod. Dev.* **2007**, *74*, 116–124. [CrossRef]
110. Yamamoto, A.; Kitamura, K.; Hihara, D.; Hirose, Y.; Katsuyama, S.; Hiraoka, Y. Spindle checkpoint activation at meiosis I advances anaphase II onset via meiosis-specific APC/C regulation. *J. Cell Biol.* **2008**, *182*, 277–288. [CrossRef]
111. Ruchaud, S.; Carmena, M.; Earnshaw, W.C. Chromosomal passengers: Conducting cell division. *Nat. Rev. Mol. Cell Biol.* **2007**, *8*, 798–812. [CrossRef]
112. Vader, G.; Cruijsen, C.W.; Van Harn, T.; Vromans, M.J.; Medema, R.H.; Lens, S.M. The chromosomal passenger complex controls spindle checkpoint function independent from its role in correcting microtubule–kinetochore interactions. *Mol. Biol. Cell* **2007**, *18*, 4553–4564. [CrossRef] [PubMed]
113. Muñoz-Barrera, M.; Monje-Casas, F. Increased Aurora B activity causes continuous disruption of kinetochore–microtubule attachments and spindle instability. *Proc. Natl. Acad. Sci. USA* **2014**, *111*, E3996–E4005. [CrossRef] [PubMed]
114. Landino, J.; Ohi, R. The timing of midzone stabilization during cytokinesis depends on myosin II activity and an interaction between INCENP and actin. *Curr. Biol.* **2016**, *26*, 698–706. [CrossRef] [PubMed]
115. Kitagawa, M.; Lee, S.H. The chromosomal passenger complex (CPC) as a key orchestrator of orderly mitotic exit and cytokinesis. *Front. Cell Dev. Biol.* **2015**, *3*, 14. [CrossRef]

116. Giet, R.; Prigent, C. Aurora/Ipl1p-related kinases, a new oncogenic family of mitotic serine-threonine kinases. *J. Cell Sci.* **1999**, *112*, 3591–3601. [CrossRef] [PubMed]
117. Petersen, J.; Paris, J.; Willer, M.; Philippe, M.; Hagan, I.M. The *S. pombe* aurora-related kinase Ark1 associates with mitotic structures in a stage dependent manner and is required for chromosome segregation. *J. Cell Sci.* **2001**, *114*, 4371–4384. [CrossRef]
118. Ducat, D.; Zheng, Y. Aurora kinases in spindle assembly and chromosome segregation. *Exp. Cell Res.* **2004**, *301*, 60–67. [CrossRef]
119. Goldenson, B.; Crispino, J.D. The aurora kinases in cell cycle and leukemia. *Oncogene* **2015**, *34*, 537–545. [CrossRef]
120. Carmena, M.; Earnshaw, W.C. The cellular geography of aurora kinases. *Nat. Rev. Mol. Cell Biol.* **2003**, *4*, 842–854. [CrossRef]
121. Yu, H.-G.; Koshland, D. The Aurora kinase Ipl1 maintains the centromeric localization of PP2A to protect cohesin during meiosis. *J. Cell Biol.* **2007**, *176*, 911–918. [CrossRef]
122. Monje-Casas, F.; Prabhu, V.R.; Lee, B.H.; Boselli, M.; Amon, A. Kinetochore orientation during meiosis is controlled by Aurora B and the monopolin complex. *Cell* **2007**, *128*, 477–490. [CrossRef] [PubMed]
123. Balboula, A.Z.; Schindler, K. Selective disruption of aurora C kinase reveals distinct functions from aurora B kinase during meiosis in mouse oocytes. *PLoS Genet.* **2014**, *10*, e1004194. [CrossRef] [PubMed]
124. Tang, C.-J.C.; Lin, C.-Y.; Tang, T.K. Dynamic localization and functional implications of Aurora-C kinase during male mouse meiosis. *Dev. Biol.* **2006**, *290*, 398–410. [CrossRef] [PubMed]
125. Demidov, D.; Lermontova, I.; Weiss, O.; Fuchs, J.; Rutten, T.; Kumke, K.; Sharbel, T.F.; Van Damme, D.; De Storme, N.; Geelen, D. Altered expression of Aurora kinases in *Arabidopsis* results in aneu-and polyploidization. *Plant J.* **2014**, *80*, 449–461. [CrossRef]
126. Niu, B.; Wang, L.; Zhang, L.; Ren, D.; Ren, R.; Copenhaver, G.P.; Ma, H.; Wang, Y. *Arabidopsis* cell division cycle 20.1 is required for normal meiotic spindle assembly and chromosome segregation. *Plant Cell* **2015**, *27*, 3367–3382. [CrossRef]
127. Zamariola, L.; Tiang, C.L.; De Storme, N.; Pawlowski, W.; Geelen, D. Chromosome segregation in plant meiosis. *Front. Plant Sci.* **2014**, *5*, 279. [CrossRef]
128. Carmena, M.; Wheelock, M.; Funabiki, H.; Earnshaw, W.C. The chromosomal passenger complex (CPC): From easy rider to the godfather of mitosis. *Nat. Rev. Mol. Cell Biol.* **2012**, *13*, 789–803. [CrossRef]
129. Komaki, S.; Takeuchi, H.; Hamamura, Y.; Heese, M.; Hashimoto, T.; Schnittger, A. Functional analysis of the plant chromosomal passenger complex. *Plant Physiol.* **2020**, *183*, 1586–1599. [CrossRef]
130. Kirioukhova, O.; Johnston, A.J.; Kleen, D.; Kägi, C.; Baskar, R.; Moore, J.M.; Bäumlein, H.; Groß-Hardt, R.; Grossniklaus, U. Female gametophytic cell specification and seed development require the function of the putative *Arabidopsis* INCENP ortholog WYRD. *Development* **2011**, *138*, 3409–3420. [CrossRef]
131. Gassmann, R.; Carvalho, A.; Henzing, A.J.; Ruchaud, S.; Hudson, D.F.; Honda, R.; Nigg, E.A.; Gerloff, D.L.; Earnshaw, W.C. Borealin: A novel chromosomal passenger required for stability of the bipolar mitotic spindle. *J. Cell Biol.* **2004**, *166*, 179–191. [CrossRef]
132. Komaki, S.; Tromer, E.C.; De Jaeger, G.; De Winne, N.; Heese, M.; Schnittger, A. Molecular convergence by differential domain acquisition is a hallmark of chromosomal passenger complex evolution. *Proc. Natl. Acad. Sci. USA* **2022**, *119*, e2200108119. [CrossRef] [PubMed]
133. Sun, S.-C.; Wei, L.; Li, M.; Lin, S.-L.; Xu, B.-Z.; Liang, X.-W.; Kim, N.-H.; Schatten, H.; Lu, S.-S.; Sun, Q.-Y. Perturbation of survivin expression affects chromosome alignment and spindle checkpoint in mouse oocyte meiotic maturation. *Cell Cycle* **2009**, *8*, 3365–3372. [CrossRef] [PubMed]
134. Wang, K.; Jiang, G.-J.; Wei, L.; Liang, X.-W.; Miao, D.-Q.; Sun, S.-C.; Guo, L.; Wang, Z.-B.; Lu, S.-S. Survivin is a critical regulator of spindle organization and chromosome segregation during rat oocyte meiotic maturation. *Zygote* **2011**, *19*, 307–313. [CrossRef] [PubMed]
135. Chelysheva, L.; Diallo, S.; Vezon, D.; Gendrot, G.; Vrielynck, N.; Belcram, K.; Rocques, N.; Márquez-Lema, A.; Bhatt, A.M.; Horlow, C. AtREC8 and AtSCC3 are essential to the monopolar orientation of the kinetochores during meiosis. *J. Cell Sci.* **2005**, *118*, 4621–4632. [CrossRef]
136. Cuacos, M.; H. Franklin, F.C.; Heckmann, S. Atypical centromeres in plants—What they can tell us. *Front. Plant Sci.* **2015**, *6*, 913. [CrossRef] [PubMed]
137. Nasmyth, K.; Haering, C.H. The structure and function of SMC and kleisin complexes. *Annu. Rev. Biochem.* **2005**, *74*, 595–648. [CrossRef]
138. Haering, C.H.; Löwe, J.; Hochwagen, A.; Nasmyth, K. Molecular architecture of SMC proteins and the yeast cohesin complex. *Mol. Cell* **2002**, *9*, 773–788. [CrossRef]
139. Hirano, T. SMC proteins and chromosome mechanics: From bacteria to humans. *Philos. Trans. R. Soc. Lond. B Biol. Sci.* **2005**, *360*, 507–514. [CrossRef]
140. Schubert, V. SMC proteins and their multiple functions in higher plants. *Cytogenet. Genome Res.* **2009**, *124*, 202–214. [CrossRef]
141. Cai, X.; Dong, F.; Edelmann, R.E.; Makaroff, C.A. The *Arabidopsis* SYN1 cohesin protein is required for sister chromatid arm cohesion and homologous chromosome pairing. *J. Cell Sci.* **2003**, *116*, 2999–3007. [CrossRef]
142. Kudo, N.R.; Anger, M.; Peters, A.H.; Stemmann, O.; Theussl, H.-C.; Helmhart, W.; Kudo, H.; Heyting, C.; Nasmyth, K. Role of cleavage by separase of the Rec8 kleisin subunit of cohesin during mammalian meiosis I. *J. Cell Sci.* **2009**, *122*, 2686–2698. [CrossRef] [PubMed]
143. Watanabe, Y.; Nurse, P. Cohesin Rec8 is required for reductional chromosome segregation at meiosis. *Nature* **1999**, *400*, 461–464. [CrossRef] [PubMed]

144. Sakuno, T.; Hiraoka, Y. Rec8 cohesin: A structural platform for shaping the meiotic chromosomes. *Genes* **2022**, *13*, 200. [CrossRef]
145. Kudo, N.R.; Wassmann, K.; Anger, M.; Schuh, M.; Wirth, K.G.; Xu, H.; Helmhart, W.; Kudo, H.; Mckay, M.; Maro, B. Resolution of chiasmata in oocytes requires separase-mediated proteolysis. *Cell* **2006**, *126*, 135–146. [CrossRef] [PubMed]
146. Martinez-Perez, E.; Schvarzstein, M.; Barroso, C.; Lightfoot, J.; Dernburg, A.F.; Villeneuve, A.M. Crossovers trigger a remodeling of meiotic chromosome axis composition that is linked to two-step loss of sister chromatid cohesion. *Genes Dev.* **2008**, *22*, 2886–2901. [CrossRef] [PubMed]
147. Xu, H.; Beasley, M.D.; Warren, W.D.; van der Horst, G.T.; McKay, M.J. Absence of mouse REC8 cohesin promotes synapsis of sister chromatids in meiosis. *Dev. Cell* **2005**, *8*, 949–961. [CrossRef]
148. Doll, E.; Molnar, M.; Cuanoud, G.; Octobre, G.; Latypov, V.; Ludin, K.; Kohli, J.R. Cohesin and recombination proteins influence the G1-to-S transition in azygotic meiosis in *Schizosaccharomyces pombe*. *Genetics* **2008**, *180*, 727–740. [CrossRef]
149. Peirson, B.N.; Bowling, S.E.; Makaroff, C.A. A defect in synapsis causes male sterility in a T-DNA-tagged *Arabidopsis thaliana* mutant. *Plant J.* **1997**, *11*, 659–669. [CrossRef]
150. Lam, W.S.; Yang, X.; Makaroff, C.A. Characterization of *Arabidopsis thaliana* SMC1 and SMC3: Evidence that AtSMC3 may function beyond chromosome cohesion. *J. Cell Sci.* **2005**, *118*, 3037–3048. [CrossRef]
151. Lhuissier, F.G.; Offenberg, H.H.; Wittich, P.E.; Vischer, N.O.; Heyting, C. The mismatch repair protein MLH1 marks a subset of strongly interfering crossovers in tomato. *Plant Cell* **2007**, *19*, 862–876. [CrossRef]
152. Mercier, R.; Grelon, M. Meiosis in plants: Ten years of gene discovery. *Cytogenet. Genome Res.* **2008**, *120*, 281–290. [CrossRef] [PubMed]
153. Mercier, R.; Mézard, C.; Jenczewski, E.; Macaisne, N.; Grelon, M. The molecular biology of meiosis in plants. *Annu. Rev. Plant Biol.* **2015**, *66*, 297–327. [CrossRef] [PubMed]
154. Hamant, O.; Ma, H.; Cande, W.Z. Genetics of meiotic prophase I in plants. *Annu. Rev. Plant Biol.* **2006**, *57*, 267–302. [CrossRef] [PubMed]
155. Shonn, M.A.; McCarroll, R.; Murray, A.W. Spo13 protects meiotic cohesin at centromeres in meiosis I. *Genes Dev.* **2002**, *16*, 1659–1671. [CrossRef]
156. Yokobayashi, S.; Watanabe, Y. The kinetochore protein Moa1 enables cohesion-mediated monopolar attachment at meiosis I. *Cell* **2005**, *123*, 803–817. [CrossRef]
157. Kim, J.; Ishiguro, K.-i.; Nambu, A.; Akiyoshi, B.; Yokobayashi, S.; Kagami, A.; Ishiguro, T.; Pendas, A.M.; Takeda, N.; Sakakibara, Y. Meikin is a conserved regulator of meiosis-I-specific kinetochore function. *Nature* **2015**, *517*, 466–471. [CrossRef]
158. Bonner, A.M.; Hughes, S.E.; Hawley, R.S. Regulation of polo kinase by matrimony is required for cohesin maintenance during *Drosophila melanogaster* female meiosis. *Curr. Biol.* **2020**, *30*, 715–722.e713. [CrossRef]
159. Yokobayashi, S.; Yamamoto, M.; Watanabe, Y. Cohesins determine the attachment manner of kinetochores to spindle microtubules at meiosis I in fission yeast. *Mol. Cell. Biol.* **2003**, *23*, 3965–3973. [CrossRef]
160. Ferrandiz, N.; Barroso, C.; Telecan, O.; Shao, N.; Kim, H.-M.; Testori, S.; Faull, P.; Cutillas, P.; Snijders, A.P.; Colaiácovo, M.P. Spatiotemporal regulation of Aurora B recruitment ensures release of cohesion during *C. elegans* oocyte meiosis. *Nat. Commun.* **2018**, *9*, 834. [CrossRef]
161. McKee, B.D.; Yan, R.; Tsai, J.-H. Meiosis in male Drosophila. *Spermatogenesis* **2012**, *2*, 167–184. [CrossRef]
162. Yu, H.-G.; Dawe, R.K. Functional redundancy in the maize meiotic kinetochore. *J. Cell Biol.* **2000**, *151*, 131–142. [CrossRef]
163. Shao, T.; Tang, D.; Wang, K.; Wang, M.; Che, L.; Qin, B.; Yu, H.; Li, M.; Gu, M.; Cheng, Z. OsREC8 is essential for chromatid cohesion and metaphase I monopolar orientation in rice meiosis. *Plant Physiol.* **2011**, *156*, 1386–1396. [CrossRef] [PubMed]
164. Bai, X.; Peirson, B.N.; Dong, F.; Xue, C.; Makaroff, C.A. Isolation and characterization of SYN1, a RAD21-like gene essential for meiosis in *Arabidopsis*. *Plant Cell* **1999**, *11*, 417–430. [CrossRef] [PubMed]
165. Bhatt, A.M.; Lister, C.; Page, T.; Fransz, P.; Findlay, K.; Jones, G.H.; Dickinson, H.G.; Dean, C. The DIF1 gene of *Arabidopsis* is required for meiotic chromosome segregation and belongs to the REC8/RAD21 cohesin gene family. *Plant J.* **1999**, *19*, 463–472. [CrossRef]
166. Zhang, L.; Tao, J.; Wang, S.; Chong, K.; Wang, T. The rice OsRad21-4, an orthologue of yeast Rec8 protein, is required for efficient meiosis. *Plant Mol. Biol.* **2006**, *60*, 533. [CrossRef] [PubMed]
167. Moens, P.B.; Kolas, N.K.; Tarsounas, M.; Marcon, E.; Cohen, P.E.; Spyropoulos, B. The time course and chromosomal localization of recombination-related proteins at meiosis in the mouse are compatible with models that can resolve the early DNA-DNA interactions without reciprocal recombination. *J. Cell Sci.* **2002**, *115*, 1611–1622. [CrossRef] [PubMed]
168. Watanabe, Y. Modifying sister chromatid cohesion for meiosis. *J. Cell Sci.* **2004**, *117*, 4017–4023. [CrossRef]
169. Altschul, S.F.; Gish, W.; Miller, W.; Myers, E.W.; Lipman, D.J. Basic local alignment search tool. *J. Mol. Biol.* **1990**, *215*, 403–410. [CrossRef]
170. Kerrebrock, A.W.; Moore, D.P.; Wu, J.S.; Orr-Weaver, T.L. Mei-S332, a *Drosophila* protein required for sister-chromatid cohesion, can localize to meiotic centromere regions. *Cell* **1995**, *83*, 247–256. [CrossRef]
171. Yao, Y.; Dai, W. Shugoshins function as a guardian for chromosomal stability in nuclear division. *Cell Cycle* **2012**, *11*, 2631–2642. [CrossRef]
172. Katis, V.L.; Galova, M.; Rabitsch, K.P.; Gregan, J.; Nasmyth, K. Maintenance of cohesin at centromeres after meiosis I in budding yeast requires a kinetochore-associated protein related to MEI-S332. *Curr. Biol.* **2004**, *14*, 560–572. [CrossRef] [PubMed]

173. Kitajima, T.S.; Kawashima, S.A.; Watanabe, Y. The conserved kinetochore protein shugoshin protects centromeric cohesion during meiosis. *Nature* **2004**, *427*, 510–517. [CrossRef] [PubMed]
174. Marston, A.L.; Tham, W.-H.; Shah, H.; Amon, A. A genome-wide screen identifies genes required for centromeric cohesion. *Science* **2004**, *303*, 1367–1370. [CrossRef] [PubMed]
175. Rabitsch, K.P.; Gregan, J.; Schleiffer, A.; Javerzat, J.-P.; Eisenhaber, F.; Nasmyth, K. Two fission yeast homologs of *Drosophila* Mei-S332 are required for chromosome segregation during meiosis I and II. *Curr. Biol.* **2004**, *14*, 287–301. [CrossRef] [PubMed]
176. Hamant, O.; Golubovskaya, I.; Meeley, R.; Fiume, E.; Timofejeva, L.; Schleiffer, A.; Nasmyth, K.; Cande, W.Z. A REC8-dependent plant Shugoshin is required for maintenance of centromeric cohesion during meiosis and has no mitotic functions. *Curr. Biol.* **2005**, *15*, 948–954. [CrossRef]
177. Gómez, R.; Valdeolmillos, A.; Parra, M.T.; Viera, A.; Carreiro, C.; Roncal, F.; Rufas, J.S.; Barbero, J.L.; Suja, J.A. Mammalian SGO2 appears at the inner centromere domain and redistributes depending on tension across centromeres during meiosis II and mitosis. *EMBO Rep.* **2007**, *8*, 173–180. [CrossRef]
178. Lee, J.; Kitajima, T.S.; Tanno, Y.; Yoshida, K.; Morita, T.; Miyano, T.; Miyake, M.; Watanabe, Y. Unified mode of centromeric protection by shugoshin in mammalian oocytes and somatic cells. *Nat. Cell Biol.* **2008**, *10*, 42–52. [CrossRef]
179. Xu, Z.; Cetin, B.; Anger, M.; Cho, U.S.; Helmhart, W.; Nasmyth, K.; Xu, W. Structure and function of the PP2A-shugoshin interaction. *Mol. Cell* **2009**, *35*, 426–441. [CrossRef]
180. Macy, B.; Wang, M.; Yu, H.-G. The many faces of shugoshin, the "guardian spirit," in chromosome segregation. *Cell Cycle* **2009**, *8*, 35–37. [CrossRef]
181. Tanno, Y.; Kitajima, T.S.; Honda, T.; Ando, Y.; Ishiguro, K.-I.; Watanabe, Y. Phosphorylation of mammalian Sgo2 by Aurora B recruits PP2A and MCAK to centromeres. *Genes Dev.* **2010**, *24*, 2169–2179. [CrossRef]
182. Clift, D.; Marston, A. The role of shugoshin in meiotic chromosome segregation. *Cytogenet. Genome Res.* **2011**, *133*, 234–242. [CrossRef] [PubMed]
183. Zamariola, L.; De Storme, N.; Tiang, C.; Armstrong, S.; Franklin, F.; Geelen, D. SGO1 but not SGO2 is required for maintenance of centromere cohesion in *Arabidopsis thaliana* meiosis. *Plant Reprod.* **2013**, *26*, 197–208. [CrossRef] [PubMed]
184. Zamariola, L.; De Storme, N.; Vannerum, K.; Vandepoele, K.; Armstrong, S.J.; Franklin, F.C.H.; Geelen, D. SHUGOSHIN s and PATRONUS protect meiotic centromere cohesion in *Arabidopsis thaliana. Plant J.* **2014**, *77*, 782–794. [CrossRef] [PubMed]
185. Cromer, L.; Jolivet, S.; Horlow, C.; Chelysheva, L.; Heyman, J.; De Jaeger, G.; Koncz, C.; De Veylder, L.; Mercier, R. Centromeric cohesion is protected twice at meiosis, by SHUGOSHINs at anaphase I and by PATRONUS at interkinesis. *Curr. Biol.* **2013**, *23*, 2090–2099. [CrossRef] [PubMed]
186. Wang, M.; Tang, D.; Wang, K.; Shen, Y.; Qin, B.; Miao, C.; Li, M.; Cheng, Z. OsSGO1 maintains synaptonemal complex stabilization in addition to protecting centromeric cohesion during rice meiosis. *Plant J.* **2011**, *67*, 583–594. [CrossRef]
187. Mahjoubi, H.; Tamari, Y.; Takeda, S.; Bouchabké-Coussa, O.; Hanin, M.; Herzog, E.; Schmit, A.-C.; Chaboute, M.-E.; Ebel, C. The wheat TdRL1 is the functional homolog of the rice RSS1 and promotes plant salt stress tolerance. *Plant Cell Rep.* **2018**, *37*, 1625–1637. [CrossRef]
188. Cromer, L.; Jolivet, S.; Singh, D.K.; Berthier, F.; De Winne, N.; De Jaeger, G.; Komaki, S.; Prusicki, M.A.; Schnittger, A.; Guérois, R. Patronus is the elusive plant securin, preventing chromosome separation by antagonizing separase. *Proc. Natl. Acad. Sci. USA* **2019**, *116*, 16018–16027. [CrossRef]
189. Ogawa, D.; Abe, K.; Miyao, A.; Kojima, M.; Sakakibara, H.; Mizutani, M.; Morita, H.; Toda, Y.; Hobo, T.; Sato, Y. RSS1 regulates the cell cycle and maintains meristematic activity under stress conditions in rice. *Nat. Com.* **2011**, *2*, 278. [CrossRef]
190. Ravi, M.; Bondada, R. Genome elimination by tailswap CenH3: In vivo haploid production in *Arabidopsis thaliana*. In *Chromosome and Genomic Engineering in Plants: Methods and Protocols*; Springer: Berlin/Heidelberg, Germany, 2016; pp. 77–99. [CrossRef]
191. Marimuthu, M.P.; Jolivet, S.; Ravi, M.; Pereira, L.; Davda, J.N.; Cromer, L.; Wang, L.; Nogue, F.; Chan, W.L.; Siddiqi, I.; et al. Synthetic clonal reproduction through seeds. *Science* **2011**, *331*, 876. [CrossRef]
192. Wang, N.; Gent, J.I.; Dawe, R.K. Haploid induction by a maize *cenh3* null mutant. *Sci. Adv.* **2021**, *7*, eabe2299. [CrossRef]
193. Lv, J.; Yu, K.; Wei, J.; Gui, H.; Liu, C.; Liang, D.; Wang, Y.; Zhou, H.; Carlin, R.; Rich, R. Generation of paternal haploids in wheat by genome editing of the centromeric histone CENH3. *Nat. Biotechnol* **2020**, *38*, 1397–1401. [CrossRef] [PubMed]
194. Sanei, M.; Pickering, R.; Kumke, K.; Nasuda, S.; Houben, A. Loss of centromeric histone H3 (CENH3) from centromeres precedes uniparental chromosome elimination in interspecific barley hybrids. *Proc. Natl. Acad. Sci. USA* **2011**, *108*, E498–E505. [CrossRef] [PubMed]

current issues in molecular biology

MDPI

Article

Transcriptome Analysis and VIGS Identification of Key Genes Regulating Citric Acid Metabolism in Citrus

Tianxin Chen, Juan Niu, Zhimin Sun, Jing Chen, Yue Wang, Jianhua Chen * and Mingbao Luan *

Institute of Bast Fiber Crops, Chinese Academy of Agricultural Sciences, Changsha 410205, China
* Correspondence: chenjianhua@caas.cn (J.C.); luanmingbao@caas.cn (M.L.)

Abstract: Citrus (*Citrus reticulata*) is one of the world's most widely planted and highest-yielding fruit trees. Citrus fruits are rich in a variety of nutrients. The content of citric acid plays a decisive role in the flavor quality of the fruit. There is a high organic acid content in early-maturing and extra-precocious citrus varieties. Reducing the amount of organic acid after fruit ripening is significant to the citrus industry. In this study, we selected a low-acid variety, "DF4", and a high-acid variety, "WZ", as research materials. Through WGCNA analysis, two differentially expressed genes, citrate synthase (CS) and ATP citrate-pro-S-lyase (ACL), were screened out, which related to the changing citric acid. The two differentially expressed genes were preliminarily verified by constructing a virus-induced gene-silencing (VIGS) vector. The VIGS results showed that the citric acid content was negatively correlated with CS expression and positively correlated with ACL expression, while CS and ACL oppositely control citric acid and inversely regulate each other. These results provide a theoretical basis for promoting the breeding of early-maturing and low-acid citrus varieties.

Keywords: citrus; citric acid; VIGS; WGCNA

Citation: Chen, T.; Niu, J.; Sun, Z.; Chen, J.; Wang, Y.; Chen, J.; Luan, M. Transcriptome Analysis and VIGS Identification of Key Genes Regulating Citric Acid Metabolism in Citrus. *Curr. Issues Mol. Biol.* **2023**, *45*, 4647–4664. https://doi.org/10.3390/cimb45060295

Academic Editors: Angel Llamas and Quan Zou

Received: 27 March 2023
Revised: 17 May 2023
Accepted: 25 May 2023
Published: 28 May 2023

1. Introduction

Citrus (*Citrus reticulata*) is one of the world's most widely planted and highest-yielding fruit trees. Citrus fruits are rich in various nutrients, such as carbohydrates, vitamins, antioxidants, and mineral elements [1]. The flavor of citrus fruits is the main selection criterion for the consumer market. Organic acids, soluble sugars, and aromatic substances are the key factors forming the fruit flavor quality. The sugar and acid content of the fruit, especially the level and proportion of sugar and acid components, ultimately determine the flavor of the fruit [2]. With the development of the citrus industry, some problems have gradually appeared in the structure of the citrus industry. Currently, the main citrus varieties in China are mainly mid-maturing, saturating the market from November to December. A few extra-precocious and early-maturing varieties mature before October, and the late-maturing varieties ripen from February to September the following year [3]. There is a problem with the high organic acid content in early-maturing and extra-precocious citrus varieties, which lengthens the fruit storage cycle and lags behind the marketing time of citrus. Therefore, reducing the content of organic acids after fruit ripening is significant when adjusting the citrus market structure.

There are many kinds of organic acids in fruits, which vary from fruit to fruit. However, most fruits contain one or two main organic acids, and a few others exist in small amounts. According to the types of organic acids that are mainly formed and accumulated during fruit growth and development, fruits can be divided into three acid types: malic acid, tartaric, and citric acid [4]. Citrus fruit is a typical citric acid fruit. Yamaki et al. determined 48 citrus varieties regarding the type and content of organic acid. The results showed that citric acid had the highest content in most citrus varieties, at a rate of about 75.40%, comprising 96.90% of the total organic acid content, followed by malic acid [5]. The changes in the accumulation of organic acids in different varieties or the growth and development stages of the same type are inconsistent, and the opposite situation can occur [6]. As a typical citric acid fruit, the change in organic acid content in fruit mainly depends on the level of synthesis and accumulation of citric acid. For most citrus fruits, rapid synthesis and the accumulation of citric acid occur at the early stage of fruit development, reaching a relatively high level. However, with the gradual ripening of the fruit, the citric acid content shows a gradual downward trend [7]. Citric acid accumulation in citrus fruits is a complex process, and its content is affected by citric acid synthesis, degradation, storage, transport, and other aspects. Previous studies and reports have agreed on the metabolic pathway, accumulation mode, and theory of citric acid in citrus [8]. However, the key metabolic pathways affecting citric acid content in fruits need further research and verification [9]. Many studies on the decomposition and consumption of citric acid in citrus fruits are available, and most reports support the idea that citric acid decomposition and utilization directly affect citric acid content. In addition to the classical tricarboxylic acid cycle (TCA) pathway, the Gamma-aminobutyric acid (GABA) pathway and citric acid cleavage pathway have also been reported to be involved in regulating citric acid content in citrus fruits, but the main regulatory pathways or regulatory genes in these pathways need to be further clarified [10]. Gene differences between varieties can be assessed by transcriptome sequencing.

Virus-induced gene silencing (VIGS) is a plant RNA-silencing technique that uses viral vectors carrying the target gene fragment to induce endogenous gene silencing. The function of the target genes can be assessed according to the phenotypic variation [11]. Compared with the traditional gene function analysis methods, VIGS can silence and analyze the target genes in infected plants, avoid plant genetic transformation, play a role in different genetic backgrounds, and analyze the gene function more thoroughly [12]. The technical operation is simple, and the research cycle is short. Generally, one test cycle can be completed in 1–3 weeks. This has been used to study the functional genes of disease resistance, growth and development, and metabolic regulation in tobacco, tomato, and other plants [13–15]. It is often used in citrus leaves but rarely used in fruits.

In this study, RNA sequencing data were used to reveal the key differential genes regulating the citric acid metabolism pathway between Dafen 4 (low-acid variety) and Weizhang (high-acid variety) at 66–136 days after flowering in six developing stages. In addition, we preliminarily verified the key differential genes using the VIGS technique. This study aims to provide a theoretical basis for understanding the molecular mechanism of citric acid metabolism.

2. Materials and Methods

2.1. Plant Material

We selected two kinds of citrus materials for this experiment: low-acid-content citrus (DF4) and high-acid-content citrus (WZ). The details of the plant materials were explained in our previous report [16,17]. The plants were cultivated in the experimental field of the Institute of Bast Fiber Crops, Chinese Academy of Agricultural Sciences (Yuanjiang, China). Organic acid metabolic and variations in sugar content were evaluated, and the sample collection time ranged from July to September 2021. The fruit was harvested on the 66 DAF (the day after flowering) and then collected every 14 days until 136 DAF. During the collection, four fruits located in the northwest and southeast were collected as a sample, and each sample had three biological repeats. The collected part was put into liquid nitrogen and stored in an ultra-low freezer (−80 °C) for transcriptome sequencing, and the other part was used for quality analysis.

2.2. Measurement of Total of Soluble Sugar (TSS) Titratable Acid (TA) in Citrus Flesh

After being mixed well, a refractometer was used to measure TSS in the collected samples. TA was measured using titrimetric analysis, using 0.1 mol/L NaOH for neutralization [18].

2.3. Quantification of Organic Acids and Soluble Sugars in Citrus Fruit Flesh

The citrus pulp was ground with liquid nitrogen and mixed with 80% ethanol for ultrasonic extraction. The citric acid and malic acid contents were determined using high-performance liquid chromatography (HPLC, Agilent 1260, Santa Clara, CA, USA) with 2% methanol and 98% K_2HPO_4 used as the mobile phase and 0.6 mL/min as the flow rate. The glucose, fructose, and sucrose content were determined with the anthrone sulfuric acid method and spectrophotometer (SV, MAPADA UV-3000, Beijing, China).

2.4. RNA Extraction and Transcriptome Sequencing Data Comparison

Total RNA was extracted from the six stages of development (66, 80, 94, 108, 122, and 136) DAF using the RNA prep Pure Plant kit (Tiangen, Beijing, China). Then, using NanoDrop Agilent 2100 bioanalyzer (Thermo Fisher Scientific, Waltham, MA, USA), the purity of the RNA sequencing of transcriptome library through Huada Gene company (Illumina, Beijing, China) was measured. Then, the remaining clean reads were aligned by reference Citrus clementina genome (Citrus clementina v1.0) using HISAT2 (v2.1.0) [19], from which unigenes were obtained.

2.5. Analysis and Enrichment of Differential Expression Genes

The differentially expressed genes (DEGs) were identified with the DESeq2 R package [20]; statistical significance was defined using Benjamini–Hochberg-adjusted p-values < 0.05. After the pairwise comparison of different varieties in the same period and different periods of the same variety, the differentially expressed genes were obtained. GO, and KEGG enrichment was performed with the Clusterprofiler R package [21].

2.6. WGCNA for Identifying Hub Genes and Weight Modules

Weighted gene co-expression network analysis (WGCNA) was performed in R to rank genes into coexpressed modules. FPKM values were normalized, and a nearness matrix was made. The quality data and FPKM were imported into R, and correlation-based associations between quality data and factor modules were calculated [22,23]. After the structure topological overlap matrix (TOM), transcripts with the same expression pattern were divided into a module, and the characteristic genes of these modules were calculated together.

2.7. Validation of Intramodular Candidates through RT-qPCR Analysis

Quantitative reverse transcription-PCR (RT-qPCR) was employed to validate the RNAseq data. Total RNA extracted from 36 samples was reverse-transcribed to cDNA using the PrimeScript RT Master Mix for qPCR (Takara, Dalian, China). The sequences of specific primers for the internal control and differentially expressed genes are shown in Table S1. Relative expression levels were calculated using the method used in the previous report [24].

2.8. Construction of the VIGS Vector

Using the candidate gene coding frame (300–500 bp) and the polyclonal restriction site of the VIGS vector with Tobacco rattle virus (TRV), primers were designed, and the recombinant DNA was constructed with cDNA as a template. The recombinant plasmids DNA and TRV2 vector were digested with a restriction endonuclease. The target fragments were recovered, ligated with T4DNA ligase, and transformed into *E. coli* competent cells. The recombinant plasmids were screened with kanamycin (Kan^+). PCR and sequencing identified the transformation results. The correct plasmids were transformed into Agrobacterium tumefaciens competent cells. The recombinant plasmids were screened using a kanamycin (Kan^+) resistance/rifampicin (rif) plate [25]. The transformation results were identified by PCR and sequencing.

2.9. Infection in Citrus Fruit with VIGS Vector

The vector transferred into *Agrobacterium* was used for the expanded culture. When the OD value of Agrobacterium tumefaciens was 2.5–2.7, 10 mL of the bacterial solution was obtained; this was centrifuged at 4500 rpm speed for 10 min, and the supernatant was discarded. Then, the sediment was mixed by resuspending (10% MES,10% $MgCl_2$,80% ddH_2O). Then, 10 μL AS (10 mg/mL) was added, TRV1 and candidate genes were mixed at 1:1, and TRV1 and TRV2 1:1 were mixed and placed in an ice box. A 5 mL disposable syringe was used to inject the top of the fruit with *Agrobacterium* tumefaciens (the fruit can be injected in multiple locations). Three biological repeats were selected for each variety, and two fruits were chosen for injection at two locations for each repetition [26]. After infection for 14 days, fruits were collected for further processing.

2.10. Validation of Quality Traits and VIGS Infection

The methods described above were used to determine the TA, citric acid, and malic acid after fruit picking. Similarly, the above-described method was used to extract RNA and reverse-transcribe it into cDNA. All the cDNAs were verified by 1% agarose gel electrophoresis with the VIGS primers. The correct cDNA was validated using RT-qPCR.

3. Results

3.1. Variations among TSS, TA, and TSS/TA during Citrus Fruit Flesh Development

During fruit development, TSS, TA, and TSS/TA can reflect the flavor quality. Figure 1A showed that, with changes in the developmental stage, the TSS of DF4 and WZ showed an overall upward trend, and both reached the maximum value at 136 DAF, but there was no significant difference between the varieties. In Figure 1B, changes in the developmental stage showed that the TA of DF4 and WZ initially increased and then decreased. The TA inflection point of DF4 was 14 days earlier than that of WZ, and the TA of DF4 was significantly lower than that of WZ. Figure 1C showed that the TSS/TA of the two varieties indicated a gradual upward trend with the developmental stage. TSS/TA in DF4 increased significantly more after 108 DAF than in WZ. The results showed that the main factor affecting quality traits in the two varieties was TA, and the TSS had no significant effect on quality traits.

3.2. Variations of Minalic Acid and Citric Acid during Citrus Fruit Flesh Development

Malic acid and citric acid mainly affected the acid content of the fruit. A significant difference was found in the citric acid content between DF4 and WZ. Figure 2A shows that the citric acid content of DF4 initially increased and then decreased as the developmental stages changed. The citric acid content of WZ was lower during early growth and higher in the later stage. As shown in Figure 2B, the malic acid content of the citrus fruits showed an overall downward trend as the plants developed. The malic acid content in WZ was higher in the early growth stage. Still, it decreased rapidly at 80 DAF, and there was a significant difference between periods. In comparison, the malic acid content in DF4 was lower in the early stage and slowly decreased at 94 DAF, and there was no significant difference between the periods. As shown in Figure 2C, citric acid, and malic acid are dominant. In the late developmental stage (108 DAF–136 DAF), the citric acid percentage of low-acid-variety DF4 was lower than that of high-acid-variety WZ, but the malic acid proportion of low-acid-variety DF4 was higher than that of high-acid-variety WZ.

3.3. Variations in Fructose, Sucrose, and Glucose during Citrus Fruit Flesh Development

The fructose content of the citrus fruit showed an upward trend as the plants developed (Figure 3A). The fructose content of the two varieties was lower in the early stage. The DF4 fructose content continued to increase at 94 DAF, while that of WZ continued to rise at 108 DAF. Figure 3B shows that, with changes in growth, the sucrose content of the citrus fruits showed an upward trend. The sucrose content of the two varieties remained low in the early stage and rapidly increased at 108 DAF. The sucrose content of DF4 and

WZ reached the maximum at 136 DAF. During the development period, the citrus fruit's sucrose content initially increased, then decreased and increased again (Figure 3C). The glucose content of the two varieties was lower in the early stage. DF4 glucose content decreased at 108 DAF and further increased at 122 DAF. The WZ glucose content decreased at 94 DAF and rose again at 108 DAF. There were significant differences in fructose at 108 DAF between the two varieties. With the continuous immaturity of fruit, the sugar content of the two varieties gradually increased, but there was no significant difference, which indicated that fructose, sucrose, and glucose were not the main factors leading to the difference between low-acid varieties and high-acid varieties.

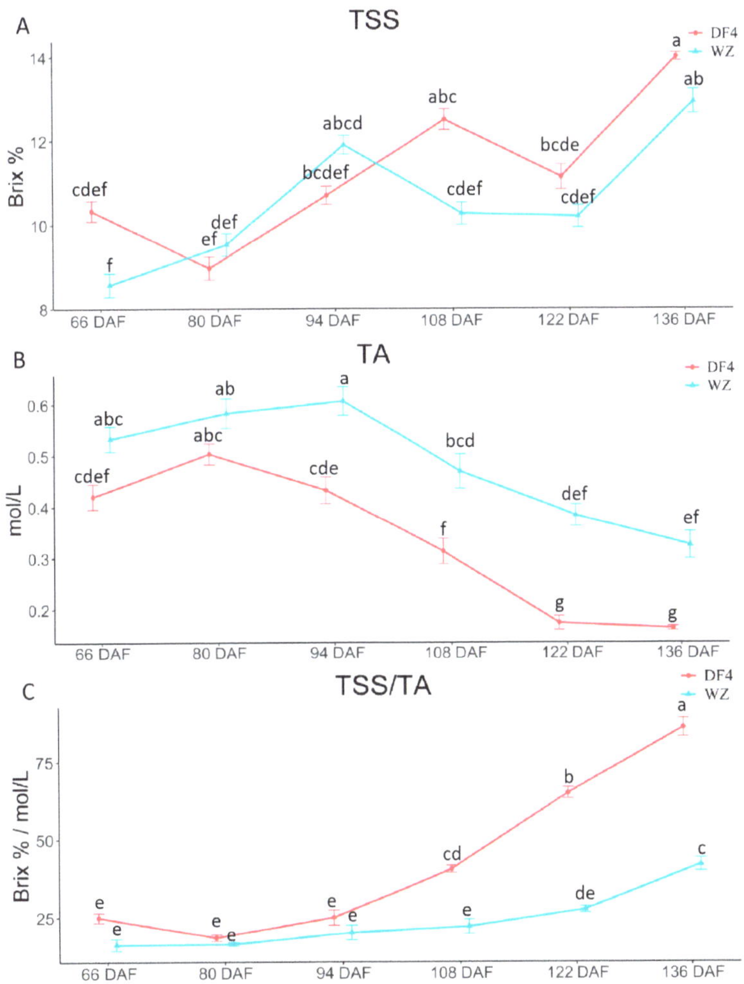

Figure 1. TSS (Brix %), TA (mol/L), and TSS/ TA in citrus fruit at 66–136 DAF. (**A**) TSS in citrus fruit; (**B**) TA in citrus fruit; (**C**) TSS/ TA in citrus fruit. "a, b, c, d, e, f" indicates the statistical significance of the data; "a" was the most significant. Other letters decrease in alphabetical order.

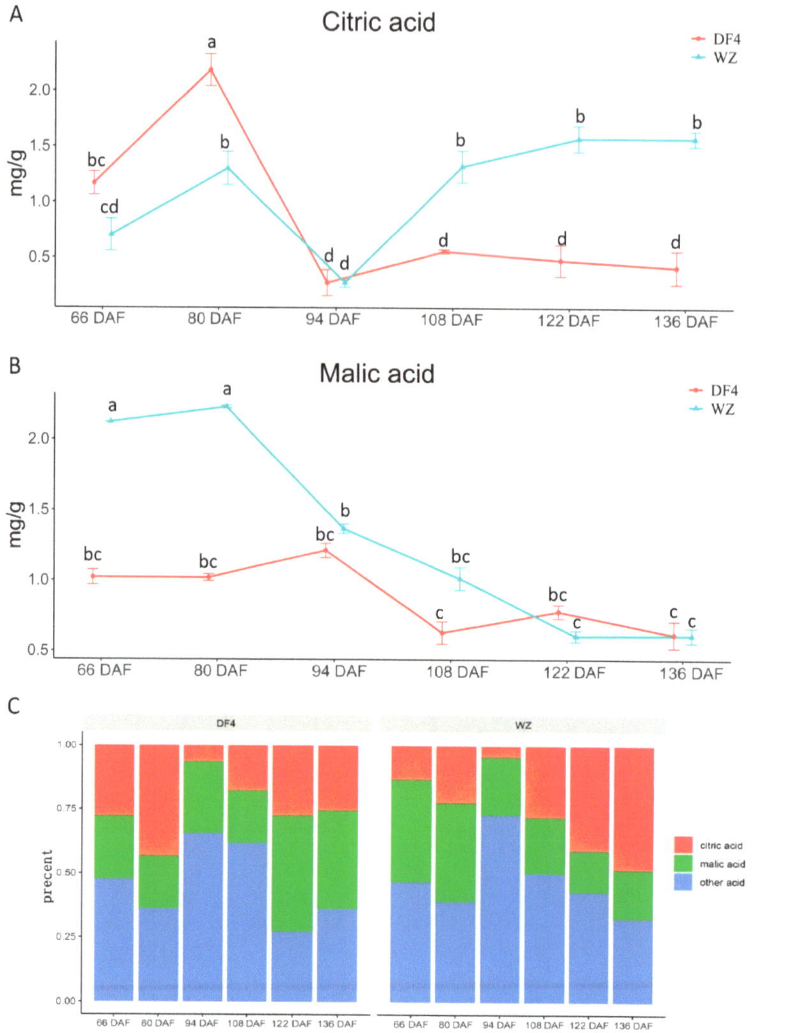

Figure 2. Malic acid and citric acid (mg/g) in citrus fruit at 66–136 DAF. (**A**) Malic acid in citrus fruit; (**B**) Citric acid in citrus fruit; (**C**) Malic acid and citric acid percentages of TA in citrus fruit. "a, b, c, d" indicates the statistical significance of the data; "a" was the most significant. Other letters decrease in alphabetical order.

3.4. Identification and Analysis of DEGs at Different Fruit Developmental Stages

After analyzing the citrus fruit quality traits, two varieties, DF4 and WZ, were selected, including six developmental stages, three biological repeats, and 36 samples for transcriptome sequencing. After removing the low-quality reads and joint sequences, 796 million clean reads were obtained, with a Q20 base of 95.75–96.67% and a GC content of 45.59–46.81% (Table S2). After using the hisat2 comparison and htseq-count quantification, 16310 gene read counts were obtained. As seen in Figure 4, a total of 7105 DEGs were identified using a pairwise comparison of different stages and the same stage in each sample.

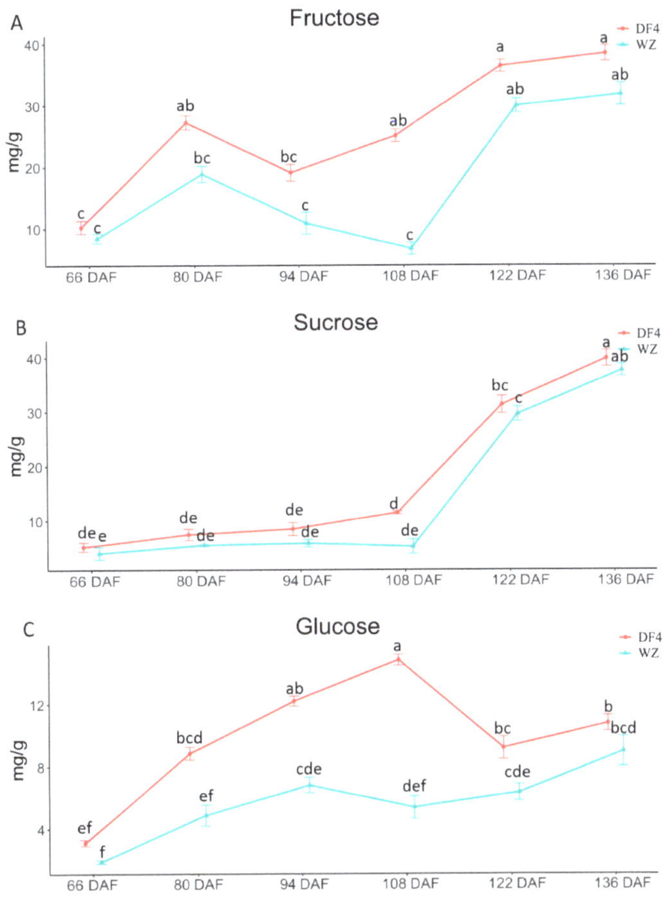

Figure 3. Fructose, sucrose, and glucose content (mg/g) in citrus fruit at 66–136 DAF. (**A**) Fructose in citrus fruit; (**B**) Sucrose in citrus fruit; (**C**) Glucose in citrus fruit. "a, b, c, d, e, f" indicates the statistical significance of the data; "a" was the most significant. Other letters decrease in alphabetical order.

3.5. Obtaining Hub Genes during Weighted Correlation Network Analysis

A correlation analysis was conducted between the 7105 DEGs obtained from the transcriptome data analysis and quality trait data. Figure 5 shows the regulatory genes related to quality traits, classified into several modules; different colors represent different modules. In Figure 6, the module MEtan is the weight module of the citric acid change, with a correlation between module MEtan and citric acid of higher than 0.7. Figure 7 shows that, after gene enrichment analysis, hub genes in the module MEtan were mainly enriched in terms of the organic acid metabolic process and the TCA cycle pathway. The hub genes that were enriched in the pathway were used as candidate genes for the next analysis. The main direction of this study was to mine the genes regulating citric acid metabolism. The tricarboxylic acid cycle pathway is the main pathway of citric acid metabolism; therefore, we screened the weight genes that were enriched in the tricarboxylic acid cycle. The screened genes are shown in Table S3. In Figure 8, the red box represents the candidate genes that were enriched in the tricarboxylic acid cycle. While enriching the candidate genes, we found that the candidate genes Ciclev10031423m.1 and Ciclev10031633m.1 are also enriched at the key enzymes of citrate metabolism (represented by the green module in Figure 8).

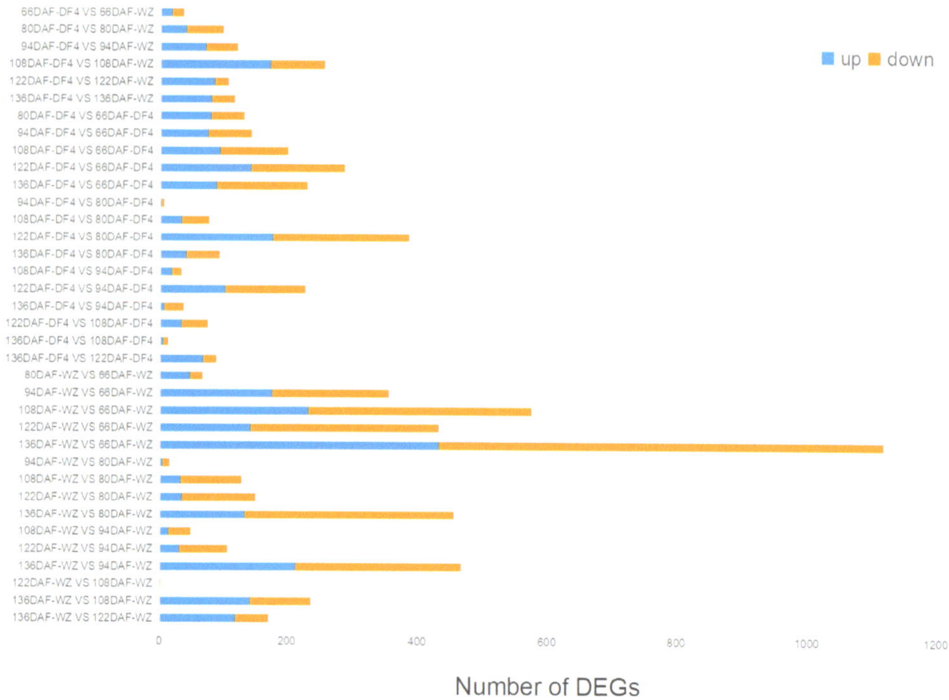

Figure 4. DEGs between six fruit developmental stages of the two samples. Pairwise comparisons of gene expression levels at different stages within each sample and gene expression levels at the same stage between the two samples.

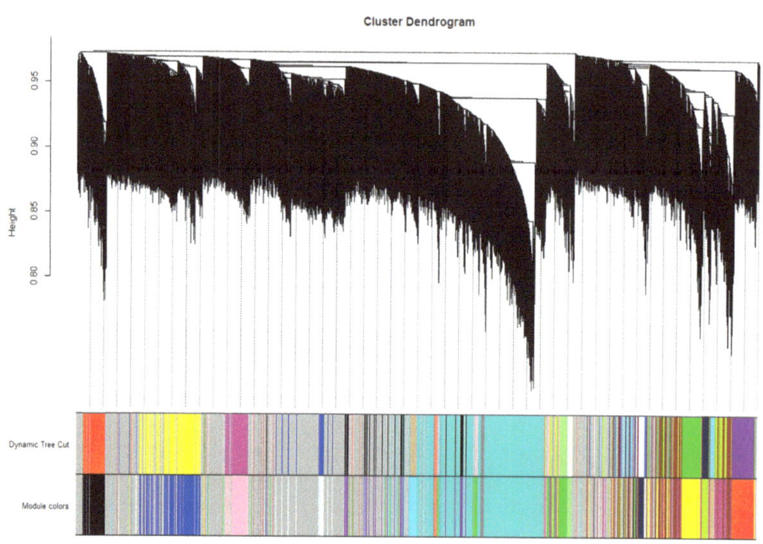

Figure 5. WGCNA quality traits Cluster Map. The result of quality data and cluster analysis, through which quality trait data were grouped into different color modules according to their clustering results as representatives.

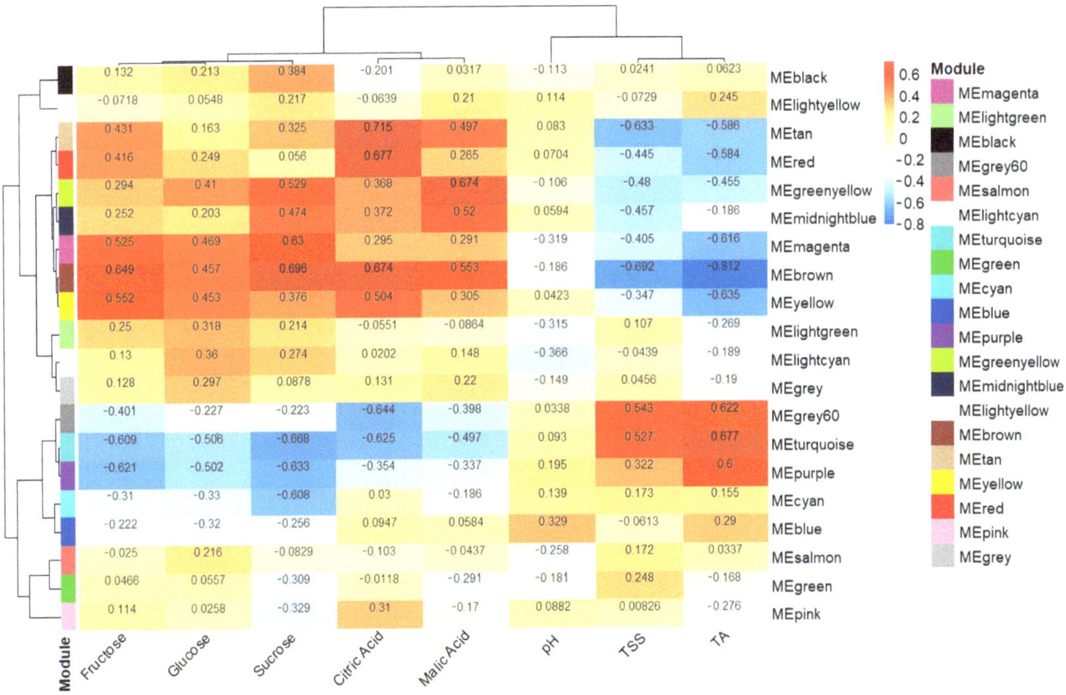

Figure 6. Heat map of correlation between WGCNA quality traits and differentially expressed genes. The module with the highest correlation degree was chosen (the color in the figure is red to indicate a correlation degree > 0.7). Figures on the heat map denote correlation and *p*-value between module and traits. In each module, the upper number represents correlation; the lower number represents *p*-value.

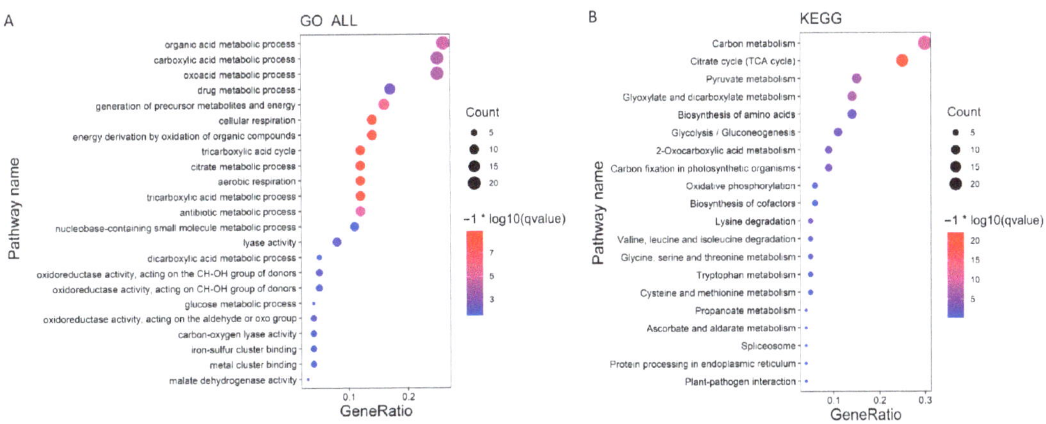

Figure 7. Module MEtan enrichment diagram. (**A**) GO enrichment analysis; (**B**) KEGG enrichment analysis. −1 * log10 means the log transformation of qvalue.

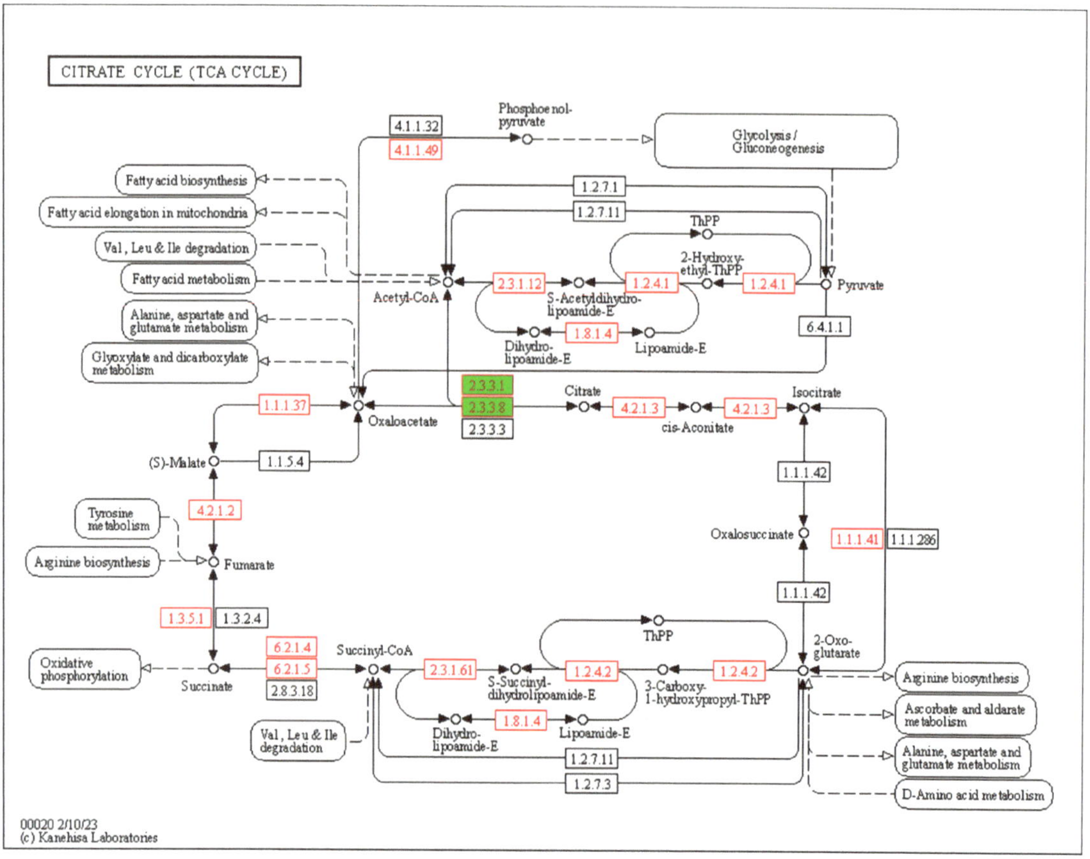

Figure 8. Candidate genes are involved in the TCA pathway metabolic processes.

3.6. Screening Candidate Genes via RT-qPCR

Bioinformatics analyses were used to enrichen differentially expressed genes in the module MEtan, which directly regulates citric acid content in the tricarboxylic acid cycle. RT-qPCR primers were designed for the differentially expressed genes and verified with RT-qPCR. After combining these with quality trait data, it was found that the RT-qPCR results of two differentially expressed genes, Ciclev10031423m.1 (CS:citrate synthase) and Ciclev10031633m.1 (ACL:ATP citrate (pro-S)-lyase), were consistent with the changing trend of FPKM. The correlation between relative expression value and FPKM was evaluated by calculating the Pearson correlation coefficient (PCC). The PCC values of these two genes were higher than 0.7. The gene expression trends of RNA-seq and RT-qPCR results were highly consistent (see Figure 9). These two genes (CS and ACL) were used as candidate genes for further analysis.

3.7. Construction of VIGS Vector and Verification of Impregnation Result

The PCR products were verified with 1% agarose gel electrophoresis. Figure 10 shows that CS and ACL were successfully amplified in DF4 and WZ with lengths of 1407 bp and 1272 bp, respectively. In this figure, ACL-DF4 is Thr, and ACL-WZ is Met. VIGS vectors were detected in all the infected fruits, indicating that VIGS vectors were successfully transferred into citrus fruits (Figure 11). After comparing the amplified sequences of the two varieties, Figure 12A shows that CS has a single base mutation at 473 bp, which changes the amino acid sequence. In this figure, CS-DF4 is Val, and CS-WZ is Ala. Figure 12B shows that ACL has a single base mutation at 863 bp, which changes the amino acid sequence.

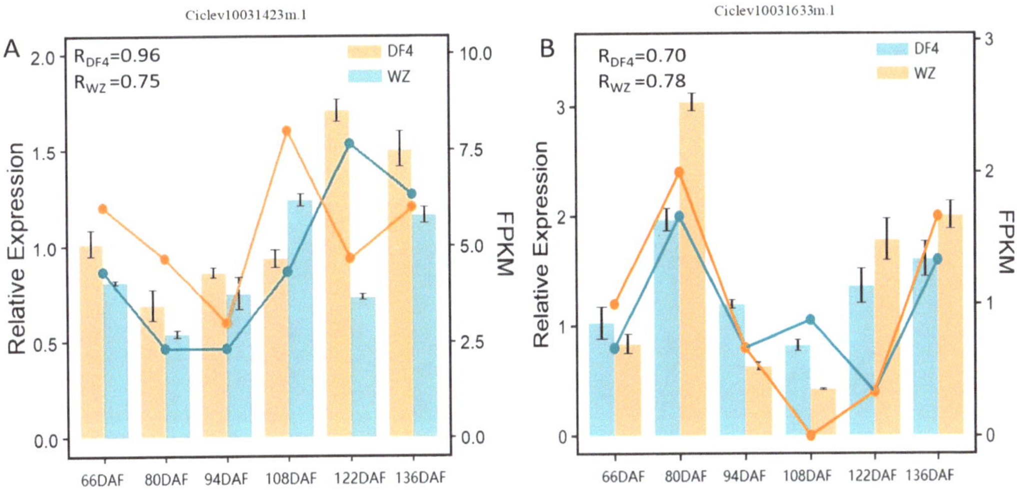

Figure 9. RT-qPCR verification and FPKM results of candidate expression genes. R means Pearson correlation coefficient (PCC) between relative expression and FPKM. (**A**) RT-qPCR verification and FPKM results of CICLE_v10031423mg (CS); (**B**) RT-qPCR verification and FPKM results of CICLE_v10031633mg (ACL). The bar graph shows gene RT-qPCR values, and the line graph shows the gene FPKM.

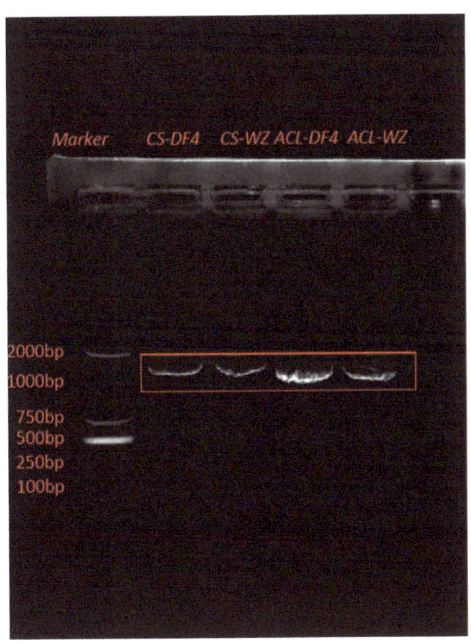

Figure 10. PCR verification with CS and ACL. The red box indicated correct site.

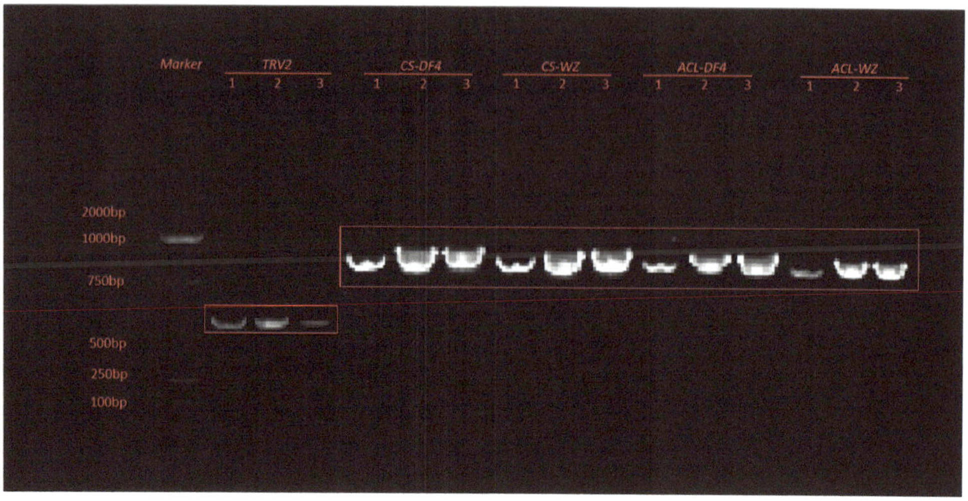

Figure 11. PCR verification with TRV2, VIGS-CS, and VIGS-ACL. The red box indicated correct site.

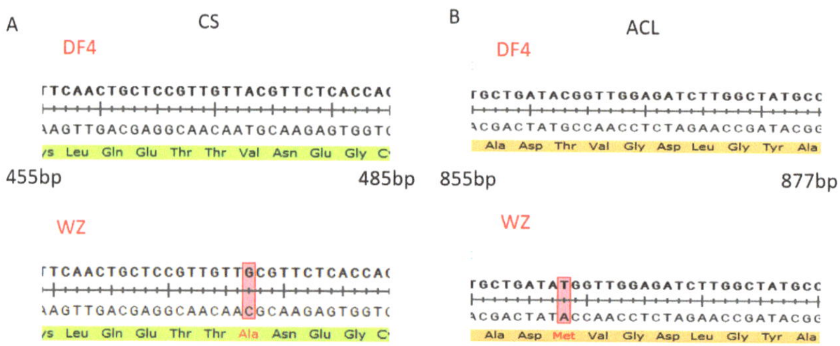

Figure 12. Sequence alignment between DF4 and WZ in CS and ACL. (**A**) Sequence alignment between DF4 and WZ in CS; (**B**) Sequence alignment between DF4 and WZ in ACL. The red box indicated the location of the base mutation.

3.8. Effect of VIGS Infection on the Content of RT-qPCR Citric Acid and TA

Both the RT-qPCR of CS and ACL were measured in VIGS-CS and VIGS-ACL fruits. Figure 13A showed that, after VIGS insertion in DF4, the expression of CS and ACL decreased compared with the control group, which indicated that VIGS had an effect. The expression of ACL increased after silencing CS, and after silencing ACL, the expression of CS significantly increased. Figure 13B shows that, after VIGS insertion in WZ, the two varieties showed the same trend. This result suggests that CS and ACL may influence each other. Figure 14A shows that silencing CS increases TA compared with the control TRV2 while silencing ACL decreases TA. Compared with the control TRV2, silencing CS increased citric acid content, while silencing ACL decreased citric acid content (Figure 14B). Both DF4 and WZ show the same trend. In sum, CS reversely regulates citric acid content for TA, while ACL positively regulates TA and citric acid content.

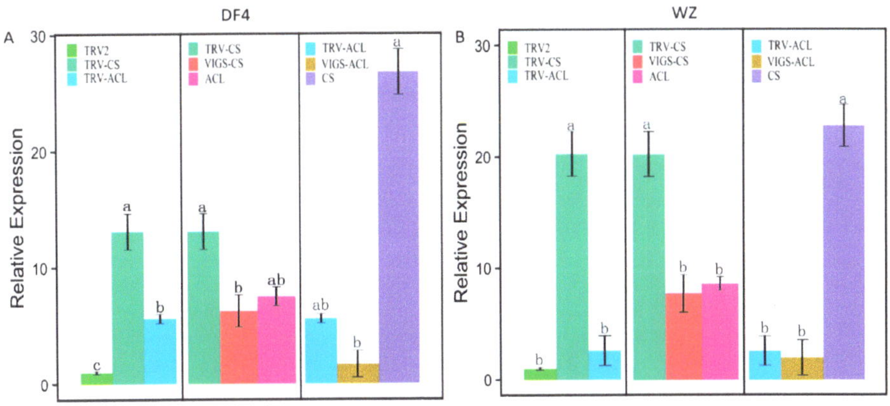

Figure 13. RT-qPCR verification of candidate gene expression with VIGS. (**A**) RT-qPCR verification of candidate gene expression in DF4; (**B**) RT-qPCR verification of candidate gene expression in WZ. TRV2 (Empty vector as a control). TRV-CS/TRV-ACL (the amount of CS/ACL gene expression in fruits without VIGS infection as a control). VIGS-CS/VIGS-ACL (the amount of CS/ACL gene expression after fruit infection). ACL (the expression of ACL gene in fruits silenced with CS). CS (the expression of CS gene in fruits with silenced ACL). "a, b, c" indicates the statistical significance of the data; "a" was the most significant. Other letters decrease in alphabetical order.

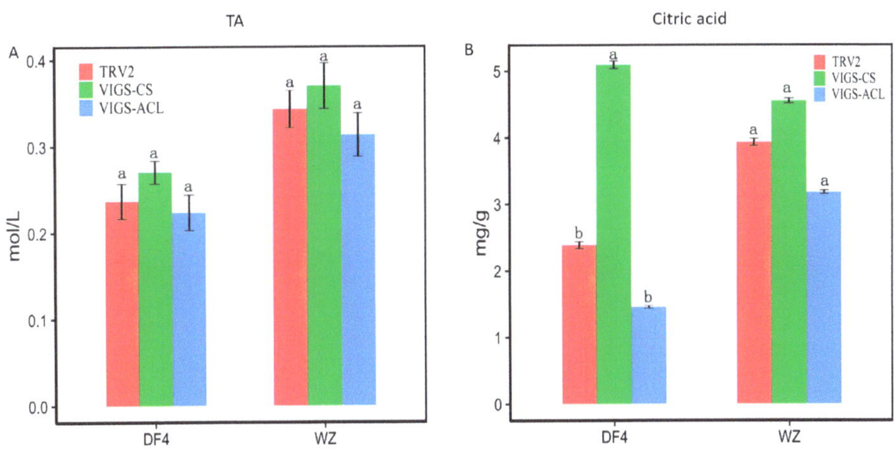

Figure 14. TA (mol/L) and Citric acid (mg/g) in citrus fruit with VIGS. (**A**) TA in citrus fruit; (**B**) Citric acid in citrus fruit. "a, b" indicates the statistical significance of the data; "a" was the most significant. Other letters decrease in alphabetical order.

4. Discussion

Soluble sugars in citrus mainly exist as fructose, sucrose, and glucose, among which the sucrose content is the highest [27,28]. Studies have shown that the ratio of sucrose, glucose, and fructose in some early- and medium-maturing citrus fruits is mainly 2:1:11 [29]. Related studies showed no significant difference in sugar content between wild and cultivated citrus fruits, indicating no substantial change in the soluble sugar content during citrus domestication [30]. During the development of the citrus fruit, the content of soluble sugar in the juice sacs gradually accumulated with fruit development, and the accumulation rate of sucrose was higher than that of glucose and fructose [31]. Great differences in soluble sugar content among different citrus varieties have been reported [32]. In this study, the soluble sugar content of DF4 and WZ gradually increased during the growth period, consistent with previous studies. At 150 DAF, the proportion of fructose, sucrose, and glucose in DF4 is about 3%, 4%, and 1%. In WZ, the ratio of fructose, sucrose, and glucose is about 4.5%, 5.3%, and 5.3%. Compared with previous studies, the ratio of the glucose content is relatively consistent, but the ratio of fructose to sucrose is different. This showed differences in the content and proportion of soluble sugars in citrus fruits among different varieties and growth stages.

While the content of soluble sugars in citrus fruit is high, the content of organic acid in the fruit has become the key factor affecting the quality of citrus fruit. In the early-ripening variety Wenzhou mandarin, citric and malic acid increased at the same rate in the initial stage of fruit growth and development, and their contents were almost the same. However, with fruit growth, the content of citric acid rapidly increased, reaching its peak in September and then gradually decreasing; its accumulation was much higher than that of malic acid [33]. The research found that the organic acid content of Clemens citrus reached its peak 137 days after flowering and then gradually decreased [34]. However, some studies have found that the change in the organic acid content of some citrus fruits differs from that of most citrus fruits with the growth and development of the fruits, but this showed the opposite trend. The organic acid content gradually increased or maintained a low level during the whole process of growth and development. For example, the citric acid content in high-acid citrus and high-acid lemon fruits gradually increased during growth and development. The citric acid content of high-acid lemon peaked 150 days after flowering, and the citric acid content of high-acid lemon reached a peak 100 days after flowering. The citric acid content of both remained at a high level during growth and development. The citric acid in acid-free citrus and acid-free lemon fruits remained stable at a low level during the growth and development process [35]. In fruit growth, development, and ripening, the citric acid content of red willow, acid-free pomelo, and other varieties remained at a low level [36]. In this study, DF4 and WZ reached their inflection point at 80 DAF and their minimum at 94 DAF. Overall, a trend of higher acid content in the early stage and lower acid content in the later stage was maintained, which was consistent with previous studies, but the acid reduction period was relatively early compared with other varieties. This showed that the acid reduction inflection point of citrus fruit was substantially different among varieties.

The citric acid in citrus fruit is the intermediate product of the TCA cycle. Studies have shown that citrus fruits can independently carry out photosynthesis before changing color [37], which can provide substrates for the tricarboxylic acid cycle. On the other hand, studies have shown that citrus leaves affect citric acid accumulation in fruits [38], but there is little direct evidence for citric acid's transport from roots or leaves into fruits. Citrate synthase (CS) is one of the most important enzymes in the tricarboxylic acid cycle, which catalyzes the combination of oxaloacetic acid (OAA) and acetyl-coenzyme A (acetyl-CoA) to form citric acid [39]. Early studies showed that spraying arsenate inhibited CS activity and decreased citric acid content in citrus fruits, indicating that CS plays an important role in citric acid accumulation in fruits [40]. However, many reports showed no significant correlation between CS and citric acid accumulation. Shi et al. [41] studied the relationship between CS gene expression and pomelo fruit acid content and concluded that it had no significant correlation with the high- and low-citric-acid phenotypes of pomelo fruit. The AN analysis of the gene expression and enzyme activity of three fruits, sweet lemon, sour lemon, and "Shamoti sweet orange", with different acidity levels also showed that the difference in citric acid content was not directly related to the change in CS [42]. Some reports showed that spray phosphate reduced CS activity and, thus, citrate content. The study of Chen et al. [43] focusing on lemon, Jin orange, Bingtang orange, and Fengjie navel orange also showed no significant relationship between a change in CS activity and the difference in citric acid content in different types of citrus fruits. Therefore, although CS is an important enzyme directly involved in citric acid synthesis, its role in citric acid accumulation in fruit remains unclear. The Ciclev10031423m.1 (CS) gene, regulating CS, was identified in this study. The expression of Ciclev10031423m.1 (CS) in DF4 and WZ initially decreased and then increased with the growth period, contrary to the citric acid content trend. To some extent, CS may reverse regulate the activity of the CS enzyme, thus affecting the change in citric acid content in citrus fruit. ATP-citrate lyase (ACL) exists in the cytoplasm, catalyzes the decomposition of citrate into acetyl-CoA and oxaloacetic acid, and also affects the accumulation of citric acid [44]. Katz et al. [45] reported that ACL was involved in citric acid cleavage during fruit-ripening in Washington navel oranges, while research showed that the downregulation of ACL gene expression could lead to citric acid accumulation, and its expression had slightly different effects on the citric acid content in different citrus fruits [46]. The Ciclev10031633m.1 (ACL) gene regulating ACL was also identified in this study. The expression of Ciclev10031633m.1 (ACL) in DF4 and WZ first increased and then decreased with the growth period, which was the same as the changing trend of citric acid content. This indicates that ACL may positively regulate the activity of ACL enzyme, thus affecting the change in citric acid content in citrus fruit.

Currently, the establishment of the VIGS system is reported more in citrus leaves than in citrus fruits. In this study, VIGS vectors for two candidate genes were constructed and successfully transferred into fruits. After silencing, the expression of the two candidate genes was downregulated, which confirmed the reliability of VIGS. Compared with the control group, the changes in the TA and citric acid content of TRV2, CS, and ACL silencing were opposite. The results further verified the regulatory effect of candidate genes on citric acid. It was found that the expression of ACL increased in CS silencing, and the expression of CS increased in ACL silencing and was consistent in DF4 and WZ. These results indicated that the expression of CS and ACL might influence and regulate each other in reverse.

At present, the main citrus varieties are still mid-maturing varieties from November to December, and relatively few early-maturing citrus varieties mature before October. In addition, the phenomenon of high citric acid content is common in early-maturing varieties, so the breeding of low-acid and early-maturing citrus varieties can not only meet the needs of consumers but also expand and regulate the citrus market and promote the healthy development of the citrus industry [47–49]. The results of this study provide theoretical support for the breeding of low-acid and early-maturing citrus varieties and the development of molecular markers.

5. Conclusions

In this study, two differentially expressed genes related to the change in citric acid content using transcriptome and quality trait data were screened. The two differentially expressed genes of CS and ACL were preliminarily verified by constructing a VIGS vector. The results showed that the two differentially expressed genes could regulate the change in citric acid content. Through VIGS, we found that CS and ACL oppositely regulated citric acid and controlled each other inversely. In this study, candidate genes for regulating citric acid content during citrus fruit development were identified and preliminarily verified via VIGS. These results provide a theoretical basis for promoting the breeding of early-maturing and low-acid citrus varieties.

Supplementary Materials: The following supporting information can be downloaded at: https://www.mdpi.com/article/10.3390/cimb45060295/s1, Table S1: Primers for candidate genes of RT-qPCR; Table S2: Quality control results of transcriptome sequencing machine data; Table S3: Candidate genes in the TCA cycle.

Author Contributions: Conceived the study and managed the projects T.C., J.C. (Jianhua Chen) and M.L.; participated in sample collection and test guidance, Z.S., J.C. (Jing Chen), J.N. and Y.W.; performed data analysis and wrote a revised article, T.C. All authors have read and agreed to the published version of the manuscript.

Funding: This work was financially supported by the Central Public-Interest Scientific Institution Basal Research Fund, grant number 1610242020010; the Agricultural Science and Technology Innovation Program (ASTIP) of CAAS, grant number 2017IBFC.

Institutional Review Board Statement: Not applicable.

Informed Consent Statement: Not applicable.

Data Availability Statement: All data needed to evaluate the conclusions in this paper are present in the paper and/or the Supplementary Materials.

Acknowledgments: This work was financially supported by the Central Public-Interest Scientific Institution Basal Research Fund, grant number 1610242020010; the Agricultural Science and Technology Innovation Program (ASTIP) of CAAS, grant number 2017IBFC.

Conflicts of Interest: The authors declare that they have no known competing financial interest or personal relationships that could have appeared to influence the work reported in this paper.

References

1. Xu, Q.; Chen, L.-L.; Ruan, X.; Chen, D.; Zhu, A.; Chen, C.; Bertrand, D.; Jiao, W.-B.; Hao, B.-H.; Lyon, M.P.; et al. The draft genome of sweet orange (*Citrus sinensis*). *Nat. Genet.* **2013**, *45*, 59–66. [CrossRef] [PubMed]
2. Kelebek, H.; Selli, S. Determination of volatile, phenolic, organic acid and sugar components in a Turkishcv Dortyol (*Citrus sinensis* L. Osbeck) orange juice. *Sci. Food Agric.* **2011**, *91*, 1855–1862. [CrossRef] [PubMed]
3. Goldenberg, L.; Yaniv, Y.; Porat, R.; Carmi, N. Mandarin fruit quality: A review. *J. Sci. Food Agric.* **2018**, *98*, 18–26. [CrossRef]
4. Wang, T.; Lin, M.; Feng, X.; Wang, P.; Cao, X.; Zhang, W. Established methods and comparison of 10 organic acids based on reversed phase chromatography and hydrophilic interaction chromatography. *CyTA J. Food* **2022**, *20*, 206–217. [CrossRef]
5. Yamaki, Y.T. Organic acids in the juice of citrus fruits. *J. Jpn. Soc. Hortic. Sci.* **1989**, *58*, 587–594. [CrossRef]
6. Lado, J.; Gambetta, G.; Zacarias, L. Key determinants of citrus fruit quality: Metabolites and main changes during maturation. *Sci. Hortic.* **2018**, *233*, 238–248. [CrossRef]
7. Emmanouilidou, M.G.; Kyriacou, M.C. Rootstock-modulated yield performance, fruit maturation and phytochemical quality of 'Lane Late' and 'Delta' sweet orange. *Sci. Hortic.* **2017**, *225*, 112–121. [CrossRef]
8. Hussain, S.B.; Shi, C.-Y.; Guo, L.-X.; Kamran, H.M.; Sadka, A.; Liu, Y.-Z. Recent Advances in the Regulation of Citric Acid Metabolism in Citrus Fruit. *Crit. Rev. Plant Sci.* **2017**, *36*, 241–256. [CrossRef]
9. Veldhuizen, M.G.; Siddique, A.; Rosenthal, S.; Marks, L.E. Interactions of Lemon, Sucrose and Citric Acid in Enhancing Citrus, Sweet and Sour Flavors. *Chem. Senses* **2018**, *43*, 17–26. [CrossRef]
10. Oliver, D.J.; Nikolau, B.J.; Wurtele, E.S. Acetyl-CoA-Life at the metabolic nexus. *Plant Sci.* **2009**, *176*, 597–601. [CrossRef]
11. Terrier, N.; Deguilloux, C.; Sauvage, F.X.; Martinoia, E.; Romieu, C. Proton pumps and anion transport in Vitis vinifera: The inorganic pyrophosphatase plays a predominant role in the energization of the tonoplast. *Plant Physiol. Biochem.* **1998**, *36*, 367–377. [CrossRef]
12. Lee, W.S.; Rudd, J.J.; Kanyuka, K. Virus induced gene silencing (VIGS) for functional analysis of wheat genes involved in *Zymoseptoria tritici* susceptibility and resistance. *Fungal. Genet. Biol.* **2015**, *79*, 84–88. [CrossRef] [PubMed]
13. Mustafa, R.; Shafiq, M.; Mansoor, S.; Briddon, R.W.; Scheffler, B.E.; Scheffler, J.; Amin, I. Virus-Induced Gene Silencing in Cultivated Cotton (*Gossypium* spp.) Using Tobacco Rattle Virus. *Mol. Biotechnol.* **2016**, *58*, 65–72. [CrossRef] [PubMed]
14. Orzaez, D.; Medina, A.; Torre, S.; Fernández-Moreno, J.P.; Rambla, J.L.; Fernández-Del-Carmen, A.; Butelli, E.; Martin, C.; Granell, A. A visual reporter system for virus-induced gene silencing in tomato fruit based on anthocyanin accumulation. *Plant Physiol.* **2009**, *150*, 1122–1134. [CrossRef] [PubMed]
15. Ramegowda, V.; Senthil-Kumar, M.; Udayakumar, M.; Mysore, K.S. A high-throughput virus-induced gene silencing protocol identifies genes involved in multi- stress tolerance. *BMC Plant Biol.* **2013**, *13*, 193–210. [CrossRef]
16. Renhui, D.; Ziniu, D.; Rencai, W.; Suming, D. Analysis of storage conditions of extra precocious Wenzhou mandarin Dafen 4. *Hunan Agric. Sci.* **2015**, *4*, 87–89, 92.
17. Deyu, L.; Jianzhong, Z.; Hong, C.; Wengjing, L.; Zuyong, X. Demonstration and Popularization of Wenzhou mandarin tail as Interstock High grafting Yi Red Orange. *Zhejiang Citrus* **2011**, *28*, 18–19.
18. Blanke, M. Non-invasive Assessment of Firmness and NIR Sugar (TSS) Measurement in Apple, Pear and Kiwi Fruit. *Erwerbs-Obstbau* **2013**, *55*, 19–24. [CrossRef]
19. Kim, D.; Paggi, J.M.; Park, C.; Bennett, C.; Salzberg, S.L. Graph-based genome alignment and genotyping with HISAT2 and HISAT-genotype. *Nat. Biotechnol.* **2019**, *37*, 907–915. [CrossRef] [PubMed]
20. Utturkar, S.; Dassanayake, A.; Nagaraju, S.; Brown, S.D. Bacterial Differential Expression Analysis Methods. In *Metabolic Pathway Engineering*; Himmel, M., Bomble, Y., Eds.; Methods in Molecular Biology; Humana: New York, NY, USA, 2020; Volume 96, pp. 89–112.
21. Yu, G.; He, Q. ReactomePA: An R/Bioconductor package for reactome pathway analysis and visualization. *Mol. BioSyst.* **2016**, *12*, 477–479. [CrossRef]

22. Umer, M.J.; Bin Safdar, L.; Gebremeskel, H.; Zhao, S.; Yuan, P.; Zhu, H.; Kaseb, M.O.; Anees, M.; Lu, X.; He, N.; et al. Identification of key gene networks controlling organic acid and sugar metabolism during watermelon fruit development by integrating metabolic phenotypes and gene expression profiles. *Hortic. Res.* **2020**, *7*, 193. [CrossRef]
23. Langfelder, P.; Horvath, S. WGCNA: An R package for weighted correlation network analysis. *BMC Bioinform.* **2008**, *9*, 559. [CrossRef] [PubMed]
24. Jawad, U.M.; Gao, L.; Gebremeskel, H.; Safdar, L.B.; Yuan, P.; Zhao, S.; Xuqiang, L.; Nan, H.; Hongju, Z.; Liu, W. Expression pattern of sugars and organic acids regulatory genes during watermelon fruit development. *Sci. Hortic.* **2020**, *265*, 109102.
25. Wang, F.; Xu, Y.; Liu, X.; Shen, W.; Zhu, S.; Zhao, X. Use of TRV-mediated VIGS for functional genomics research in citrus. *Plant Cell Tiss. Organ Cult.* **2019**, *139*, 609–613. [CrossRef]
26. Fantini, E.; Giuliano, G. Virus-Induced Gene Silencing as a Tool to Study Tomato Fruit Biochemistry. *Plant Signal. Transduct.* **2016**, *1363*, 65–78.
27. Avi, S.; Lyudmila, S.; Itzhak, K.; Eduardo, B. Primary Metabolism in Citrus Fruit as Affected by Its Unique Structure. *Front. Plant Sci.* **2019**, *10*, 1167.
28. Hussain, S.B.; Guo, L.X.; Shi, C.Y.; Khan, M.A.; Bai, Y.-X.; Du, W.; Liu, Y.-Z. Assessment of sugar and sugar accumulation-related gene expression profiles reveal new insight into the formation of low sugar accumulation trait in a sweet orange (*Citrus sinensis*) bud mutant. *Mol. Biol. Rep.* **2020**, *47*, 2781–2791. [CrossRef]
29. Pangbom, R.M. Relative taste of selected sugars and organic acids. *Food Sci.* **1963**, *28*, 726–733. [CrossRef]
30. Wang, L.; He, F.; Huang, Y.; He, J.; Yang, S.; Zeng, J.; Deng, C.; Jiang, X.; Fang, Y.; Wen, S.; et al. Genome of wild mandarin and domestication history of mandarin. *Mol. Plant* **2018**, *11*, 1024–1037. [CrossRef]
31. Yehudit, A.; Candace, H.H.; Sarah, J.; Melody, W.; Deborah, P.D. A membrane-associated form of sucrose synthase and its potential role in synthesis of cellulose and callose in plants. *Proc. Natl. Acad. Sci. USA* **1995**, *92*, 9353–9357.
32. Komatsu, A.; Takanokura, Y.; Moriguchi, T.; Omura, M.; Akihama, T. Differential expression of three sucrose-phosphate synthase isoforms during sucrose accumulation in citrus fruits (*Citrus unshiu* Marc.). *Plant Sci.* **1999**, *140*, 169–178. [CrossRef]
33. Albertini, V.; Carcouet, E.; Pailly, O.; Gambotti, C.; Luro, F.; Berti, L. Changes in organic acids and sugars during early stages of development of acidic and acidless citrus fruit. *Agric. Food Chem.* **2006**, *54*, 8335–8339. [CrossRef]
34. Shi, L.; Tu, B.P. Acetyl-CoA and the regulation of metabolism: Mechanisms and consequences. *Curr. Opin. Cell Biol.* **2015**, *33*, 125–131. [CrossRef]
35. Loka, D.A.; Oosterhuis, D.M.; Baxevanos, D.; Noulas, C.; Hu, W. Single and combined effects of heat and water stress and recovery on cotton (*Gossypium hirsutum* L.) leaf physiology and sucrose metabolism. *Plant Physiol. Biochem.* **2020**, *148*, 166–179. [CrossRef]
36. Ruan, Y.L. Sucrose Metabolism: Gateway to Diverse Carbon Use and Sugar Signaling. *Annu. Rev. Plant Biol.* **2014**, *65*, 33–67. [CrossRef]
37. Hiratsuka, S.; Suzuki, M.; Nishimura, H.; Nada, K. Fruit photosynthesis in Satsuma mandarin. *Plant Sci.* **2015**, *241*, 65–69. [CrossRef] [PubMed]
38. Lu, X.; Cao, X.; Li, F.; Li, J.; Xiong, J.; Long, G.; Cao, S.; Xie, S. Comparative transcriptome analysis reveals a global insight into molecular processes regulating citrate accumulation in sweet orange(*Citrus sinensis*). *Physiol. Plant.* **2016**, *158*, 463–482. [CrossRef] [PubMed]
39. Li, S.; Yin, X.; Xie, X.; Allan, A.C.; Ge, H.; Shen, S.; Chen, K.-S. The Citrus transcription factor, CitERF13,regulates citric acid accumulation via a protein-protein interaction with the vacuolar proton pump, CitVHA-c4. *Sci. Rep.* **2016**, *6*, 20151. [CrossRef]
40. Lin, Q.; Li, S.; Dong, W.; Feng, C.; Yin, X.; Xu, C.; Sun, C.; Chen, K. Involvement of CitCHX and CitDIC in developmental-related and postharvest-hot-air driven citrate degradation in citrus fruits. *PLoS ONE* **2015**, *10*, 94–110. [CrossRef]
41. Shi, C.; Song, R.; Hu, X.; Liu, X.; Jin, L.; Liu, Y. CitrusPH5-like H^+-ATPase genes: Identification and transcript analysis to investigate their possible relationship with citrate accumulation in fruits. *Front. Plant Sci.* **2015**, *6*, 135. [CrossRef]
42. Liu, X.; Hu, X.; Jin, L.; Shi, C.; Liu, Y.; Peng, S. Identification and transcript analysis of two glutamate decarboxylase genes, CsGAD1 and CsGAD2,reveal the strong relationship between CsGAD1 and citrate utilization in citrus fruit. *Mol. Biol. Rep.* **2014**, *41*, 6253–6262. [CrossRef] [PubMed]
43. Chen, M.; Xie, X.; Lin, Q.; Chen, J.; Grierson, D.; Yin, X.; Sun, C.; Chen, K. Differential expression of organic acid degradation-related genes during fruit development of Navel Oranges (*Citrus sinensis*) in two habitats. *Plant Mol. Biol. Rep.* **2013**, *31*, 1131–1140. [CrossRef]
44. Chyper, M.; Zaidi, N.; Smans, K. ATP-citrate lyase: A mini-review. *Biochem. Biophys. Res. Commun.* **2012**, *422*, 1–4. [CrossRef] [PubMed]
45. Katz, E.; Fon, M.; Lee, Y.; Phinney, B.S.; Sadka, A.; Blumwald, E. The citrus fruit proteome: Insights into citrus fruit metabolism. *Planta* **2007**, *226*, 989–1005. [CrossRef]
46. Rata, J.; Czak, R. Structure, function and regulation of the plant vacuolar H^+-translocating ATPase. *Biochim. Biophys. Acta Biomembr.* **2000**, *1465*, 17–36.
47. Usman, M.; Fatima, B. Mandarin (*Citrus reticulata* Blanco) Breeding. In *Advances in Plant Breeding Strategies: Fruits*; Al-Khayri, J., Jain, S., Johnson, D., Eds.; Springer: Cham, Germany, 2018.

48. Cuenca, J.; Garcia-Lor, A.; Navarro, L.; Aleza, P. Citrus Genetics and Breeding. In *Advances in Plant Breeding Strategies: Fruits*; Al-Khayri, J., Jain, S., Johnson, D., Eds.; Springer: Cham, Germany, 2018.
49. Khan, I.A.; Kender, W.J. Citrus breeding: Introduction and objectives. In *Citrus Genetics, Breeding and Biotechnology*; CABI Books; CABI International: Wallingford, UK, 2007.

current issues in molecular biology

MDPI

Article

Exploring Genetic and Epigenetic Changes in Lingonberry Using Molecular Markers: Implications for Clonal Propagation

Umanath Sharma [1,2], Arindam Sikdar [1,2], Abir U. Igamberdiev [1] and Samir C. Debnath [2,*]

1 Department of Biology, Memorial University of Newfoundland, 45 Arctic Avenue, St. John's, NL A1C 5S7, Canada; usharma@mun.ca (U.S.); asikdar@mun.ca (A.S.); igamberdiev@mun.ca (A.U.I.)
2 St. John's Research and Development Centre, Agriculture and Agri-Food Canada, 204 Brookfield Road, St. John's, NL A1E 0B2, Canada
* Correspondence: samir.debnath@agr.gc.ca

Abstract: Lingonberry (*Vaccinium vitis-idaea* L.) is an important and valuable horticultural crop due to its high antioxidant properties. Plant tissue culture is an advanced propagation system employed in horticultural crops. However, the progeny derived using this technique may not be true-to-type. In order to obtain the maximum return of any agricultural enterprise, uniformity of planting materials is necessary, which sometimes is not achieved due to genetic and epigenetic instabilities under in vitro culture. Therefore, we analyzed morphological traits and genetic and epigenetic variations under tissue-culture and greenhouse conditions in lingonberry using molecular markers. Leaf length and leaf width under greenhouse conditions and shoot number per explant, shoot height and shoot vigor under in vitro conditions were higher in hybrid H1 compared to the cultivar Erntedank. Clonal fidelity study using one expressed sequence tag (EST)—polymerase chain reaction (PCR), five EST—simple sequence repeat (SSR) and six genomic (G)—SSR markers revealed monomorphic bands in micropropagated shoots and plants in lingonberry hybrid H1 and cultivar Erntedank conforming genetic integrity. Epigenetic variation was studied by quantifying cytosine methylation using a methylation-sensitive amplification polymorphism (MSAP) technique. DNA methylation ranged from 32% in greenhouse-grown hybrid H1 to 44% in cultivar Erntedank under a tissue culture system. Although total methylation was higher in in vitro grown shoots, fully methylated bands were observed more in the greenhouse-grown plants. On the contrary, hemimethylated DNA bands were more prominent in tissue culture conditions as compared to the greenhouse-grown plants. The study conclude that lingonberry maintains its genetic integrity but undergoes variable epigenetic changes during in vitro and ex vitro conditions.

Keywords: DNA methylation; epigenetic variation; greenhouse-grown plants; in vitro culture; molecular markers; shoot proliferation

Citation: Sharma, U.; Sikdar, A.; Igamberdiev, A.U.; Debnath, S.C. Exploring Genetic and Epigenetic Changes in Lingonberry Using Molecular Markers: Implications for Clonal Propagation. *Curr. Issues Mol. Biol.* **2023**, *45*, 6296–6310. https://doi.org/10.3390/cimb45080397

Academic Editor: Shimeles Tilahun

Received: 9 June 2023
Revised: 20 July 2023
Accepted: 26 July 2023
Published: 28 July 2023

1. Introduction

Plants respond to changes in the environment by altering their growth, physiology and reproductive processes. The molecular basis of such changes is based on the alteration in the underlying DNA or its plastic modification, including DNA methylation [1]. Commercial micropropagation is performed in a unique optimized environment containing various inorganic nutrients and growth hormones, controlled light, humidity and osmotic conditions. However, such an artificial environment may create a stressful situation for the plant material, resulting in genetic or epigenetic changes [2]. Among epigenetic mechanisms, DNA methylation is the most important phenomenon affecting plant phenotype [3]. However, for the sake of uniformity, both genetic and epigenetic variations are not desired in the commercial micropropagation of plants, including that of lingonberry (*Vaccinium vitis-idaea* L., family *Ericaceae*).

Lingonberry is a berry fruit bearing shrub commonly found in the Northern Hemisphere [4,5]. The lingonberry plant bears red edible fruits; both fruits and leaves are rich in nutrients and bioactive compounds, such as sugars, organic acids, vitamins, minerals, dietary fiber, and polyphenolics [4,6]. While lingonberries are cultivated in some parts of the world, the majority of the fruits are collected from wild natural habitats and are consumed fresh and frozen. A substantial part of a harvest is processed into food products bearing longer shelf-life and pharmaceutical products [4]. Lingonberry possesses strong antiviral, antimicrobial, antioxidant, anti-inflammatory, and neuroprotective potential [7]. Therefore, eating fresh or processed lingonberries may reduce the risk or eliminate the development of gastrointestinal, metabolic, cardiovascular, renal, and neurodegenerative disorders [8].

The propagation of lingonberry is generally performed by vegetative methods using the rhizome because, being genetically heterozygous, progeny derived from lingonberry seeds are not true-to-type. Although vegetative propagation retains the genetic characteristics, this method is not economically viable in lingonberry due to its short life span and poor rhizome production [9]. Commercial production requires a large number of uniform plants, which can be achieved using in vitro propagation techniques. Propagation of lingonberry by tissue culture is much faster than traditional methods [10], but occasional variations in the tissue-cultured progeny, termed somaclonal variations, have been reported in several crops [11]. Therefore, we assessed the genetic and epigenetic stability of lingonberry in tissue culture medium. Genetic variation in tissue culture can arise due to point mutations, chromosomal rearrangements, relocation of mobile genetic elements, or changes in the ploidy level [12]. Although epigenetic mechanisms stabilize cell identity and maintain tissue organization, they entail a variety of reversible biochemical modifications that can occur on the underlying DNA, its interacting proteins, or both, modifying chromatin structure and resulting in an altered phenotype [13]. At the molecular level, such epigenetic phenomena are moderated by reversible mechanisms such as histone modifications, DNA methylation, and small RNAs, thus affecting the regulatory states of genes [13,14].

In plants, DNA methylation of cytosine base is a widespread epigenetic mechanism that contributes to the regulation of gene expression, maintenance of genomic integrity, cellular differentiation, and plant response to biotic and abiotic stresses [3,13,15]. DNA methylation also plays a vital role in many crucial biological processes, such as genomic imprinting, transposable element silencing, maintenance of heterochromatin, and inactivation of X-chromosome [16]. DNA methylation occurs when a methyl group is transferred from *S*-adenosyl methionine (SAM) to the fifth carbon of cytosine residue of DNA by DNA methyltransferases [1]. In plants, DNA methylation occurs at symmetric mCG and mCHG, or asymmetric mCHH contexts, where mC = methylated cytosine, and H = A, T or C [17]. In *Arabidopsis thaliana*, 55% of methylated cytosines are reported in CG sites, while cytosine methylation accounts for 23% and 22% at CHG and CHH sites, respectively [18]. The gene's function is reported to be affected by the position where the methylation of cytosines has occurred, for example, in the regions of transposons or the promoter regions of the gene [19]. Because of the heritable nature of variation, DNA methylation marks are useful in sexually and asexually propagated crops. Given the significant role that DNA methylation has in the regulation of gene expression [20], it is relevant to investigate how different growth conditions affect cytosine methylation.

Methylation-sensitive amplification polymorphism (MSAP) is a modified amplified fragment length polymorphism (AFLP) method commonly used to study DNA cytosine methylation [21,22]. In the place of the use of a frequent-cutter restriction enzyme, *Mse*I, in the AFLP technique, DNA is cleaved using two different enzymes, *Hpa*II and *Msp*I. The recognition sequence for both of these restriction enzymes is CCGG; however, they cleave the DNA fragment based on the particular pattern of methylated cytosines. This method is popular as it offers several advantages, especially in non-model plants such as lingonberry: because the obtained loci cover the information on the whole genome, obtaining a general idea about the DNA methylation is relatively quick, and the method is cost-effective compared to other techniques such as whole genome bisulfite sequencing [21].

For commercial micropropagation, genetic and epigenetic stability is necessary for phenotypic integrity. As a super food, lingonberry is increasing in popularity day by day and, therefore, holds huge potential for commercialization. In this context, we studied the clonal fidelity and global DNA methylation in micropropagated and greenhouse-grown lingonberry leaves using molecular markers. Information obtained through this investigation is expected to contribute to the commercialization of lingonberry as a medicinally important crop.

2. Materials and Methods

2.1. Plant Material, Growth Conditions and Morphological Data

Lingonberry cultivar Erntedank and a selected hybrid designated as H1, developed at St. John's Research and Development Centre, Agriculture and Agri-Food Canada, St. John's, Newfoundland and Labrador, Canada [6], were used for this study. Ten plants from each genotype were used to record the morphological data. For the genetic analysis, five plants were randomly selected, chopped, mixed and sampled for DNA analysis. Each experiment was replicated three times.

In vitro cultures were initiated using transversely segmented leaf explants from cultivar Erntedank and hybrid H1 following Arigundam et al. [10]. Surface sterilized explants were inoculated on a semi-solid medium on Fisherbrand™ Petri dishes covered with clear lids. The 25 cm^3 sterilized basal medium in each Petri dish contained 3/4 micro salts and macro salts [10] supplemented with 20 g dm^{-3} sucrose, 1.25 g dm^{-3} gelrite and 3.5 g dm^{-3} Sigma A 1296 agar, pH 5.0. The plant growth regulator (PGR) added to the medium was zeatin at a concentration of 1 mg L^{-1}. The cultures were kept in dark in a growth chamber at 20 $\pm$ 2 °C and relative humidity of 60–70%. After 2 weeks, the cultures were then exposed to cool white fluorescent lamps emitting PPFD of 30 µmol m^{-2} s^{-1}. After 4 weeks of inoculation, culture-initiated explants were transferred to Sigma bottles containing the same medium [10] where shoot regeneration was obtained. Leaf length and leaf width measurements were taken from 2-year-old greenhouse-grown lingonberry genotypes. Other morphological data were taken from 2-month-old cultures of cultivar Erntedank (Figure 1) and hybrid H1. Leaf length, leaf width and shoot height were measured in cm. The shoots per explant and the leaves per shoot were counted, and shoot vigor was assessed using the visual scale of 1–8, 1 being the poorest and 8 being the best looking shoot [10]. The shoots were then sampled for DNA analysis.

Figure 1. Two-month-old shoots of lingonberry cultivar Erntedank in in vitro semi-solid medium in a Sigma bottle (**left**) and greenhouse-grown plants in a plastic pot (**right**) containing peat and perlite medium.

From mother plants, young leaves were taken for DNA extraction from the cultivar Erntedank (Figure 1) and hybrid plants that were maintained in a greenhouse under natural light conditions having photosynthetic photon flux density (PPFD) of 90 μmol m^{-2} s^{-1}, temperature of 20 ± 2 °C and relative humidity of 85% maintained using automatic control systems. Plants were grown and maintained in 10 cm plastic pots containing peat and perlite in the ratio of 2:1 (v/v).

2.2. DNA Isolation

Actively growing lingonberry leaves from greenhouse plants and shoots from tissue culture plants (100 mg) were sampled, and genomic DNA was isolated using DNeasy Plant Mini Kits (Qiagen GmbH, Hilden, Germany) following the manufacturer's instructions. Briefly, the sampled lingonberry leaves were shock-frozen in liquid nitrogen immediately after collection and stored at −80 °C in a freezer until DNA isolation. Lysis was performed in 2 mL centrifuge tubes containing 600 μL AP1 buffer (80% ethanol, 100 mM NaCl and 10 mM Tris-HCl, pH 7.5) using two ceramic beads in a FastPrep 24 tissue and cell homogenizer (MP Biomedicals, Irvine, CA, USA). Then, 20 μL proteinase K (20 mg/mL) was added to the mixture and incubated for 1 h at 65 °C. Subsequently, 4.5 μL RNAse (100 mg/mL) was added to the mix and incubated for 15 min under the same conditions at 65 °C. Neutralization was performed using 425 μL P3 buffer, incubated at −20 °C for 18 min. DNA was separated from the mixture using a QIAshredder Mini Spin column placed in a 2 mL collection tube and centrifuged at 20,000× *g*. The lysate was washed using 1.5 volume of AW1 and 500 μL AW2 in a DNeasy mini spin column, where the DNA was trapped in the DNeasy membrane, which was eluted using 50 μL AE buffer. The DNA concentration of 50 ng/μL was maintained across the samples using 1 × TE buffer. The quality of the DNA was assessed using the absorbance ratio of A260 to A280 in the range of 1.8–1.9 and absorbance ratio of A260 to A230 in the range of 2.0–2.2 using a nanodrop spectrophotometer. The DNA was used to assess global DNA methylation in tissue culture and greenhouse-grown plants using the methylation-sensitive amplification polymorphism technique.

2.3. Clonal Fidelity Experiment

DNA samples from cv. Erntedank were diluted to 10 ng μL^{-1} using 1 × TE buffer. Amplification of the DNA regions containing markers was carried out using 12 molecular markers including one expressed sequence tag (EST)—PCR, five EST—simple sequence repeats (SSR) and six genomic simple sequence repeats (GSSR) [23] that were proven effective on *Vaccinium* species. Amplification reactions were carried out in a 25 μL reaction mixture containing 2.5 μL DNA template (25 ng DNA per reaction), 2.5 μL PCR buffer (1.5 mM $MgCl_2$), 0.1 μL Taq DNA polymerase (5U μL^{-1} stock), 0.5 μL dNTP (10 mM stock), and 0.5 μL primer (10 μL stock), and the final volume was adjusted with PCR-grade water (Sigma Chemical Co., St. Louis, MO, USA). DNA amplification was performed in Mastercycler ep Gradient S (Eppendorf AG, 22331 Hamburg, Germany). Initial denaturation was carried out at 94 °C for 10 min. The reaction was run for 40 cycles. Each cycle of amplification reaction consisted of denaturation of template DNA at 92 °C for 40 s. Primer extension was attained at 72 °C for 2 min. The reaction was completed with the final extension allowed to incubate at 72 °C for 10 min.

Separation of DNA fragments was performed by gel electrophoresis in 1.6% agarose (Agarose 3:1 HRB™, Amresco, Solon, OH, USA) gel pre-casted in a solution containing Tris-borate EDTA buffer (TBE) and GelRed nucleic acid stain (Biotium Inc., Hayward, CA, USA) in the ratio of 2:1.

A 100 bp Low-Ranger and a 50 bp Mini sizer DNA ladder (Norgen Bioteck Corp., Thorold, ON, Canada) was used as size marker. The gel was run for 1.2 h at 100 V. DNA bands were photographed digitally under UV light using a gel documentation system (InGenius 3; Syngene, Beacon House, Cambridge, UK). Visual observation of presence or absence of bands was recorded for further interpretation.

2.4. Methylation-Sensitive Amplification Polymorphism (MSAP) Assay

DNA samples from the cultivar Erntedank and a hybrid lingonberry were used to determine cytosine methylation. The AFLP technique for DNA fingerprinting [24], modified to the MSAP technique by [21], was adopted in the experiment. The MSAP assay was performed in the following steps, using MSAP adapters and primers (Table 1).

Table 1. List of adapter sequences, preamplification primers and selective amplification primers for methylation-sensitive amplification polymorphism analysis of in vitro-propagated and greenhouse-grown lingonberry genotypes.

Oligo Name	Function	Nucleotide Sequences
Ad. *Eco*RI	Forward adaptor	5′-CTG TAG ACT GCG TAC C-3′
Ad. *Eco*RI	Reverse adaptor	3′-CAT CTG ACG CAT GGT TAA-5′
Ad. *Msp*I/*Hpa*II	Forward adaptor	5′-GAT CAT GAG TCC TGC T-3′
Ad. *Msp*I/*Hpa*II	Reverse adaptor	3′-AGT ACT CAG GAC GAG C-5′
*Eco*RI (E)	Preselective amplification primer	5′-GAC TGC GTA CCA ATT CA-3′
*Msp*I/*Hpa*II (MH)	Preselective amplification primer	5′-ATC ATG AGT CCT GCT CGG-3′
E-TT	Selective amplification primer	5′-GAC TGC GTA CCA ATT CAT T-3′
E-TG	Selective amplification primer	5′-GAC TGC GTA CCA ATT CAT G-3′
MH-ATG	Selective amplification primer	5′-ATC ATG AGT CCT GCT CGG ATG-3′
MH-AAC	Selective amplification primer	5′-ATC ATG AGT CCT GCT CGG AAC-3′
MH-AAG	Selective amplification primer	5′-ATC ATG AGT CCT GCT CGG AAG-3′
MH-ACA	Selective amplification primer	5′-ATC ATG AGT CCT GCT CGG ACA-3′
MH-ATT	Selective amplification primer	5′-ATC ATG AGT CCT GCT CGG ATT-3′
MH-TCC	Selective amplification primer	5′-ATC ATG AGT CCT GCT CGG TCC-3′
MH-AAT	Selective amplification primer	5′-ATC ATG AGT CCT GCT CGG AAT-3′
MH-TCG	Selective amplification primer	5′-ATC ATG AGT CCT GCT CGG TCG-3′

2.4.1. Digestion

Digestion of DNA (900–1100 ng) was performed with *Eco*RI, *Msp*I, and *Hpa*II restriction endonuclease (Thermo Scientific, Waltham, MA, USA). Genomic DNA in a 75 μL reaction volume containing 3× FastDigest buffer was cleaved using 3 U of *Eco*RI at 37 °C for 1.5 h. Inactivating *EcoRI*, the reaction was stopped by incubating the mixture at 65 °C for 10 min. The *Eco*RI-digested DNA was allocated into three distinct aliquots and subjected to three separate reactions in a 50 μL reaction volume containing 1× the corresponding buffer. One of the aliquots was digested with 2 U *Msp*I, another with 2 U *Hpa*II, and the remaining aliquot was treated with 2 U each of *Msp*I and *Hpa*II restriction endonucleases and incubated for 3 h at 37 °C. The restriction enzyme reaction were stopped by denaturing the enzymes and incubating for 15 min at 65 °C.

2.4.2. Ligation

The digested DNA (50 μL) was ligated to adapters using 5 U of T4 DNA ligase, 10 μL 1 × T4 DNA ligase buffer, 1 μL of 10 μM *Eco*RI adapter, 1 μL of 100 μM *Msp*I/*Hpa*II adapter and 2 μL polyethylene glycol (50% *w*/*v*). The final volume was adjusted to 100 μL using PCR water and incubated at 23 °C for 5 h. The reaction of the enzymes was stopped by placing the mixture at 65 °C for 10 min.

2.4.3. Preamplification

The DNA fragments (4 μL) ligated to the adaptors were amplified by PCR using *EcoRI* (E) as forward and *MspI-HpaII* (MH) as reverse primers (Table 1). A total volume of 50 μL pre-selective amplification was carried out containing a final concentration of 200 μM of each dNTP (Amresco LLC, Solon, OH, USA), 1× PCR buffer (Qiagen Inc., Toronto, ON, Canada), 1 U of Top Taq DNA polymerase (Qiagen) and 0.2 μM of E and MH primers. PCR amplifications were carried out in an Eppendorf Mastercycler Gradient thermocycler (Eppendorf AG, 22331 Hamburg, Germany). Pre-selective amplification products showed a 100 to 1000 bp smear in 1.8% agarose gel. PCR products were diluted seven times using 0.1 × TE buffer for selective amplification.

2.4.4. Selective Amplification

Diluted pre-amplified products from the previous step were selectively amplified with 16 primer combinations in total. The two *EcoRI* selective amplification primers consisted of two particular extra bases (TT and TG) of preamplification *EcoRI* (E) as forward primers, and eight *Msp*I-*Hpa*II primers consisting of three different base overhangs (ATG, AAC, AAG, ACA, ATT, TCC, AAT, TCG) of *Msp*I-*Hpa*II (MH) preamplification primers used as reverse primers. A total volume of 25 µL PCR amplification reaction contained 1× PCR buffer, 400 µM dNTPs, 0.4 µM of each selective primer, 1 U of Top Taq DNA polymerase, and 4 µL pre-amplified PCR product. PCR amplification was performed using the touch-down cycles with the following conditions: initially heated to 94 °C for 5 min; then, 13 cycles were run for 30 s at 94 °C, 1 min at 65 °C, which was reduced by 0.7 °C per cycle, and 72 °C for 2 min, followed by 23 cycles of 30 s at 94 °C, 1 min at 56 °C and 2 min at 72 °C with a final extension step of 72 °C for 10 min. The selective amplification products were separated on 6% denaturing polyacrylamide gel electrophoresis (PAGE).

2.4.5. Polyacrylamide Gel Electrophoresis (PAGE)

Denaturing formamide dye solution was prepared using 10 mM EDTA of pH 8.0, 98% formamide, 0.01% (*w*/*v*) xylene cyanol and 0.01% (*w*/*v*) bromophenol blue. Selective amplification products were denatured by mixing them with equal volumes of denaturing formamide dye and heating for 4 min at 95 °C, then cooling immediately for 5 min and keeping at −20 °C. The 6% PAGE gels were pre-run at 90 V for about 1 h to clean the wells. Denatured selective amplification products (10 µL) were loaded in the gels, and a potential difference of 95 V was applied for 3 h. The DNA fragments separated in gels were stained for 30 min in the dark in 1% PAGE GelRed™ (Biotium Inc., Hayward, CA, USA) with gentle agitation and visualized using InGenius 3 gel documentation system (Syngene, Frederick, MD, USA). The 1 kb and 50 bp DNA ladders (Norgen Biotek, Thorold, ON, Canada) were used as a molecular size marker. The experiments were repeated twice, and the reproducible results were used to score for further analysis.

2.4.6. Profiling Scoring and Data Analysis

The methylation status at tetranucleotide restriction sites (5′-CCGG-3′) was detected by comparing the DNA profiles, based on the presence or absence of DNA bands by the reaction of restriction enzymes *Eco*RI+*Msp*I, *Eco*RI+*Hpa*II and *Eco*RI+*Msp*I+*Hpa*II. In the absence of methylation at 5′-CCGG-3′, both the isoschizomers *Msp*I and *Hpa*II cleave the DNA fragments at this site. Therefore, the DNA bands present in all three lanes were considered non-methylation (first case). DNA bands identified in *Eco*RI+*Msp*I and *Eco*RI+*Msp*I+*Hpa*II lanes but absent from the *Eco*RI+*Hpa*II lane were considered as fully methylated internal cytosine, as *Hpa*II cannot cleave at fully methylated internal cytosine (second case). The bands that were present in the lanes of both *Eco*RI+*Hpa*II and *Eco*RI+*Msp*I+*Hpa*II but not in *Eco*RI+*Msp*I were considered hemimethylation of external cytosine (5′-mCCGG-3′, third case). In another case, bands absent from the *Eco*RI+*Msp*I+*Hpa*II lane but present in the *Eco*RI+*Hpa*II lane indicated the existence of a site for internal 5′-CmCGG-3′ [25] and were scored accordingly (fourth case). In aggregate, the number of bands present in the second, third, and fourth cases resulted in the total number of methylated bands.

$$\text{Total methylation} = \frac{\text{Methylated band numbers}}{\text{Total band numbers}} \times 100\% \quad (1)$$

$$\text{Fully methylated percentage} = \frac{\text{Fully methylated band numbers}}{\text{Total methylated band numbers}} \times 100\% \quad (2)$$

$$\text{Hemimethylated percentage} = \frac{\text{Hemimethylated band numbers}}{\text{Total methylated band numbers}} \times 100\% \quad (3)$$

2.4.7. Statistical Analysis

Morphological statistical data analysis was performed through one-way ANOVA in jamovi [26] software (3rd generation) followed by standard error (SE) and mean.

3. Results

3.1. Morphological Characteristics

Among the greenhouse-grown lingonberry samples, the average leaf length of hybrid lingonberry H1 was 2.44 cm; this was significantly longer than that of the lingonberry cultivar, which was only 1.98 cm (Figure 2). Similarly, the average leaf width of hybrid lingonberry was 1.3 cm, which was significantly wider than that of the cultivar, for which the average width of leaf was 1.04 cm under greenhouse conditions. In tissue cultures, the average shoots per explant in hybrid was 4, which was significantly more than that for the cultivar with an average of 3 shoots per explant. Similarly, shoot height was significantly higher in the hybrid (3.67 cm) as compared to the cultivar (3.36 cm). Plants of hybrid lingonberry showed significantly better vigor (average 6.4) as compared to the cultivar (average 5.9). However, there was no any significant difference in the number of shoots between the studied lingonberry cultivar and the hybrid.

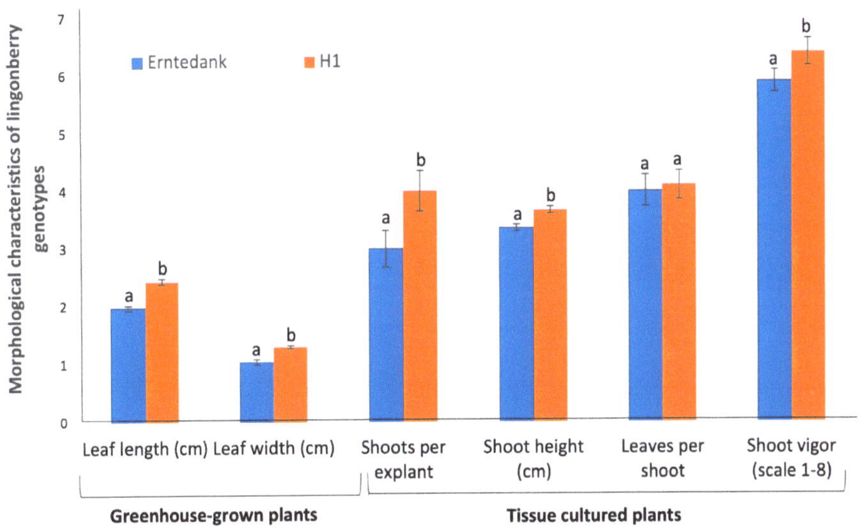

Figure 2. Morphological characteristics (means ± standard error) of lingonberry cultivar Erntedank and hybrid H1. Data on leaf length and leaf width were recorded from 2-year-old greenhouse-grown plants. Shoot number per explant, shoot height, leave number per shoot and shoot vigor (scale 1–8; 1 being the poorest, and 8 being the best) were taken from 2-month-old tissue cultures. Bars, within the same group, followed by same letter are not significantly different according to Tukey's range test at $p = 0.05$.

3.2. Clonal Fidelity

Clonal fidelity in tissue-cultured lingonberry cultivar Erntedank was assessed using 12 molecular markers including one expressed sequence tag PCR (EST PCR), five expressed sequence tag—simple sequence repeats (EST SSR) and six genomic simple sequence repeats (GSSR). Altogether, 31 monomorphic DNA bands were present from 12 molecular markers, yielding an average of 2.5 bands per primer (Table 2). A representative figure including seven markers is shown in Figure 3.

Table 2. List of PCR primers including the primer type, primer name, sequence information, annealing temperature, the number of bands present and the size of amplified alleles in the in vitro propagated shoots and greenhouse-grown plants of lingonberry cultivar Erntedank.

Primer Type	Primer Name	Primer Sequence	Annealing Temperature	Bands Present (No.)	Size of Amplified Alleles (bp)
EST PCR	CA21	F:TCCGATAACCGTTACCAAGC R:TATACAGCGACACGCCAAAA	54	2	110, 230
EST SSR	CA23	F:GAGAGGGTTTCGAGGAGGAG R:GTTTAGAAACGGGACTGTGAGACG	60	2	100, 175
EST SSR	CA169	F:TAGTGGAGGGTTTTGCTTGG R:GTTTATCGAAGCGAAGGTCAAAGA	54	2	260, 350
EST SSR	CA421	F:TCAAATTCAAAGCTCAAAATCAA R:GTTTAAGGATGATCCCGAAGCTCT	60	2	175, 250
EST SSR	NA398	F:TCCTTGCTCCAGTCCTATGC R:GTTTCCTTCCACTCCAAGATGC	60	2	145, 200
EST SSR	NA1040	F:GCAACTCCCAGACTTTCTCC R:GTTTAGTCAGCAGGGTGCACAA	56	3	150, 210, 350
GSSR	VCCB3	F:CCTTCGATCTTGTTCCTTGC R:GTTTGATGCAATTGAGGTGGAGA	62	3	125, 270, 300
GSSR	VCCI2	F:AGGCGTTTTTGAGGCTAACA R:TAAAAGTTCGGCTCGTTTGC	62	3	130, 300, 325
GSSR	VCCJ9	F:GCGAAGAACTTCCGTCAAAA R:GTGAGGGCACAAAGCTCTC	60	3	75, 120, 135
GSSR	VCCJ1	F:CTCATGGGTTCCCATAGACAA R:TGCAGTGAGGCAAAAGATTG	62	3	275, 300, 350
GSSR	VCCK4	F:CCTCCACCCCACTTTCATTA R:GCACACAGGTCCAGTTTTTG	62	3	100, 140, 150
GSSR	VCCS10	F:ATTTGGTGTGAAACCCCTGA R:GTTTGCGGCTATATCCGTGTTTGT	60	3	150, 175, 215

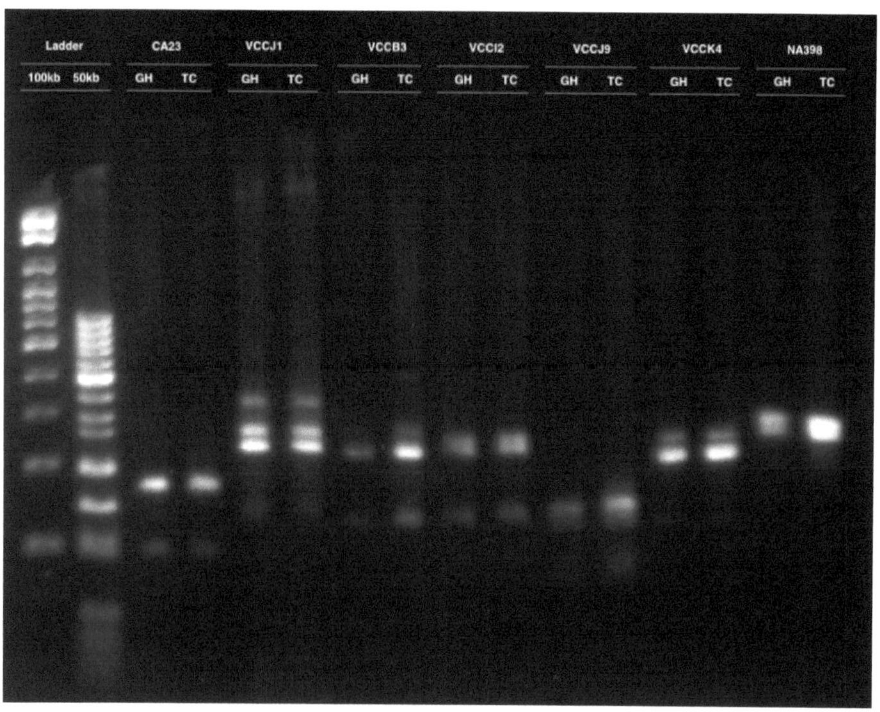

Figure 3. Agarose gel image showing DNA banding pattern in greenhouse (GH)- and growth chamber (TC)-grown lingonberry cultivar Erntedank. The first two lanes are the 100 bp and 50 bp DNA ladder. Each of the consecutive two lanes represent the DNA bands with the molecular marker, the first sample being from the greenhouse-grown plants and the second from the tissue-culture shoots in a growth chamber.

3.3. DNA Methylation Pattern

DNA methylation profiles were explained based on the polymorphism of the fragments digested with *EcoRI* and one or both of the isoschizomers *MspI/HpaII*, resulting in three lanes. When DNA fragments appeared in all three lanes, they represented non-methylation at the 5′-CCGG-3′ site. *MspI*-specific fragments appeared in the *EcoRI+MspI* and *EcoRI+MspI+HpaII* lanes by the digestion of methylated internal cytosine (5′-CmCGG-3′).

In contrast, *HpaII*-specific fragments appeared in the *EcoRI+HpaII* and *EcoRI+MspI+HpaII* lanes that resulted from cleavage at hemimethylated external cytosine (5′-mCCGG-3′). With *HpaII*-specific fragments, lanes only present on *HpaII* were counted as methylation, as it accounts for internal cytosine methylation [25]. The latter three conditions were considered as methylation at the 5′-CCGG-3′ site. A gel image shows the non-methylated, hemimethylated, and methylated cytosines at the 5′-CCGG-3′ site in Figure 4.

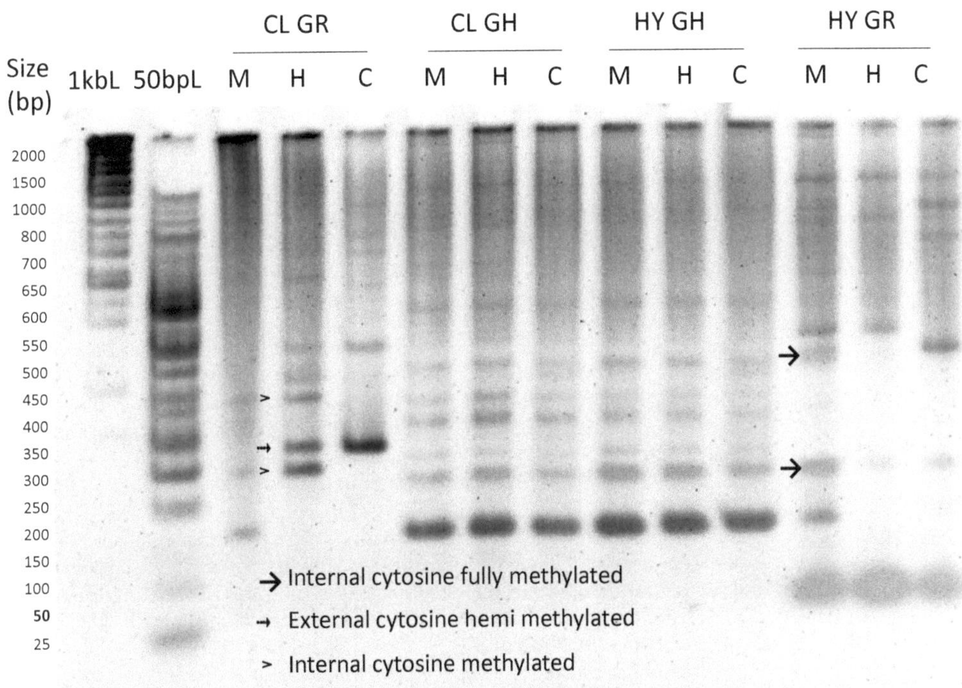

Figure 4. Polyacrylamide gel image showing cytosine methylation in the greenhouse (GH) and in vitro grown in semi-solid media in growth chamber (GR) grown lingonberry genotypes, cultivar Erntedank (CL) and a hybrid H1 (HY). CL GR = cultivar growth chamber, CL GH = cultivar greenhouse, HY GR = hybrid growth chamber, and HY GH = hybrid greenhouse. DNA methylation pattern was detected in lingonberry cultivar and hybrid using methylation-sensitive amplification polymorphism (MSAP) assay. Arrows show fully methylated internal cytosine, and arrows with a broken line show hemimethylated external cytosine. The arrowheads represent internal cytosine methylation.

In Erntedank, 344 bands were obtained in the greenhouse-grown plants, and 329 bands were obtained in the tissue-cultured plants (Table 3). In the same genotype, 116 (33.72%) bands were found to be methylated in greenhouse-grown plants, whereas 140 (20%) bands were obtained in tissue-cultured plants. In the hybrid lingonberry, 353 bands were obtained in greenhouse plants, while 364 bands were obtained in tissue-cultured plants. In hybrid lingonberry, 113 (32.01%) of the bands were methylated at the CCGG site, while 160 (43.96%) bands were methylated in tissue cultures. Variation was also observed if the cytosine was fully or hemimethylated. Full methylation ranged between 42.86% to 57.52%, with the highest observed in greenhouse-grown hybrid lingonberry and the lowest on the tissue-cultured lingonberry cultivar Erntedank (Table 3). Hemimethylation was highest (57.14%) in tissue-cultured lingonberry cultivar and lowest in greenhouse-grown hybrid (42.48%). In general, in vitro grown plants showed higher DNA methylation as compared to the greenhouse-grown plants.

Table 3. Cytosine methylation in greenhouse- and growth chamber-grown lingonberries.

DNA Bands	Erntedank		Hybrid (H1 = HY GH, HY GR)	
	Greenhouse	Tissue Culture	Greenhouse	Tissue Culture
Type 1	228	189	240	204
Type 2	54	80	48	86
Type 3	46	48	46	50
Type 4	16	12	19	24
Total analyzed bands	344	329	353	364
Total methylated bands	116	140	113	160
Fully methylated bands	62	60	65	74
Fully methylated percentage	53.45%	42.86%	57.52%	46.25%
Hemimethylated bands	54	80	48	86
Hemimethylated percentage	46.55%	57.14%	42.48%	53.75%
MSAP percentage	33.72%	42.55%	32.01%	43.96%

4. Discussion

Tissue culture is a rapid propagation technique used to propagate transgenic crops and clonally born plants. However, since the tissue culture process bypasses the normal developmental events in the tissue culture microenvironment, it may be stressful for plant tissue, resulting in genetic and epigenetic instabilities. These variations are known as somaclonal variation [22]. Tissue culture-induced variations or their effect in morphological and biochemical response to different stresses have been reported in various plant species [10,27,28]. Morphological characteristics compared between Erntedank and H1 reflected that the selected hybrid H1 had bigger leaves as compared to the cultivar Erntedank. In the tissue cultures also, there were better shoot height, shoot vigor and number of shoots per explant. Phenotypic variations in any organism independent of DNA sequence variation are epigenetic modifications. Although there is no change in the DNA sequence, transcription of the gene is effectively altered by epigenetic factors. Therefore, epigenetic factors are important mediators of gene expression [20]. For the organisms whose genome sequence information is not available, the MSAP technique is widely used to detect epigenetic variations due to DNA methylation in tissue cultures [29,30]. This method has been utilized to determine the epigenetic variation in banana [31], grapevine [32], oil palm [29], blueberry [2] and lingonberry [33].

Variations are a source of novel traits in breeding programs. Phenotypic plasticity is another phenomenon in plant cells that helps the tissues to cope with environmental variation. However, commercial sustainability of the in vitro regeneration systems depends upon the maintenance of genetic integrity. The tissue culture system did not alter the genetic integrity in the lingonberry cultivar in our study. In line with these results, no genetic differences were found in tissue culture-derived plants in blueberry [34], lingonberry [10], or *Jatropha* [35]. However, some genetic variations have been reported in tissue-cultured gerbera [36].

Researchers have reported various levels of cytosine methylation in plants. In our study in lingonberry, cytosine methylation was found to be between 32.01% and 43.96%. However, in tissue-cultured potato, cytosine methylation was found to be very low (0–3.4%) [37]. Methylation in the range of 64.36–67.00% was reported in pepper (*Capsicum* sp.) [38]. Therefore, it can be said that there exists a wide range of variation in the DNA methylation profile in the plant kingdom. Moreover, the effect of micropropagation on cytosine methylation in lingonberries was similar to that in banana [31], orchid [14] and blueberry [2] plants (Table 3), where in vitro conditions resulted in higher cytosine methylation. The organized tissues in the presence of growth regulators dedifferentiate to the undifferentiated mass of totipotent cells called callus and further re-differentiate to produce plant organs [39]. This alteration in the differentiation status might have resulted in higher cytosine methylation in tissue-cultured lingonberries.

The elusive aspect of epigenetic mechanisms is their variable inheritance. During mitosis cell division, such as in asexual propagation, such variations are frequently heritable, sometimes for multiple generations [39]. However, during sexual reproduction, epigenetic marks are partially reset during meiosis and partially transmitted through meiosis. For example, the epigenetic marks at locus *FLC* affecting the vernalization response are changed during the meiotic process, although transposon methylation is firmly maintained [40]. In another study with a methylation-deficient *Arabidopsis* mutant, it was revealed that methylated CpG are fundamental to epigenetic memory throughout generations [41]. Unlike genetic alleles, epialleles have a tendency to react more frequently to change in the environment, are reversible, and can be retained for a number of generations only [42]

Environmental factors such as exposure to stress change DNA methylation patterns in plants [43]. This adaptive mechanism of plants under altered ecological conditions affects gene expression, including the genes that are involved in the synthesis of biochemicals that have a significant role in abiotic stress tolerance [20]. In an experiment with high-temperature exposure in in vitro conditions, 60% of grapevine somaclones retained the altered DNA methylation pattern even one year after the treatment [32]. In the current study, 16 selective primers revealed lower levels of methylated loci in greenhouse-grown plants than in tissue-cultured plants (Table 3). This trend was similar for both of the genotypes used in the study. In *Oryza sativa*, DNA methylation in the promoter regions of genes has been shown to reduce their expression levels, thus affecting the phenotype [44]. Li et al. [45] also established a relationship between DNA methylation in promoter regions and gene expression, which showed a negative correlation between gene expression and cytosine methylation levels in the 2 kb regions of the promoter [44]. In our experiments, although the location of the DNA methylation in the lingonberry chromosome was not known, if it occurs in the functional DNA region such as a promoter, it could cause significant physiological changes, including those affecting antioxidant properties.

DNA methylation plays a vital role in regulating dedifferentiation and redifferentiation phases in the tissue culture system. In the tissue culture, each plant cell of an explant experiences the medium differentially, and their altered responses result in polymorphism in the methylation pattern [35]. In lingonberry genotypes, higher variation was observed if the cytosine was fully methylated or hemimethylated, which was found irrespective of the percentage of DNA methylation. However, although full methylation was higher in greenhouse-grown plants than in in vitro-grown shoots, the hemimethylated DNA bands were more prominent under in vitro conditions (Table 3). Somaclonal variations (genetic and epigenetic) take place in plants due to environmental stress. MET1, the main cause of methylation, is present in tissue culture plants. The PGRs zeatin and indole-3-butyric acid were used for shoot proliferation in vitro and rooting of microshoots, respectively [10]. The expression of the AUXIN RESPONSE FACTOR3 (ARF3) gene is inhibited by hyper-methylation, giving rise to apical dominance of micropropagated plants, and hypo-methylation enhances expression of the ARF3 gene. In the current investigation, the variable DNA methylation (both hypo- and hyper-methylation) might have been due to a number of factors, including culture conditions, media type, and type and concentration of PGR during lingonberry propagation under ex vitro and in vitro conditions [33]. Under tissue culture conditions, variation in the epigenetic pattern was genotype-specific (Table 3). Similar results have been obtained in blueberry [2], where tissue-cultured blueberry showed higher methylation in a genotype-specific manner. This process may also be linked to alteration in the plant material's hormonal balance and hormone signaling pathway [45,46]. The plant tissue culture process involves the action of PGRs and a complex network of interactions among them. In *Arabidopsis thaliana*, adding indole-3-butyric acid and zeatin in a shoot induction medium resulted in hypo-methylation, which enhanced the expression of the gene encoding ARF3 [47]. Tissue-cultured blueberry in PGR-containing medium exhibited higher cytosine methylation than the cutting or donor counterparts, respectively [2]. In this line, zeatin (1 mg L^{-1}) added in the tissue-cultured media seemed to cause lingonberry's increased DNA methylation.

Lingonberries are characterized as a superfood because of their high antioxidant properties. Several secondary metabolites have antioxidant properties and are regulated by DNA methylation by modulating the expression of key genes involved in this process [48]. Several experiments have shown the involvement of increased DNA methylation to suppress the genes' function, thus resulting in reduced products of those genes. For example, in red sage (*Salvia miltiorrhiza*), *S*-adenosyl methionine (SAM), a donor for DNA methylation, dramatically inhibited accumulation of the phenolic compound [49]. In contrast, the expressions of key genes involved in phenolic acid biosynthesis were downregulated [49], whereas 5-azacytidine, an inhibitor of methylation, significantly enhanced the accumulation of phenolic compounds with a significant upregulation of the key gene expressions involved in phenolic acid biosynthesis [49]. In *Arabidopsis thaliana*, hypomethylation resulted in a substantial increase in the production of a protein involved in growth regulation [50]. CG methylation in *Arabidopsis* is maintained by a conserved METHYLTRANSFERASE1 (*MET1*), a protein homologous to animal *DNMT1* [48,51]. However, methylation in the CHG context is maintained by CHROMOMETHYLASE3 (*CMT3*), and CHH context is maintained by the plant-specific CHROMOMETHYLASE2 (*CMT2*) [51,52]. Analysed expression patterns of *MET1*, *CMT* methyltransferases exhibited higher expression levels in fast-growing calli, and regenerated plants were hypermethylated [52]. In another experiment, genes *MET1* and *CMT3* that code for DNA methylases during somatic embryo formation were found to be upregulated, while genes encoding DNA demethylases were downregulated [51,53]. Therefore, global DNA methylation seemed to affect the transcriptional activity of coding genes, ultimately affecting several physiological processes, including the production of secondary metabolites. In *Arabidopsis thaliana*, a statistical model predicted that 65% of the variance in plant height was the result of DNA methylation [51]. The significance of DNA methylation to the amount of antioxidant production could be particularly important in this medicinally important crop, lingonberry.

5. Conclusions

Although no genetic distinction was observed in the lingonberry cultivar Erntedank, variation in DNA methylation patterns in leaf tissues of tissue-cultured and greenhouse-grown lingonberries suggested the effect of in vitro propagation. These variations in turn might affect the expression of the genes involved in vital processes. Given the elusive nature of variation, it would be interesting to evaluate the effects on the vegetative and reproductive stages of mature plants to determine if the tissue culture-induced variation is transient or permanent. This study might be a valuable consideration for the use of commercial micropropagation of lingonberry. Another exciting research direction could be to investigate the effect of epigenetic marks on vital physiological processes, such as the pathways for production of antioxidant compounds.

Author Contributions: U.S.: Conducted all of the experiments and wrote the original manuscript draft. A.S.: Helped in conducting experiments. A.U.I.: Co-supervised the research and helped in research planning and execution. S.C.D.: Had roles in conceptualization, supervision, research activity planning and execution, funding acquisition, manuscript writing, and reviewing and editing. All authors have read and agreed to the published version of the manuscript.

Funding: This research received no external funding.

Institutional Review Board Statement: Not applicable.

Informed Consent Statement: Not applicable.

Data Availability Statement: Original data presented in the study are included in the main text, and further inquiries can be directed to the corresponding author.

Acknowledgments: The authors thank Darryl Martin for his excellent technical help. All of the experiments were performed at St. John's Research and Development Centre (SJRDC), Agriculture and Agri-Food Canada and in the Department of Biology, Memorial University of Newfoundland. This work is an SJRDC contribution.

Conflicts of Interest: The authors declare that they have no known competing financial interests or personal relationships that could have appeared to influence the work reported in this paper.

References

1. Rohde, A.; Junttila, O. Remembrances of an Embryo: Long-Term Effects on Phenology Traits in Spruce. *New Phytol.* **2008**, *177*, 2–5. [CrossRef] [PubMed]
2. Ghosh, A.; Igamberdiev, A.U.; Debnath, S.C. Tissue Culture-Induced DNA Methylation in Crop Plants: A Review. *Mol. Biol. Rep.* **2021**, *48*, 823–841. [CrossRef] [PubMed]
3. Chinnusamy, V.; Zhu, J.K. Epigenetic Regulation of Stress Responses in Plants. *Curr. Opin. Plant Biol.* **2009**, *12*, 133–139. [CrossRef] [PubMed]
4. Bujor, O.C.; Ginies, C.; Popa, V.I.; Dufour, C. Phenolic Compounds and Antioxidant Activity of Lingonberry (*Vaccinium Vitis-Idaea* L.) Leaf, Stem and Fruit at Different Harvest Periods. *Food Chem.* **2018**, *252*, 356–365. [CrossRef]
5. Jin, Y.; Liu, Z.; Liu, D.; Shi, G.; Liu, D.; Yang, Y.; Gu, H.; Yang, L.; Zhou, Z. Natural Antioxidant of Rosemary Extract Used as an Additive in the Ultrasound-Assisted Extraction of Anthocyanins from Lingonberry (*Vaccinium Vitis-Idaea* L.) Pomace. *Ind. Crops Prod.* **2019**, *138*, 111425. [CrossRef]
6. Debnath, S.C.; Arigundam, U. In Vitro Propagation Strategies of Medicinally Important Berry Crop, Lingonberry (*Vaccinium Vitis-Idaea* L.). *Agronomy* **2020**, *10*, 744. [CrossRef]
7. Kelly, E.; Vyas, P.; Weber, J.T. Biochemical Properties and Neuroprotective Effects of Compounds in Various Species of Berries. *Molecules* **2018**, *23*, 26. [CrossRef]
8. Nile, S.H.; Park, S.W. Edible Berries: Bioactive Components and Their Effect on Human Health. *Nutrition* **2014**, *30*, 134–144. [CrossRef]
9. Holloway, P.S. Rooting of Lingonberry, Vaccinium Vitis-Idaea, Stem Cuttings. *Plant Prop.* **1985**, *31*, 7–9.
10. Arigundam, U.; Variyath, A.M.; Siow, Y.L.; Marshall, D.; Debnath, S.C. Liquid Culture for Efficient In Vitro Propagation of Adventitious Shoots in Wild Vaccinium Vitis-Idaea Ssp. Minus (Lingonberry) Using Temporary Immersion and Stationary Bioreactors. *Sci. Hortic.* **2020**, *264*, 109199. [CrossRef]
11. Larkin, P.J.; Scowcroft, W.R. Somaclonal Variation—A Novel Source of Variability from Cell Cultures for Plant Improvement. *Theor. Appl. Gen.* **1981**, *60*, 197–214. [CrossRef]
12. Azman, A.S.; Mhiri, C.; Tam, S.M. Transposable Elements and the Detection of Somaclonal Variation in Plant Tissue Culture: A Review. *Malaysian App. Biol.* **2014**, *43*, 1–12.
13. Hirsch, S.; Baumberger, R.; Grossniklaus, U. Epigenetic Variation, Inheritance, and Selection in Plant Populations. In *Cold Spring Harbor Symposia on Quantitative Biology*; Cold Spring Harbor Laboratory Press: New York, NY, USA, 2012. [CrossRef]
14. Park, S.Y.; Murthy, H.N.; Chakrabarthy, D.; Paek, K.Y. Detection of Epigenetic Variation in Tissue-Culture-Derived Plants of Doritaenopsis by Methylation-Sensitive Amplification Polymorphism (MSAP) Analysis. *In Vitro Cell. Dev. Biol.-Plant* **2009**, *45*, 104–108. [CrossRef]
15. Bräutigam, K.; Vining, K.J.; Lafon-Placette, C.; Fossdal, C.G.; Mirouze, M.; Marcos, J.G.; Fluch, S.; Fraga, M.F.; Guevara, M.Á.; Abarca, D.; et al. Epigenetic Regulation of Adaptive Responses of Forest Tree Species to the Environment. *Ecol. Evol.* **2013**, *3*, 399–415. [CrossRef] [PubMed]
16. Hatorangan, M.R.; Laenen, B.; Steige, K.A.; Slotte, T.; Köhler, C. Rapid Evolution of Genomic Imprinting in Two Species of the Brassicaceae. *Plant Cell* **2016**, *28*, 1815–1827. [CrossRef] [PubMed]
17. Henderson, I.R.; Jacobsen, S.E. Epigenetic Inheritance in Plants. *Nature* **2007**, *447*, 418–424. [CrossRef] [PubMed]
18. Lister, R.; O'Malley, R.C.; Tonti-Filippini, J.; Gregory, B.D.; Berry, C.C.; Millar, A.H.; Ecker, J.R. Highly Integrated Single-Base Resolution Maps of the Epigenome in *Arabidopsis*. *Cell* **2008**, *133*, 523–536. [CrossRef] [PubMed]
19. Maunakea, A.K.; Chepelev, I.; Cui, K.; Zhao, K. Intragenic DNA Methylation Modulates Alternative Splicing by Recruiting MeCP2 to Promote Exon Recognition. *Cell Res.* **2013**, *23*, 1256–1269. [CrossRef]
20. Gallusci, P.; Dai, Z.; Génard, M.; Gauffretau, A.; Leblanc-Fournier, N.; Richard-Molard, C.; Vile, D.; Brunel-Muguet, S. Epigenetics for Plant Improvement: Current Knowledge and Modeling Avenues. *Trends Plant Sci.* **2017**, *22*, 610–623. [CrossRef] [PubMed]
21. Reyna-López, G.E.; Simpson, J.; Ruiz-Herrera, J. Differences in DNA Methylation Patterns Are Detectable during the Dimorphic Transition of Fungi by Amplification of Restriction Polymorphisms. *Mol. Gen. Genet.* **1997**, *253*, 703–710. [CrossRef] [PubMed]
22. Alonso, C.; Pérez, R.; Bazaga, P.; Medrano, M.; Herrera, C.M. MSAP Markers and Global Cytosine Methylation in Plants: A Literature Survey and Comparative Analysis for a Wild-Growing Species. *Mol. Ecol. Resour.* **2016**, *16*, 80–90. [CrossRef] [PubMed]
23. Boches, P.S.; Bassil, N.V.; Rowland, L.J. Microsatellite Markers for Vaccinium from EST and Genomic Libraries. *Mol. Ecol. Notes* **2005**, *5*, 657–660. [CrossRef]
24. Vos, P.; Hogers, R.; Bleeker, M.; Reijans, M.; van de Lee, T.; Hornes, M.; Friters, A.; Pot, J.; Paleman, J.; Kuiper, M.; et al. AFLP: A New Technique for DNA Fingerprinting. *Nucleic Acids Res.* **1995**, *23*, 4407–4414. [CrossRef] [PubMed]

25. Fulneček, J.; Kovařík, A. How to Interpret Methylation Sensitive Amplified Polymorphism (MSAP) Profiles? *BMC Genet.* **2014**, *15*, 2. [CrossRef]
26. Şahin, M.; Aybek, E. Jamovi: An easy to use statistical software for the social scientists. *Int. J. Assess. Tools Educ.* **2019**, *6*, 670–692. [CrossRef]
27. Mohan Jain, S. Tissue Culture-Derived Variation in Crop Improvement. *Euphytica* **2001**, *118*, 153–166. [CrossRef]
28. Machczyńska, J.; Zimny, J.; Bednarek, P.T. Tissue Culture-Induced Genetic and Epigenetic Variation in Triticale (× *Triticosecale* Spp. Wittmack Ex A. Camus 1927) Regenerants. *Plant Mol. Biol.* **2015**, *89*, 279–292. [CrossRef]
29. Jaligot, E.; Beulé, T.; Rival, A. Methylation-Sensitive RFLPs: Characterisation of Two Oil Palm Markers Showing Somaclonal Variation-Associated Polymorphism. *Theor. Appl. Genet.* **2002**, *104*, 1263–1269. [CrossRef]
30. Miguel, C.; Marum, L. An Epigenetic View of Plant Cells Cultured *In Vitro*: Somaclonal Variation and Beyond. *J. Exp. Bot.* **2011**, *62*, 3713–3725. [CrossRef]
31. Peraza-Echeverria, S.; Herrera-Valencia, V.A.; Kay, A.J. Detection of DNA Methylation Changes in Micropropagated Banana Plants Using Methylation-Sensitive Amplification Polymorphism (MSAP). *Plant Sci.* **2001**, *161*, 359–367. [CrossRef]
32. Baránek, M.; Čechová, J.; Raddová, J.; Holleinová, V.; Ondrušíková, E.; Pidra, M. Dynamics and Reversibility of the DNA Methylation Landscape of Grapevine Plants (Vitis Vinifera) Stressed by *In Vitro* Cultivation and Thermotherapy. *PLoS ONE* **2015**, *10*, e0126638. [CrossRef]
33. Sikdar, A.; Sharma, U.; Barua, R.; Igamberdiev, A.U.; Debnath, S.C. Epigenomic Insight of Lingonberry and Health-Promoting Traits during Micropropagation. *Sci. Rep.* **2022**, *12*, 12487. [CrossRef] [PubMed]
34. Goyali, J.C.; Igamberdiev, A.U.; Debnath, S.C. Propagation Methods Affect Fruit Morphology and Antioxidant Properties but Maintain Clonal Fidelity in Lowbush Blueberry. *HortSci* **2015**, *50*, 888–896. [CrossRef]
35. Rathore, M.S.; Jha, B. DNA Methylation and Methylation Polymorphism in Genetically Stable *In Vitro* Regenerates of Jatropha Curcas L. Using Methylation-Sensitive AFLP Markers. *Appl. Biochem. Biotechnol.* **2016**, *178*, 1002–1014. [CrossRef] [PubMed]
36. Bhatia, R.; Singh, K.P.; Jhang, T.; Sharma, T.R. Assessment of Clonal Fidelity of Micropropagated Gerbera Plants by ISSR Markers. *Sci. Hortic.* **2009**, *119*, 208–211. [CrossRef]
37. Tiwari, J.; Saurabh, S.; Chandel, P.; Singh, B.; Bhardwaj, V. Analysis of Genetic and Epigenetic Variation in *in Vitro* Propagated Potato Somatic Hybrid by AFLP and MSAP Markers. *Electr. J. Biotechnol.* **2013**, *16*, 5. [CrossRef]
38. Xu, X.; Li, T.; Li, Y.; Wang, H. Variation of DNA Cytosine Methylation Patterns Among Parent Lines and Reciprocal Hybrids in Hot Pepper. *Chem. Eng. Trans.* **2015**, *46*, 1345–1350. [CrossRef]
39. Richards, E.J. Inherited Epigenetic Variation—Revisiting Soft Inheritance. *Nat. Rev. Genet.* **2006**, *7*, 395–401. [CrossRef]
40. Bond, D.M.; Finnegan, E.J. Passing the Message on: Inheritance of Epigenetic Traits. *Trends Plant Sci.* **2007**, *12*, 211–216. [CrossRef]
41. Mathieu, O.; Reinders, J.; Čaikovski, M.; Smathajitt, C.; Paszkowski, J. Transgenerational Stability of the *Arabidopsis* Epigenome Is Coordinated by CG Methylation. *Cell* **2007**, *130*, 851–862. [CrossRef]
42. Kalisz, S.; Purugganan, M.D. Epialleles via DNA Methylation: Consequences for Plant Evolution. *Trends Ecol. Evol.* **2004**, *19*, 309–314. [CrossRef] [PubMed]
43. Mirouze, M.; Paszkowski, J. Epigenetic Contribution to Stress Adaptation in Plants. *Curr. Opin. Plant Biol.* **2011**, *14*, 267–274. [CrossRef] [PubMed]
44. Akimoto, K.; Katakami, H.; Kim, H.-J.; Ogawa, E.; Sano, C.M.; Wada, Y.; Sano, H. Epigenetic Inheritance in Rice Plants. *Ann. Bot.* **2007**, *100*, 205–217. [CrossRef] [PubMed]
45. Li, J.; Wang, M.; Li, Y.; Zhang, Q.; Lindsey, K.; Daniell, H.; Jin, S.; Zhang, X. Multi-Omics Analyses Reveal Epigenomics Basis for Cotton Somatic Embryogenesis through Successive Regeneration Acclimation Process. *Plant Biotechnol. J.* **2019**, *17*, 435–450. [CrossRef]
46. Ikeuchi, M.; Ogawa, Y.; Iwase, A.; Sugimoto, K. Plant Regeneration: Cellular Origins and Molecular Mechanisms. *Development* **2016**, *143*, 1442–1451. [CrossRef]
47. Li, W.; Liu, H.; Cheng, Z.J.; Su, Y.H.; Han, H.N.; Zhang, Y.; Zhang, X.S. DNA Methylation and Histone Modifications Regulate de Novo Shoot Regeneration in *Arabidopsis* by Modulating WUSCHEL Expression and Auxin Signaling. *PLoS Genet.* **2011**, *7*, e1002243. [CrossRef]
48. Law, J.A.; Jacobsen, S.E. Establishing, Maintaining and Modifying DNA Methylation Patterns in Plants and Animals. *Nat Rev. Genet.* **2010**, *11*, 204–220. [CrossRef]
49. Yang, D.; Huang, Z.; Jin, W.; Xia, P.; Jia, Q.; Yang, Z.; Hou, Z.; Zhang, H.; Ji, W.; Han, R. DNA Methylation: A New Regulator of Phenolic Acids Biosynthesis in Salvia Miltiorrhiza. *Ind. Crops Prod.* **2018**, *124*, 402–411. [CrossRef]
50. Hu, Y.; Morota, G.; Rosa, G.J.; Gianola, D. Prediction of Plant Height in *Arabidopsis thaliana* Using DNA Methylation Data. *Genetics* **2015**, *201*, 779–793. [CrossRef] [PubMed]
51. Grzybkowska, D.; Morończyk, J.; Wójcikowska, B.; Gaj, M.D. Azacitidine (5-AzaC)-Treatment and Mutations in DNA Methylase Genes Affect Embryogenic Response and Expression of the Genes That Are Involved in Somatic Embryogenesis in *Arabidopsis*. *Plant Growth Regul.* **2018**, *85*, 243–256. [CrossRef]

52. Rival, A.; Jaligot, E.; Beule, T.; Finnegan, E.J. Isolation and Expression Analysis of Genes Encoding MET, CMT, and DRM Methyltransferases in Oil Palm (*Elaeis Guineensis* Jacq.) in Relation to the 'Mantled' Somaclonal Variation. *J. Exp. Bot.* **2008**, *59*, 3271–3281. [CrossRef] [PubMed]
53. De-la-Peña, C.; Nic-Can, G.I.; Galaz-Ávalos, R.M.; Avilez-Montalvo, R.; Loyola-Vargas, V.M. The Role of Chromatin Modifications in Somatic Embryogenesis in Plants. *Front. Plant Sci.* **2015**, *6*, 635. [CrossRef] [PubMed]

MDPI AG
Grosspeteranlage 5
4052 Basel
Switzerland
Tel.: +41 61 683 77 34

Current Issues in Molecular Biology Editorial Office
E-mail: cimb@mdpi.com
www.mdpi.com/journal/cimb

www.ingramcontent.com/pod-product-compliance
Lightning Source LLC
Chambersburg PA
CBHW041425210926
51850CB00013B/3360
* 9 7 8 3 7 2 5 8 4 1 3 5 6 *